McMurtrie's Human Anatomy Coloring Book

McMURTRIE'S
HUMAN ANATOMY COLORING BOOK

*A Systemic Approach to the Study of
the Human Body: Thirteen Systems*

HOGIN McMURTRIE

STERLING PUBLISHING CO., INC.
NEW YORK

To my angel— during physical life, my wife—Denise—for her unwavering love,
support and courage from flagfall to finish, and all the years after ...
to my pride and joy, daughter Megan and son Joshua, whose love will never fail ...
and to J.B., wherever you are, my friend, for that plane flight back from San Francisco.

PRODUCED BY THE REFERENCE WORKS, INC.

Harold Rabinowitz, Director
Pamela Adler, Managing Editor
Elizabeth O'Sullivan, Assistant Editor
Barbara Arnstein, Copy Editor
Allen McCormack, Proofreader
David L. Kulak, Consulting Editor
Shulamit Roditi-Kulak, Consulting Editor
Jonathan Wiesen, Consulting Editor
Jennifer Hron, Editorial Intern
Cynthia Hamilton, Editorial Intern

Library of Congress Cataloging-in-Publication Data Available Upon Request

10 9 8 7 6 5 4 3 2 1

Published by Sterling Publishing Co., Inc.
387 Park Avenue South, New York, NY 10016
© 2006 by Hogin McMurtrie, The Reference Works, Inc. and
    Sterling Publishing Co., Inc.
Distributed in Canada by Sterling Publishing
c/o Canadian Manda Group, 165 Dufferin Street
Toronto, Ontario, Canada M6K 3H6
Distributed in the United Kingdom by
GMC Distribution Services
Castle Place, 166 High Street, Lewes, East Sussex,
England BN7 1XU
Distributed in Australia by Capricorn Link (Australia) Pty. Ltd.
P.O. Box 704, Windsor, NSW 2756, Australia

Manufactured in the United States of America

Sterling ISBN-13: 978-1-4027-3788-6
        ISBN-10: 1-4027-3788-2

For information about custom editions, special sales, premium
and corporate purchases, please contact Sterling Special Sales
Department at 800-805-5489 or special sales@sterlingpub.com.

The Author has made appropriate attempts necessary to contact
copyright holders for consideration of material and ideas that
might be construed as borrowed. If any have inadvertently been
overlooked, the Author will be pleased to immediately make the
necessary arrangements upon receipt of inquiry.

Acknowledgments

The original McMurtrie professional work published fifteen plus years ago, upon which this smart, condensed Sterling trade revision is based, was molded and nurtured along by many people and their devoted staffs to whom I will always be indebted for their assistance, contributions and reviews, and who have previously been acknowledged—in particular my colleague and good friend, James Krall Rikel, Ph.D.

The many years of my life spent authoring, designing and illustrating the work have been fruitful, and supported by the professionalism of many others.

It has taken me many years of perseverance to reach the stage where I now stand. I have been very fortunate in cultivating first, the friendship, and second, the professional relationship, with Harold Rabinowitz of The Reference Works, whose steadfast and arduous work has finally helped make possible my lifelong dream of placing my human anatomy coloring book into the major retail bookstore markets for the layman, for which it was originally intended. It is Harold's efforts and his dreams of forging a new experiment in the publishing world that I must first and foremost mention.

Also great thanks to the dedication, skill, and smooth editorial direction of Pamela Adler, whose command of considerable details and communication skills under extensive pressure and target deadlines is to be highly commended; to the tasty design talents of Jordan Rosenblum for his sharp attention to detail and efficiency; to Barbara Berger, our editor at Sterling, who lent her experience and expertise—and her sharp sense of design—to the project; to my good buddy Deborah Broide, who always kept me in good spirits and helped show me the lay of the land; and to those other publishing staff members and reviewers whom I have not yet had the pleasure of meeting, whose advice, comments and expertise will make major contributions to this project.

To close, I am very grateful to Sterling Publishing, with whom Harold forged my project, and in particular to Charles Nurnberg, Sterling's publisher and CEO, who took the time and saw the vision where others did not, who aligned with Harold's and my panoramic views on this and other future projects, and supported them throughout, resulting in an end product based on quality, first-rate support, and facilitation of the educational process.

TABLE OF CONTENTS

PREFACE

For the curious layman seeking a simple way to find out more about how the amazing human body is put together—through specific structures or overall organization—or the beginning human anatomy student facing the daunting task of comprehending the body's complexities in coursework (oftentimes in a very short period of time), *McMurtrie's Human Anatomy Coloring Book* offers time-tested effective assistance in many ways. This condensed, revised Sterling trade edition is based on over fifteen years of successful teaching to hundreds of thousands of students through McMurtrie's professional supplemental human anatomy text.

FACILITATING LEARNING

The pedagogical approach chosen for this book is a simple but comprehensive study of the human body via the body systems or a systemic anatomy. Under this approach, the book's design facilitates learning by providing useful tools in an effective format.

The first main tool is the grouping of the thirteen body systems into four wider and easy-to-understand parts related to the human body's overall function: support and movement (four systems), integration and control (two systems), regulation and maintenance (six systems), and species continuation (one system). This device aids in putting all the systems into an easily remembered proper perspective.

The second main tool is the breaking down of the entire summary of the human body into a consistent format of two-page subject spreads, placed within each system chapter. A connected left page and right page are dedicated to one subject/topic. This compartmentalization of subject matter has proven very helpful for retention.

The third main tool—the one with which most people are familiar and comfortable—is the specific activity of learning through the process of color association (color linkage), an active visualization technique which has many proven benefits over rote memorization. *The simple act of spending time with a subject through the manual process of coloring labels and then coloring objects, planes, and surfaces in matching colors fosters familiarity and ultimately aids in the retention of both detail and the context of that detail.* This effort, which is a pleasant and fun experience, makes the completed colored book a personalized document of one's knowledge of human anatomy, and can continue to serve as a useful integrated tool for future study, coursework, or home reference.

TIPS ON HOW TO USE THIS BOOK

For students and laymen alike, it is recommended that Part I: Orientation and Organization (Chapters 1-3) be studied first, as they establish a foundation for the basic terminology, structural relationships, and levels of complexity to come in the following four parts of body systems (Chapters 4-16—See Note at the end of the Preface).

For students, each of the body systems may be studied in any order, adapting to any course sequence. However, the presentation of each chapter's subject matter follows a logical sequence, and should be studied from beginning to end.

Each system chapter begins as an introductory spread, containing a table of contents, an overview of the system, a simple system illustration, a system components and system functions box, and chapter coloring guidelines. Also, included in the front of this book is a removable cardboard tablet to use as a hard drawing surface.

Each system chapter proceeds from a general overview, followed by spreads of specific areas, usually beginning with the most superior or upper region (i.e. head, neck), and ending with the most inferior or lower region (i.e. leg, foot).

Within a chapter, each two-page layout, in general, contains descriptive text, and the names of structures (enveloped by rectangular boxes) and illustrations. This visual separation is valuable in organizing learning and also enables the pages to be used as "flash cards."

THE COLORING PROCESS

When looking at two pages, the name of a structure is found within a rectangular box. Choose a color and color within the borders of the box, around the name (or over the name if the text remains visible after coloring). A reference number precedes the box. (Latin terms, popular names and useful descriptive material appears outside the box.) Then locate the same reference number in the illustration. Use the same color from the box and color the structure. The name and the structure are now visually related through color linkage, and when colored, whole illustrations and whole body systems take on an expression of your own individuality, enhancing your retention of the subject matter.

Using colored pencils is best, because felt-tip pens can bleed and wax crayons are too broad and coarse. Plus, with pencils, you can create different densities by varying pencil pressure.

Color suggestions are provided in boxes with the heading "Color Guidelines," and further suggestions regarding coloring methods and procedures are provided in boxes with the heading "Coloring Notes." Readers will no doubt make many of their own choices more suited to them. Color conventions used in many books on the subject are: red for arteries, blue for veins, green for lymphatic vessels, reddish-brown for muscles and yellow for fat. In general, for best results, color larger areas with lighter colors, and smaller areas with darker colors.

ADDITIONAL REVIEW MATERIAL AND CONTENT

Visit www.MCMURTRIESANATOMY.COM for free additional study and review charts. View it as an extension to this book. You can also post suggestions for improvement or write comments on the website.

Students will find here invaluable system review pages, including comprehensive reviews of the whole skeletal system, the articular system, the muscular system, the central nervous system, the cardiovascular system, and the reproductive system (summary comparisons of mitosis and meiosis, plus menstrual and hormonal cycles).

New content will be added on an ongoing basis, including specific articular joints, more in-depth nervous system information, mechanics of respiration, embryology subjects and updated discussions and information on controversial issues in society (human cloning and embryonic stem cell research).

Note: There are thirteen body systems (others consolidate them into as little as few systems). Articulations are treated as the articular system. The circulatory system is divided into the cardiovascular (blood vascular) system and lymph vascular system. The lymphoid immune system is its own separate system.

CHAPTER 1: ORIENTATION TO THE BODY

CONTENTS

BODY PLANES (SECTIONS) AND DIRECTIONAL TERMS

A helpful way of gaining a mental orientation when viewing an object such as a part of the body (or the whole body) is to visualize the body as an object inside a transparent, three-dimensional box. The faces or sides of the box become the VIEWING PLANES through which one views the object from different positions around it. Thus, for example, in the illustrations in this chapter, viewing the body through the FRONT or ANTERIOR VIEWING PLANE of the box provides an ANTERIOR VIEW, which is the view that appears on page 11. When viewing the body through the back of the body (the dotted plane in the drawing on page 11), one has a BACK or POSTERIOR VIEW as one looks through the POSTERIOR VIEWING PLANE.

The body itself can be imagined to have planes through it that help locate structures, organs, etc. The plane that divides the body into symmetrical halves is the MIDASIGITTAL PLANE—alongside that plane are planes that divide into non-symmetrical portions and separate the left and right portion of the body—these are the PARASAGITTAL PLANES. A plane perpendicular to the sagittal planes divides the body into (non-symmetrical) front and back halves—this is the FRONTAL or CORONAL PLANE. Finally, a plane perpendicular to both the sagittal and the frontal planes divides the body into upper and lower halves—this is known as the HORIZONTAL, TRANSVERSE or CROSS-SECTIONAL PLANE.

DIRECTIONAL TERMS AS "VIEWING PLANES"

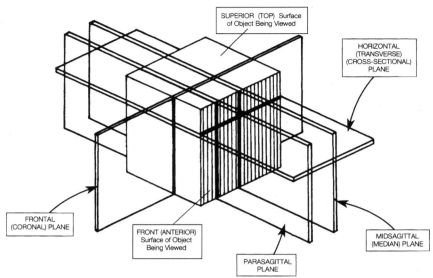

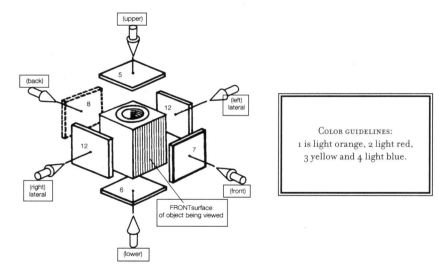

COLOR GUIDELINES:
1 is light orange, 2 light red, 3 yellow and 4 light blue.

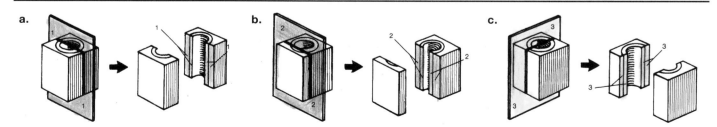

a.   **b.**   **c.**

In panels (a) through (e) above and on the next page, the body planes are shown as both imaginary planes through an object (left) and as cross-sections of the object (right). Panel (a) shows the midsagittal plane and cross-section; (b) shows the parasagittal plane and cross-section; (c) shows frontal plane and cross-section; (d) shows the transverse plane and cross-section. A plane may also pass through the body obliquely and be used to locate organs and structures. Panel (e) shows the OBLIQUE PLANE (left) and cross-section (right).

Directional terms and body planes are given in reference to the STANDARD ANATOMICAL POSITION (or simply the ANATOMICAL POSITION). The standard anatomical position is when the body is:

· erect directed forward, facing the observer; with
· eyes directed forward, toward the observer;
· arms at the sides of the body, palms of hands turned forward; and
· feet are placed flat on the floor, parallel to each other.

BODY PLANES:
Imaginary flat surfaces that pass through the body.

1	Midsagittal (median)
2	Parasagittal
3	Frontal (coronal)
4	Horizontal (transverse/cross-sectional)

DIRECTIONAL TERMS to explain the exact location of various body structures relative to each other.

5	Superior (Craniad) Cephalic Upper
6	Inferior (Caudad) Lower
7	Anterior (Ventral) Front
8	Posterior (Dorsal) Back
9	Proximal
10	Distal
11	Medial
12	Lateral

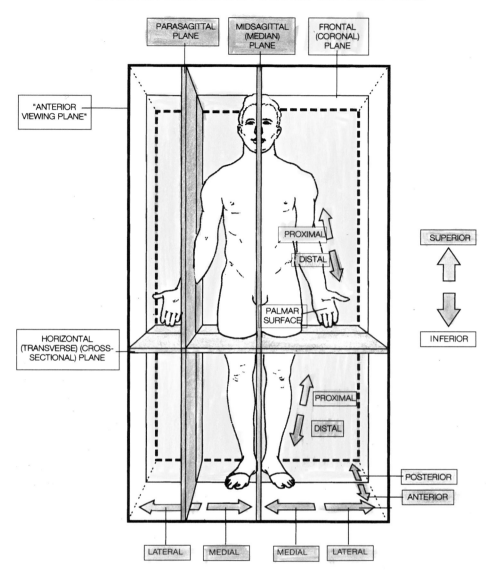

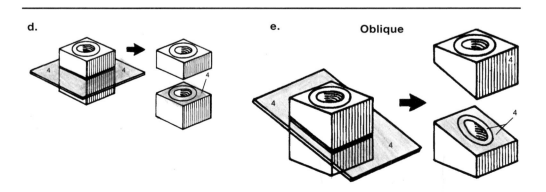

d.

e. **Oblique**

COLORING NOTES:
Begin by coloring the viewing planes and the body planes on page 10 and the bottom of page 11. This will serve to orient you to the visualization of the body. Then color the viewing planes and body planes using the labels and the drawing of the Standard Anatomical View on page 11. Then, color the planes and surfaces of the Right Lateral View of the body on page 12. Finally, color the directional arrows and labels on pages 10 through 13.

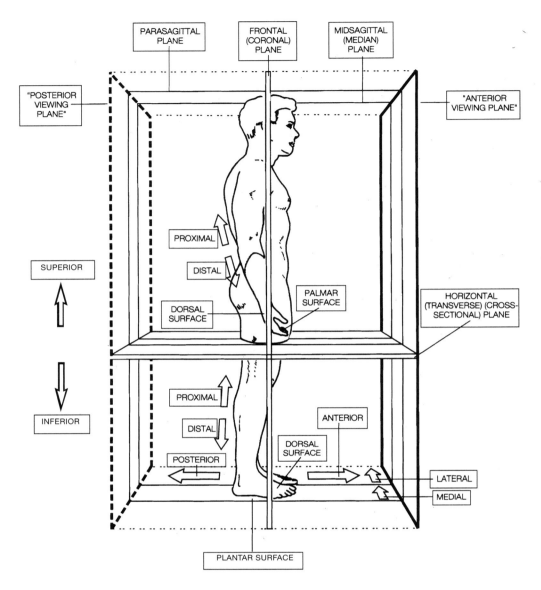

The body directions are used to explain the exact location of various body structures relative to each other. Learning to refer to the position of organs and structures using the directional terms and the viewing and body planes is essential for dealing with and for discussing the complex anatomy of the human body. THE RIGHT LATERAL VIEW, shown above, displays the same planes through the body as the anterior view on the previous page, only from the side. The ANTERIOR VIEW has a mirror image in the form of the POSTERIOR VIEW (which is the body seen from the rear), and the right lateral view has a mirror image in the form of the LEFT LATERAL VIEW (which is the body viewed from the left side).

DIRECTIONAL TERMS are used to pinpoint body organs and structures relative to the body as a whole or relative to other structures and organs. Any structure that is above another structure (nearer the head) is said to be SUPERIOR (CRANIAD) to it. Any structure that is lower than another (toward the feet) is INFERIOR (CAUDAD). Any structure in front of another is ANTERIOR (VENTRAL). POSTERIOR (DORSAL) refers to a structure that is in back of another, toward the back. MEDIAL structures are closer to the midsagittal plane. LATERAL structures are farther away from the midsagittal plane, towards the sides. IPSILATERAL structures are on the same side of the body, while CONTRALATERAL structures are on opposite sides. INTERMEDIATE structures are between a medial structure and a lateral structure, or between two structures. INTERNAL (DEEP) structures are away from the body surface relative to another structure. EXTERNAL (SUPERFICIAL) structures are more toward the body surface in reference to another structure. VISCERAL refers to the internal covering of an internal organ (viscus) and PARIETAL refers to the outer covering of an organ or the inner covering of a body cavity.

DIRECTIONAL TERMS FOR THE HUMAN BODY

TERM	DEFINITION	EXAMPLE
SUPERIOR	Toward the head; above another structure	The heart is superior to the liver
CRANIAL; CEPHALIC	Of or near the skull (cranium) or head	The brain is a cranial organ
INFERIOR	Away from the head; relatively lower	The intestines are inferior to the lungs
CAUDAL	Toward the feet; lower in the body	The legs are a caudal part of the anatomy
ANTERIOR (VENTRAL)	Toward the front (center); in front of	The sternum is anterior to the heart
POSTERIOR (DORSAL)	Toward the rear (posterior); in back of	The kidneys are posterior to the intestines
MEDIAL	Toward the midsagittal plane or middle	The heart is medial to the lungs
LATERAL	Toward the side or further from the middle	The ears are lateral to the brain
IPSILATERAL	On the same side of the body	The gallbladder and the ascending colon of the large intestine are ipsilateral
CONTRALATERAL	On the opposite side of the body	The ascending and descending colons of the large intestine are contralateral
INTERMEDIATE	In between two structures	The ring finger is intermediate to the middle finger and the little finger
INTERNAL (DEEP)	Away from the surface of the body	The brain is internal to the cranium
EXTERNAL (SUPERFICIAL)	On or toward the surface of the body	The skin is external to the muscles
PROXIMAL	Toward the main mass of the body	The knee is proximal to the foot
DISTAL	Further from the main mass of the body	The hand is distal to the elbow
VISCERAL	Relating to an internal organ	The visceral pleura covers the lungs
PARIETAL	Relating to the wall of a body cavity	The parietal pleura is the inside lining of the thoracic cavity

DIRECTIONAL TERMS OF THE LIMBS

Directional terms for limbs of the body are dictated by their relationship with the torso, so that the upper arm is proximal relative to, say the elbow, though it is further from the midline of the body in the standard anatomical position. The same is true of the structures of the legs. Shown at right are the ANTERIOR and POSTERIOR VIEWS of the limbs for the left limbs only. Similar terms and arrangements are true of the right limbs. For limbs, any structure closer to the body mass (the torso) than another structure is PROXIMAL to the other structure; and any structure further from the body mass than another structure is DISTAL to that structure.

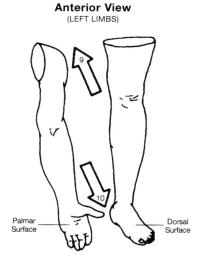

Anterior View
(LEFT LIMBS)

Palmar Surface

Dorsal Surface

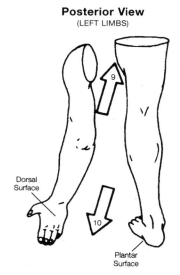

Posterior View
(LEFT LIMBS)

Dorsal Surface

Plantar Surface

Chapter 2: Systems and Regions of the Body

This study of the structural relationships of the parts of the body in a particular region is emphasized mostly in laboratory courses based on regional dissections (see www.mcmurtriesanatomy.com for more information). Surface anatomy's aim is to visualize structures that lie beneath the skin and are palpable, observing the surface of the body and the structures under it, studying the living body at rest and in action.

The clinical anatomy approach, not covered in this book, incorporates both systemic and regional approaches, and stresses clinical applications, problems and specific case studies important in the practice of medicine and other allied health sciences. The clinical application of surface anatomy is the physical examination of a person.

Approaches to the Study of Human Anatomy

There are three main approaches to studying human gross anatomy (the examination of body structures perceived without the aid of a microscope): SYSTEMIC, REGIONAL (including SURFACE) and CLINICAL. In this book, the main pedagogical approach presented for the study of the complex human physical body is via the body systems, or SYSTEMIC ANATOMY. The ten essential life processes are delegated to the thirteen body systems, aiding our conceptualization of how the body is wholly organized, both structurally and functionally. Further assisting our learning task in this book is the visual division of each system into two-page spreads, each with its own unique identity, "dissecting" so to speak each system into smaller pedagogical "compartments." Ease of comprehension has been further accomplished by grouping the thirteen body systems into four wider functional units:

· support and movement (four systems);
· integration and control (two systems);
· regulation and maintenance (six systems); and
· species continuation (one system).

As we shall see in the next chapter on organization, levels of increasing complexity within the human body, beginning with the single cell, finally lead up to the complete human organism. Aggregations of similar specialized cells constitute tissues.

There are four basic tissue types which integrate in various ways to form all the body's organs and structures, each with a definite form and function. All organs are grouped into two main categories in the human body: VISCERAL ORGANS/STRUCTURES (many of them hollow organs) found within the body cavities, and the BODY WALL (SOMATIC) ORGANS/STRUCTURES found outside the body cavities.

A system is an association of organs/structures working together to perform a common function. Within any system, its defined organs/structures are not necessarily confined to a certain location or grouped together closely, and thus may be found occupying various regions throughout the body. No one system works independently from the others. All thirteen body systems functioning together, working as a whole, constitute one living individual, the human organism. Health and well-being are dependent on the coordinated, harmonious effort of each and every system.

There is some disagreement in classification as to the number of body systems (some view as few as ten systems), but for our purposes, a total of thirteen body systems have been chosen—treating articulations as the articular system, dividing the circulatory system into the blood vascular and lymph vascular systems, and the lymphoid immune system as separate from the lymph vascular system.

Another way to look at the general design of the body is to conceptualize the systems as being grouped into "MASTER" tissues and "VEGETATIVE" systems. The two highly specialized and modified "MASTER" tissues—NERVOUS TISSUE and SKELETAL MUSCLE TISSUE—are those which specialize in receiving stimuli from the external and internal environments and reacting to them. The "vegetative" systems (VASCULAR, IMMUNE, RESPIRATORY, DIGESTIVE, URINARY, and ENDOCRINE [regulating]) are those which serve the basic utilities of life by dealing with the source of energy—FOOD—and breaking it down and transporting it around the body, making it available to the cells for the release of energy, growth and repair, reproduction and movement, and to get rid of waste.

The activities of these "vegetative" systems are regulated partly through a section of the "master" tissues—the autonomic nervous system (ANS) of the peripheral nervous system (PNS. See Nervous System, chapter 8). Its two complimentary divisions—parasympathetic and sympathetic—carry opposite (antagonistic) messages to visceral organs/structures. These message impulses are rapidly adjusted and balanced by the central nervous system (CNS) brain centers to integrate the actions of the "vegetative" systems to the constantly changing needs of the whole human body.

Body Systems: 13 Systems

Support and Movement (Four Systems)

There are four systems concerned principally with body support and movement. The INTEGUMENTARY (SKIN) and SKELETAL SYSTEMS chiefly support and protect the body, with assistance from the ARTICULAR and MUSCULAR SYSTEMS. The structural and functional relationships of the latter three systems coordinate the activities with respect to bodily movement. The skeletal, articular and muscular systems structure's (bones, joints, ligaments, muscles) work together in the limbs to constitute the major portion of the locomotor system, which produces overall locomotion of the body. (Other organs/structures within the LOCOMOTOR SYSTEM are those from the blood vascular system [arteries and veins] which supply nutrients and oxygen, and remove waste; and from the nervous system [nerves], which stimulates action.)

LIFE PROCESS	SYSTEM
Protection from environment	Integumentary
Contractility (movement)	Locomotor (skeletal, articular, muscular)

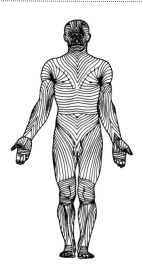 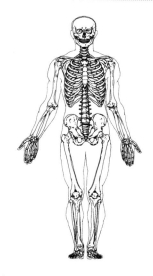 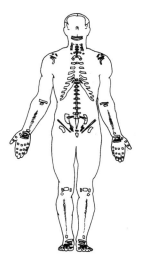 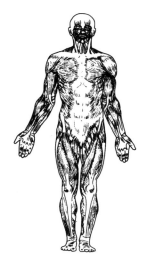

INTEGUMENTARY SYSTEM

The layered integument, the SKIN—an extensive sensory organ, the largest organ in the body—in addition to supporting hair, nails, sweat glands, vessels, immune cells and antibodies, and sensory receptors, forms a protective covering and container for the body, resisting pathogens and injury from the external environment through layers of cells and keratin. It is also very important in maintaining homeostasis in the internal environment.

SYSTEM COMPONENTS:
The SKIN and its structural derivatives (such as HAIR, NAILS, SWEAT and OIL GLANDS).

SYSTEM FUNCTIONS:
External support of the body
Regulation of body temperature
Protection of the body
Elimination of wastes
Reception of certain stimuli (such as temperature, pressure, pain, touch and vibration)

SKELETAL SYSTEM

The skeletal system consists of bones (including their covering the periosteum) and cartilage. The bones have diagnostic features and perform the mechanical functions of support, protection and movement. The skeletal system is what the muscular system acts on to produce movement. The fibrous fasciae that ensheath the skeletal muscles (and body wall), contributing to bodily structural stability, may be included here.

The skeletal system also protects vital organs (heart, lungs, pelvic organs), and performs the metabolic functions of hemopoiesis and mineral storage.

SYSTEM COMPONENTS:
The bones of the body and their associated cartilages.

SYSTEM FUNCTIONS:
Internal support of the body
Protection of the body
Body leverage
Production of blood cells
Storage of minerals

ARTICULAR SYSTEM

The articular system consists of joints (both movable and fixed), and their associated ligaments (that secure and connect the bony parts of the skeletal system at joints).

Also included are joint related structures such as joint capsules, synovial membranes, and discs/menisci. The joints and ligaments provide the sites at which movements occur. The joints are classified both structurally and functionally.

SYSTEM COMPONENTS:
Joints and their associated ligaments.

SYSTEM FUNCTIONS:
Flexible fibrous connective tissue at points of contact between bones or cartilage and bones.

MUSCULAR SYSTEMS

The muscular system consists of muscles that contract to move parts of the body. (Based on mass, the largest portion of muscle in the body belongs to the highly specialized skeletal muscle tissue, which moves the skeletal bones that articulate at joints and give form to the body.) The many skeletal muscles are adapted to contract to electrochemical stimuli from the nerves in order to carry out the functions of motion, heat production, posture, and bodily support. Also included are the facial muscles, the cardiac muscle of the heart walls, and the smooth muscle in many visceral walls, blood and lymphatic vessel walls, and in the skin.

SYSTEM COMPONENTS:
The muscle tissue of the body (including skeletal, cardiac and smooth contractile tissue).

SYSTEM FUNCTIONS:
Movement
Heat production
Maintenance of body posture
Maintenance of body temperature

Body Systems, continued

Integration and Control (Two Systems)

There are two integration and control body systems, the nervous system and the endocrine system, that regulate the activities of all the other eleven body systems. Specific organs/structures from these two systems help a person to respond with appropriate action to both external and internal stimuli, and to play their part in the maintenance of balanced metabolic functions of whole bodily HOMEOSTASIS (maintenance of a relatively stable internal physiological condition in the presence of fluctuating environmental conditions).

LIFE PROCESS	SYSTEM
Irritability and control	Nervous
Metabolism and growth	Endocrine

Regulation and Maintenance (Six Systems)

There are six systems concerned primarily with regulation and maintenance of the body. They are sometimes referred to loosely as the unconscious "vegetative" systems, whose principal function is to maintain a condition of balanced, healthy homeostasis in the visceral organs/structures, by, for example, maintaining stable concentrations of various substances in the body's fluids. The lymphoid immune system further maintains homeostasis by acting to protect the body from many diseases.

The circulatory system is divided into the blood vascular (cardiovascular) system and the lymph vascular (lymphatic) system, which both function in parallel.

LIFE PROCESS	SYSTEM
Transport	Blood vascular lymph vascular
Defense against disease	Lymphoid immune
Respiration	Respiratory
Nutrition	Digestive
Excretion	Excretory (digestive, urinary)

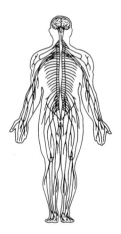

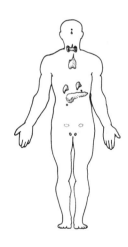

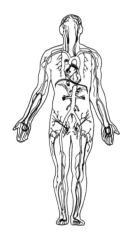

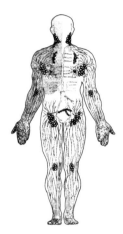

Nervous System

The nervous system, whose highly specialized conductive tissue generates and transmits electrochemical impulses, is organized into the central nervous system (CNS)—the brain and spinal cord—and the peripheral nervous system (PNS)—the cranial and spinal nerves—together with their motor and sensory endings, and the sensory receptor organs. The nervous system relates the body to the environment and controls and coordinates the functions of organs.

SYSTEM COMPONENTS:
The brain, the spinal cord, the nerves and the sense organs (i.e., eye and ear).

SYSTEM FUNCTION:
Regulation of body activities through the generation and conductive transmission of nerve impulses

Endocrine System

The endocrine system is a regulating system consisting of ductless glands that secrete regulatory chemical agents (molecules called hormones) into the blood and tissue fluids, where they are carried by the circulatory system to all parts of the body (target tissues, specific bonding receptor proteins) and elicit physiological responses. The endocrine glands assist in maintaining metabolism and growth functions throughout the body's systems, and are under partial control by the CNS (hypothalamus of the brain).

SYSTEM COMPONENTS:
The glands and tissues of hormone production.

SYSTEM FUNCTION:
Regulation of body activities through transportation of hormones by the blood vascular system

Cardiovascular System

The blood vascular system consists of the heart and the blood vessels that propel and conduct blood throughout the body. Life-sustaining oxygen, nutrients, gases and molecular material are brought to the tissue cells in the "red blood" via arteries, arterioles and the finest capillaries; metabolic wastes are carried away from the tissue cells by the venous end of the capillaries, and veins return the "blue blood" from the tissue cells back to the heart.

SYSTEM COMPONENTS:
The blood, heart and blood vessels.

SYSTEM FUNCTION:
Distribution of oxygen and nutrients to the cells
Carrying of carbon dioxide and wastes from the cells
Maintenance of body acid-base balance
Protection against disease
Prevention of hemorrhage (blood clot formation)
Regulation of body temperature

Lymph Vascular System

The lymph vascular system is a network of "start-closed" non-blood lymphatic vessels structurally end-connected to the superior vena cavae system which assists the inadequacy of the veins regarding full demand of tissue drainage in recovering much of the body's tissue fluids and returning them to the heart. Excess tissue fluid called lymph (basically blood plasma) drained from the interstitial (intercellular) fluid compartment, is filtered through lymph nodes (found in the lymphatic vessels and elsewhere throughout the body) and returned to the bloodstream.

SYSTEM COMPONENTS:
Lymph, lymph nodes, vessels, ducts and glands (i.e.: spleen, tonsils and thymus gland).

SYSTEM FUNCTION:
Return of protein and fluid to the blood vascular system
Transportation of fats from the digestive system to the blood vascular system
Filtration of blood
Production of white blood cells
Protection against disease

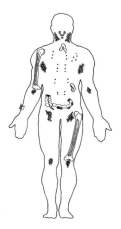

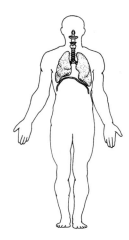

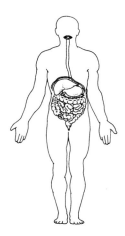

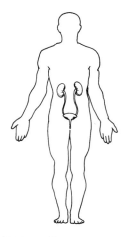

LYMPHOID IMMUNE SYSTEM

The lymphoid immune system consists of lymphatic organs/tissue (in various stages of encapsulation) concerned with nonspecific and specific defenses to protect the body from the invasion of disease-causing agents. Primary lymphoid organs are bone marrow and thymus; secondary lymphoid organs are lymph nodes, spleen, tonsils, appendix, and M.A.L.T. (small pocketed aggregations of lymphoid tissue). Within various lymphoid organs/tissues, B-lymphocytes provide antibody-mediated immunity, and T-lymphocytes provide cell-mediated immunity.

SYSTEM COMPONENTS:
Primary organs (bone marrow and thymus), secondary organs (spleen [filters blood], lymph nodes [filter lymph], tonsils/adenoids, appendix, M.A.L.T.) and immune cells (B-lymphocytes and T-lymphocytes).

SYSTEM FUNCTION:
Resistance to disease-causing agents
Removal of damaged/abnormal cells
B-lymphocyte antibody-mediated immunity
T-lymphocyte cell-mediated immunity

RESPIRATORY SYSTEM

The respiratory system takes oxygen into the body and eliminates carbon dioxide out of it via the air conduction tracts (upper and lower) and the lungs; it also helps to maintain the acid-base balance of the body. After the air traverses through the tract airways, gases are only ultimately exchanged at a microscopic level between the lumen space of the lung alveoli (air cells) and the blood-filled lumen space of the fine lung capillaries, called the alveolar-capillary membrane ("air-blood barrier").

SYSTEM COMPONENTS:
The lungs and a series of passageways in and out of the lungs.

SYSTEM FUNCTION:
Supplies oxygen
Elimination of carbon dioxide
Regulation of body acid-base balance

DIGESTIVE SYSTEM

The digestive system prepares food for cellular utilization through a one-way transportation system (mouth to anus) involving processes from the organs of the gastrointestinal (GI) tract: ingestion, mastication (chewing), deglutition (swallowing), peristalsis, chemical digestion, assimilation (absorption), and defecation (elimination) of residual feces (solid wastes) which remain after nutrient absorption. Assisting the digestive processes are abdominal accessory organs (glands), including the liver, pancreas and biliary system (gallbladder and related ducts).

SYSTEM COMPONENTS:
A tubular passageway and associated organs (i.e.: salivary glands, liver, gallbladder and pancreas).

SYSTEM FUNCTION:
Physical and chemical breakdown of food for cell usage
Elimination of solid wastes

URINARY SYSTEM

The urinary system functions mainly through the two kidneys in the maintenance and regulation of acid-base balance, chemical composition, volume and conservation of the extracellular body fluids. In the kidneys, metabolic wastes (especially urea) from the cells (carried to them via the blood) are filtered out and eventually eliminated from the body in the form of residual fluid (urine); the urine from the kidneys is excreted through two ureters to an expanding "holding organ"—the urinary bladder—then discharge to the outside world via the urethra. The kidneys also return water, electrolyzes and nutrients to the blood.

SYSTEM COMPONENTS:
The organs of urine production, collection and elimination

SYSTEM FUNCTION:
Regulation of blood chemical composition
Elimination of wastes
Regulation of fluid/electrolyte balance and volume
Maintenance of body acid-base balance

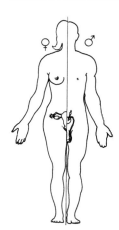

CONTINUANCE OF SPECIES (ONE SYSTEM)

The reproductive (genital) system is concerned with the reproduction and perpetuation of the human species through the male and female reproductive systems, resulting in a constant stream of new and genetically unique individual human beings.

LIFE PROCESS	SYSTEM
Reproduction	Reproductive

REPRODUCTIVE SYSTEM

The organs of both systems, male (testes) and female (ovaries), are responsible for the secretion of their own gender's sex hormones (which have regulatory functions within the body) and production, maintenance and transportation of reproductive cells (sperm and ova respectively). The male system assists in the transport of the sperm germ cells to the female genital tract; the female system elicits the receipt and transport of the male sperm germ cells to the fertilization site where the ova are fertilized, implant, and begin embryonic development; in addition, the female system maintains the developing embryo/fetus, is adapted for delivery (birth) and is responsible for sustaining (nursing) the newborn during infancy.

SYSTEM COMPONENTS:
The organs (testes and ovaries) for production of reproductive cells (sperm and ova) and organs for transportation and storage of those cells.

SYSTEM FUNCTION:
Reproduction of the organism
Perpetuation of the species
Passage of genetic material from generation to generation

BODY REGIONS
GENERALIZED BODY REGIONS

The human body is divided into five major regions, all identifiable on the surface: HEAD, NECK, TRUNK (TORSO), UPPER EXTREMITY (TWO) and LOWER EXTREMITY (TWO). They can be subdivided into more generalized and specific localized areas. Each region contains internal structures with specific anatomical names. Learning the specific regional terminology provides a solid basis for later learning the names of the underlying structures and identifying their location.

The HEAD (CAPUT) contains the BRAIN and SPECIAL SENSE ORGANS. It provides openings into the RESPIRATORY SYSTEM and DIGESTIVE SYSTEM.

The complexity of NECK (COLLUM) musculature provides a wide variety of movements for the head, and connects the head to the THORAX. Four neck regions contain major organs. The ANTERIOR REGION (CERVIX) contains portions of the RESPIRATORY TRACT (TRACHEA) and DIGESTIVE TRACT (ESOPHAGUS), the LARYNX or VOICE BOX, MAJOR BLOOD VESSELS to and from the head, HYOID BONE, NERVES and THYROID and PARATHYROID GLANDS. The RIGHT and LEFT LATERAL REGIONS contain major neck muscles and LYMPH NODES. The POSTERIOR REGION (NUCHA) contains the SPINAL CORD and CERVICAL VERTEBRAE.

The TRUNK (TRUNCUS) or TORSO is frequently divided into an upper THORAX (PECTUS) or chest, and a lower ABDOMEN (VENTER), separated by a muscular internal DIAPHRAGM. The torso is a major site of vital visceral organs.

The thorax contains the bony RIB CAGE within which lies continuations of the trachea and esophagus, the LUNGS, HEART, MAJOR BLOOD and LYMPHATIC VESSELS and THYMUS GLAND (outside the rib cage are the BREASTS and LYMPH NODES).

The ABDOMEN contains the lower ESOPHAGUS, STOMACH, SMALL INTESTINE, LARGE INTESTINE (major portion), LIVER, GALLBLADDER, SPLEEN, PANCREAS, KIDNEYS and ADRENAL GLANDS.

Two other divisions of the torso include the BACK (dorsum) and the PELVIS, which contains the large intestine (terminal portion only), URINARY BLADDER and INTERNAL REPRODUCTIVE ORGANS. The floor of the pelvis is called the PERINEUM and includes the EXTERNAL GENITALIA.

The upper extremities are adaptive for freedom of movement. The lower extremities are important for locomotion and bearing weight.

COLORING NOTES:
The five major regions are identified by dot and line patterns, but you may wish to color these areas nonetheless along with the label boxes below. Use cool colors for the extremities, warm colors for the trunk and yellows for head and neck.

HEAD (CAPUT) REGION:

1. CRANIUM – Skull, Braincase
2. FACE – Facies

NECK (COLLUM) REGION:

3. ANTERIOR NECK – Cervical Region (Cervix)
4. POSTERIOR NECK – Nuchal Region (Nucha)

TRUNK (TORSO) REGION:

5. THORAX – Pectus (Chest/Pectoral Region)
6. ABDOMEN – Venter
7. PELVIS
5+6. BACK (DORSUM) and PERINEUM

Posterior Thorax and Posterior Abdomen

UPPER EXTREMITY

8. SHOULDER – Omos (Deltoid Region)
9. BRACHIUM – Upper Arm (Humeral Region)
10. ELBOW
11. ANTEBRACHIUM – Forearm
12. WRIST – Carpus
13. HAND – Manus

LOWER EXTREMITY

14. BUTTOCK
15. THIGH – Upper Leg (Femur/Femoral Region)
16. KNEE
17. LEG – Crus
18. ANKLE – Tarsus
19. FOOT – Pes

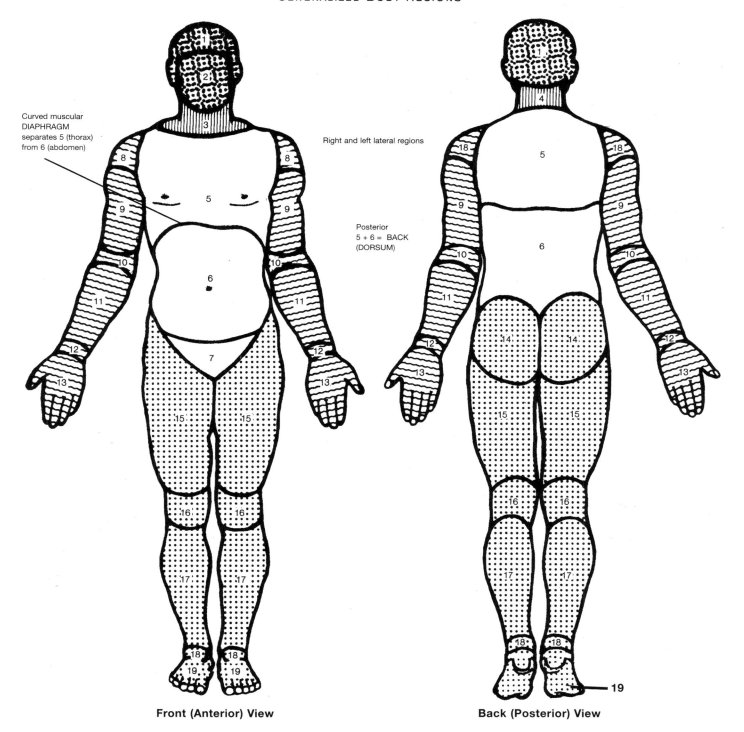

HEAD [CAPUT]	NECK [COLLUM]	TRUNK [TRUNCUS]	UPPER EXTREMITY (2)	LOWER EXTREMITY (2)

GENERALIZED BODY REGIONS

Curved muscular DIAPHRAGM separates 5 (thorax) from 6 (abdomen)

Right and left lateral regions

Posterior
5 + 6 = BACK
(DORSUM)

Front (Anterior) View

Back (Posterior) View

BODY REGIONS, CONTINUED
SPECIFIC BODY REGIONS AND ABDOMINOPELVIC QUADRANTS

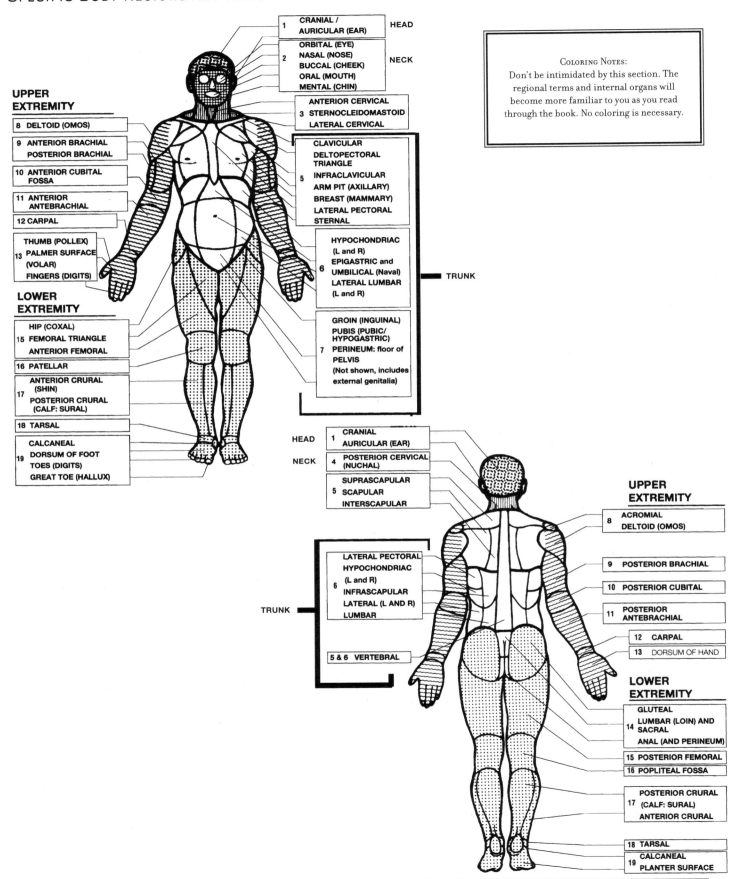

HEAD
1 CRANIAL / AURICULAR (EAR)

2 ORBITAL (EYE)
NASAL (NOSE)
BUCCAL (CHEEK)
ORAL (MOUTH)
MENTAL (CHIN)

NECK
3 ANTERIOR CERVICAL
STERNOCLEIDOMASTOID
LATERAL CERVICAL

UPPER EXTREMITY

8 DELTOID (OMOS)

9 ANTERIOR BRACHIAL
POSTERIOR BRACHIAL

10 ANTERIOR CUBITAL FOSSA

11 ANTERIOR ANTEBRACHIAL

12 CARPAL

13 THUMB (POLLEX)
PALMER SURFACE (VOLAR)
FINGERS (DIGITS)

LOWER EXTREMITY

15 HIP (COXAL)
FEMORAL TRIANGLE
ANTERIOR FEMORAL

16 PATELLAR

17 ANTERIOR CRURAL (SHIN)
POSTERIOR CRURAL (CALF: SURAL)

18 TARSAL

19 CALCANEAL
DORSUM OF FOOT
TOES (DIGITS)
GREAT TOE (HALLUX)

5 CLAVICULAR
DELTOPECTORAL TRIANGLE
INFRACLAVICULAR
ARM PIT (AXILLARY)
BREAST (MAMMARY)
LATERAL PECTORAL
STERNAL

6 HYPOCHONDRIAC (L and R)
EPIGASTRIC and
UMBILICAL (Naval)
LATERAL LUMBAR (L and R)

7 GROIN (INGUINAL)
PUBIS (PUBIC/ HYPOGASTRIC)
PERINEUM: floor of PELVIS
(Not shown, includes external genitalia)

TRUNK

Coloring Notes:
Don't be intimidated by this section. The regional terms and internal organs will become more familiar to you as you read through the book. No coloring is necessary.

HEAD
1 CRANIAL
AURICULAR (EAR)

NECK
4 POSTERIOR CERVICAL (NUCHAL)

5 SUPRASCAPULAR
SCAPULAR
INTERSCAPULAR

TRUNK
6 LATERAL PECTORAL
HYPOCHONDRIAC (L and R)
INFRASCAPULAR
LATERAL (L AND R)
LUMBAR

5 & 6 VERTEBRAL

UPPER EXTREMITY

8 ACROMIAL
DELTOID (OMOS)

9 POSTERIOR BRACHIAL

10 POSTERIOR CUBITAL

11 POSTERIOR ANTEBRACHIAL

12 CARPAL

13 DORSUM OF HAND

LOWER EXTREMITY

14 GLUTEAL
LUMBAR (LOIN) AND SACRAL
ANAL (AND PERINEUM)

15 POSTERIOR FEMORAL

16 POPLITEAL FOSSA

17 POSTERIOR CRURAL (CALF: SURAL)
ANTERIOR CRURAL

18 TARSAL

19 CALCANEAL
PLANTER SURFACE

20   ORIENTATION AND ORGANIZATION

1   Right Hypochondriac

Liver (right lobe)
Gallbladder
Right Kidney (upper 1/3)
Right Adrenal Gland

2   Epigastric

Liver (left lobe, medial part of right lobe)
Stomach (pyloric part and lesser curvature)
Small Intestine (superior and descending duodenum)
Pancreas (body and upper head)
Two Adrenal Glands

3   Left Hypochondriac

Stomach (body and fundus)
Large Intestine (Left
Colic or Splenic Flexure)
Spleen
Pancreas (tail)
Left Kidney (upper 2/3)
Left Adrenal Gland

4   Right Lateral | Right Lumbar

Small Intestine (Jejunum, Ileum)
Large Intestine (Right Colic or Hepatic Flexure, Ascending Colon,
Superior Cecum)
Right Kidney (lower 2/3, lateral portion)

5   Umbilical

Small Intestine (Inferior Duodenum, Jejunum, Ileum)
Large Intestine (middle of Transverse Colon)
Bifurcation of Abdominal Aorta and Inferior Vena Cava
Kidneys (Hilar regions)

6   Left Lateral | Left Lumbar

Small Intestine (Jejunum, Ileum)
Large Intestine (Descending Colon)
Left Kidney (lower 1/3)

7   Right Inguinal | R. Iliac

Small Intestine (Ileum)
Large Intestine (lower end of Cecum)
Appendix

8   Hypogastric | Pubic

Small Intestine (Ileum)
Large Intestine (part of Sigmoid Colon and Rectum)
Urinary Bladder (when full)

9   Left Inguinal | L. Iliac

Small Intestine (Ileum)
Large Intestine (junction of Descending and Sigmoid Colon)

ABDOMINOPELVIC REGIONS

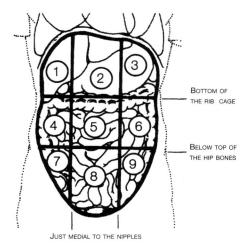

BOTTOM OF THE RIB CAGE

BELOW TOP OF THE HIP BONES

JUST MEDIAL TO THE NIPPLES

ABDOMINOPELVIC QUADRANTS

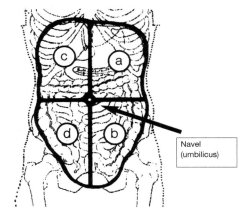

Navel (umbilicus)

a.   Left Upper Quadrant

b.   Left Lower Quadrant

c.   Right Upper Quadrant

d.   Right Lower Quadrant

Body Cavities: Divisions and Linings/Membranes
The Two Principal Body Cavities: Cavity Divisions and Body Membranes

BODY CAVITIES are CONFINED SPACES within the body that house and support organs. They serve to segregate or confine organs and systems that have related functions. The organs within the body cavities are PROTECTED, SUPPORTED and COMPARTMENTALIZED (separated) by associated MEMBRANES. The two principal body cavities are the DORSAL BODY CAVITY and VENTRAL BODY CAVITY.

The dorsal body cavity is comprised of two cavities: the CRANIAL CAVITY and VERTEBRAL CAVITY (CANAL), housing the BRAIN and SPINAL CORD, respectively. The major portion of the nervous system (the central nervous system or CNS) occupies the dorsal cavity. The covering and protective membranes are the three-layered MENINGES.

During embryological development, a body cavity is formed within the trunk called the COELOM, which is lined with a membrane secreting lubricating fluid. In time, the coelom is partitioned by the MUSCULAR DIAPHRAGM into an upper THORACIC (CHEST) CAVITY and a lower ABDOMINOPELVIC cavity. The organs contained within the coelom are called VISCERA or VISCERAL organs.

SEROUS membranes line the THORACIC and ABDOMINOPELVIC CAVITIES as PARIETAL MEMBRANES and cover the VISCERA as VISCERAL MEMBRANES. Serous membranes secrete a watery SEROUS FLUID for lubrication and they compartmentalize visceral organs so that infections and diseases cannot spread from one compartment to another.

The THORACIC CAVITY contains the principal organs of the RESPIRATORY and CIRCULATORY SYSTEMS: two pleural cavities contain the right and left lungs and one PERICARDIAL CAVITY contains the heart. The area between the two lungs within the thoracic cavity is the MEDIASTINUM. The PARIETAL PLEURA lines both pleural cavities, and the VISCERAL PLEURA covers both lungs. The potential space between them is called the PLEURAL CAVITY; the pericardial cavity is lined with the PARIETAL PERICARDIUM (PERICARDIAL SAC) and the heart is covered by the VISCERAL PERICARDIUM (EPICARDIUM) and the potential space between them is called the PERICARDIAL SPACE.

The abdominopelvic cavity consists of two cavities. The ABDOMINAL CAVITY hosts the primary organs of digestion and the PELVIC CAVITY hosts the internal reproductive organs.

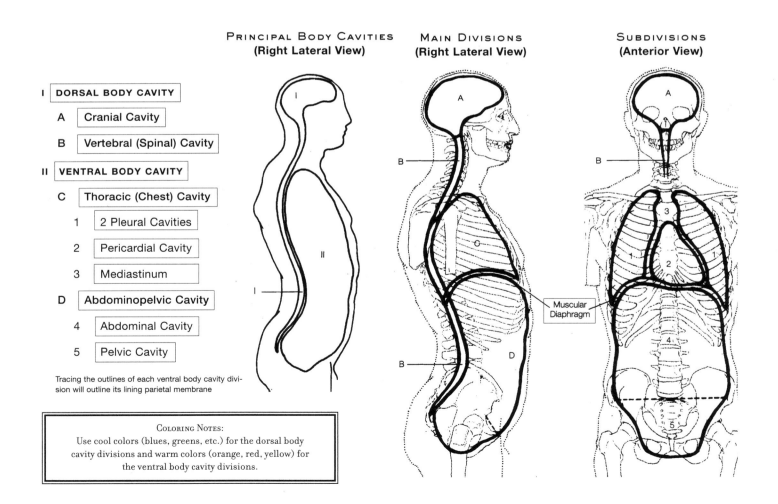

PRINCIPAL BODY CAVITIES
(Right Lateral View)

MAIN DIVISIONS
(Right Lateral View)

SUBDIVISIONS
(Anterior View)

I	DORSAL BODY CAVITY
A	Cranial Cavity
B	Vertebral (Spinal) Cavity
II	VENTRAL BODY CAVITY
C	Thoracic (Chest) Cavity
1	2 Pleural Cavities
2	Pericardial Cavity
3	Mediastinum
D	Abdominopelvic Cavity
4	Abdominal Cavity
5	Pelvic Cavity

Tracing the outlines of each ventral body cavity division will outline its lining parietal membrane

Muscular Diaphragm

> COLORING NOTES:
> Use cool colors (blues, greens, etc.) for the dorsal body cavity divisions and warm colors (orange, red, yellow) for the ventral body cavity divisions.

The two basic types of body membranes are SEROUS MEMBRANES and MUCOUS MEMBRANES. Both are composed of thin layers of connective tissue and epithelial tissue. Mucous membranes secrete a thick, viscous MUCUS for lubrication and protection. Serous membranes line the principal body cavities and their divisions, called CLOSED CAVITIES. Mucous membranes, associated with what are called the OPEN VISCERAL CAVITIES (mainly the tracts of the digestive, respiratory and genitourinary systems) line the LUMINA (hollow tubular canals) of the esophagus, stomach, intestines, trachea, uterus, oral and nasal cavities.

Several smaller cavities exist within the head. The ORAL (BUCCAL) CAVITY contains the TEETH and TONGUE primarily for DIGESTION (secondarily for RESPIRATION). The nasal cavity is for respiration. Two ORBITAL CAVITIES house the eyeballs and associated muscles, nerves and blood vessels for the SENSATION OF VISION. Two MIDDLE EAR CAVITIES house the ear ossicles (bones) for the SENSATION OF HEARING.

COMPARTMENTILIZATION
Cavities lined by membranes compartmentalize organs.
Four distinct compartments in the THORACIC CAVITY house organs.
Two PLEURAL CAVITIES house the LUNGS.
One MEDIASTINUM (anterior, posterior, superior and middle).
One PERICARDIAL CAVITY houses the HEART (middle mediastininum).

THE MEDIASTINUM

THE MEDIASTINUM

The space between the lungs extending from the STERNUM (breastbone) to the VERTEBRAL COLUMN (in the back) is called the mediastinum. The mediastinum houses ALL the contents of the thoracic cavity EXCLUDING the LUNGS and PLEURAE (serous membranes of the lungs).

S Superior Mediastinum

M Middle Mediastinum

The pericardial cavity, heart and pericardium comprise the middle mediastinum.

P Posterior Mediastinum

A Anterior Mediastinum

a Visceral Pleurae

b Parietal Pleurae

c Pleural Cavity

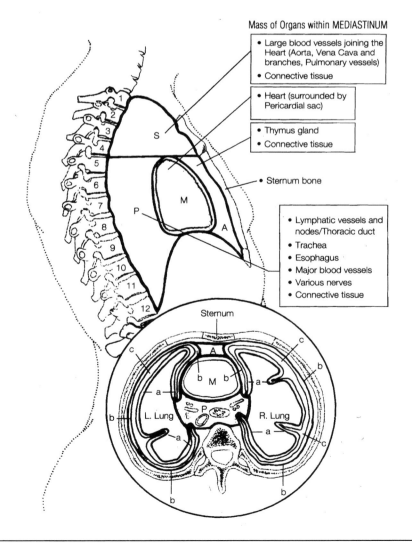

Mass of Organs within MEDIASTINUM

- Large blood vessels joining the Heart (Aorta, Vena Cava and branches, Pulmonary vessels)
- Connective tissue

- Heart (surrounded by Pericardial sac)

- Thymus gland
- Connective tissue

- Sternum bone

- Lymphatic vessels and nodes/Thoracic duct
- Trachea
- Esophagus
- Major blood vessels
- Various nerves
- Connective tissue

Chapter 3: Organization of the Body: Cells and Tissues

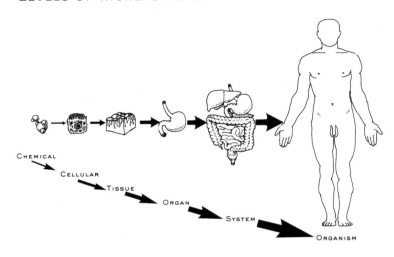

LEVELS OF INCREASING COMPLEXITY

CHEMICAL
CELLULAR
TISSUE
ORGAN
SYSTEM
ORGANISM

Basic Types of Interdependent Bodily Phenomena Requisite for Life

The entire constituents of the human body are living CELLS organized into TISSUES and ORGANS (and SYSTEMS), nonliving CONNECTIVE TISSUE FIBERS (elongated strand-like products of cells), and FLUID (nonliving secretions of cells/organs). Environmental factors upon which human life is totally dependent: oxygen, food, heat, pressure, protection (clothing and shelter), gravity, and size of the body in relation to the size of the earth.

Levels of Increasing Complexity

Each level represents an association of units from the preceding level. The CHEMICAL LEVEL is composed of atoms joined together to form various molecular structures. All chemical substances are essential for life maintenance. The basic structural and functional units of the organism are at the CELLULAR LEVEL, which is composed of chemical substances called PROTOPLASM. Small protoplasmic ORGANELLES within the cell carry out specific functions. Tissue consists of aggregated groups or layers of similar specialized cells (and their intercellular substance) whose individual, unique structure is directly related to the performance of a specific function. Similar cells secrete a nonliving MATRIX (liquid, semisolid, or solid) which spaces them uniformly and binds them together as tissue. (Secretory cells, however, do not have a binding matrix, and are solitary amidst another tissue.) Structures of definite form and function are in the ORGAN LEVEL. They are composed of two or more tissues joined together to form a structure that has a particular function. The association of organs working together to perform a

common function creates a SYSTEM. An organism has all body systems functioning together to constitute one LIVING INDIVIDUAL or ORGANISM. The systems do not work independently, but depend on a highly coordinated effort.

The ORGANIZATIONAL PHENOMENA are related to the structural complexity of the organism and include five types:

DEVELOPMENT—Cellular differentiation during morphogenesis changes unspecialized tissues into structurally and functionally distinct organs.

ORGANIZATION—Each body structure is genetically determined and adapted to specific functions.

ADAPTATION—Narrow tolerance levels result in highly specialized structures with low adaptability.

SYNTHESIS—Process of forming a complex substance from simple elements or compounds (such as protein, DNA, RNA, and fats).

SECRETION—Specific chemical compounds synthesized within specialized cells for intercellular communication and other purposes.

METABOLIC PHENOMENA are the sum total of all dynamic energy and material transformations that occur continuously within the living organism and its cells. ENERGY CHANGES involve the release of potential energy of food and its conversion into mechanical work (movement) or heat. MATERIAL CHANGES of substances occur during the life periods of GROWTH, MATURITY and SENESCENCE (old age).

ANABOLISM—Building-up process that converts ingested substances into the constituents of protoplasm.

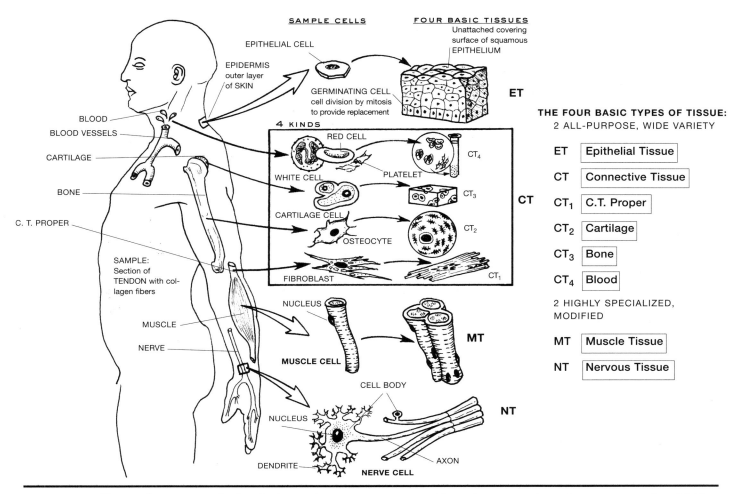

SAMPLE CELLS

EPITHELIAL CELL

EPIDERMIS
outer layer
of SKIN

GERMINATING CELL
cell division by mitosis
to provide replacement

4 KINDS

BLOOD

BLOOD VESSELS

CARTILAGE

BONE

C. T. PROPER

RED CELL

WHITE CELL

PLATELET

CARTILAGE CELL

OSTEOCYTE

SAMPLE:
Section of
TENDON with col-
lagen fibers

FIBROBLAST

MUSCLE

NERVE

NUCLEUS

MUSCLE CELL

CELL BODY

NUCLEUS

DENDRITE

AXON

NERVE CELL

FOUR BASIC TISSUES
Unattached covering
surface of squamous
EPITHELIUM

ET

CT_4

CT_3

CT_2

CT_1

CT

MT

NT

THE FOUR BASIC TYPES OF TISSUE:
2 ALL-PURPOSE, WIDE VARIETY

ET	Epithelial Tissue
CT	Connective Tissue
CT_1	C.T. Proper
CT_2	Cartilage
CT_3	Bone
CT_4	Blood

2 HIGHLY SPECIALIZED,
MODIFIED

| MT | Muscle Tissue |
| NT | Nervous Tissue |

CATABOLISM—Tearing-down process that breaks down substances into simpler substances, whose end products are usually excreted.

GROWTH—Growth processes occur both at the cellular and systems level. Newly formed daughter cells need to grow and mature to function. A genetically determined growth process of the organism arises when body structures reflect metabolic needs.

CELLS, TISSUES, ORGANS AND SYSTEMS

Because of their specialized nature within this multicellular organism, living CELLS, while the structural and functional units of the human body, do not function independently of one another. Aggregations of similar cells that perform specific functions are called TISSUES, usually similarly shaped and attached together. Non-living intercellular products/secretions (matrix fiber and fluid) of similar cells space them uniformly and bind them together to form tissue. An ORGAN is composed of two or more tissues joined together to form a structure that has a particular overall function; rising to higher levels of complexity, a SYSTEM is an association of organs working together to form a common function. All thirteen body systems functioning in a highly coordinated effort (none independent) comprise the human bodily ORGANISM.

All of the human body's organs and structures (which make up the thirteen systems), are developed from only four basic categories of tissue, determined by structure and function: two all-purpose, wide-variety: EPITHELIAL and CONNECTIVE; and two highly specialized and modified (packaged in groups): MUSCLE and NERVOUS. Each organ/structure is formed from an array of intricate combinations of these four tissues.

Basically, the epithelial tissue (EPITHELIUM) serves as a working surface for the organ or outer covering of body and organ surfaces, inner lining body cavities and lumen cavities (organs, vessels and ducts) and canals, and forms various glands. The connective tissue serves to hold the organ together, binding, supporting and protecting body parts. The muscle tissue contracts for organ movement or control of blood flow; and nervous tissue initiates and transmits nerve impulses from one body part to another to regulate activity and respond to stimuli.

Of the four categories of tissue, connective tissue (C.T.) offers the widest variety of different cell types, residing within either a solid, semisolid, or fluid matrix (the matrix mostly containing differing fiber structure). Dispersed among the fibers are FIBROBLASTS, the most numerous of connective tissue cells, whose job is the production of all the various fibers found in C.T. There are four kinds of connective tissue: CONNECTIVE TISSUE PROPER (five types in a semisolid matrix/collagen-elastic-reticular fibers), CARTILAGE (three types in a solid elastic matrix/various fibers), BONE (two types in a solid rigid matrix/calcium-magnesium salts) and BLOOD (fluid matrix). (See pages 32–37.)

THE CELLULAR LEVEL
THE GENERALIZED CELL

THE CELL: THE BASIC STRUCTURAL AND FUNCTIONAL UNIT OF LIFE

The phenomena that distinguish living from non-living things are exemplified in the basic structural and functional activities of highly organized microscopic units called CELLS. The human being is composed of between 60 and 100 trillion living cells, with wide variation in structure and function. Only a few hundred specific kinds exist. However, all cells have certain features in common. All cells are composed entirely of chemical substances called PROTOPLASM. A generalized cell consists of a relatively fixed percentage of protoplasmic compounds that are arranged into small functional structures called ORGANELLES, each one a specific working component of the cell. Protoplasm is divided into a single NUCLEUS and the CYTOPLASM. The nucleus is the hereditary and control center of the cell, without which the cell dies. The cytoplasm is all the protoplasm of the cell outside the nucleus.

PRINCIPAL INTERIOR ORGANELLES:
Function Compartmentalization within the Watery Hyaloplasm

1 | Endoplasmic Reticulum (ER)

2 | Ribosomes | — Microsomes

3 | Golgi Apparatus | —Golgi Complex

4 | Mitochondria

5 | Lysosomes | — Contain Enzymes

6 | Vacuoles

7 | Peroxisomes

8 | Centrosome | — Centrosphere

N | Nucleus

  N1 | Nucleolus

CYTOPLASMIC LATTICE

9 | Microtubules

10 | Microfilaments | – Microfibrils

COLOR GUIDELINES:
N = bright yellow; a = flesh; c = light blue;
8 = green; 4 = orange; 1 = light purple;
other organelles = your choice

CILIUM-FLAGELLUM

Nine bundles in each cilium (2 microtubules per bundle) + 2 in middle

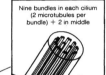

8-CENTROSOME: TWO CENTRIOLES

Nine bundles in each centriole (3 microtubules per bundle)

90°

CYTOPLASMIC LATTICE within the CYTOPLASM

3-GOLGI APPARATUS

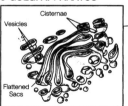

Cisternae
Vesicles
Flattened Sacs

1-ENDOPLASMIC RETICULUM

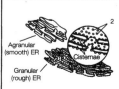

Agranular (smooth) ER
Cisternae
Granular (rough) ER

Size of microtubules in relation to other parts is exaggerated

(Size of All Parts Exaggerated)

Vesicles produced by Golgi apparatus

EXOCYTOSIS

MICROVILLI

CILIA-FLAGELLA

PINOCYTOSIS

INERT INCLUSION

4-MITOCHONDRION

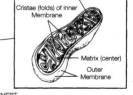

Inner Membrane
Cristae (folds) of Inner Membrane
Matrix (center)
Outer Membrane

a-CELL (PLASMA) MEMBRANE

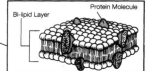

Bi-lipid Layer
Protein Molecule

BASIC CONSTITUENTS OF PROTOPLASM

Protoplasm is made up of certain ELEMENTS in CHEMICAL COMBINATION. There are four elements that make up most of the body. OXYGEN (O) is 62.2%, CARBON (C) 18.0%, HYDROGEN (H) 9.4% and NITROGEN (N) 2.6%. Most of the (O) and (H) elements combine to form water. The combination of (O) + (H) + (C) makes CARBOHYDRATES and FATS, which are the main sources of energy in living protoplasm. PROTEINS are formed by (O) + (H) +(C) + (N). They are the main building constituent of protoplasm. The basic function of all cells is the production of protein. The general fixed percentage of protoplasmic compounds of the cell is 80% water, proteins 15%, lipids (fats) 3% and the 1% carbohydrates, MINERALS and NUCLEIC acids.

Cells arise only from preexisting cells. New cells arise from cell division, either by MITOSIS or MEIOSIS (see WWW.MCMURTRIESANATOMY.COM for more information). Growth and development arise from increase and numbers of cells and differentiation of cells into different types of tissues.

The structure and form of a cell are closely correlated with its functions. Cells of one tissue differ from those of other tissues, depending on the specialized function they perform (in accordance with their modified structure).

CELL JUNCTIONS are cell-to-cell attachments, channels of communication and cellular bonding between close epithelial tissue cells.

A cell has the power of exercising the vital processes of life required for the maintenance of the life of the human organism as a whole. In that sense, the cell can be viewed as a MICROCOSM of the MACROCOSMIC world of humans. The essential life processes that are carried on at the cellular level are, in the human organism, delegated to the separate whole systems, which function together to constitute the living individual. (See chapter 16, Reproductive System for more information.) Before entering into a detailed chapter-by-chapter journey through the separate body systems (and the organs grouped into those systems), we will familiarize ourselves with the cellular level, how the tissues are derived from three embryonic germ layers and a discussion of the organization of the four primary types of body tissues and their principal subdivisions.

The essential life processes of the human organism involved with metabolic phenomena active at the cellular level are delegated to the separate systems level. For example, at the cellular level, CONTRACTILITY and MOVEMENT (life processes) are delegated at the systems level to the MUSCULAR SYSTEM. Likewise, INTEGRATON and CONTROL (as wells as IRRITABILITY and RESPONSIVENESS) at the cellular level are delegated to the NERVOUS SYSTEM and ENDOCRINE SYSTEM (as well as the INTEGUMENTARY SYSTEM; CIRCULATION to the CARDIOVASCULAR and LYMPH VASCULAR SYSTEMS; cellular RESPIRATION delegated to the RESPIRATORY SYSTEM; cellular NUTRITION (INGESTION, DIGESTION and ABSORPTION) to the DIGESTIVE SYSTEM; cellular EXCRETION delegated to the URINARY SYSTEM (as well as the RESPIRATORY and INTEGUMENTARY SYSTEMS); REPRODUCTION (cellular division) to the REPRODUCTIVE SYSTEM; DEFENSE AGAINST DISEASE to the LYMPHOID IMMUNE SYSTEM; METABOLISM and GROWTH (related to the cellular MITOCHONDRIA and NUCLEUS) to the REPRODUCTIVE and NERVOUS SYSTEMS. Other organizational phenomena at the cellular level can be viewed at the systems level: PROTECTION FROM THE ENVIRONMENT (the CELL MEMBRANE) is delegated to the INTEGUMENTARY SYSTEM; the interior cellular CYTOPLASMIC LATTICE can be viewed as the SKELETAL SYSTEM; the NUCLEUS can be imagined as the CENTRAL NERVOUS SYSTEM or BRAIN of the cell.

REPRODUCTION is a function of the reproductive system. The INTEGUMENTARY SYSTEM is delegated to CELL (PLASMA) MEMBRANE. CYPTOPLASMIC LATTICE is a system of the SKELETAL SYSTEM. The CENTRAL NERVOUS SYSTEM (BRAIN) is at the NUCLEUS.

CELL JUNCTIONS
The junction completely encircles each cell that is joined.

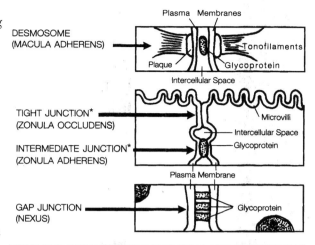

PRINCIPAL PARTS OF A GENERALIZED CELL

Protoplasm

N | Nucleus |
The largest organelle, hereditary and control center

Cytoplasm

ENVELOPING ORGANELLE

a | Cell (Plasma) Membrane |

b | Interior Organelles |

PARAPLASTIC BODIES

• | Inert Inclusions |

CLEAR, VISCOUS MATRIX

c | Hyaloplasm |

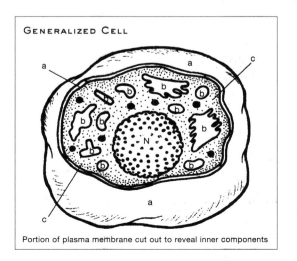

GENERALIZED CELL

Portion of plasma membrane cut out to reveal inner components

CELL DIVISION: MITOSIS

During the life of the human organism between birth and death, the body is constantly shedding and replacing cells, dividing at the rate of millions per second. In a twenty-four-hour period, trillions of cells are lost by the average adult from all parts of the body (through damage, disease, wear-and-tear, or death). Obviously, the SOMATIC (body) cell must contain within itself a self-duplicating process for continual cell replacement, by which the body grows, increases its body cells, and maintains its gradually changing form.

Each and every human body cell during its non-dividing activity carries within its NUCLEUS twenty-three matched pairs (forty-six) of DNA CHROMOSOMES. When it divides, a PARENT BODY CELL duplicates itself exactly into two new DAUGHTER CELLS (i.e., where there was one cell, now there are two). This entire duplication process consists of both MITOSIS (division of the nucleus where main cellular changes occur) and CYTOKINESIS (division of the CYTOPLASM). Each of the resulting two daughter cells contain the same number (forty-six) and kind (twenty-three matched pairs) of DNA chromosomes as the original parent cell (thus, hereditarily and genetically identical to each other). This "normal" method of cell division is used by all the cells in one's body except for those involved in producing SPERM and EGGS (see Chapter 16).

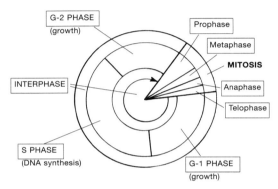

BODY CELL LIFE CYCLE

FREQUENCY OF MITOSIS:

The frequency of CELL LOSS is directly associated with the frequency of CELL REPLACEMENT via MITOSIS. Some cells are being lost continuously (cornea of the eye, outer epidermal layer of the skin [renewed about every two weeks], the stomach lining [renewed about every two to three days]); others not so frequently.

A collection of cells with a common function (usually similar in shape and attached together) is defined as a TISSUE. In the four types of tissue cells in the body, mitosis is very frequent in the two all-purpose wide-variety types, EPITHELIAL and CONNECTIVE tissues; in the two modified highly specialized types, mature striated MUSCLE tissue cells not very frequently at all; and in the nerve tissue cells, very rarely throughout the life of the organism.

Uncontrolled mitosis or CANCER, which is invasive, spreads directly into surrounding tissues, and tends to METASTASIZE to new sites disseminated via the BLOOD and LYMPH VASCULAR SYSTEMS. There are two types of cancer: CARCINOMAS have their origin in epithelial tissues while SARCOMAS develop from connective tissues (and structures whose origin was in EMBRYONIC MESODERMAL TISSUES).

The Nucleus is the supervisor of cell activity, exercising control of cellular development and function, the vital body in the protoplasm of the cell, containing the chromosomes and the essential agent in growth, metabolism, reproduction, and transmission of cell characteristics.

1 | Double Nuclear Membrane

Double layer of protein that acts as a boundary for the nucleus and as a semipermeable membrane.

2 | Perinuclear Cisterna

Space between layers of the nuclear membrane (envelope).

3 | Nuclear Pores

The two membranes of the nuclear envelope join at regular intervals to form nuclear pores for migration of ribosomal RNA into the hyaloplasm.

4 | Nucleoplasm

Also called kairyolymph, the specialized protoplasmic substance of the nucleus.

5 | Nucleolus

Spheroid body within the nucleus where RNA (ribonucleic acid), especially ribosomal RNA, is produced and stored.

6 | Chromatin

Substance containing the genetic material DNA (deoxyribonucleic acid), bistones, and other proteins. It is present in the nondividing or normal metabolic phase of the cell (interphase). It is a dispersed form of the chromosomes, a threadlike extended mass of DNA.

7 | Prophase "Chromosomes"

Rodlike, "packaged" DNA that becomes microscopically visible in the dividing phase of the cell (mitosis); the "replication" of the DNA molecule in preparation for cell division takes place during mid-INTERPHASE (S-phase). On the chromosomes are the GENES, the carriers of inheritable characteristics.

SISTER CHROMATIDS

8 | Chromatid 1

9 | Chromatid 2

10 | Centromore

11 | Split Centromere

12 | Centrosome

13 | Centriole

14 | Aster

15 | Spindle Fiber Apparatus

16 | Cell Membrane

CELL LIFE CYCLE PHASES
INTERPHASE

G-1 | Growth

S | DNA Synthesis Replication

G-2 | Growth

MITOSIS

P | Prophase

M | Metaphase

A | Anaphase

T | Telophase

> COLORING NOTES:
> Choose warm colors for the membranes and cool colors for the centrioles and asters. Use neutral colors for the cytoplasm and nucleoplasm. Start coloring at INTERPHASE and repeat colors appropriately in succeeding phases.

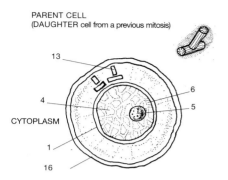

PARENT CELL
(DAUGHTER cell from a previous mitosis)

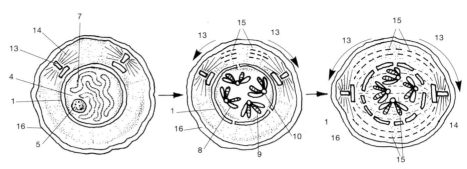

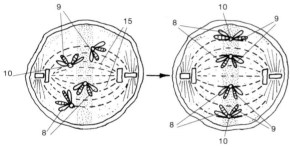

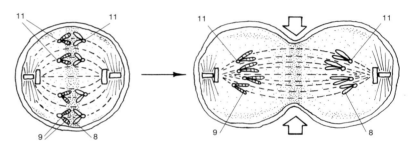

INTERPHASE (I), THE NON-DIVIDING PHASE

The body cell, when it is growing or carrying out its everyday life functions (the majority of its life) is in a long, non-dividing phase called INTERPHASE (METABOLIC or QUIESCENT PHASE). This phase follows a preceding mitosis (M) phase, if a cell is predisposed to divide.

Sandwiched between two growth/life phases (G-1, G-2) is the S-PHASE, where DNA synthesis (or duplication of new DNA) occurs within the nucleoplasm in preparation for cell division. The replicated DNA is now in its extended form, called CHROMATIN, a diffuse threadlike network of fine fibril strands (viewable in an electron microscope). The DNA chromatin material has now doubled, exactly replicating itself. The cell membrane, nucleus, double nuclear membrane and nucleolus are intact.

MITOSIS (M), FOUR DIVIDING PHASES: I. PROPHASE:

This is where mitosis begins. Each pair of centrioles now project a series of radiating fibers (microtubules) called ASTERS. The centriole pairs now separate and move to opposite poles of the cell (on opposite sides of the nucleus), each with its own aster. They both will stay "locked" in their respective positions throughout the dividing process. Having duplicated in the S-Phase, the chromatin DNA fibrils now all coil into condensed, compact rod-like structures called PROPHASE CHROMOSOMES (viewable in an ordinary light microscope). As the prophase chromosomes become shorter and thicker, it is noted that each one of the forty-six is actually made up of two subunits called CHROMATIDS. Each chromatid being attached to its chromatid pair by a small spherical body called a CENTROMERE. The nuclear membrane begins to disappear; the nucleolus is no longer visible.

2. METAPHASE:

The centromere of each chromatid pair attaches itself to a spindle fiber. The chromatid pairs now line up on the equatorial plane of the spindle fiber. The apparatus has now moved into the open area of the nucleus and moves to the center of the cell, half (forty-six) on one side, half (forty-six) on the other. By now, the nuclear membrane has disappeared completely.

3. ANAPHASE:

All centromeres divide, leaving a sister centromere still attached to each chromatid, which now tends to separate from its pair. There is a movement now of complete identical sets of chromatids (now officially called chromosomes as they are now isolated) to opposite poles of the cell. Each sister centromere (attached to a spindle fiber) seems to drag the trailing parts of its chromosome to its ipsilateral (same side) pole of the cell. As the anaphase ends, the sister chromosomes (identical sets) reach their respective poles (forty-six on each pole, four on each pole in the illustration) and the cell begins to be pinched in two.

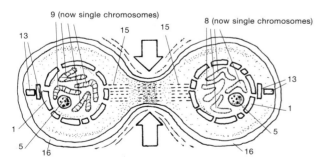

9 (now single chromosomes)
8 (now single chromosomes)

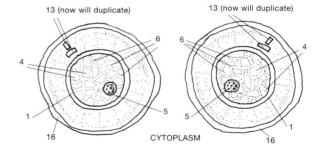

13 (now will duplicate)
13 (now will duplicate)

CYTOPLASM

4. TELOPHASE:

As a result of CYTOKINESIS (cell division), the cytoplasm and organelles have duplicated during the mitotic phases, and now contribute and flow into the newly forming cells. The telophase can be viewed as nearly a reverse of prophase. The spindle fiber apparatus and asters begin to disappear, while new double nuclear membranes begin to form to enclose the sister chromosomes in each new cell and nucleoli begin to reappear in the newly constituted nuclei while the centromeres disappear.

TWO NEW DAUGHTER CELLS

Each cell now duplicates its single centriole pair (not shown), so that each cell now has two centriole pairs (two nonmembranous centrosomes) ready for the next division. When what was the parent cell completely cleaves itself into the two new daughter cells, the process of mitosis is complete, and each new cell immediately enters a new cycle, beginning with interphase, beginning the cyclic process described above all over again.

THE TISSUE LEVEL: EMBRYONIC GERM LAYERS

TISSUE DERIVATIVES

Tissues, organs, and systems derivce from the primary embryonic germ layers. The four basic tissue groups, and all their variations, are established during the development of the EMBRYO, beginning at a time when three specific cellular embryonic germ layers (comprising the embryonic disc) have been laid down and differentiated from the earlier, primordial embryonic stem cells (inner cell mass). All the tissues of the body develop from just these three basic germ layers: ECTODERM, MESODERM and ENDODERM. The boundaries of each germ layer are not specifically fixed in any rigid manner: each layer's cells divide, migrate, aggregate and differentiate in predictable precise patterns to form the various organs and systems. As overall differentiation continues within the embryo, the organs of the body are formed (each of which is composed of a specific arrangement of tissues). Through a process called INDUCTION, one germ layer may cause the cells of another layer to become specialized or differentiate to form specific organs or tissues.

Most tissues are composed of combinations of these three germ layers, but many major tissues, organs and systems can trace their primary origin back to a specific germ layer.

In simplest terms, the ectoderm is divided into two areas, the SURFACE ECTODERM (from whence is derived the outer covering surface epithelial tissue); and a specialized region called the NEUROECTODERM, which also gives rise to numerous tissues, including the nervous tissue. Major mesoderm derivatives are connective tissue (all four kinds—C.T. proper, cartilage, bone, blood) and all muscle tissue: smooth and striated). Endoderm derivatives are the inner lining epithelial tissues of open visceral cavities.

PRE-EMBRYONIC PERIOD

Day 1: Within the entire gestational period— a span of thirty-eight weeks (266 days)— the prenatal (before birth) development of the human being grows from one single cell into a complex organization of over 200 million cells at birth. The development of the eventual human organism is initiated at CONCEPTION, when a female sex cell—an OVULATED OVUM (egg)—is fertilized by a male sex cell—a SPERMATOZOA (sperm). The resulting single ZYGOTE (fertilized egg) contains in its nucleus the combined DNA material from the pronuclei (half the amount of normal chromosomes) of each sex cell. The now normal twenty-three pairs of chromosomes contain all the genetic information required for the development and differentiation of all body structures.

Day 3: We each spend about thirty hours as a single cell. After that time, the cleavage process begins with the zygote's first mitotic division resulting in two identical daughter cells called BLASTOMERES, surrounded by a translucent protective protein layer, the ZONA PELLUCIDA. More cleavages occur as the structure passes down the uterine tube of the mother and by the third day, enters the uterine cavity. Composed of sixteen cells, it is now called the morula (about the same size as the original zygote due to the absence of nutrients in the cells necessary for growth).

Day 7 (End of Week 1): In about three more days, the morula mass remains unattached in the uterine cavity, drawing in fluid from the cavity in isolated spaces. As the spaces fuse they form a large central fluid-filled cavity called the BLASTOCOELE, the zona pellucida begins to disappear, and two distinct groups of cells form: An outer

SPHERICAL FORMATIONS

1 Zygote

2 Morula

3 Blastocyst

2-LAYERED EMBRYONIC DISC

A1 Epiblast | Embryonic Ectoderm

B1 Hypoblast | Embryonic Endoderm

DEVELOPING STRUCTURES

D Trophoblast

D1 Extraembryonic Mesoderm

D2 Connecting Body Stalk

E Inner Cell Mass
Embryoblast

CAVITIES

F Amniotic Cavity

Secondary Yolk Sac:

G Blastocyst Cavity

H Gut Cavity

I Extraembryonic Coelom
Chorionic Cavity

3-LAYERED EMBRYONIC DISC

A2 Ectoderm

B2 Endoderm

C1 Mesoderm

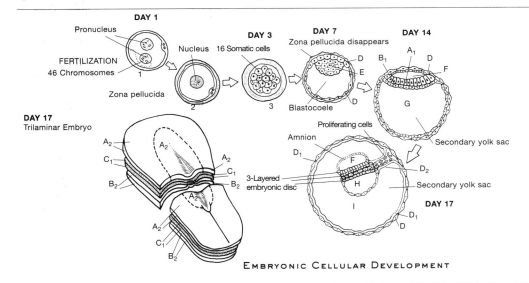

DAY 1
Pronucleus
FERTILIZATION
46 Chromosomes
Zona pellucida
Nucleus
DAY 3
16 Somatic cells
DAY 7
Zona pellucida disappears
Blastocoele
DAY 14
Secondary yolk sac
Proliferating cells
Amnion
3-Layered embryonic disc
Secondary yolk sac
DAY 17
DAY 17
Trilaminar Embryo

EMBRYONIC CELLULAR DEVELOPMENT

flattened wall (single layer of cells) called the TROPHOBLAST—the future placenta (embryonic contribution)—and on one end of the ball a small, inner aggregation of cells called the INNER CELL MASS (embryoblast)—the future human embryo. The latter, composed of from 50-150 cells, are also referred to contemporarily as EMBRYONIC STEM CELLS. These primordial cells are totipotent, meaning they are able to differentiate into all derivatives of the three primary germ layers, capable of developing into each of the more than 200 cell types of the adult body. As these two groups of cells form, the free-floating morula becomes known as the BLASTOCYST.

Between the fifth and seventh day, implantation of the blastocyst into the uterine wall occurs. The post-migrating epiblast becomes the ectoderm during days fourteen to seventeen.

A GUT CAVITY (secondary yolk sac) has also been completed bordering the hypoblast and a connecting body stalk (future umbilical cord) filled with proliferating cells is formed from the extraembryonic mesoderm (membrane) derived from the trophoblast and is attached between the trophoblast and one end of the embryonic coelom. A large cavity, the EXTRAEMBRYONIC COELOM (which will become the chorionic cavity), has been formed from fused pocketing spaces within the extraembryonic mesoderm and surrounds the amniotic cavity, the gut cavity and the disc (except where the connecting stalk is located).

At this stage, the three primary germ layers are officially constituted, and the three-layered embryonic disc is formed, consisting of ECTODERM, MESODERM and ENDODERM. The trilaminar embryo is complete, and this demarks the end of the pre-embryonic period.

EMBRYONIC PERIOD

Day 18 to End of Week 8: Day 18 marks the beginning of the embryonic period, which lasts until two months from conception have been completed. (Often, the embryonic period is referred to as lasting from the fourth week to the eighth week (a total of five weeks), but it does actually begin earlier, during the 3rd week (Day 18 on, when the central nervous system [neuroectoderm] and cardiovascular system [mesoderm] begin to form.) A week after Day 18 (Day 24) the flat embryonic disc has perceptibly given way to a rounded form within the amniotic cavity (via a process called EMBRYONIC FOLDING) with a definite head end and a tail end.

This embryonic period is a crucial one for human development, because this is the time for the developmental beginnings of all the major external and internal structures of the body. At the end of the embryonic period, at the end of the eighth week, when the human embryo is the size of a raspberry, all the main organs and organ systems have begun to develop, and all their foundations have been laid down—although at this time their functions are minimal at best. At this point, having gone through what many would describe as strange and bizarre morphological changes, the embryo begins to take on a striking human appearance.

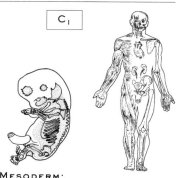

C₁

MESODERM: MUSCLE TISSUE AND CONNECTIVE TISSUE

Smooth Muscle:
Viscera, vessels and ducts

Striated Skeletal Muscle
Head and neck
Trunk and back
Extremities
Cardiac muscle (heart)

Connective Tissue:
C.T. Proper
Cartilage
Bone
Blood

Cardiovascular and Lymphatic Systems:
Blood and lymph cells, bone marrow
Blood and lymphatic vessels
Lymphoid immune tissue

Inner Lining Epithelial Tissue of:
Joint cavities
Body cavities

Serous Membranes:
Pleura
Pericardium
Peritoneum

Urogenital System:
Kidneys and ureters
Gonads (ovaries and testes)
Genital ducts and accessory glands

Dermis (skin inner layer)
Dentin of teeth
Outer adrenal cortex
Spleen

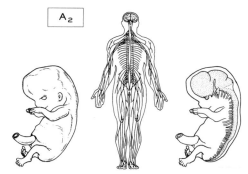

A₂

ECTODERM: SURFACE ECTODERM: OUTER COVERING EPITHELIAL TISSUE EPIDERMIS (SKIN OUTER LAYER)

Skin Derivatives:
Hair
Nails
Cutaneous (skin) glands

Head Region:
Portions of sensory organs:
Sensory epithelia of eye, ear, nose
Lens of eye
Inner ear
Enamel of teeth
Anterior lobe of pituitary gland
Linings of oral and nasal cavities
Mammary glands
Linings of anal and vaginal cavities
(continuous with the skin)

NEUROECTODERM: NERVOUS TISSUE AND SENSORY RELATED TISSUE

Neural Tube:
Central Nervous System (CNS)
Brain
Spinal cord
Retina of eye
Posterior lobe of pituitary gland
Pineal body

Neural Crest:
Coverings of the CNS
Peripheral Nervous System
Cranial, spinal and autonomic ganglia
    and nerves
Ensheathing cells of PNS
Inner adrenal medulla
Pigment cells of dermis (skin inner
    layer)
Branchial arch cartilage origin of
    connective tissues, bone and
    muscle
Head mesenchyme

B₂

ENDODERM: INNER LINING EPITHELIAL TISSUE OF OPEN VISCERAL CAVITIES:

Gastrointestinal (GI) Tract

Respiratory Tract:
Larynx
Trachea
Bronchi

Epithelium of:
Liver
Pancreas

Epithelial Parts of:
Pharynx
Pharyngotympanic tube
Tympanic cavity
Tonsils
Thyroid and parathyroid glands
Thymus
Lungs
Urinary bladder
Urethra and vagina
Urachus

The Four Basic Tissue Groups

Specialized and Modified Tissues: Muscle and Nervous Tissues

Of the four principal groups of tissues, two highly specialized groups are the MUSCLE TISSUE and the NERVOUS TISSUE. There are also two all-purpose, wide-variety groups: EPITHELIAL TISSUE and CONNECTIVE TISSUE.

Muscle tissue is derived from MESODERM. The outstanding characteristic of muscle tissue is the ability to shorten or contract, effecting movement of the body or an organ within the body for movement of materials. They are so specialized for contraction that the tissue cannot replicate after birth. Muscle cells (fibers) are elongated in the direction of contraction. There is little intercellular matrix with the fibers lying close together. The three kinds are SMOOTH, CARDIAC and SKELETAL. The properties of action consist of contractability, conductivity, irritability and elasticity.

Nervous tissue is derived from ECTODERM (neuroembryonic ectoderm). It initiates and conducts impulses for coordination of body activities. The two kinds are NEURONS and NEUROGLIA. Neurons are specialized in irritability, conductivity and integration and are incapable of mitosis. Neuroglia are five times as abundant as neurons and they support and assist neurons. They are nonreceptive and nonconductive. There are six types of neuroglia (See Chapter 8: Nervous System). Master tissues of the body are the nervous tissue and skeletal muscle. These tissues work together in receiving stimuli from the external and internal environment and then in reacting to them. They are the most highly specialized tissues in the body.

TWO HIGHLY SPECIALIZED AND MODIFIED GROUPS

> Muscle Tissue

> Nervous Tissue

TWO ALL-PURPOSE, WIDE-VARIETY GROUPS

> Epithelial Tissue

> Connective Tissue

Components of Tissue

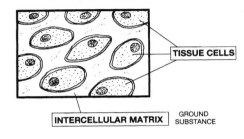

TISSUE CELLS

INTERCELLULAR MATRIX — GROUND SUBSTANCE

> **Coloring Notes:**
> Color the names of the four principal groups of tissues, then use those same colors as names repeat in further pages. Try reddish brown for muscle, gray for nervous, pinkish for epithelial and light blue for connective tissue.

Two Kinds of Nervous Tissue:

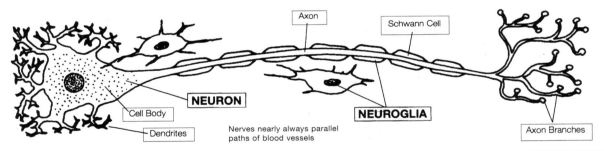

Axon

Schwann Cell

NEURON

NEUROGLIA

Cell Body

Dendrites

Nerves nearly always parallel paths of blood vessels

Axon Branches

Three Kinds of Muscle Tissue:

Smooth muscle is spindle-shaped, nonstriated and has a single nucleus. It is the least specialized and it has a slow, rhythmical contraction and no voluntary control (involuntary). It is found in walls of hollow internal organs (VISCERA and BLOOD VESSELS).

CARDIAC MUSCLE has striations and is more highly specialized. It is characterized by rapid, rhythmical contraction that spreads through the whole muscle mass. There is no voluntary control (involuntary) and it is found only in the HEART WALL (myocardium).

INTERCALATED DISCS adhere cells to cells. Branching is present, but no crossing of protoplasm.

SKELETAL MUSCLE is cylinder-shaped and multinucleated. This is the most highly specialized kind of muscle tissue. It has marked striations and very rapid, powerful contractions of individual fibers and allows for conscious control (voluntary). It is found in all skeletal muscles (head, trunk, limbs), tongue, pharynx, and upper esophagus. It is formed in thick bundles with no branching.

SMOOTH MUSCLE
(VISCERAL, PLAIN)

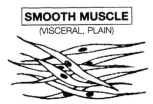

CARDIAC MUSCLE

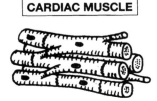

SKELETAL MUSCLE

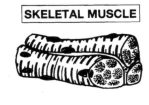

Two Kinds of All-Purpose Tissues: Epithelial and Connective

Of the four principal groups of tissue, there are two all-purpose, wide variety groups: EPITHELIAL TISSUE and CONNECTIVE TISSUE. Here we will discuss the two kinds of epithelial tissue: CLOSED/LINING EPITHELIUM (SIMPLE and STRATIFIED TYPES) and GLANDULAR EPITHELIA. Epithelia is derived from ECTODERM, MESODERM and ENDODERM (the three layers of the embryo). It covers body and organ surfaces and lines body cavities and lumina. Epithelia forms various glands, the EPIDERMIS of the skin and the surface layer of mucous and serous membranes. It consists of one or more cellular layers. One layer is called SIMPLE EPITHELIUM, and multilayered is known as STRATIFIED EPITHELIUM. It has tightly packed cells with little INTERCELLULAR MATRIX between them. It is typically avascular (without blood vessels) and rests on a basement membrane. The two kinds are COVER and LINING EPITHELIUM and GLANDULAR. The general functions are protection, absorption, excretion and secretion. Specialized functions consist of movement of substances through ducts, production of spermatozoa, reception of stimuli and an excellent ability to regenerate (as frequently as every twenty-four hours).

CONNECTIVE TISSUE is derived from the MESODERM. All other connective tissues differentiate from an EMBRYONIC connective tissue called MESENCHYME: MESENCHYMAL CELLS in a semisolid, jellylike matrix with fine fibrils. The connective tissues are located throughout the body. The four kinds of connective tissues are connective tissue proper, cartilage, bone and blood. Its general functions are to support and bind other tissues, and tissues and parts. It stores nutrients and manufactures protective and regulatory materials. The bulk of connective tissue consists of an EXTRACELLULAR MATRIX, whose nature gives each type of connective tissue its unique properties. The cells of connective tissue are few in number compared with the matrix substance. Connective tissues are highly vascular (except for cartilage).

The five types of connective tissue proper are LOOSE (areolar), ADIPOSE (fatty), DENSE FIBROUS (regular and irregular), ELASTIC and RETICULAR. Two dense connective tissues are CARTILAGE and BONE. The only type of connective tissue free-floating in a fluid matrix is BLOOD.

Two Kinds of Epithelia
Cover/Lining and Glandular

| Simple | Single Layer |

Capillary

| Stratified | Multilayered |

| Glandular |

Basement membrane: glycoprotein from the epithelial cells, collagen and reticular fiber mesh from underlying connective tissue

Connective Tissue Proper (Five Types)

Four types of cells in a semisolid matrix with thick collagenous and elastic fibers, one type of cell in a semisolid matrix with a fine network of extracellular reticular fibers:

| Loose Areolar | | Adipose Fatty | | Dense Fibrous | | Elastic | | Reticular |

Dense Connective Tissue (Two Types)
Cartilage and Bone

| Cartilage | (three varieties)

Cells in a solid elastic matrix (with fibers)

Bone (two varieties)

· Cells in a solid rigid matrix impregnated with calcium and magnesium salts
· Forms rigid body framework
· COMPACT and SPONGY varieties

Non-Fibrous Free-Floating Connective Tissue (One Type)

| Blood | · Cells in a fluid matrix (fibrils not present until clotting occurs)
· Cells float free in matrix

HYALINE   FIBROCARTILAGE   ELASTIC

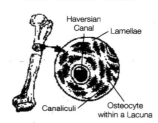

Haversian Canal   Lamellae

Canaliculi   Osteocyte within a Lacuna

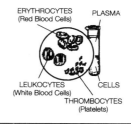

ERYTHROCYTES (Red Blood Cells)   PLASMA

LEUKOCYTES (White Blood Cells)   CELLS

THROMBOCYTES (Platelets)

Epithelial Tissue Detailed

Two General Kinds of Epithelial Tissue

1. COVERING AND LINING EPITHELIA

I. SIMPLE (ONE LAYER)

II. PSEUDOSTRATIFIED**
(ONE LAYER)

III. STRATIFIED
(TWO OR MORE LAYERS)

**Pseudostratified may be regarded as simple epithelium because each cell is in contact with the basement membrane.

LOWER SURFACE SUPPORT OF COVERING AND LINING EPITHELIA

a | Basement Membrane

b | Underlying Supportive Connective Tissue

2. GLANDULAR EPITHELIA

I. | Exocrine Glands

II. | Endocrine Glands

COMPONENT STRUCTURE OF GLANDULAR EPITHELIA:

c | Duct of Gland

d | Secretory Portion

Covering and Lining Epithelia

I. SIMPLE (1 LAYER)

1 | Simple Squamous

2 | Simple Cuboidal

3 | Simple Columnar Cylindrical

II. PSEUDOSTRATIFIED (1 LAYER)**

4 | Pseudostratified Ciliate Columnar with Goblet Cells

III. STRATIFIED (MULTILAYERED)

5 | Stratified Squamous*

6 | Stratified Cuboidal

7 | Stratified Columnar

8 | Stratified Transitional

**In the skin, stratified squamous tissue cells contain the pigment keratin. Simple squamous epithelium is classified either as endothelium (which lines the blood and lymph vessels and the heart) or mesothelium (which lines the serous body cavities and visceral organs).

Glandular Epithelia

I. EXOCRINE GLANDS
Secretion reaches an epithelial surface either directly or through a duct.

STRUCTURAL CLASSIFICATION:

9 | Unicellular | Goblet Cell

Multicellular

10 | Simple

Tubular
Branched Tubular
Coiled Tubular
Acinar
Tubuloacinar

11 | Compound

Tubular
Acinar
Tubuloacinar

FUNCTIONAL CLASSIFICATION:

12 | Holocrine

Gland and/or its secretion consists of altered cells of the same gland.

13 | Merocrine

Glandular cell remains intact during elaboration and discharge of secretion.

14 | Apocrine

Secretory cells contribute part of their protoplasm to the material secreted.

II. ENDOCRINE GLANDS:
Secretion is directly into the bloodstream.

The EPITHELIUM is the layer of cells forming the epidermis of the skin and the surface layer of mucous and serous membranes. The cells rest on a BASEMENT MEMBRANE and lie close to each other with little intercellular material between them. The epithelium may be either simple (consisting of a single layer) or stratified (consisting of several layers). Epithelium cells may be flat (SQUAMOUS), cube-shaped (CUBOIDAL) or cylindrical (columnar). Modified forms of epithelium include CILIATED, PSEUDOSTRATIFIED, GLANDULAR and NEUROEPITHELIUM.

Neuroepithelium is a specialized epithelial structure forming the termination of a nerve of special sense (gustatory cells, olfactory cells, inner ear hair cells, and rods and cones of the retina—

see chapter 8: Nervous System). The glandular epithelium lies in clusters deep to the covering and lining epithelium. A GLAND may consist of one cell (unicellular GOBLET CELLS, which secrete mucus) or two groups of highly specialized multicellular epithelium that secrete substances into ducts or into the blood: the EXOCRINE GLANDS and the ENDOCRINE GLANDS (see chapter 9: Endocrine System). Exocrine glands secrete their products into ducts or tubes that empty at the surface of the covering and lining epithelium (at the skin surface or into the cavity or lumen of a hollow organ): enzymes, oil or sweat. Endocrine glands secrete their products (HORMONES) directly into the blood.

I. SIMPLE (1 Layer)

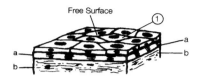

Free Surface

①
a
a
b
b

②
a
b
a
b

Unicellular
Exocrine Gland:
Goblet Cell

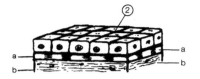

Ciliated
Nonciliated
③
③
③
g
a
a
b
b

II. PSEUDOSTRATIFIED (1 layer)

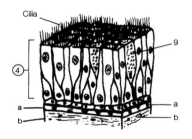

Cilia
9
④
a
b

II. ENDOCRINE GLANDS

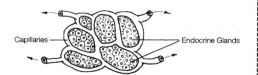

Capillaries
Endocrine Glands

III. STRATIFIED (2 or more layers)

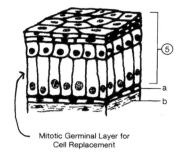

⑤
a
b

Mitotic Germinal Layer for
Cell Replacement

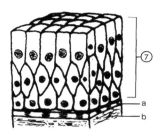

⑦
a
b

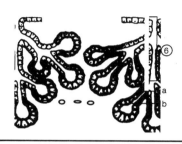

⑥
a
b

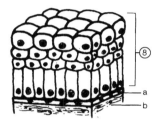

⑧
a
b

GLANDULAR EPITHELIA

I. EXOCRINE GLANDS

c
d

TUBULAR
BRANCHED TUBULAR
ACINAR
TUBULOACINAR

COILED TUBULAR

c   d

SIMPLE

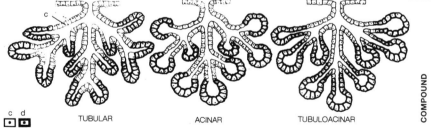

c

c   d

TUBULAR
ACINAR
TUBULOACINAR

COMPOUND

STRUCTURAL CLASSIFICATION

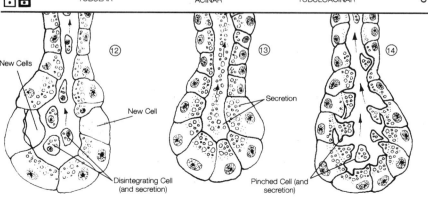

New Cells

⑫

New Cell

⑬

Secretion

⑭

Disintegrating Cell
(and secretion)

Pinched Cell (and
secretion)

FUNCTIONAL CLASSIFICATION

Connective Tissue Detailed

MESENCHYME—EMBRYONIC CONNECTIVE TISSUE:

The four kinds of connective tissue (CONNECTIVE TISSUE PROPER, CARTILAGE, BONE and BLOOD) are all derived from embryonic mesoderm, from which it migrates and gives rise to all other kinds of connective tissue. The cytoplasmic processes of the mesenchymal cells are drawn out to touch neighboring cells. The MUCOUS CONNECTIVE TISSUE (not shown) is a type found only in the umbilical cord of the fetus.

[In this section, we will only discuss connective tissue proper and cartilage. BONE (OSSEOUS) TISSUE, the hardest and most rigid connective tissue, is discussed in chapter 5. BLOOD (VASCULAR) TISSUE, specialized, viscous connective tissue, the only connective tissue with a fluid matrix, is discussed in chapter 11.]

I. CONNECTIVE TISSUE PROPER: SEMISOLID MATRIX

Common elements of connective tissue proper are collagen (white) fibers, elastic (yellow) fibers, reticular fibers, fibroblasts (fibrocytes), macrophages (histiocytes), plasma cells, fat cells (adipocytes) and matrix.

1 Loose (Areolar)

It is also called LOOSE FIBROUS PACKING tissue between organs. It forms sheaths around muscles, nerves, and blood vessels.

2 Adipose (Fatty)

It provides a protective cushion for organs and is the insulating layer in deep skin. The cytoplasm of fat cells is displaced by stored fat reserves.

3 Dense Fibrous

It is very strong and inelastic, yet pliable (e.g., ligaments, capsules of joints, heart valves).

4 Elastic

It is very strong, flexible and extensible (e.g., in walls of blood vessels and air passages).

5 Reticular

The reticular cells are phagocytic (capable of ingesting particles). Reticular fibers form a three-dimensional net for the framework of the spleen, lymph nodes, and bone marrow.

II. CARTILAGE: SOLID ELASTIC MATRIX

Common elements of cartilage are collagen (white fibers), elastic (yellow) fibers, lacuna, chondrocytes and matrix.

6 Hyaline (Gristle)

It is firm, yet resilient and is the transition stage in bone formation (e.g., in air passages, ends of bones at joints).

7 Fibrocartilage

It is also called white fibrocartilage. It is tough, resistant to stretching and has a shock-absorbing function between vertebrae.

8 Elastic

It is also called yellow fibrocartilage. It is flexible and resilient (e.g., in the larynx and ear).

I. CONNECTIVE TISSUE PROPER

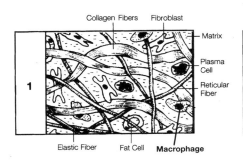

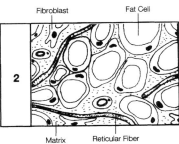

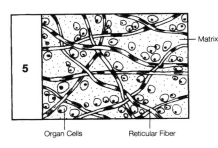

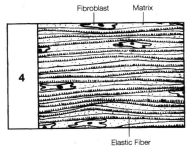

REGULAR AND IRREGULAR CONNECTIVE TISSUE

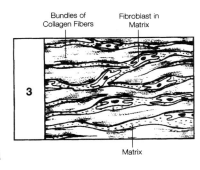

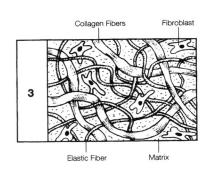

II. CARTILAGE

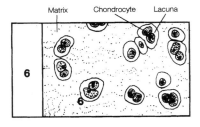

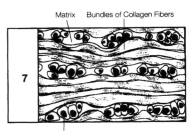

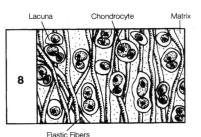

COLORING NOTES:
Color the names of the nine types of connective tissue and then use the same colors in the numbered boxes alongside the drawings—or use pencil to try to continue the drawing into each numbered box.

Somatic and Visceral Tissue Integration

Viewed regionally, the physical composition of the human body as a whole, interior to and including the skin, can be described as three major areas:
- The BODY CAVITY (all closed hollow confined spaces and open passageways combined) lined internally with associated membranes/linings that face the spaces.
- The visceral organs/structures lying within or passing through the hollow confined body cavity spaces (organs may or may not contain inner open cavities); or surrounding or containing the body cavity's open passageways.
- The BODY WALL comprising all the somatic organs/structures other than the visceral organs/structures and lying outside the body cavity—the musculo-skeletal framework of the body—examples: the upper and lower extremities, the perineum, the thoracic and abdominopelvic walls, the back wall, the neck wall, and the face-head wall—and covered by the skin, which separates the human body from the outside physical environment.

THE BODY CAVITY

The body cavity can be divided into two main cavity types: CLOSED BODY CAVITIES (more akin to the body wall, closed to the outside world); and OPEN VISCERAL CAVITIES: TRACTS (found within various visceral organs or structures that are open to the outside world).

**Special Note: Other minor cavity types to be considered would be: smaller cavities within the head region and the inner central lumen spaces of blood and lymphatic vessels (as well as the heart and various other organs) that do not have secreting membranes lining them. They have a form of non-secreting flat squamous lining epithelium called endothelium that face their inner lumens.

THE CLOSED BODY CAVITIES

The DORSAL body cavity and VENTRAL body cavity are sometimes referred to as the "shut sacs." They are not open to the outside of the body and do not communicate with the air. If any organ residing innately within (or passing through) these closed cavities possesses an interior cavity, central lumen or canal (open passageway) within itself, these internal spaces do not open into these larger closed cavities.

The membranes of the dorsal body cavity are non-secretory (containing no epithelia) and composed of thick, tough, fibrous connective tissue called DURA MATER; those of the ventral body cavity are thin, watery serous membranes, which contain lymph-like-secreting epithelia in addition to thin connective tissue.

The dorsal body cavity is divided into two cavities: the upper spherical cranial cavity, which houses the brain, and the lower stalk-like vertebral (spinal) cavity or canal, which houses the spinal cord—the cavities connecting to one another through the circular FORAMEN MAGNUM opening at the base of the skull (the brain and spinal cord together also express a continuous organic structure).

The dorsal body cavity is lined with a thick, tough white fibrous connective tissue membrane, the dura mater, which is in contact, for the most part, with the bone. The cranial dura mater is double-layered, and fused over most of the brain. A thicker, outer PERISOTEAL layer adheres to the cranial PERIOSTEUM of the skull. A thinner, inner MENINGEAL layer turns inward at specific points to follow the general contour of the brain. The spinal dura mater has only one layer, similar to and continuous with the meningeal layer of the cranial dura mater. It forms a tubular dural sheath as it continues into the vertebral canal and surrounds the spinal cord. This dural sheath on its exterior does not connect directly to the vertebrae forming the vertebral canal. Between them is a highly vas-cular epidural space containing areolar and adipose connective tissue, forming a protective pad around the spinal cord.

The ventral body cavity is divided into two cavities: The upper thoracic (chest) cavity which houses the visceral organs of the respiratory and circulatory systems (lungs, heart and various structures [blood and lymphatic vessels, nerves, tubular airways of the respiratory and digestive systems]), and the larger, lower abdominopelvic cavity which houses the primary and associated visceral organs of digestion, urination, and reproduction, along with large numbers of blood and lymphatic vessels, glands and nerves.

The roof of the thoracic cavity is membranous. The floor is the musculomembranous thoracic diaphragm. The surrounding thoracic wall is composed of alternating muscle tissue and skeletal bones of the thoracic (rib) cage allowing it capacity for expansion and contraction.

The thoracic cavity has four distinct compartments that house organs: two (left and right) pleural cavities house the left and right lungs respectively. One central partition space, the mediastinum—which separates the thoracic cavity into the two distinct left and right halves that create the pleural cavities—houses the major blood vessels joining the heart, large blood vessels, lymphatic vessels, nerves, thymus gland, trachea, esophagus, and connective tissue; the pericardial cavity houses the heart (and is considered part of the mediastinum as the middle mediastinum).

Serous membranes line the internal surface of the two pleural cavities (as pleura) and the pericardial cavity (as pericardium). All serous membranes consist of a single epithelial cell layer (simple squamous epithelium) called mesothelium (which secretes a lymph-like watery fluid) with a thin underlying vascular connective tissue layer. The epithelial cells always face towards the cavity.

We will now compare the overall structure of two of the three major areas of the human body, a portion of the somatic body wall (which has no cavities) and a sample visceral organ/structure (with an open visceral cavity). By examining a section of each, we can learn important similarities and differences in the integration (order, arrangement, and form) of the body's four basic tissues that comprise them: epithelial, connective, muscle, and nerve. The overall construction of each structure is designed as a reflection of its overall functions, and how each individual tissue within each structure functions in relation to the overall function of the latter.

Bear in mind the section of the body wall structure would be viewed with the naked eye and is obviously much larger and thicker (composed mostly of musculoskeletal mass) than the microscopic section of the thinner visceral small intestine wall we are examining. And the vessels presented (arteries and veins in the body wall section, and arterioles, venules, capillaries and lymphatic vessels in the intestinal wall section) are made up of more than one basic tissue, and all have inner lumen spaces.

Both sections present a general common order of tissue layers: epithelial tissue layer (secreting) always on surface (top), followed immediately by a connective tissue layer(s), then underneath that, muscle tissue.

In the body wall, the secreting/glandular surface epithelial layer is the outer layer, facing the open "space" of the outside world, farthest away from the center of the somatic structure, with the subsequent tissue layers moving towards the center. In the case of the visceral organ (wall), we have a reversal of sorts: the secreting/glandular epithelial layer is the inner layer, facing the lumen (and thus facing the center of the organ) with the same subsequent layers moving away from the center to the outer regions; however here, it ends in another secreting outer epithelial layer, preceded by connective tissue.

THE BODY WALL SOMATIC STRUCTURE

EPITHELIAL TISSUE:

A1 | Epidermis | Skin Outer Layer

CONNECTIVE TISSUE:

B1 | Dermis | Skin Deep Layer

B2 | Superficial Fascia

B3 | Deep Fascia

B4 | Ligament

B5 | Bone

B6 | Periosteum

MUSCLE TISSUE:

C1 | Skeletal Muscle

NERVOUS TISSUE:

D1 | Nerve

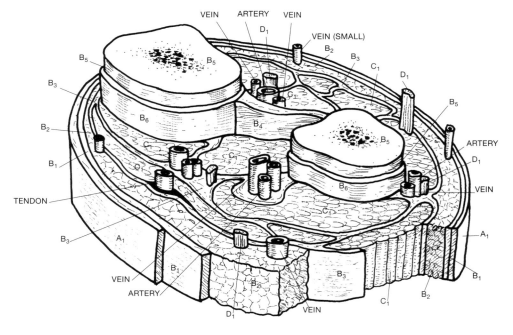

VISCERAL STRUCTURE (OPEN VISCERAL CAVITY)

EPITHELIAL TISSUE:

A_2 | Mucosal Lining |

A_3 | Gland |

A_4 | Serosa Outer Layer |

CONNECTIVE TISSUE:

B_7 | Lamina Propria |

B_8 | Submucosa |

B_9 | Serosa | Inner Layer

MUSCLE TISSUE:

C_2 | Smooth Muscle |

NERVOUS TISSUE:

D_2 | Nerve Cells/Plexus |

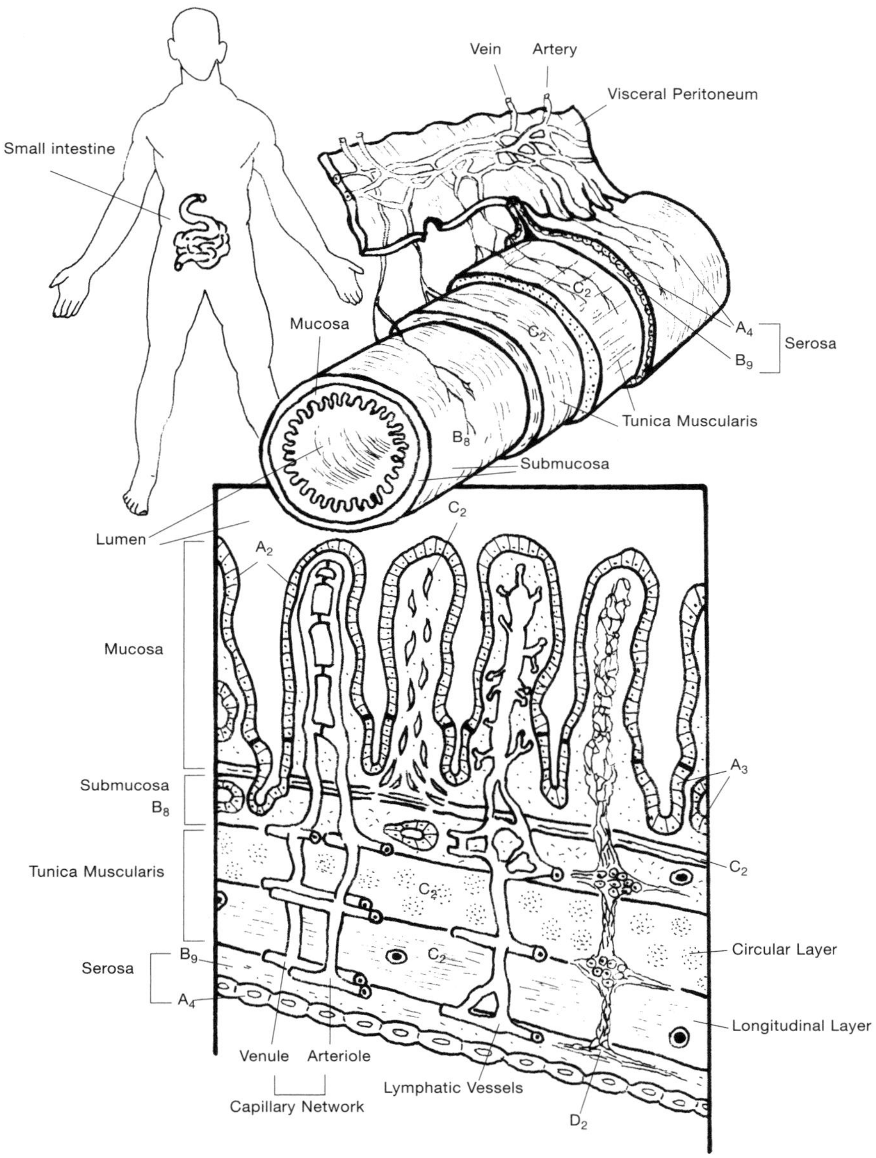

COLOR GUIDELINES:
Membranes = light, cool colors;
muscle = reddish-brown,
nerves = greys; bones = whitish;
connective tissue = light yellow and orange pastels.
As you color the vessels, keep in mind that they
are composed of more than one tissue:
arteries, arterioles = reds;
veins, venules = blues; capillaries = purples;
lymphatics = greens.

Vein   Artery

Visceral Peritoneum

Small intestine

Mucosa

C_2

C_2

A_4
B_9 } Serosa

B_8

Tunica Muscularis

Submucosa

Lumen

A_2

C_2

Mucosa

A_3

Submucosa
B_8

C_2

Tunica Muscularis

C_2

C_2

Circular Layer

Serosa

B_9

A_4

Longitudinal Layer

Venule   Arteriole

Capillary Network

Lymphatic Vessels

D_2

CHAPTER 4: INTEGUMENTARY SYSTEM

The INTEGUMENTARY SYSTEM accounts for close to seven percent of the total body weight. It consists of the skin (INTEGUMENT OR CUTIS) and its associated structures (HAIR, NAILS, SWEAT and OIL GLANDS), integrated with millions of sensory receptors from the nervous system and an incredibly complex vascular network from the blood vascular system. The skin is the largest organ in the body and is essential for the survival of the human organism. It is often considered a simple covering that basically protects the body and keeps all of its parts together, but it is much more than that. It is an amazingly complex, dynamic organ that interlaces between the constantly changing external environment and the internal environment; its metabolic activity must maintain homeostasis.

The skin is often thought to be the same everywhere around the body, but a simple inspection shows that there are significant variations in texture and thickness. Texture varies from rough and calloused (elbow, heel, knuckle joints) to soft and sensitive (eyelids, genitalia, ear drum, nipple area). The general thickness through the body is on average 1.0-2.0 mm. The epidermis ranges from 0.07 mm to 0.12 mm in thickness. The thickest skin is found on the soles of the feet (6.0 mm, with the epidermis 1.4 mm). The thinnest skin is that found on the eyelids—0.5 mm.

Elastic and collagen fibers within the dermis create lines of tension on the skin (shown below). These are of particular interest to surgeons because incisions made parallel to these lines of tension heal more rapidly and create less scar tissue than transverse incisions.

The total surface area is 3,000 sq. inches (7,620 sq. cm). The skin is considered to be an organ because it is comprised of tissues that are joined together to form a structure that has a particular function.

In general, the skin is thicker on the dorsal surface of a part of the body than on the ventral surface (except in the feet and hands, where the ventral portions are thicker due to wear and abrasion).

CONTENTS

THE INTEGUMENT (SKIN)

SYSTEM COMPONENTS

The skin and its structural derivatives, such as hair, nails, sweat and oil glands.

SYSTEM FUNCTIONS

External support of the body, regulation of body temperature, protection of the body, elimination of wastes and reception of certain stimuli, such as temperature, pressure, pain, touch and vibration.

COLOR GUIDELINES:
You may want to choose your color palette for this chapter by comparing colored illustrations from your main textbook.
Usual choices are: arteries = red; , veins = blue;
epidermis = fleshy, warm pinks, or browns; dermis = light grays; fatty tissue = yellow; sweat and oil glands = light or pale greens; hair = light browns.

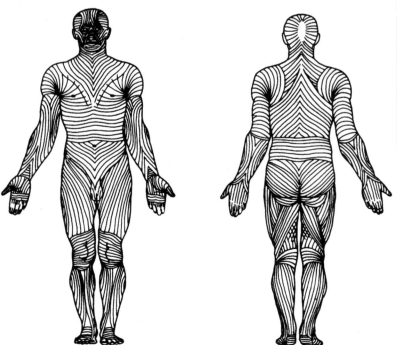

Anterior View   **Posterior View**

THE MAJOR BODY REGIONS AS A PERCENTAGE OF SKIN SURFACE

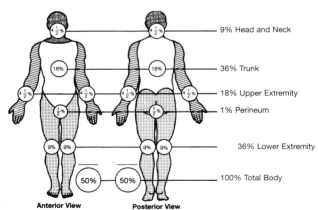

9% Head and Neck
36% Trunk
18% Upper Extremity
1% Perineum
36% Lower Extremity
100% Total Body

Anterior View   Posterior View

The Integument: Skin

The Three Cutaneous Layers:
Epidermis, Dermis, and Hypodermis

The three principal layers of the skin are established by the eleventh week of embryonic development. The EPIDERMIS is the thinner, outermost layer (CUTICLE), and is composed of four to five sublayers, which are continually replenished with cells generated in inner layers (STRATUM GERMINATIVUM to STRATUM CORNEUM) every six to eight weeks. The epidermis and associated structures are derived from the ECTODERM GERM LAYER. The considerably thicker dermis lies directly under the epidermis and consists of two sublayers. The deepest layer is the HYPODERMIS and consists of a single layer. The dermis and hypodermis are derived from the MESODERM GERM LAYER.

The epidermis is thirty to fifty cells thick (mostly STRATUM CORNEUM). The stratification of cells forms a dense protective barrier. It is keratinized and cornified. The innermost STRATUM GERMINATIVUM constantly divides mitotically to replace the rest of the epidermis as it wears away.

The DERMIS (CORIUM) is the true skin. The upper papillary layer is in contact with the EPIDERMIS and the lower reticular layer is in contact with the HYPODERMIS. The loose connective tissue contains numerous capillaries, lymphatics, nerve endings, hair follicles, sebaceous glands, sweat glands and their ducts, and smooth muscle fibers. The HYPODERMIS is the deepest subcutaneous layer. Loose fibrous connective tissue contains adipose (fatty) cells, interlaced with blood vessels. It binds the DERMIS to underlying organs, stores lipids, insulates and cushions the body, regulates temperature, and aids sexual attraction in mature females (soft contouring). Skin color is produced as a combination of the pigment MELANIN, produced as a protection against the sun's ultraviolet rays and CAROTENE in the skin and hemoglobin in the blood.

PRINCIPAL SKIN LAYERS

I.   EPIDERMIS

a   STRATUM CORNEUM

b   STATUM LUCIDUM

c   STRATUM GRANULOSUM

d   STRATUM SPINOSUM

e   STRATUM GERMINATIVUM

II. DERMIS

f   PAPILLARY LAYER

g   RETICULAR LAYER

III. HYPODERMIS

PRINCIPAL SKIN LAYERS

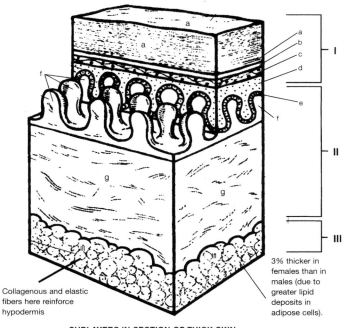

Collagenous and elastic fibers here reinforce hypodermis

3% thicker in females than in males (due to greater lipid deposits in adipose cells).

SUBLAYERS IN SECTION OF THICK SKIN
(sole of foot or palm of hand)

EPIDERMAL SUBLAYERS

	CELL ROWS	CELL TYPE	NUCLEI	SUBSTANCE FORMATION
a	25–30	FLAT, SCALY, DEAD, CORNIFIED	NONE	KERATIN (Keratinization; Waterproofing)
b	3–5 (Only in soles and palms)	FLAT, DEAD, CLEAR. TRANSLUCENT	NOT VISIBLE, DEGENERATE	ELEDIN (Keratinization; Waterproofing)
c	1–5 (Living part of epidermis)	FLAT, GRANULAR	SHRIVELED, DEGENERATE	KERATOHYALIN (Keratinization; Waterproofing)
d	8-10 (Living part of epidermis)	POLYHEDRAL; SPINE-LIKE PROJECTIONS	CENTERED, OVAL, LARGE	MELANOCYTES (Produce melanin, a proteinaceous pigment that protects against the sun's ultraviolet rays)
e	1 (Living part of epidermis)	CUBOIDAL, COLUMNAR	MITOTIC	

The Hair and Nails

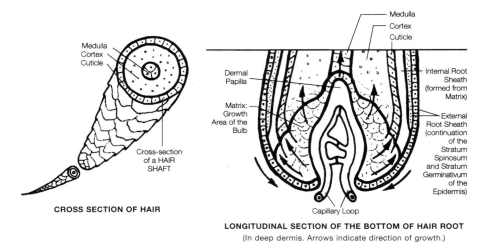

Medulla
Cortex
Cuticle

Medulla
Cortex
Cuticle

Cross-section of a HAIR SHAFT

CROSS SECTION OF HAIR

Dermal Papilla

Matrix: Growth Area of the Bulb

Internal Root Sheath (formed from Matrix)

External Root Sheath (continuation of the Stratum Spinosum and Stratum Germinativum of the Epidermis)

Capillary Loop

LONGITUDINAL SECTION OF THE BOTTOM OF HAIR ROOT
(In deep dermis. Arrows indicate direction of growth.)

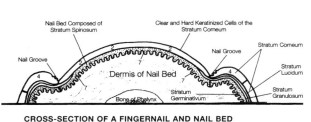

Nail Bed Composed of Stratum Spinosum

Clear and Hard Keratinized Cells of the Stratum Corneum

Nail Groove

Nail Groove

Stratum Corneum

Dermis of Nail Bed

Stratum Lucidum

Bone of Phalynx

Stratum Germinativum

Stratum Granulosum

CROSS-SECTION OF A FINGERNAIL AND NAIL BED

Visible Proximal End of Nail Body

(9) Lunula of the Nail Body (2)

Root

Root Hidden in Nail Groove

Free Edge Projects beyond Distal End of Digit

PARTS OF A SINGLE NAIL

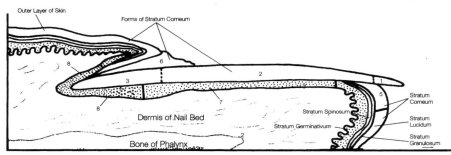

Outer Layer of Skin

Forms of Stratum Corneum

Dermis of Nail Bed

Stratum Germinativum

Stratum Spinosum

Bone of Phalynx

Stratum Corneum

Stratum Lucidum

Stratum Granulosum

SAGITTAL SECTION OF A FINGERNAIL AND NAIL BED

I. EPIDERMIS

a Stratum Corneum
b Stratum Lucidum
c Stratum Granulosum
d Stratum Spinosum
e Stratum Germinativum

HAIR

1 Shaft
2 Root
3 Follicle
4 Bulb Of Follicle

Matrix + Dermal Papilla

NAIL (ON PAGE 42)

1 Free Edge
2 Nail Body
3 Nail Root
4 Nail Fold
5 Hyponychium (quick)
6 Eponychium (cuticle)
7 Nail Bed
8 Matrix
9 Lunula

INTEGUMENTARY GLANDS (ON PAGE 43)

5 Sebaceous Gland | Oil Gland

SUDORIFEROUS GLANDS

6 Apocrine Sweat Gland
7 Eccrine Sweat Gland

II. DERMIS

PAPILLARY LAYER

8 Papillae

RETICULAR LAYER

8 Collagenous, Elastic, Reticular Fibers
10 Arrector Pili Muscle

III. HYPODERMIS

11 Loose Fibrous Connective Tissue
12 Adipose Tissue
13 Artery
14 Vein
15 Lymphatic Vessel
16 Motor Nerve

Hair and integumentary glands form from the EPIDERMIS and along with nails they constitute the three epidermal skin derivatives. Hair has a limited functional value. It is for protection, to distinguish individuals, and serves as sexual ornament of attraction. During midfetal life, the epidermis develops downgrowths into the dermis called FOLLICLES, from which the actual hair grows (root and shaft). The follicles surround the hair. Each single hair develops from stratum germinativum cells within the bulb of the follicle, which is the enlarged base of the root. Nutrients from dermal blood vessels in the dermal papilla of the bulb foster hair growth. As the hair grows and cells divide, they are pushed away from the blood supply. Results are death of the hair cells and keratinization. Each hair lost is replaced by a new hair from the bulb of the follicle that pushes the old hair out. The life span of an eyelash is three to four months, while scalp hair lives as long as three to four years.

Nails serve mainly to protect the digits and for enhanced grasping. The main visible part of the nail—the NAIL BODY—rests on a NAIL BED and appears pink due to the underlying vascular tissue. Growth of the nails begins in the NAIL ROOT as new cells are pushed forward at a rate of about 1 mm per week.

THE INTEGUMENTARY GLANDS

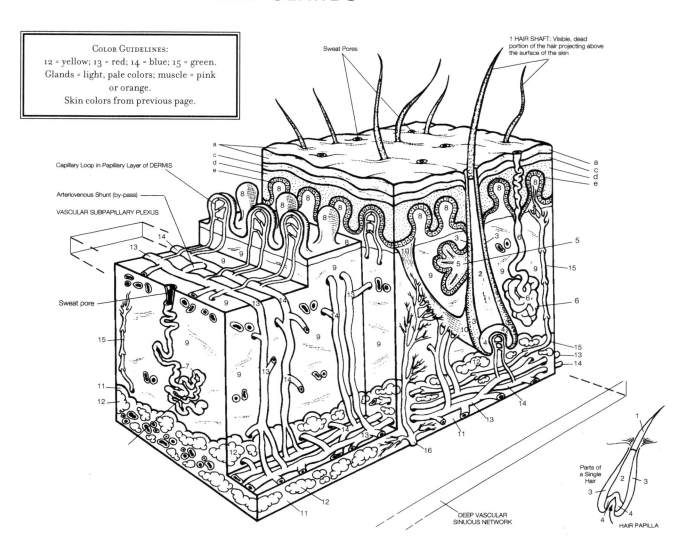

All of the glands of the skin are located in the dermis, even though they are derived from the epidermis. The integumentary glands have extremely important functional value: body defense, perspiration (sweat) for excretion of wastes and evaporative cooling, as well as the maintenance of homeostasis. The sebaceous oil gland almost always associates with and opens into a hair follicle. It is a branched alveolar gland. The holocrine-type gland has a secretion, sebum, that results from cell disintegration in the alveoli pouches.

Apocrine sweat glands are localized in the axillary-pubic regions. It is associated with hair follicles and not functional until puberty.

The endocrine sweat gland is widely distributed throughout the body, mostly in palms and soles, forehead, and back. It has an evaporative cooling function and is formed totally before birth. Relaxed eccrine perspiration is about 300 to 800 ml per day of water loss due to external temperature and humidity. Active perspiration can equal five liters per day of water loss due to active exercise.

There are also specialized sudoriferous glands throughout the body, such as the eruminous sweat glands in external auditory meatus, which produce cerumen (earwax) when secretions combine with oil from sebaceous gland; and mammary glands in the breasts, which secrete milk during lactation periods, under the influence of pituitary and ovarian hormones.

The oily sebaceous glands develop from the follicular epithelium of the hair (external root sheath, see illustration). These holocrine (secretory) glands and specialized smooth muscles, the arrector pili muscles which raise the hair under the influence of cold or fright, are both attached to the hair follicle.

Note the complex vascular system in the hypodermis and the upper reticular layer of the dermis. Also note the two major locations for capillary loop: in the papillary layer of the dermis and in the dermal papilla of each hair root. There is also an abundance of sensory nerve receptors in the skin that convey sensory impulses through the peripheral nervous system to the sensorium portion of the brain. (See chapter 8: The Nervous System. For more material, visit WWW.MCMURTRIESANATOMY.COM.)

CHAPTER 5: SKELETAL SYSTEM

CONTENTS

The SKELETAL SYSTEM in the adult human organism consists of 206 individual bones (and related cartilage) uniquely arranged into the flexible but strong framework of the body. Each bone is an "organ" unto itself and all bones work together to provide the total functioning of the skeletal system. In addition to the principal components of BONE (OSSEUS TISSUE) and CARTILAGE, every bone consists of other kinds of tissues—VASCULAR (BLOOD and LYMPH VESSELS) and NERVOUS—that all coordinate in elaborating each bone's structure and function. The skeletal system should not be considered an isolated system, but one that supports and protects all of the systems of the body, and in particular has close association with the CARDIOVASCULAR, MUSCULAR, and RESPIRATORY SYSTEMS. It helps serve the circulatory system by producing blood cells in well-protected areas of bone. It helps serve the muscular system in a twofold manner. Bony projections provide attachments for muscles that span movable joints, optimizing the muscles' power to move the bones; the skeletal system serves as the basis for the complexity of body movement through space. Bones act as a storehouse for the calcium needed for muscle contraction

> ### SYSTEM COMPONENTS
> *The bones of the body and their associated cartilages*
> *(see also Chapter 6: Articular System).*
>
> ### SYSTEM FUNCTIONS
> *Internal support and protection of the body, body leverage, the production*
> *of blood cells and the storage of minerals.*

and serve the respiratory system by forming passageways in the nasal cavity that help clean, moisten, and warm inhaled air. Bones of the thorax are specially shaped and positioned to serve the chest in expansion during inhalation.

Bones also serve as landmarks for students of anatomy (as well as surgeons). Frequent, misleading conceptions of the hard, dry, lifeless material of laboratory study must be properly replaced by a conception of bone as vital living substance and as a site of enormous metabolic activity, performing many crucial body functions, including support, protection, form and shape, body movement, blood cell production, and storage of mineral nutrients.

For purposes of study, the skeleton is divided into two portions: the AXIAL SKELETON and the APPENDICULAR SKELETON. The axial skeleton consists of the bones (skull, breastbone/sternum, backbone/vertebral column and ribs) related to the axis of the body that support and protect the head, neck, and trunk. The axis is an imaginary longitudinal line that runs through the head and down to the space between the feet. This straight line runs vertically along the body's center of gravity and is considered the center of the human body. The appendicular skeleton consists of the bones of the upper and lower extremities and the girdles that anchor them to the axial skeleton (the UPPER LIMBS with the PECTORAL GIRDLE and the LOWER LIMBS with the PELVIC GIRDLE).

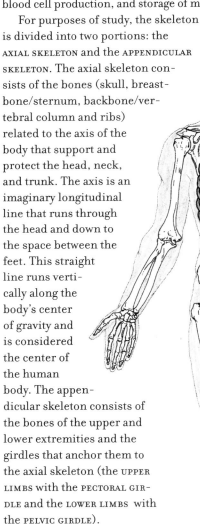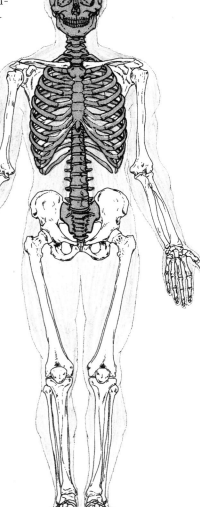

The drawing at right marks the distinction between the Axial Skeleton (shaded) and the Appendicular Skeleton (unshaded).

Anterior View

GENERAL SKELETAL ORGANIZATION
SUPERFICIAL SURVEY OF THE SKELETON

The skeleton system consists of two of the four general kinds of connective tissue: CARTILAGE and BONE (OSSEOUS TISSUE). Bone is the most rigid of all connective tissue, and, unlike cartilage, is the site of great metabolic activity enhanced by a large vascular supply. Bone and bone tissue should not be viewed as the dry, lifeless material studied in the anatomy laboratory.

The process of bone formation is called OSSIFICATION or OSTEOGENESIS. Bone is formed from the human embryo "skeleton" by transformation of either fibrous membranes or hyaline cartilage. In the young child, bone tissue has completely replaced cartilage in the bones, except in two regions, the articular cartilage and the epiphyseal line. The ARTICULAR CARTILAGE covers the articular surfaces of the two EPIPHYSES, the proximal and distal ends of a typical long bone. The EPIPHYSEAL LINE is a remnant of a cartilaginous epiphyseal plate between each epiphysis and the DIAPHYSIS (the shaft, or long portion, of a typical bone). During growth, the EPIPHYSEAL PLATE is a region of mitotic activity responsible for elongation of bone.

COLOR GUIDELINES AND COLORING NOTES:
Axial Skeleton = warm colors;
Appendicular Skeleton = cool colors.
Use the same colors for cranial and facial bones on all pages.
Observe the direct correlation between the bones of the upper limb and the lower limb in the appendicular skeleton (humerus-femur; ulna-tibia, etc.). You may want to color correlate bones the same color to enhance your understanding of these relationships.
Axial skeleton = gray;
Appendicular skeleton = light blue.

The skeleton at right provides a superficial picture of the human skeleton taken as a whole.

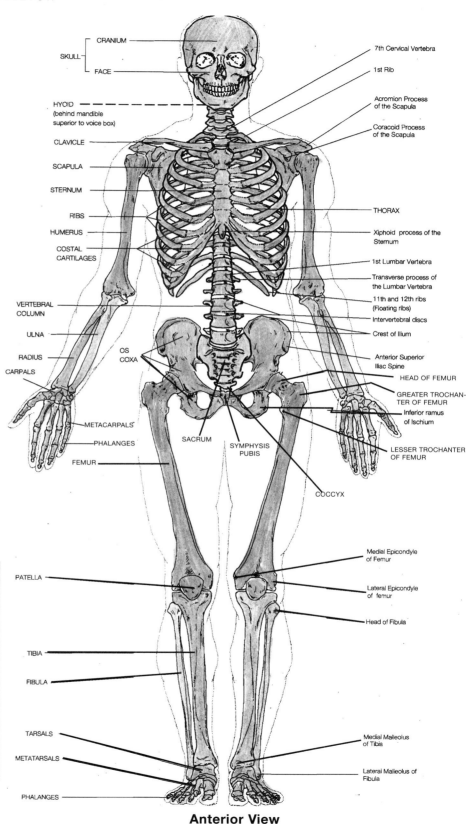

Anterior View

Gross and Microscopic Bone Anatomy

The PERIOSTEUM is a dense, white fibrous tissue covering the surface of the bone not covered by articular cartilage. It is essential for diametric (width) bone growth, repair, and nutrition. It is the point of attachment for ligaments and muscle tendons. The two layers of the periosteum are the INNER OSTEOGENIC LAYER and the OUTER FIBROUS LAYER. The inner osteogenic layer covers compact bone and is composed of elastic fibers (not shown) and blood vessels. The OSTEOBLASTS are specialized cells that form new bone and help in repair. The outer fibrous layer is made of fibrous connective tissue and blood vessels; lymphatic vessels; and nerves that pass into the bone.

Along the diaphysis are nutrient foramina that allow passage of nutrient vessels into the bone. In the diaphysis of a typical long bone, the COMPACT BONE forms a cylinder that surrounds a central cavity, called the MEDULLARY CAVITY. Also called the MARROW CAVITY, it contains fatty YELLOW MARROW, which consists of fat cells and a few scattered blood cells. The medullary cavity is lined with a layer of connective tissue called the ENDOSTEUM and contains OSTEOBLASTS. In addition, scattered OSTEOCLASTS aid in bone removal.

There are two types of bone tissue, and most bones have both types: CANCELLOUS SPONGY BONE and COMPACT, DENSE BONE. The spongy bone contains many large spaces filled with RED MARROW, which is responsible for producing red blood cells, some white blood cells and platelets. It is the highly porous, highly vascular inner portion of bone and makes bone lighter. It makes up most of the epiphyses of long bones and almost all of short, flat, irregularly-shaped bones. Compact bone contains few spaces and is deposited in a layer over spongy bone. This dense layer is thicker along the diaphysis than around the epiphyses. A hard outer layer provides protection, durable strength and support, and resists weight stress. In adults, compact bone has a concentric ring structure (spongy bone does not).

Recall that bone tissue consists of widely scattered separated cells surrounded by a very large amount of intercellular substance—mainly composed of CALCIUM PHOSPHATE and CALCIUM CARBONATE. These salts are referred to as the HYDROXYAPATITES. As bone grows, its framework is formed by COLLAGENOUS FIBERS. As the salts are deposited into this fiber framework of the intercellular substance, the tissue hardens (becomes OSSIFIED), giving bone its rigid character.

For a more detailed analysis of gross and microscopic bone anatomy, see WWW.MCMURTRIESANATOMY.COM.

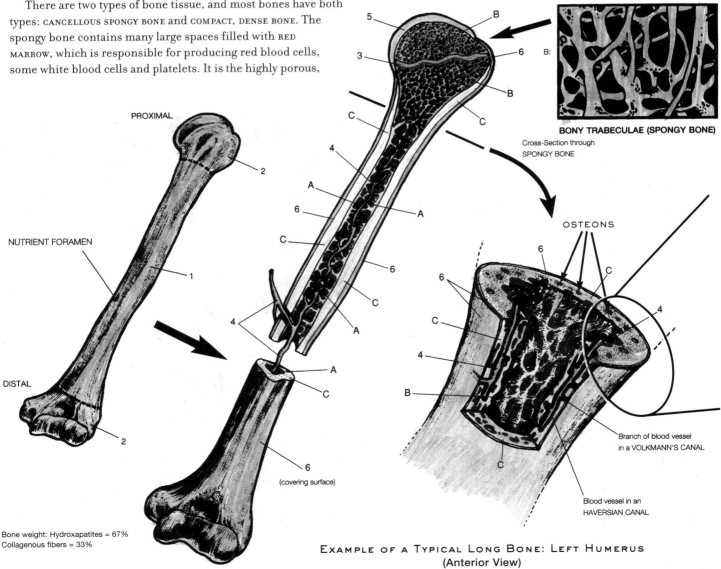

BONY TRABECULAE (SPONGY BONE)

Cross-Section through SPONGY BONE

PROXIMAL

NUTRIENT FORAMEN

DISTAL

OSTEONS

Branch of blood vessel in a VOLKMANN'S CANAL

Blood vessel in an HAVERSIAN CANAL

(covering surface)

Bone weight: Hydroxapatites = 67%
Collagenous fibers = 33%

Example of a Typical Long Bone: Left Humerus
(Anterior View)

TYPICAL STRUCTURE OF A LONG BONE

1　**Diaphysis** Shaft

　A　Medullary Cavity/Yellow Marrow

2　**Epiphyses**

　　Proximal Epiphysis and Distal Epiphysis
　B　Cancellous Spongy Bone/Red Marrow

3　**Epiphyseal Line/Plate**

4　**Nutrient Artery**

5　**Articular Cartilage**

6　**Periosteum**

TWO TYPES OF BONE TISSUE, BASED ON POROSITY

C　**Compact, Dense Bone**

　　Outer Layer

B　**Cancellous, Spongy Bone**

　　Inner Layer

MICROSCOPIC BONE ANATOMY

OSSEOUS TISSUE CELL TYPES

a　**Osteocytes**　(within lamelleae of of compact bone)

a　**Osteoblasts**　(within inner layer of periosteum and within endosteum)

c　**Osteoclasts**　(within endosteum)

PERIOSTEUM:

6a　**Outer Fibrous Layer**

6b　**Inner Osteogenic Layer**

SPACES CONTAINING OSTEOCYTES:

7　**LACUNAE**　(singular: "lacuna")

8　**CANALICULI**

Calcified Intercelluar Substance:

9　**HAVERSIAN LAMELLAE** (concentric)

10　**INTERSTITIAL LAMELLAE** (irregular)

11　**ENDOSTEUM**

HAVERSIAN SYSTEMS

(Within Osteons in Compact Bone)

12　**VOLKMANN'S CANAL**

13　**BLOOD VESSEL**

14　**HAVERSIAN CANAL**

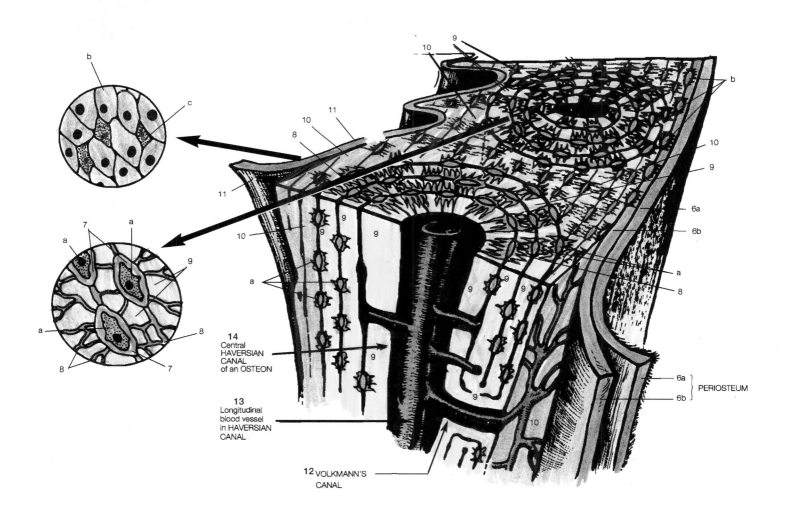

14
Central
HAVERSIAN
CANAL
of an OSTEON

13
Longitudinal
blood vessel
in HAVERSIAN
CANAL

12 VOLKMANN'S
CANAL

PERIOSTEUM

The adult skull contains twenty-two bones: eight cranial and fourteen facial

EIGHT CRANIAL BONES (NUMBER OF BONES IN THE CRANIUM)

THE NEUROCRANIUM

1 | Frontal | (1)
2 | Parietal | (2)
3 | Temporal | (2)
4 | Occipital | (1)
5 | Sphenoid | (1)
6 | Ethmoid | (1)
H | Hyoid Bone |

FOURTEEN FACIAL BONES (NUMBER OF EACH KIND OF BONE)

THE FACIAL SKELETON

7 | Maxilla Upper Jaw | (2)
8 | Palatine 2 | (2)
9 | Zygomatic Malar | (Cheekbone) (2)
10 | Lacrimal | (2)
11 | Nasal | (2)
12 | Inferior Nasal Concha | (2)
13 | Vomer | (1)
14 | Mandible Lower Jaw | (1)

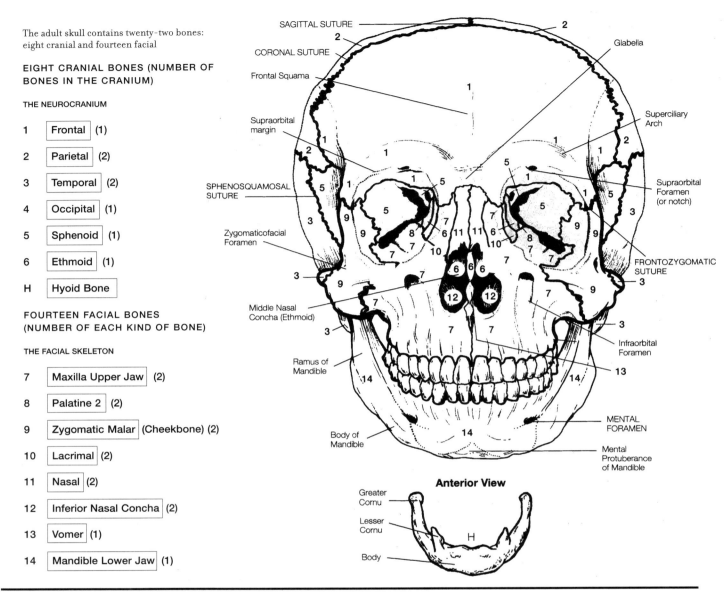

SAGITTAL SUTURE
CORONAL SUTURE
Frontal Squama
Supraorbital margin
SPHENOSQUAMOSAL SUTURE
Zygomaticofacial Foramen
Middle Nasal Concha (Ethmoid)
Ramus of Mandible
Body of Mandible
Glabella
Superciliary Arch
Supraorbital Foramen (or notch)
FRONTOZYGOMATIC SUTURE
Infraorbital Foramen
MENTAL FORAMEN
Mental Protuberance of Mandible

Anterior View

Greater Cornu
Lesser Cornu
Body

The adult human skull contains twenty-two bones. Eight are cranial (called the NEUROCRANIUM) and fourteen facial (the FACIAL SKELE-TON). Not included among these bones are the single HYOID BONE or the six OSSICLES of the inner ear. The adult human skull rests on the superior end of the VERTEBRAL COLUMN. Each bone of the skull articulates with adjacent bones. The skull contains several cavities that house the BRAIN and SENSORY ORGANS.

The neocranium encloses and protects the brain as well as the sensory organs of sight, hearing, and balance (see Chapter 8: NERVOUS SYSTEM). They are all interlocked by SUTURES. The FRONTAL BONE forms the forehead, the roof of the nasal cavity, the superior arch of the BONY ORBITS (which house the eyeballs) and the anterior part of the cranial floor (after birth, the left and right halves of the frontal bone are united by a suture, which usually disappears by age six). Two PARIETAL BONES form the superior (upper) sides of the CRANIUM and the roof of the crani-

um (CRANIAL CAVITY). Two TEMPORAL BONES form the inferior (lower) sides of the cranium and part of the CRANIAL FLOOR (structurally, each temporal bone has four parts). The OCCIPITAL BONE forms the posterior (back) portion of the skull and a prominent portion of the base of the cranium. The occipital bone contains a large hole, the FORAMEN MAGNUM, through which the spinal cord attaches to the brain; on either side of this hole are oval processes with convex surfaces—the OCCIPITAL CONDYLES, which articulate the atlas (see page 51). The SPHENOID BONE forms the anterior base of the cranium, situated at the middle part of the base of the skull, as well as part of the floor and sidewalls of the eye sockets. It is the "keystone" of the cranial floor and it articulates with all other cranial bones. The inferior pterygoid processes help form the lateral walls of the NASAL CAVITY. The ETHMOID BONE forms the roof of the nasal cavity, situated in the anterior part of the floor of the cranium between the orbits.

A middle inferior projection, the PERPENDICULAR PLATE, forms a major part of the NASAL SEPTUM that separates the nasal cavity into two chambers.

The HYOID BONE is unique among all bones of the skeletal system in that it is the only bone that does not articulate directly with any other bone. It is suspended from the STYLOID PROCESS of the temporal bone by the STYLOHYOID LIGAMENTS. It is located in the neck between the mandible and larynx (voice box), just superior to the larynx and it supports the TONGUE and provides attachments for several tongue and NECK MUSCLES and LIGAMENTS.

The fourteen facial bones form the framework and basic shape of the face, support the teeth, and serve as attachments for various muscles that move the jaw and cause facial expressions. They are the bones of the skull not in contact with the brain. Except the MANDIBLE or LOWER JAW, all are firmly interlocked by sutures with one another and the CRANIAL BONES. The MAXILLA BONE articulates with every bone of the face (except the MANDIBLE). The two ZYGOMATIC BONES form the cheekbones of the face and part of the outer wall and floor of the orbits (the posterior projection called the TEMPORAL PROCESS articulates with the anterior projection of the temporal bone, forming the ZYGOMATIC ARCH). The two LACRIMAL BONES form the anterior part of the medial wall of each orbit. The smallest bones of the face, they are thin and shaped like a fingernail. (Tear ducts pass into the nasal cavity through the LACRIMAL FOSSAE.) The bridge of the nose (superior part) is formed by the two nasal bones (small, oblong bones fused together).

The mandible forms the lower jaw. It is the only movable bone of the skull and does not form from the fusion of two bones. The body of the mandible has a horseshoe-shaped front and two horizontal lateral sides. The RAMI of the mandible are two vertical extensions from the posterior portion of the body, and the angles of the mandible are formed where each ramus meets the body. The remaining facial bones are the two PALATINE BONES, the two INFERIOR NASAL CONCHAE, and the VOMER BONE.

Not considered in the typical adult count of 206 total bones are WORMIAN, or SUTURAL bones, located within the joints (sutures) of the skull, whose number varies considerably from person to person. Sutures are immovable articulations that are found only between skull bones. Most of the names of the sutures are descriptive of the bones they connect with. There are four prominent skull sutures. The CORONAL SUTURE is between the frontal bone and the two parietals. The SAGITTAL SUTURE is between the two parietal bones. The LAMBDOIDAL SUTURE is between the occipital bone and the two parietals, and the SQUAMOSAL SUTURE (the thin, large expanded portion of the temporal bone) is between the two temporal bones and the two parietals.

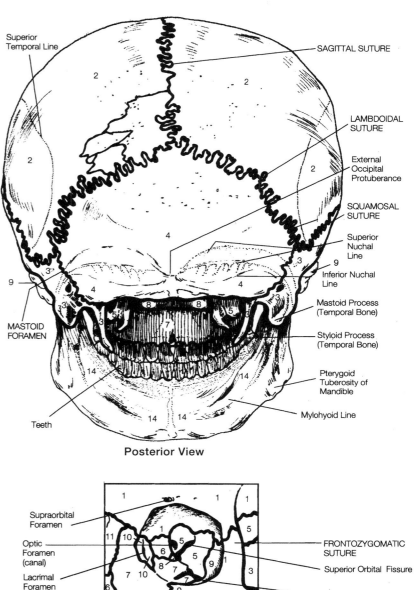

Posterior View

DETAIL OF LEFT ORBIT
(Anterior View)

COLOR GUIDELINES:
1 = pink; 2 = blue; 3 = orange;
4 = red; 5 = yellow; 6 = light blue;
7 = flesh; 8 = light orange; 9 = purple;
10 = yellow-orange; 11 = light brown;
12 = blue-green; 13 = light red; 14 = green;
15 = cream; wormian bones = grey.

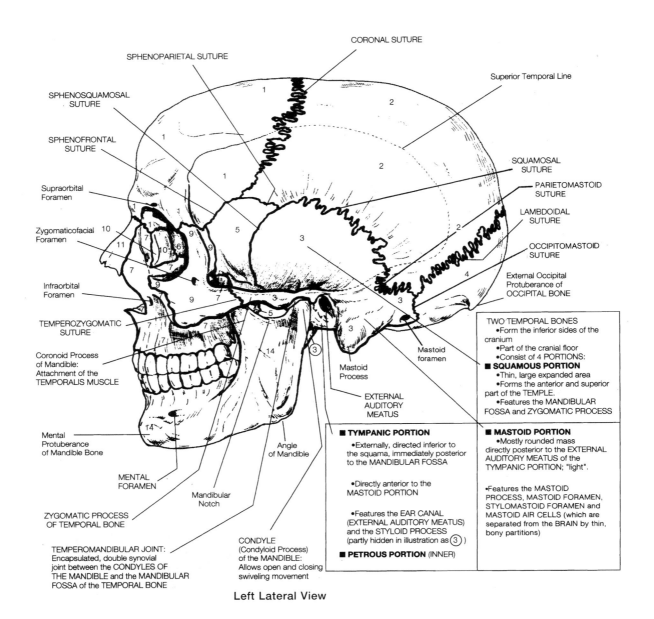

CORONAL SUTURE

SPHENOPARIETAL SUTURE

Superior Temporal Line

SPHENOSQUAMOSAL SUTURE

SPHENOFRONTAL SUTURE

SQUAMOSAL SUTURE

PARIETOMASTOID SUTURE

Supraorbital Foramen

LAMBDOIDAL SUTURE

Zygomaticofacial Foramen

OCCIPITOMASTOID SUTURE

External Occipital Protuberance of OCCIPITAL BONE

Infraorbital Foramen

TEMPEROZYGOMATIC SUTURE

Mastoid foramen

Coronoid Process of Mandible: Attachment of the TEMPORALIS MUSCLE

Mastoid Process

EXTERNAL AUDITORY MEATUS

Mental Protuberance of Mandible Bone

MENTAL FORAMEN

Mandibular Notch

Angle of Mandible

ZYGOMATIC PROCESS OF TEMPORAL BONE

TEMPEROMANDIBULAR JOINT: Encapsulated, double synovial joint between the CONDYLES OF THE MANDIBLE and the MANDIBULAR FOSSA of the TEMPORAL BONE

CONDYLE (Condyloid Process) of the MANDIBLE: Allows open and closing swiveling movement

TWO TEMPORAL BONES
 •Form the inferior sides of the cranium
 •Part of the cranial floor
 •Consist of 4 PORTIONS:
■ **SQUAMOUS PORTION**
 •Thin, large expanded area
 •Forms the anterior and superior part of the TEMPLE.
 •Features the MANDIBULAR FOSSA and ZYGOMATIC PROCESS

■ **TYMPANIC PORTION**
 •Externally, directed inferior to the squama, immediately posterior to the MANDIBULAR FOSSA

 •Directly anterior to the MASTOID PORTION

 •Features the EAR CANAL (EXTERNAL AUDITORY MEATUS) and the STYLOID PROCESS (partly hidden in illustration as ③)

■ **PETROUS PORTION** (INNER)

■ **MASTOID PORTION**
 •Mostly rounded mass directly posterior to the EXTERNAL AUDITORY MEATUS of the TYMPANIC PORTION; "light".

•Features the MASTOID PROCESS, MASTOID FORAMEN, STYLOMASTOID FORAMEN and MASTOID AIR CELLS (which are separated from the BRAIN by thin, bony partitions)

Left Lateral View

TWENTY-TWO BONES OF THE SKULL

NEUROCRANIUM: FACIAL SKELETON

EIGHT CRANIAL BONES (NUMBER OF EACH BONE)

1 Frontal (1)

2 Parietal (2)

3 Temporal (2)

4 Occipital (1)

5 Sphenoid (1)

6 Ethmoid (1)

FOURTEEN FACIAL BONES (NUMBER OF EACH BONE)

7 Maxilla (2)

8 Palatine (2)

9 Zygomatic (2)

10 Lacrimal (2)

11 Nasal (2)

12 Inferior Nasal Concha (2)

13 Vomer (1)

14 Mandible (1)

H Hyoid Bone

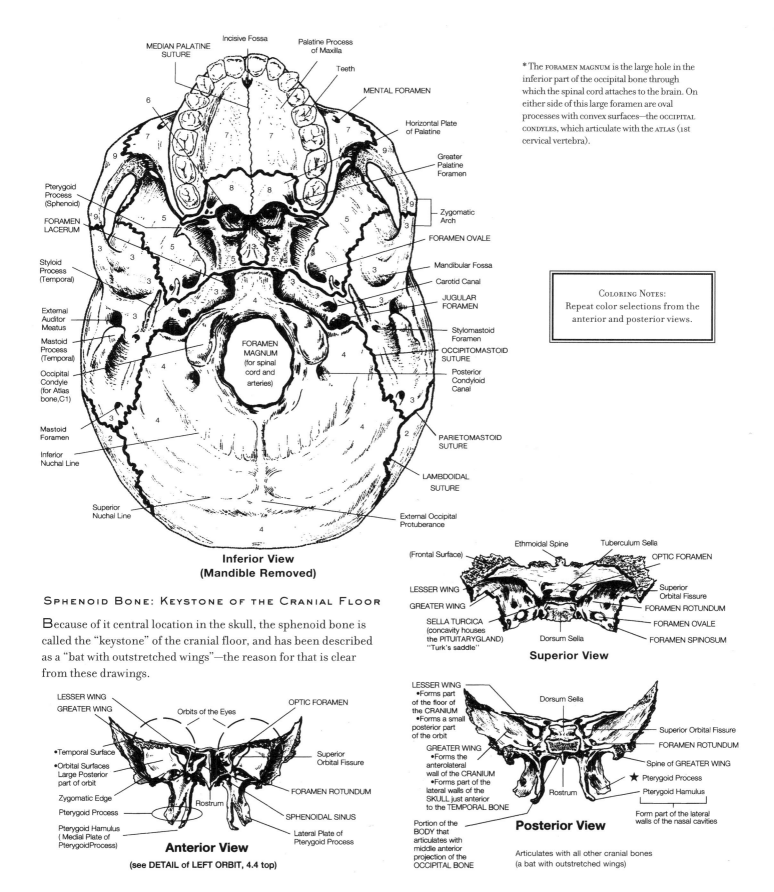

MEDIAN PALATINE SUTURE

Incisive Fossa

Palatine Process of Maxilla

Teeth

MENTAL FORAMEN

Horizontal Plate of Palatine

Greater Palatine Foramen

Zygomatic Arch

FORAMEN OVALE

Mandibular Fossa

Carotid Canal

JUGULAR FORAMEN

Stylomastoid Foramen

OCCIPITOMASTOID SUTURE

Posterior Condyloid Canal

PARIETOMASTOID SUTURE

LAMBDOIDAL SUTURE

External Occipital Protuberance

Pterygoid Process (Sphenoid)

FORAMEN LACERUM

Styloid Process (Temporal)

External Auditor Meatus

Mastoid Process (Temporal)

Occipital Condyle (for Atlas bone,C1)

Mastoid Foramen

Inferior Nuchal Line

Superior Nuchal Line

FORAMEN MAGNUM (for spinal cord and arteries)

Inferior View
(Mandible Removed)

* The FORAMEN MAGNUM is the large hole in the inferior part of the occipital bone through which the spinal cord attaches to the brain. On either side of this large foramen are oval processes with convex surfaces—the OCCIPITAL CONDYLES, which articulate with the ATLAS (1st cervical vertebra).

COLORING NOTES:
Repeat color selections from the anterior and posterior views.

SPHENOID BONE: KEYSTONE OF THE CRANIAL FLOOR

Because of it central location in the skull, the sphenoid bone is called the "keystone" of the cranial floor, and has been described as a "bat with outstretched wings"—the reason for that is clear from these drawings.

(Frontal Surface)

Ethmoidal Spine

Tuberculum Sella

OPTIC FORAMEN

LESSER WING

GREATER WING

SELLA TURCICA (concavity houses the PITUITARYGLAND) "Turk's saddle"

Dorsum Sella

Superior Orbital Fissure

FORAMEN ROTUNDUM

FORAMEN OVALE

FORAMEN SPINOSUM

Superior View

LESSER WING
GREATER WING

Orbits of the Eyes

OPTIC FORAMEN

•Temporal Surface

•Orbital Surfaces Large Posterior part of orbit

Zygomatic Edge

Pterygoid Process

Pterygoid Hamulus (Medial Plate of PterygoidProcess)

Rostrum

Superior Orbital Fissure

FORAMEN ROTUNDUM

SPHENOIDAL SINUS

Lateral Plate of Pterygoid Process

Anterior View

(see DETAIL of LEFT ORBIT, 4.4 top)

LESSER WING
•Forms part of the floor of the CRANIUM
•Forms a small posterior part of the orbit

GREATER WING
•Forms the anterolateral wall of the CRANIUM
•Forms part of the lateral walls of the SKULL just anterior to the TEMPORAL BONE

Portion of the BODY that articulates with middle anterior projection of the OCCIPITAL BONE

Dorsum Sella

Superior Orbital Fissure

FORAMEN ROTUNDUM

Spine of GREATER WING

★ Pterygoid Process

Pterygoid Hamulus

Rostrum

Form part of the lateral walls of the nasal cavities

Posterior View

Articulates with all other cranial bones
(a bat with outstretched wings)

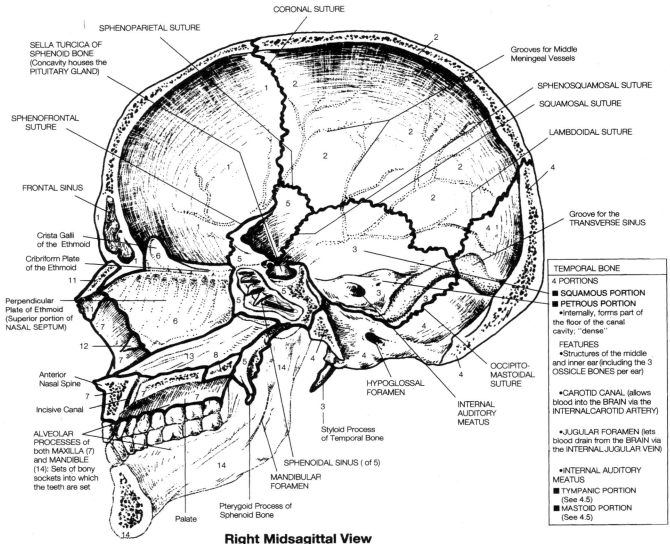

CORONAL SUTURE

SPHENOPARIETAL SUTURE

SELLA TURCICA OF SPHENOID BONE (Concavity houses the PITUITARY GLAND)

SPHENOFRONTAL SUTURE

FRONTAL SINUS

Crista Galli of the Ethmoid

Cribriform Plate of the Ethmoid

Perpendicular Plate of Ethmoid (Superior portion of NASAL SEPTUM)

Anterior Nasal Spine

Incisive Canal

ALVEOLAR PROCESSES of both MAXILLA (7) and MANDIBLE (14): Sets of bony sockets into which the teeth are set

Grooves for Middle Meningeal Vessels

SPHENOSQUAMOSAL SUTURE

SQUAMOSAL SUTURE

LAMBDOIDAL SUTURE

Groove for the TRANSVERSE SINUS

OCCIPITO-MASTOIDAL SUTURE

INTERNAL AUDITORY MEATUS

HYPOGLOSSAL FORAMEN

Styloid Process of Temporal Bone

SPHENOIDAL SINUS (of 5)

MANDIBULAR FORAMEN

Pterygoid Process of Sphenoid Bone

Palate

TEMPORAL BONE

4 PORTIONS
■ SQUAMOUS PORTION
■ PETROUS PORTION
• Internally, forms part of the floor of the canal cavity; "dense"

FEATURES
• Structures of the middle and inner ear (including the 3 OSSICLE BONES per ear)

• CAROTID CANAL (allows blood into the BRAIN via the INTERNALCAROTID ARTERY)

• JUGULAR FORAMEN (lets blood drain from the BRAIN via the INTERNAL JUGULAR VEIN)

• INTERNAL AUDITORY MEATUS
■ TYMPANIC PORTION (See 4.5)
■ MASTOID PORTION (See 4.5)

Right Midsagittal View
(View of Inner Right Cranial Cavity)

NEUROCRANIUM
TWENTY-TWO BONES OF THE SKULL: EIGHT CRANIAL BONES (NUMBER OF EACH BONE)

1 Frontal (1)

2 Parietal (2)

3 Temporal (2)

4 Occipital (1)

5 Sphenoid (1)

6 Ethmoid (1)

FACIAL SKELETON
FOURTEEN FACIAL BONES (NUMBER OF EACH BONE)

7 Maxilla (2)

8 Palatine (2)

9 Zygomatic (2)

10 Lacrimal (2)

11 Nasal (2)

12 Inferior Nasal Concha (2)

13 Vomer (1)

14 Mandible (1)

COLORING NOTES:
Repeat color selections from the anterior and posterior views.

Superior View (Transverse Cut)
(Floor of Cranial Cavity/Base of Skull)

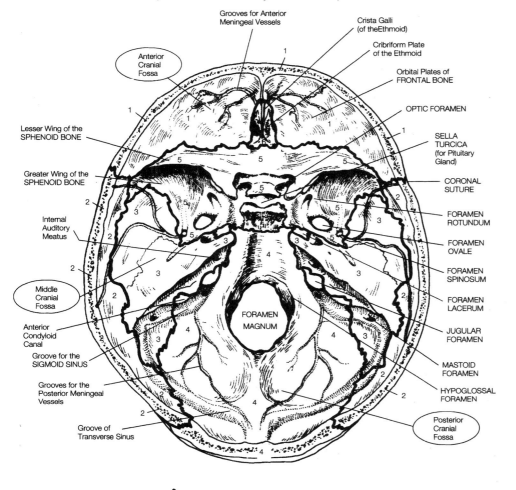

Grooves for Anterior Meningeal Vessels

Crista Galli (of the Ethmoid)

Cribriform Plate of the Ethmoid

Orbital Plates of FRONTAL BONE

OPTIC FORAMEN

SELLA TURCICA (for Pituitary Gland)

CORONAL SUTURE

FORAMEN ROTUNDUM

FORAMEN OVALE

FORAMEN SPINOSUM

FORAMEN LACERUM

JUGULAR FORAMEN

MASTOID FORAMEN

HYPOGLOSSAL FORAMEN

Posterior Cranial Fossa

Anterior Cranial Fossa

Lesser Wing of the SPHENOID BONE

Greater Wing of the SPHENOID BONE

Internal Auditory Meatus

Middle Cranial Fossa

Anterior Condyloid Canal

Groove for the SIGMOID SINUS

Grooves for the Posterior Meningeal Vessels

Groove of Transverse Sinus

FORAMEN MAGNUM

The Canals and Cavities of the Skull

The TEMPORAL BONE has four portions: SQUAMOUS, TYMPANIC, MASTOID and PETROUS. The dense petrous portion is internal and forms part of the floor of the canal cavity. It features structures of the middle and inner ear including the three ossicle bones in each ear. It also contains the CAROTID CANAL (allows blood into the brain via the inter-carotid artery); the JUGULAR FORAMEN (lets blood drain from the brain via the internal jugular vein) and the internal auditory meatus.

The CRANIAL CAVITY is the largest cavity of the skull and houses the brain; it is formed by cranial bones. The NASAL CAVITY is partitioned into two chambers (nasal fossae) by a nasal septum; it is formed by both cranial and facial bones. The PARANASAL SINUSES are four paired sets of sinus cavities set within their respective bones which surround the nasal area and opening into the nasal cavities. The two MIDDLE and INNER EAR CHAMBERS are inferior to the cranial cavity. They house organs of hearing and balance. The two ORBITS are the orbital cavities of the eyeballs, open to the face. They are formed by both cranial and facial bones. The ORAL, or BUCCAL, CAVITY is the cavity of the mouth, totally within the facial region, and only partially formed by bone.

Ethmoid Bone: Principal Supporting Structure of the Nasal Cavity

Perpendicular Plate

Crista Galli

Ethmoidal Air "Cells" (Ethmoidal sinus or labyrinth)

Orbital Surface (Portion of left eyeball socket)

Ala of Crista Galli

Cribriform Plate (Contains numerous perforations called OLFACTORY FORAMINA through which the OLFACTORY NERVES (I) pass from nasal cavity)

Orbital Surface (Portion of right eyeball socket)

Superior View

CRISTA GALLI: An upper spine of the PERPENDICULAR PLATE, projects superiorly into the cranial cavity. It is an attachment for the MENINGES covering the BRAIN.

Anterior Ethmoidal "Cells": (Labyrinth)

Ala

(Orbital Surface: Concavity where the left eye lies)

Perpendicular Plate

Unicate Process

Middle Nasal Concha

Left Lateral View

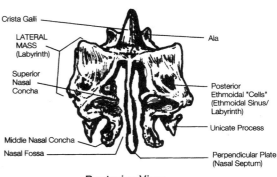

Crista Galli

LATERAL MASS (Labyrinth)

Superior Nasal Concha

Middle Nasal Concha

Nasal Fossa

Ala

Posterior Ethmoidal "Cells" (Ethmoidal Sinus/ Labyrinth)

Unicate Process

Perpendicular Plate (Nasal Septum)

Posterior View

Axial Skeleton, continued
Bones of the Adult Skull:
Lateral Walls and Medial Nasal Septum of the Nasal Cavity/External Nose

THE NASAL CAVITY

The nasal cavity is the cavity in the skull between the floor of the cranium and the roof of the mouth. The ethmoid bone is located in the anterior portion of the floor of the cranium between the orbits, where it forms the roof of the nasal cavity. A major portion of the right and left lateral walls of the nasal cavity is composed of the inner portion of the lateral masses (labyrinths) of the ethmoid bone (see page 53). Portions of the frontal and sphenoid cranial bones, along with portions of many facial bones, help to complete the lateral walls of the nasal cavity. Two scroll-shaped plates from both inner lateral walls of the ethmoid bone project into the nasal cavity, the superior and middle nasal conchae. An inferior middle projection of the ethmoid bone, the perpendicular plate, helps to form a medial septum that splits the nasal cavity into two separate chambers (the nasal septum and the external nose).

The nasal septum is the medial wall of the nasal cavity. It partitions the nasal cavity into two chambers or nasal fossae, and is formed of both BONE and CARTILAGE. There are several parts of the supporting framework of the nasal septum. The median COLUMNA (SEPTAL CARTILAGE) of the nose extends inward to form the anterior part of the septum. The thin, plowshare-shaped VOMER BONE forms the inferior and posterior part of the nasal septum. The perpendicular plate of the ethmoid bone articulates with the VOMER bone and columna, and forms the superior portion of the nasal septum. An inner portion of GREATER ALAR CARTILAGE, NASAL and VOMERONASAL CARTILAGE, and NASAL CRESTS of the nasal, sphenoid, palatine, and maxilla bones complete the nasal septum.

THE EXTERNAL NOSE

The framework of the external nose is formed of both bone and cartilage, but mostly cartilage. Two nasal bones fuse together to form the superior part of the bridge of the nose that lies between the orbits. The orifices of the nose, the nostrils or nares, are separated from each other by a median septum called the columna (septal cartilage).

Although not part of the nose, the frontal processes of the two fused maxillae bones form the posterolateral support of the nose near the nasal bones.

The fragile inferior nasal conchae are not part of the ethmoid bone, as are the superior and middle nasal concha. They are all scroll-shaped bones that form portions of the lateral walls of the nasal cavities and project into the cavities. All of these three nasal conchae allow for circulation and filtration of air before it passes into the lungs.

The paired maxillary bones unite to form the single upper jawbone, the maxilla bone. It articulates with every bone of the face except the mandible. The fusion is complete before birth. The formations of the maxilla are part of the lateral walls and floor of the nasal cavities, which are part of the roof of the mouth, comprising most of the hard palate and part of the floors of the orbits.

The palatine bones are the L-shaped bones whose horizontal processes unite to form a nasal crest. These horizontal plates form the posterior portion of the hard palate, which separates the upper nasal cavity from the lower oral cavity. The palatine bones consist of a small part of the lateral walls and floor of the nasal cavities and a small part of the floors of the orbit.

RIGHT AND LEFT LATERAL WALLS OF THE NASAL CAVITY

ASSOCIATED CRANIAL BONES

1	Frontal Bone
5	Sphenoid Bone
6	Ethmoid Bone

ASSOCIATED FACIAL BONES

7	Maxilla Bone Upper Jaw
8	Palatine Bone
10	Lacrimal Bone
11	Nasal Bone
12	Inferior Nasal Concha

NASAL SEPTUM (MEDIAL WALL OF THE NASAL CAVITY)

ASSOCIATED SKULL BONES

5	Nasal Crest of the Sphenoid Bone
6	Perpendicular Plate of the Ethmoid Bone*
11	Nasal Crest of the Nasal Bones
13	Vomer Bone *

ASSOCIATED CARTILAGES

15	Septal Cartilage* Columna
16	Greater Alar Cartilage
19	Nasal Cartilage
20	Vomeronasal Cartilage

EXTERNAL NOSE

11	Nasal Bones

Although not part of the nose, the frontal processes of the two fused maxillae bones form the posterolateral support of the nose near the nasal bones.

F	Fibro-Fatty Tissue

CARTILAGES OF EXTERNAL NOSE

15	Septal Cartilage* Columna
16	Greater Alar Cartilage
17	Lesser Alar Cartilage
18	Lateral Nasal Cartilage

*Major components of the nasal septum.

EXTERNAL NOSE

(SKIN REMOVED)

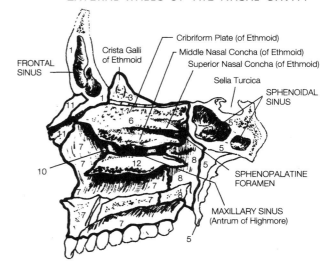

Frontal Process of Maxilla

Frontal Bone

11

15

18 18

16 16

17

18

16

15

F

F

F

Philtrum

NOSTRILS
(External Nares)

Anterior View

Left Lateral View

LATERAL WALLS OF THE NASAL CAVITY

Cribriform Plate (of Ethmoid)

Crista Galli of Ethmoid

Middle Nasal Concha (of Ethmoid)

Superior Nasal Concha (of Ethmoid)

Sella Turcica

FRONTAL SINUS

SPHENOIDAL SINUS

SPHENOPALATINE FORAMEN

MAXILLARY SINUS
(Antrum of Highmore)

See Chapter 13: Respiratory System for details on the openings of the paranasal sinuses into the nasal cavities.

COLOR GUIDELINES:
1 = pink;
5 = yellow;
6 = light blue;
7 = flesh;
8 = light orange;
10 = yellow-orange;
11 = light brown;
12 = blue-green;
13 = light red.
Cartilage can be any color except cream, green, or grey.

MEDIAL NASAL SEPTUM OF NASAL CAVITY

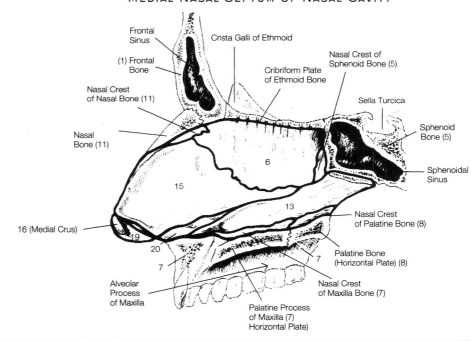

Frontal Sinus

Crista Galli of Ethmoid

Nasal Crest of Sphenoid Bone (5)

(1) Frontal Bone

Cribriform Plate of Ethmoid Bone

Sella Turcica

Nasal Crest of Nasal Bone (11)

Nasal Bone (11)

Sphenoid Bone (5)

Sphenoidal Sinus

15

6

13

Nasal Crest of Palatine Bone (8)

16 (Medial Crus)

19

20

7

Palatine Bone (Horizontal Plate) (8)

Alveolar Process of Maxilla

Nasal Crest of Maxilla Bone (7)

Palatine Process of Maxilla (7) Horizontal Plate

MAXILLA BONES

Orbital Surface
(Portion of the right eyeball socket)

Lacrimal Groove

Frontal Process

MAXILLARY SINUS

Middle Meatus

Inferior Meatus

Zygomatic Process

INFRAORBITAL FORAMEN

Nasal Notch

Greater Palatine Groove

INCISIVE CANAL

Palatine Process

Horizontal Plate

Alveolar Process

R. Lateral View

R. Medial View

PALATINE BONES

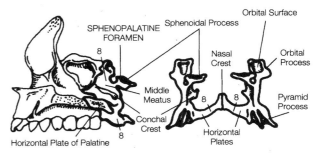

Orbital Surface

SPHENOPALATINE FORAMEN

Sphenoidal Process

Nasal Crest

Orbital Process

8

Middle Meatus

8 8

Pyramid Process

Conchal Crest

Horizontal Plate of Palatine

8

Horizontal Plates

In Situ

Posterior View
(both bones)

Axial Skeleton, continued
The Vertebral Column and Vertebrae

The VERTEBRAL COLUMN is composed of twenty-four movable VERTEBRAE (it is actually composed of thirty-three individual vertebrae, but five are fused into the SACRUM and four are more or less fused into the COCCYX). The vertebral column is commonly called the backbone (or spine). The backbone together with the SPINAL CORD makes up the SPINAL COLUMN. The twenty-six movable vertebrae are composed (from below upwards) of a COCCYXASACRUM and twenty-four VERTEBRAE. The latter have FIBROCARTILAGINOUS INTERVERTEBRAL DISCS. (The drum-shaped body or CENTRUM of each vertebra is in contact with the discs at each end.)

Of the twenty-four, there are seven CERVICAL VERTEBRAE, twelve THORACIC VERTEBRAE, and five LUMBAR VERTEBRAE. The SACRUM develops as five distinct bones that fuse together. The coccyx consists of three to five (usually four) more or less fused rudimentary tailbones.

One function of the spine (spinal column) is to support the head and the thorax by serving as a point of attachment for the ribs (and the muscles of the back). It also supports the girdles (pectoral and pelvic girdles that attach the appendages to the trunk). It houses and protects the spinal cord and permits passage of the spinal nerves. The spine is a strong, flexible rod that permits freedom of movement (anteriorly, posteriorly and laterally). The INTERVERTEBRAL DISCS lend flexibility and absorb stress of movement. Linkage of the vertebrae is accomplished by intervertebral discs, interlocking processes and binding ligaments. This permits limited movement between vertebrae and extensive movements of the entire vertebral column.

There are several openings of the vertebral column. The VERTEBRAL FORAMEN of each vertebra add up to form the SPINAL CANAL, through which the spinal cord traverses. Between the vertebrae are openings called INTERVERTEBRAL FORAMINA, through which spinal nerves traverse as they branch off from the spinal cord. TRANSVERSE FORAMINA in the transverse processes of the CERVICAL VERTEBRAE allow passage of the VERTEBRAL VESSELS, which transfer blood to and from the brain.

The fetal spine has only one curve (concave anterior). The THORACIC and PELVIC CURVATURES in the adult spine are designated as PRIMARY CURVES; the opposite CERVICAL and LUMBAR CURVATURES are designated as SECONDARY CURVES.

The ATLAS and the AXIS are two specialized cervical vertebrae. The atlas has a ring of bone with ANTERIOR and POSTERIOR ARCHES. It possesses two large LATERAL MASSES. The SUPERIOR ARTICULAR SURFACES of the lateral masses receive the OCCIPITAL CONDYLES of the skull, thus supporting the head and permitting nodding. It has a body and no spinal process. Its TUBERCLES are for the attachment of the TRANSVERSE LIGAMENT. The axis has the DENS (ODONTOID PROCESS) attached to its superior aspect. The dens is regarded as what remains of the atlas body that has become attached to the axis. The transverse ligament of the atlas passes behind the dens to create a PIVOT JOINT, permitting rotation of the skull.

General Structure of a Typical Vertebra

The body (centrum) is the weight-bearing part of a vertebra, adapted to withstand compression. Intervertebral discs contact on each end of it; the NEURAL ARCH is affixed to its posterior surface. It is composed of two short, thick PEDICLES and two ARCHED LAMELLAE. The vertebral foramen is a hollow space formed by the body and the neural arch. The spinal cord passes through all the vertebral foramina. The seven processes of the neural arch are one spinal process and two each of transverse, superior articular and inferior articular processes. The superior articular process (and facet) interlocks with the inferior articular process of the adjacent anterior vertebra. The intervertebra fora-

General Structure of a Typical Vertebra

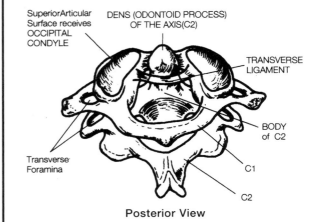

Posterior View

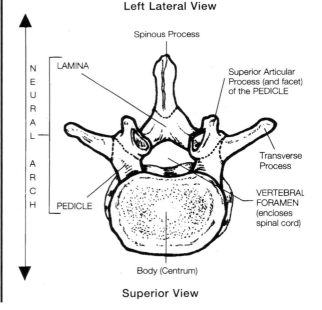

Left Lateral View

Superior View

men allows passage for spinal nerves branching off from the spinal cord.

Intervertebral disks are thicker and larger in the lumbar area, thinner and higher in the column. They are thicker anteriorly than posteriorly. Each disk has a central soft center, the nucleus pulposus, and a tough, fibrous outer capsule called the annulus fibrosus.

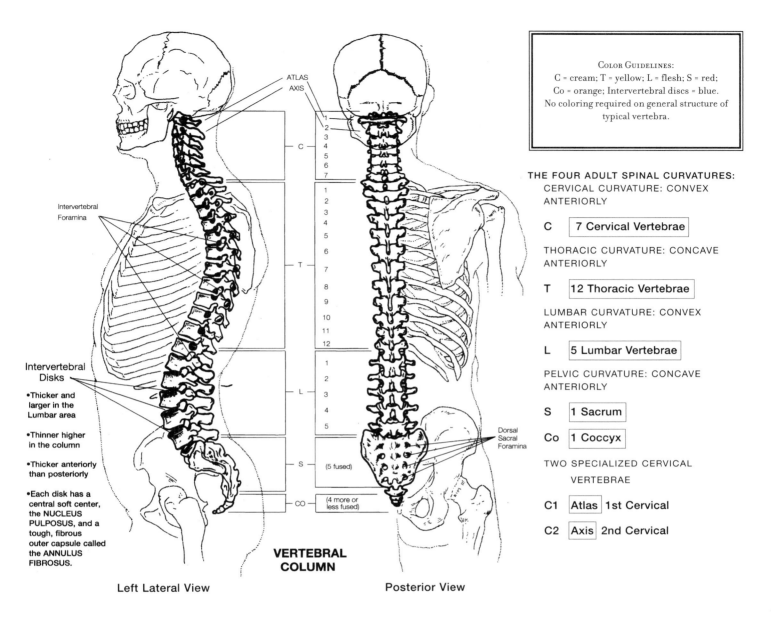

ATLAS
AXIS

THE FOUR ADULT SPINAL CURVATURES:
CERVICAL CURVATURE: CONVEX ANTERIORLY

C | 7 Cervical Vertebrae |

THORACIC CURVATURE: CONCAVE ANTERIORLY

T | 12 Thoracic Vertebrae |

LUMBAR CURVATURE: CONVEX ANTERIORLY

L | 5 Lumbar Vertebrae |

PELVIC CURVATURE: CONCAVE ANTERIORLY

S | 1 Sacrum |

Co | 1 Coccyx |

TWO SPECIALIZED CERVICAL VERTEBRAE

C1 | Atlas | 1st Cervical

C2 | Axis | 2nd Cervical

Intervertebral Foramina

Intervertebral Disks

- Thicker and larger in the Lumbar area
- Thinner higher in the column
- Thicker anteriorly than posteriorly
- Each disk has a central soft center, the NUCLEUS PULPOSUS, and a tough, fibrous outer capsule called the ANNULUS FIBROSUS.

Dorsal Sacral Foramina

(5 fused)

(4 more or less fused)

VERTEBRAL COLUMN

Left Lateral View

Posterior View

STRUCTURE OF A THORACIC VERTEBRA

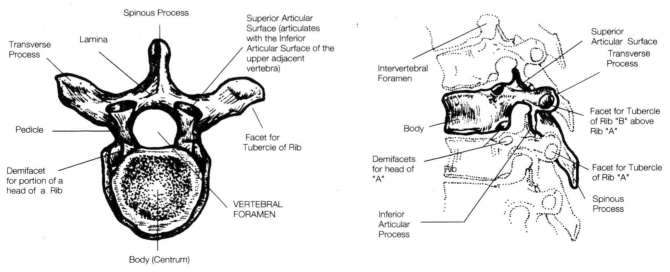

Spinous Process

Superior Articular Surface (articulates with the Inferior Articular Surface of the upper adjacent vertebra)

Transverse Process

Lamina

Pedicle

Facet for Tubercle of Rib

Demifacet for portion of a head of a Rib

VERTEBRAL FORAMEN

Body (Centrum)

Superior View (Posterior aspect on top)

Intervertebral Foramen

Superior Articular Surface

Transverse Process

Body

Facet for Tubercle of Rib "B" above Rib "A"

Demifacets for head of "A"

Rib

Facet for Tubercle of Rib "A"

Inferior Articular Process

Spinous Process

Left Lateral View

The rib (thoracic) cage—the skeletal portion of the THORAX—has a conical shape with a narrow superior inlet and a broad inferior outlet. It is compressed ANTEROPOSTERIORLY (meaning, curved from front to back). It encloses and protects the thoracic viscera. (Recall that the THORAX, or CHEST, is that region of the body—the trunk—between the base of the neck superiorly and the diaphragm inferiorly). It is very flexible and involved directly in the mechanics of breathing. It supports the PECTORAL GIRDLE and the UPPER EXTREMITIES.

The STERNUM (BREASTBONE) is an elongated, flattened bony plate that forms the ANTERIOR MIDLINE of the UPPER THORAX. It is 15 cm (6 in.) in length and attaches to the COSTAL CARTILAGES of the first through the seventh ribs.

Each of the twelve pairs of ribs attaches posteriorly to a THORACIC VERTEBRA. Anteriorly, the first seven pairs of ribs are called TRUE RIBS because they attach (with their costal cartilages) directly to the STERNUM. The last five pairs of ribs are called FALSE RIBS because they have no direct attachment or no attachment at all. Anteriorly, the first three pairs of false ribs (the eighth, ninth and tenth) attach to the costal cartilage of the seventh pair. The last two pairs of false ribs (eleventh and twelfth) have no association with the sternum at all, and are called FLOATING RIBS.

The general morphology of a typical rib is composed of a head, neck, tubercle, and body (shaft). The head articulates with the bodies of two adjacent vertebrae by two DEMIFACETS separated by an INTERARTICULAR CREST. The tubercle is the knoblike struc-ture lateral to the head. The large articulating portion attaches to the facet of the transverse process of the thoracic vertebra. The small nonarticulating portion serves as the attachment for the COSTOTRANSVERSE LIGAMENT, connecting the rib to the transverse process of the thoracic vertebra. The body is composed of parts of different curvature meeting at the angle. It curves downward and forward and cannot be flattened on a surface. The superior surface is blunt, or rounded, and the inferior surface is sharper. Both surfaces afford attachment for INTERCOSTAL MUSCLES (except the first rib).

The three basic portions of the sternum are the UPPER MANUBRIUM, the CENTRAL BODY and the LOWER XIPHOID PROCESS (ENSIFORM PROCESS). The upper manubrium is the superior, triangular portion. The JUGULAR (SUPRASTERNAL) NOTCH is a depression on its superior surface. The CLAVICULAR NOTCHES located on each side of the jugular notch articulate with the medial ends of the CLAVICLES. The upper manubrium articulates with the first and second ribs.

The central body is the middle, largest portion (the GLADIOLUS). It attaches to the COSTAL CARTILAGES of the second through tenth ribs. The STERNAL ANGLE (ANGLE OF LOUIS) is a depression between the MANUBRIUM and the body at the level of the second rib. The lower xiphoid process (ensiform process) is the inferior, smallest portion and is attached to some abdominal muscles, but not to any ribs. The COSTAL ANGLE is formed where the two costal margins meet at the xiphoid process.

THE RIB (THORACIC) CAGE

S	Sternum Breastbone
R	12 Paired Ribs = 24
CC	Costal Cartilages
T	12 Thoracic Vertebrae

THE STERNUM

a	Manubrium
b	Body
c	Xiphoid Process (cartilage in youth)

THE RIBS (A RIB = A COSTA)

There are a total of 34 ribs with 12 paired. Thus there are 14 true ribs and 10 false ribs. Of these 10 false ribs, the last 4 are floating false ribs (2 pairs).

TR	7 Paired True Ribs
FR	5 Paired False Ribs

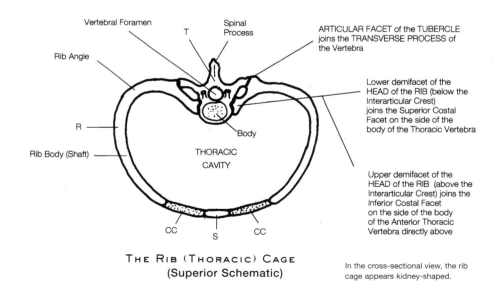

THE RIB (THORACIC) CAGE
(Superior Schematic)

Vertebral Foramen

Spinal Process

T

ARTICULAR FACET of the TUBERCLE joins the TRANSVERSE PROCESS of the Vertebra

Rib Angle

Lower demifacet of the HEAD of the RIB (below the Interarticular Crest) joins the Superior Costal Facet on the side of the body of the Thoracic Vertebra

R

Body

Rib Body (Shaft)

THORACIC CAVITY

Upper demifacet of the HEAD of the RIB (above the Interarticular Crest) joins the Inferior Costal Facet on the side of the body of the Anterior Thoracic Vertebra directly above

CC   S   CC

In the cross-sectional view, the rib cage appears kidney-shaped.

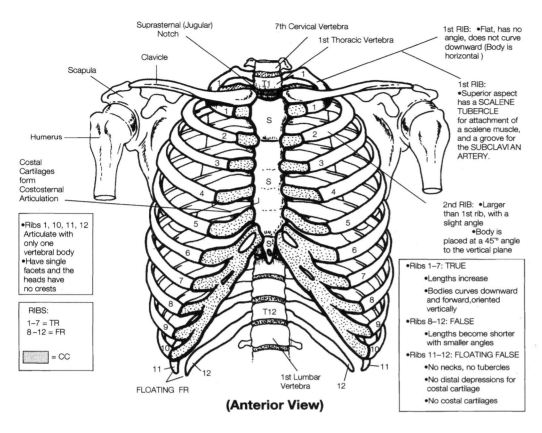

Suprasternal (Jugular) Notch

7th Cervical Vertebra

1st Thoracic Vertebra

Clavicle

Scapula

Humerus

Costal Cartilages form Costosternal Articulation

1st RIB: •Flat, has no angle, does not curve downward (Body is horizontal)

1st RIB: •Superior aspect has a SCALENE TUBERCLE for attachment of a scalene muscle, and a groove for the SUBCLAVIAN ARTERY.

2nd RIB: •Larger than 1st rib, with a slight angle •Body is placed at a 45° angle to the vertical plane

•Ribs 1, 10, 11, 12 Articulate with only one vertebral body •Have single facets and the heads have no crests

RIBS:
1–7 = TR
8–12 = FR

 = CC

11   12

FLOATING FR

1st Lumbar Vertebra   12

(Anterior View)

•Ribs 1–7: TRUE
•Lengths increase
•Bodies curves downward and forward, oriented vertically
•Ribs 8–12: FALSE
•Lengths become shorter with smaller angles
•Ribs 11–12: FLOATING FALSE
•No necks, no tubercles
•No distal depressions for costal cartilage
•No costal cartilages

THE STERNUM

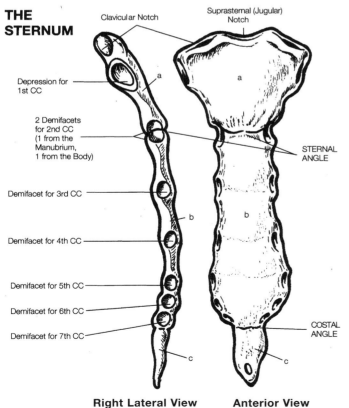

Clavicular Notch

Suprasternal (Jugular) Notch

Depression for 1st CC

2 Demifacets for 2nd CC (1 from the Manubrium, 1 from the Body)

Demifacet for 3rd CC

Demifacet for 4th CC

Demifacet for 5th CC

Demifacet for 6th CC

Demifacet for 7th CC

STERNAL ANGLE

COSTAL ANGLE

Right Lateral View   **Anterior View**

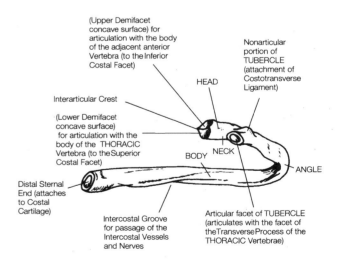

(Upper Demifacet concave surface) for articulation with the body of the adjacent anterior Vertebra (to the Inferior Costal Facet)

Nonarticular portion of TUBERCLE (attachment of Costotransverse Ligament)

HEAD

Interarticular Crest

(Lower Demifacet concave surface) for articulation with the body of the THORACIC Vertebra (to the Superior Costal Facet)

BODY   NECK

ANGLE

Distal Sternal End (attaches to Costal Cartilage)

Intercostal Groove for passage of the Intercostal Vessels and Nerves

Articular facet of TUBERCLE (articulates with the facet of the Transverse Process of the THORACIC Vertebrae)

A Typical Rib Viewed from Below and Behind

APPENDICULAR SKELETON
BONES OF THE UPPER LIMB: PECTORAL GIRDLE, ARM AND FOREARM

The PECTORAL GIRDLE consists of four bones that connect the bones of the upper extremities to the axial skeleton. Each of the two SHOULDER GIRDLES consist of one CLAVICLE and one SCAPULA. The main function of the pectoral girdle is to provide for the attachment of many muscles that move the arm (brachium) and the forearm. The pectoral girdle is not a complete girdle because it attaches to the axial skeleton at only one point, the STERNUM (STRERNOCLAVICULAR JOINT)—at the CLAVICULAR NOTCHES of the MANUBRIUM with the medial ends of the clavicles. The pectoral girdle has no attachment to the vertebral column. It is not weight-bearing and the joints are delicate and not very stable. This allows great mobility of the appendages and freedom of movement in many directions.

The CLAVICLE (anterior component of the girdle) is long, slender and S-shaped. It binds the shoulder to the axial skeleton and is responsible for maintaining the normal height of the shoulder. The clavicle allows freedom of movement by positioning the shoulder joint away from the trunk.

**THE PECTORAL GIRDLE—TWO CLAVICLES AND
TWO SCAPULAE: TWO SHOULDER GIRDLES**

1 | Clavicle | collarbone

2 | Scapula | shoulder blade

LONGER BONES OF THE UPPER EXTREMITIES:
SINGULAR BONE OF THE BRACHIUM (ARM)

3 | Humerus |

TWO SKELETAL STRUCTURES OF THE FOREARM

4 | Ulna | (Medial)

5 | Radius | (Lateral)

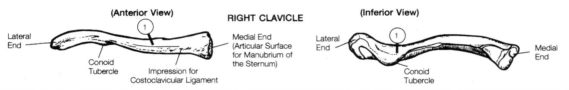

(Anterior View) **RIGHT CLAVICLE**

Lateral End
Conoid Tubercle
Impression for Costoclavicular Ligament
Medial End (Articular Surface for Manubrium of the Sternum)

(Inferior View)

Lateral End
Conoid Tubercle
Medial End

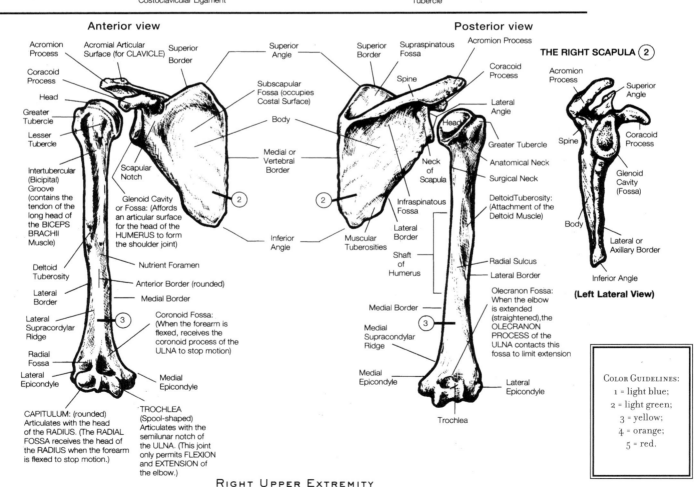

Anterior view

Acromion Process
Acromial Articular Surface (for CLAVICLE)
Coracoid Process
Head
Greater Tubercle
Lesser Tubercle
Intertubercular (Bicipital) Groove (contains the tendon of the long head of the BICEPS BRACHII Muscle)
Deltoid Tuberosity
Lateral Border
Lateral Supracordylar Ridge
Radial Fossa
Lateral Epicondyle
CAPITULUM: (rounded) Articulates with the head of the RADIUS. (The RADIAL FOSSA receives the head of the RADIUS when the forearm is flexed to stop motion.)

Superior Border
Superior Angle
Subscapular Fossa (occupies Costal Surface)
Body
Medial or Vertebral Border
Inferior Angle

Scapular Notch
Glenoid Cavity or Fossa: (Affords an articular surface for the head of the HUMERUS to form the shoulder joint)
Nutrient Foramen
Anterior Border (rounded)
Medial Border
Coronoid Fossa: (When the forearm is flexed, receives the coronoid process of the ULNA to stop motion)
Medial Epicondyle
TROCHLEA (Spool-shaped) Articulates with the semilunar notch of the ULNA. (This joint only permits FLEXION and EXTENSION of the elbow.)

Posterior view

Superior Border
Superior Angle
Supraspinatous Fossa
Spine
Lateral Angle
Head
Neck of Scapula
Infraspinatous Fossa
Lateral Border
Muscular Tuberosities
Shaft of Humerus
Medial Border
Medial Supracondylar Ridge
Medial Epicondyle

Acromion Process
Coracoid Process
Greater Tubercle
Anatomical Neck
Surgical Neck
DeltoidTuberosity: (Attachment of the Deltoid Muscle)
Radial Sulcus
Lateral Border
Olecranon Fossa: When the elbow is extended (straightened), the OLECRANON PROCESS of the ULNA contacts this fossa to limit extension
Lateral Epicondyle
Trochlea

THE RIGHT SCAPULA ②

Acromion Process
Superior Angle
Spine
Coracoid Process
Glenoid Cavity (Fossa)
Body
Lateral or Axillary Border
Inferior Angle

(Left Lateral View)

COLOR GUIDELINES:
1 = light blue;
2 = light green;
3 = yellow;
4 = orange;
5 = red.

RIGHT UPPER EXTREMITY

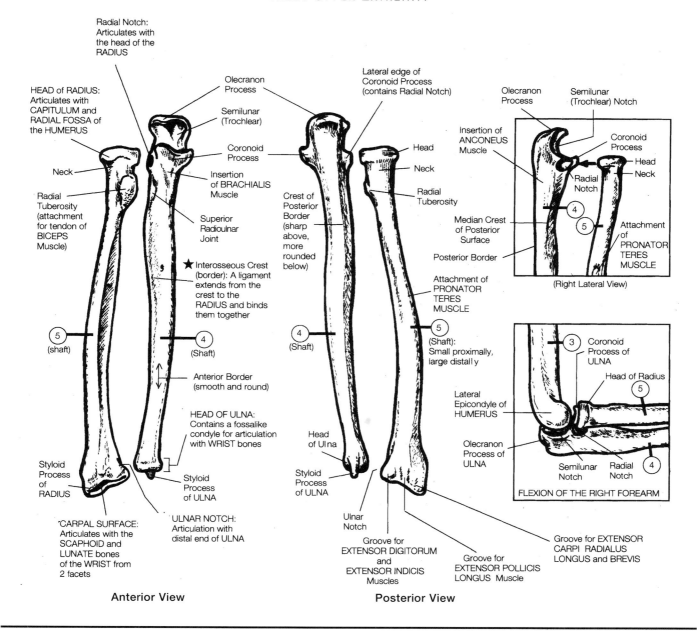

Radial Notch:
Articulates with
the head of the
RADIUS

HEAD of RADIUS:
Articulates with
CAPITULUM and
RADIAL FOSSA of
the HUMERUS

Neck

Radial
Tuberosity
(attachment
for tendon of
BICEPS
Muscle)

Olecranon
Process

Semilunar
(Trochlear)

Coronoid
Process

Insertion
of BRACHIALIS
Muscle

Superior
Radioulnar
Joint

★ Interosseous Crest
(border): A ligament
extends from the
crest to the
RADIUS and binds
them together

5
(shaft)

4
(Shaft)

Anterior Border
(smooth and round)

HEAD OF ULNA:
Contains a fossalike
condyle for articulation
with WRIST bones

Styloid
Process
of
RADIUS

Styloid
Process
of ULNA

CARPAL SURFACE:
Articulates with the
SCAPHOID and
LUNATE bones
of the WRIST from
2 facets

ULNAR NOTCH:
Articulation with
distal end of ULNA

Anterior View

Lateral edge of
Coronoid Process
(contains Radial Notch)

Head

Neck

Radial
Tuberosity

Crest of
Posterior
Border
(sharp
above,
more
rounded
below)

Attachment of
PRONATOR
TERES
MUSCLE

4
(Shaft)

5
(Shaft):
Small proximally,
large distally

Head
of Ulna

Styloid
Process
of ULNA

Ulnar
Notch

Groove for
EXTENSOR DIGITORUM
and
EXTENSOR INDICIS
Muscles

Groove for
EXTENSOR POLLICIS
LONGUS Muscle

Posterior View

Olecranon
Process

Semilunar
(Trochlear) Notch

Insertion of
ANCONEUS
Muscle

Coronoid
Process

Head

Neck

Radial
Notch

Median Crest
of Posterior
Surface

Posterior Border

4

5

Attachment
of
PRONATOR
TERES
MUSCLE

(Right Lateral View)

3 Coronoid
Process of
ULNA

Head of Radius

5

Lateral
Epicondyle of
HUMERUS

Olecranon
Process of
ULNA

Semilunar
Notch

Radial
Notch

4

FLEXION OF THE RIGHT FOREARM

Groove for EXTENSOR
CARPI RADIALUS
LONGUS and BREVIS

The SCAPULA (posterior component of the girdle) is large, flat and triangular. Some fifteen muscles attach to its processes and fossae. It lies over the posterosuperior aspect of the thorax and has three borders.

The UPPER EXTREMITIES consist of two appendages (right and left) that attach at the shoulder; each appendage consists of an ARM, a FOREARM, a WRIST and a HAND. Together the right and left upper extremities contain sixty bones. Therefore, each upper extremity contains thirty bones (twenty-seven of which comprise the wrist and hand). The three remaining are the longer bones of the upper extremity: the HUMERUS bone in the UPPER ARM (BRACHI-UM) and the ULNA and RADIUS in the forearm.

The humerus is the longest, largest bone of the upper extremity. It articulates proximally with the GLENOID FOSSA (CAVITY) of the scapula by a large, rounded HEAD, and it articulates distally with the ulna and radius by two rounded condyles. The medial condyle is the TROCHLEA, which is spool-shaped and articulates with the ulna with a pulley-like surface. The lateral condyle is the CAPITULUM, a rounded knob that receives the RADIUS.

The ulna is the medial bone of the forearm. It is located on the small finger side. Proximally it contains two processes (a prominent posterior OLECRANON PROCESS and a pointed anterior CORONOID PROCESS). Between these two processes is the SEMILU-NAR NOTCH that articulates with the trochlea of the humerus. Distally, an expanded head is separated from the wrist bones by a disc of fibrocartilage.

The radius is the lateral bone of the forearm, located on the thumb side. Proximally, a disc-shaped head articulates with the CAPITULUM of the humerus and the radial notch of the ulna. Distally, an expanded concave CARPAL SURFACE articulates with the wrist bones.

The wrist and the hand make up the distal portion of the upper extremity. One wrist and one hand contain twenty-seven bones: eight carpal (wrist) bones and nineteen hand bones. There are five metacarpal (palm) bones, and fourteen phalanges. Thus, both wrists and hands contain fifty-four bones: sixteen carpals, ten metacarpals and twenty-eight phalanges.

The wrist has eight carpal bones, arranged in two transverse rows of four bones each. The proximal row presents a curved surface that articulates with the distal and concave ends of the radius and ulna. From lateral (thumb—radial side) to medial (small finger—ulnar side) the bones are the scaphoid (shaped like a boat), lunate (half-moon shaped surface), triquetrum (shaped like a right triangle) and the pisiform (shaped like a pea). The pisiform bone forms in a tendon as a sesamoid bone.

The distal row of the wrist bears surfaces for articulation with the metacarpal bones of the hand. From the lateral (radial) to medial (ulnar) side, the bones are the trapezium (a "little table"), trapezoid (table-shaped), capitate (has a rounded head) and hamate, which has a curved, hooklike process called the hamulus.

The hand is the expression of incredible structural organization and is able to sustain a great amount of abuse, notwithstanding its complexity. Sometimes the wrist is included in the definition of the hand, which is attached to the forearm at the wrist. The palm or metacarpus (osssa metacarpalia) consists of the body of the palm of the hand and five metacarpal bones, numbered I – V from the thumb (lateral) side. Each metacarpal bone has an enlarged proximal base with a concave surface for articulation with the carpals. Distal round heads of the bones form the knuckles when the digits are flexed and articulate with the proximal phalanges. The fourteen phalanges form a total of five digits (fingers). Four fingers each carry three phalanges (proximal or first, middle or second, and distal or third or terminal). The thumb has only two phalanges.

THE WRIST (CARPUS)—EIGHT CARPAL BONES:
PROXIMAL ROW

a   Scaphoid

b   Lunate

c   Triquetrum

d   Pisiform

DISTAL ROW

e   Trapezium   greater multangular (os trapezium)

f   Trapezoid   lesser multangular

g   Capitate

h   Hamate

THE HAND (MANUS)—NINETEEN BONES:
PALM (METACARPUS)

i   5 Metacarpal Bones (I-V)

FINGERS (DIGITS)

j   14 Phalanges   (single = phalanx)

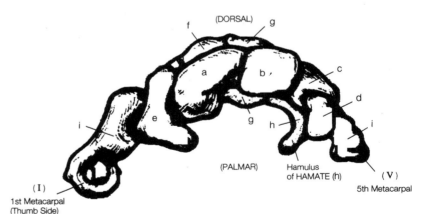

Superior View of Right Wrist

(DORSAL)

(PALMAR)

Hamulus of HAMATE (h)

(I)
1st Metacarpal (Thumb Side)

(V)
5th Metacarpal

Bones of the Right Wrist and Hand

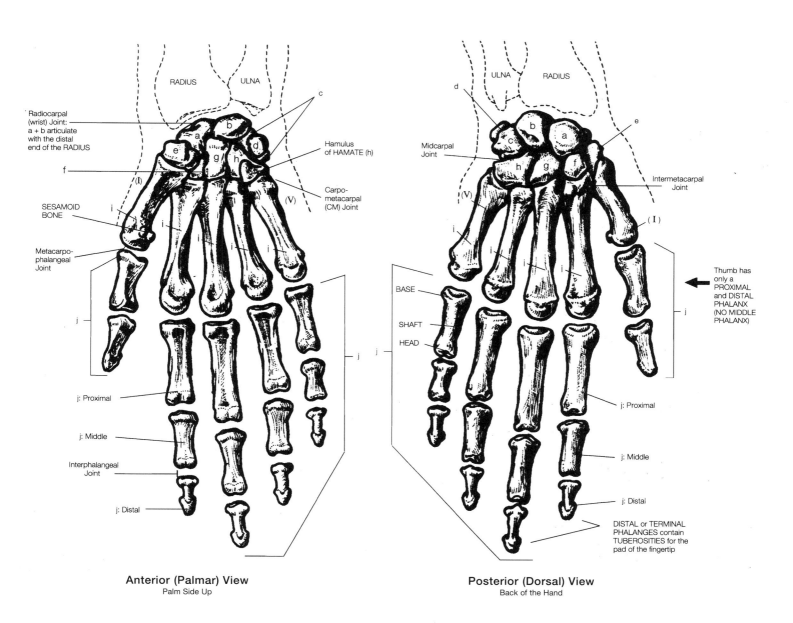

RADIUS ULNA

Radiocarpal
(wrist) Joint:
a + b articulate
with the distal
end of the RADIUS

c

a
b
e
d
g
h
f

(I)

(V)

SESAMOID
BONE

i

Metacarpo-
phalangeal
Joint

i
i
i
i

Hamulus
of HAMATE (h)

Carpo-
metacarpal
(CM) Joint

j

j: Proximal

j: Middle

Interphalangeal
Joint

j: Distal

Anterior (Palmar) View
Palm Side Up

ULNA RADIUS

d

Midcarpal
Joint

c
b
a
h
g
f

e

(V)

(I)

i

i
i
i

j

Intermetacarpal
Joint

Thumb has
only a
PROXIMAL
and DISTAL
PHALANX
(NO MIDDLE
PHALANX)

BASE

SHAFT

HEAD

j

j: Proximal

j: Middle

j: Distal

DISTAL or TERMINAL
PHALANGES contain
TUBEROSITIES for the
pad of the fingertip

Posterior (Dorsal) View
Back of the Hand

Color Guidelines
A = blue; B = green;
C = purple; D= cream;
E = flesh; f = ochre;
g = brown; h = red;
i= yellow; j = orange.

The PELVIS is a basinlike bony structure composed of four structures: PELVIC GIRDLE (TWO OSSA COXAE), the SACRUM, the COCCYX (of vertebral column) and the SYMPHYSIS PUBIS (fibrocartilage and other uniting ligaments).

The pelvic girdle is formed by the union of two ossa coxae (also called COXAL BONES, INNOMINATE BONES or HIP BONES). Anteriorly, the pubic portion of each coxal bone presents a roughened SYMPHYSEAL SURFACE that forms a joint with the opposite bone held by fibrocartiblage called the SYMPHYSIS PUBIS. Posteriorly, the iliac portion of each coxal bone presents a medial roughened auricular surface that forms the SACROILIAC JOINTS with the sacrum of the vertebral column.

Fetally, each os coxa develops as three separate bones. The ISCHIOPUBIC RAMUS is a combination of the RAMUS of the ISCHIUM (ISCHIAL RAMUS) and the INFERIOR PUBIC RAMUS. The ILIUM is the large superior bone. The ISCHIUM is inferior and posterior; the PUBIS is inferior and anterior. These three bones are fused together in the adult, but they are considered separately for descriptive purposes.

The three bones ossify (on the LATERAL SURFACE of each os coxa) at a large, circular depression (socket) called the ACETABULUM, which receives the head of the femur to form the hip joint. The three acetabular features are the acetabular fossa, lunate surface and acetabular notch (inferior).

Among the functions of the pelvic girdle is to provide strong, stable support for the lower extremities, which carry the weight of the body. It supports this body weight through the vertebral column. In its attachment to the vertebral column it permits an upright posture and locomotion on two legs (bipedal), in contrast to the four-legged locomotion of other mammals. It supports and protects the lower viscera, including the urinary bladder, the reproductive organs, and the developing fetus during pregnancy.

A SINGLE OS COXA (HIPBONE) CONSISTS OF THREE FUSED BONES IN THE ADULT

1 | os coxa | =

a | Ilium | +

b | Ischium | +

c | Pubis |

ATTACHMENTS TO THE PELVIC GIRDLE: ONE ANTERIOR ATTACHMENT (JOINT)

s.p. | Symphysis Pubis |

TWO POSTERIOR ATTACHMENTS (SACROILIAC JOINTS)

S | Sacrum |

Attachments are the two sacral auricular surfaces

Co | Coccyx |

Attached to the bottom of the sacrum

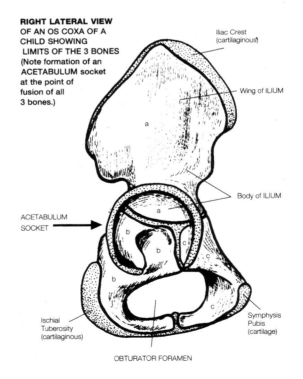

RIGHT LATERAL VIEW OF AN OS COXA OF A CHILD SHOWING LIMITS OF THE 3 BONES (Note formation of an ACETABULUM socket at the point of fusion of all 3 bones.)

Iliac Crest (cartilaginous)

Wing of ILIUM

Body of ILIUM

ACETABULUM SOCKET

Symphysis Pubis (cartilage)

Ischial Tuberosity (cartilaginous)

OBTURATOR FORAMEN

Four Views of the Right Male Adult Os Coxa

(Note: Dark Fusion Lines not actually visible.)
The gluteal muscles attach to the ilium between
three gluteal lines (posterior, anterior and inferior).

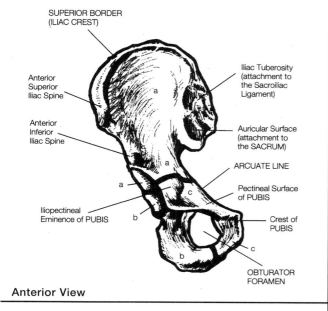

Anterior View

SUPERIOR BORDER (ILIAC CREST)

Anterior Superior Iliac Spine

Anterior Inferior Iliac Spine

Iliopectineal Eminence of PUBIS

Iliac Tuberosity (attachment to the Sacroiliac Ligament)

Auricular Surface (attachment to the SACRUM)

ARCUATE LINE

Pectineal Surface of PUBIS

Crest of PUBIS

OBTURATOR FORAMEN

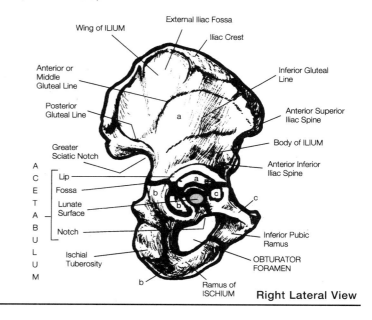

Right Lateral View

Wing of ILIUM

External Iliac Fossa

Iliac Crest

Anterior or Middle Gluteal Line

Posterior Gluteal Line

Greater Sciatic Notch

A C E T A B U L U M

Lip

Fossa

Lunate Surface

Notch

Ischial Tuberosity

Inferior Gluteal Line

Anterior Superior Iliac Spine

Body of ILIUM

Anterior Inferior Iliac Spine

Inferior Pubic Ramus

OBTURATOR FORAMEN

Ramus of ISCHIUM

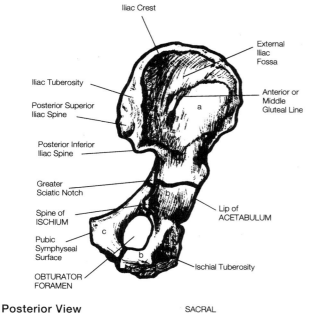

Posterior View

Iliac Crest

Iliac Tuberosity

Posterior Superior Iliac Spine

Posterior Inferior Iliac Spine

Greater Sciatic Notch

Spine of ISCHIUM

Pubic Symphyseal Surface

OBTURATOR FORAMEN

External Iliac Fossa

Anterior or Middle Gluteal Line

Lip of ACETABULUM

Ischial Tuberosity

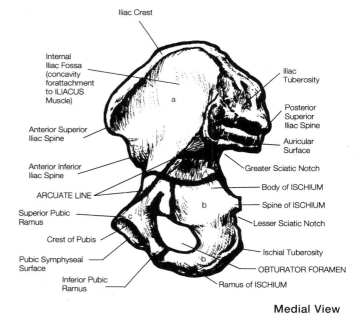

Medial View

Iliac Crest

Internal Iliac Fossa (concavity for attachment to ILIACUS Muscle)

Anterior Superior Iliac Spine

Anterior Inferior Iliac Spine

ARCUATE LINE

Superior Pubic Ramus

Crest of Pubis

Pubic Symphyseal Surface

Inferior Pubic Ramus

Iliac Tuberosity

Posterior Superior Iliac Spine

Auricular Surface

Greater Sciatic Notch

Body of ISCHIUM

Spine of ISCHIUM

Lesser Sciatic Notch

Ischial Tuberosity

OBTURATOR FORAMEN

Ramus of ISCHIUM

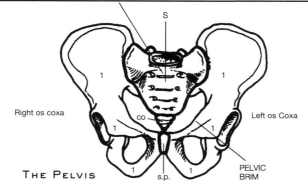

THE PELVIS

SACRAL PROMONTORY

S

Right os coxa

Left os Coxa

co

PELVIC BRIM

s.p.

COLOR GUIDELINES:
A = light green; B = light purple;
C = light blue; 1 = yellow-green;
S.P. = cream; S = red;
Co = orange; Cartilage = gray.

APPENDICULAR SKELETON, CONTINUED
BONES OF THE LOWER LIMB: MALE AND FEMALE PELVES

"Heart-Shaped" Male Pelvis

"Kidney-Shaped" Female Pelvis

CLINICAL DIVISIONS OF THE PELVIS:

FP | Greater, or False, Pelvis
Superior

TP | Lesser, or True, Pelvis
Inferior

DIVIDING CURVED, BONY RIM BETWEEN
THE FALSE AND TRUE PELVES:

PB | Pelvic Brim

DIAMETERS OF THE PELVIS

X | Transverse Diameter of
the Pelvic Inlet

Y | Anteroposterior/True Conjugate
Diameter

COMPONENTS OF THE PELVIS

2 OSSA COXAE (2 HIPBONES)

a | Ilium 2

b | Ischium 2

c | Pubis 2

s.p. | Symphysis pubis

S | Sacrum

Co | Coccyx

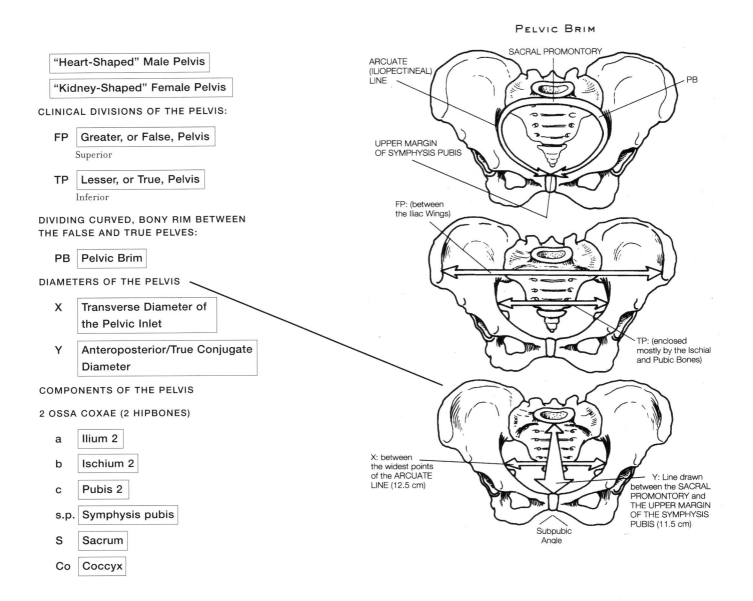

PELVIC BRIM

SACRAL PROMONTORY

ARCUATE
(ILIOPECTINEAL)
LINE

PB

UPPER MARGIN
OF SYMPHYSIS PUBIS

FP: (between
the Iliac Wings)

TP: (enclosed
mostly by the Ischial
and Pubic Bones)

X: between
the widest points
of the ARCUATE
LINE (12.5 cm)

Y: Line drawn
between the SACRAL
PROMONTORY and
THE UPPER MARGIN
OF THE SYMPHYSIS
PUBIS (11.5 cm)

Subpubic
Angle

The MALE and FEMALE PELVES (singular pelvis) have significant structural differences, particularly in relation to pregnancy and childbirth. Here are some of the general characteristics of the female pelvis that differ from the male pelvis: the joints of the pelvic girdle stretch during pregnancy and parturition (childbirth). The entire pelvic girdle is tilted forward. The GREATER (FALSE) PELVIS is shallower and the distance between the anterior superior iliac spine is wider. The LESSER (TRUE) PELVIS is spherical and wider. The pelvic inlet is larger and more oval; the pelvic outlet is comparatively large.

The obturator foramen of the PUBIS is more oval and triangular (rather than round). The symphysis pubis is shallower. The pubic arch is greater then ninety degrees, wider, and more rounded. The inferior ramus of the pubis does not have an everted surface.

The ILIUM is less vertical, and the iliac fossa is shallow (the male iliac fossa is deep). The iliac crest is less curved, and the greater sciatic notch of the ilium is wide (the male's is narrow).

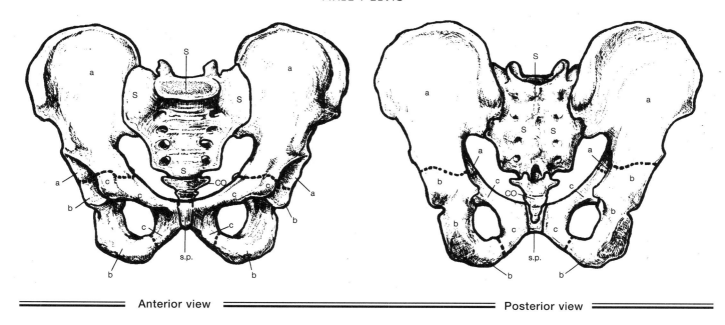

MALE PELVIS

Anterior view ═══════════════════════════════ Posterior view

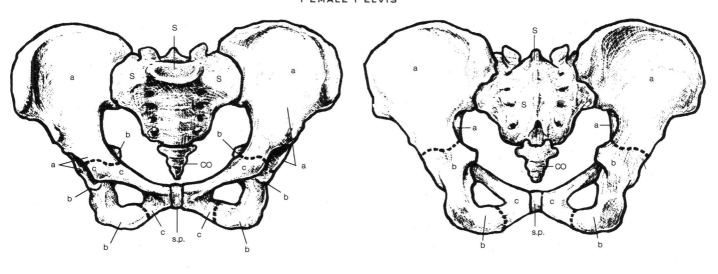

FEMALE PELVIS

In the female ischium, the ischial spine is turned inward less than the male's, and the ischial tuberosity is turned outward while the male's is turned inward.

The superior surface of the SACRUM only spans one-third of the width of the sacrum (the male's spans one-half). The auricular surface of the sacrum only extends down to S3 (the male extends well beyond S3). The sacrum is short, wide, flat and curving forward in the lower part (the male's has a long, narrow, smooth concavity).

The acetabulum and joint surfaces are small while the male's are large. The muscle attachments are indistinct while the male's are well-marked and the general bone structure is lighter and thinner (in males, it is heavy and thick).

The PELVIC INLET is the upper pelvic entrance of the LESSER (TRUE) PELVIS and the PELVIC OUTLET is the lower pelvic opening (border) of the lesser (true) pelvis.

The PELVIC BRIM is a curved, bony rim that passes inferiorly from the SACRAL PROMONTORY to the UPPER MARGIN OF THE SYMPHYSIS PUBIS. It divides the pelvis into two portions, a greater and lesser pelvis. It surrounds the pelvic inlet of the lesser pelvis as a boundary. The circumference of the pelvic brim bounds the pelvic inlet of the lesser (true) pelvis. The lower circumference of the lesser pelvis bounds the pelvic outlet, which is formed by the COCCYX, ISCHIAL PROTUBERANCES, ASCENDING RAMI OF THE ISCHIA, DESCENDING RAMI OF THE OSSA PUBIS and the SACROSCIATIC LIGAMENTS.

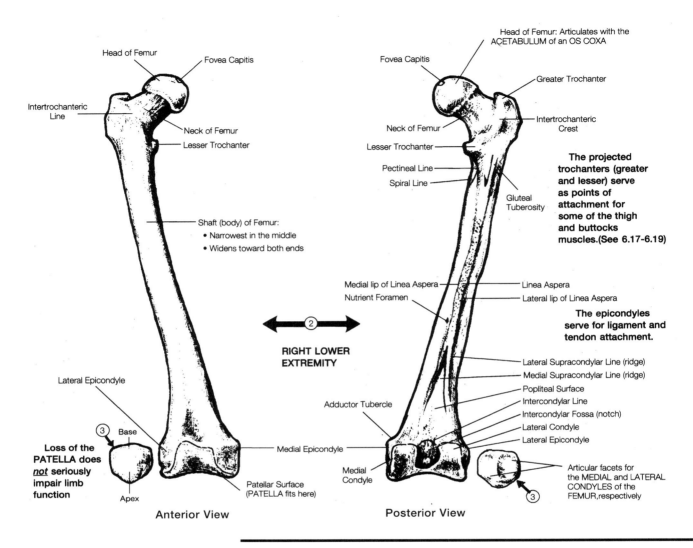

Head of Femur
Fovea Capitis
Intertrochanteric Line
Neck of Femur
Lesser Trochanter
Shaft (body) of Femur:
• Narrowest in the middle
• Widens toward both ends

Lateral Epicondyle

③ Base

Loss of the PATELLA does _not_ seriously impair limb function

Apex

Medial Epicondyle

Patellar Surface (PATELLA fits here)

Anterior View

②
RIGHT LOWER EXTREMITY

Head of Femur: Articulates with the ACETABULUM of an OS COXA
Fovea Capitis
Greater Trochanter
Neck of Femur
Intertrochanteric Crest
Lesser Trochanter
Pectineal Line
Spiral Line
Gluteal Tuberosity

The projected trochanters (greater and lesser) serve as points of attachment for some of the thigh and buttocks muscles.(See 6.17-6.19)

Medial lip of Linea Aspera
Nutrient Foramen
Linea Aspera
Lateral lip of Linea Aspera

The epicondyles serve for ligament and tendon attachment.

Adductor Tubercle
Lateral Supracondylar Line (ridge)
Medial Supracondylar Line (ridge)
Popliteal Surface
Intercondylar Line
Intercondylar Fossa (notch)
Lateral Condyle
Lateral Epicondyle

Medial Condyle

Articular facets for the MEDIAL and LATERAL CONDYLES of the FEMUR,respectively

③

Posterior View

THE SINGULAR BONE OF THE THIGH: PROXIMAL PORTION OF THE LOWER EXTREMITY

2 | Femur | Thighbone

SESAMOID BONE OF THE KNEE:

3 | Patella | Kneecap

TWO SKELETAL STRUCTURES OF THE LEG, BETWEEN THE KNEE AND THE ANKLE:

4 | Tibia | Shinbone

5 | Fibula | "Splint Bone"

The longer bones of the LOWER EXTREMITY are the FEMUR, TIBIA and FIBULA. The singular bone of the thigh is the proximal portion of the lower extremity. The femur is the longest, heaviest and strongest bone of the body. It extends from the HIP to the KNEE. The proximal, rounded HEAD articulates with the ACETABULUM of the COXAL BONE. The point of attachment for the LIGAMENTUM TERES (which helps support the head against the acetabulum) is the FOVEA CAPITIS, a roughened, shallow pit on the lower center of the rounded head. It also carries blood vessels into the bone. The SHAFT of the FEMUR bows medially to approach the femur of the opposite thigh. This convergence results in the knee joints being brought nearer to the body's line of gravity (greater in women).

It contains several important structures for muscle attachment. Its distal end is expanded for articulation with the TIBIA and includes the MEDIAL CONDYLE and LATERAL CONDYLE.

The sesamoid bone of the knee is called the PATELLA, or kneecap. It is positioned on the anterior side of the knee joint and is small and triangular. Being a sesamoid bone, it forms in the tendon of the QUARDRICEPS FEMORIS MUSCLE as a response to stress. It protects, but is not part of, the knee joint. Loss of the patella does not seriously impair limb function. It strengthens the QUADRICEPS TENDON and increases leverage of the QUADRICEPS FEMORIS MUSCLE; it also changes the direction of the muscle pull on the tibia to a more efficient angle.

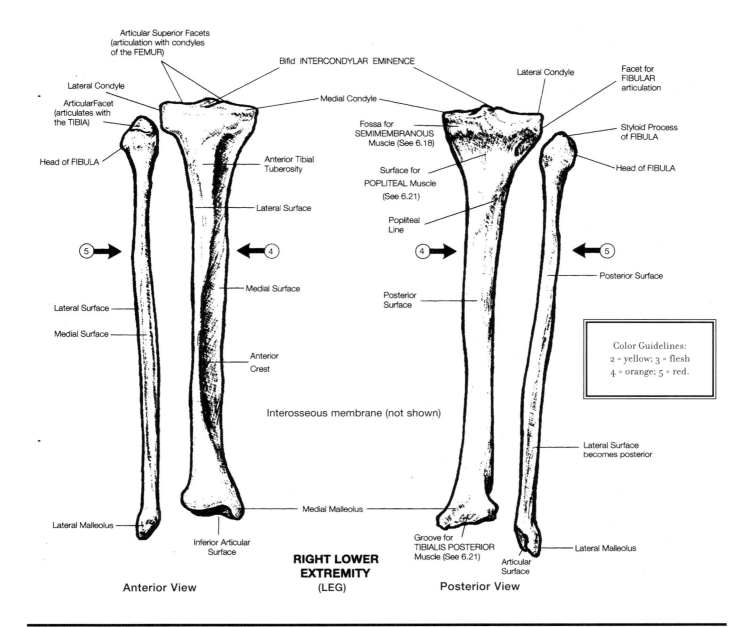

Articular Superior Facets (articulation with condyles of the FEMUR)

Bifid INTERCONDYLAR EMINENCE

Lateral Condyle

Medial Condyle

ArticularFacet (articulates with the TIBIA)

Head of FIBULA

Anterior Tibial Tuberosity

Lateral Surface

⑤→   ←④

Medial Surface

Lateral Surface

Medial Surface

Anterior Crest

Interosseous membrane (not shown)

Lateral Malleolus

Inferior Articular Surface

Medial Malleolus

RIGHT LOWER EXTREMITY (LEG)

Anterior View

Lateral Condyle

Facet for FIBULAR articulation

Fossa for SEMIMEMBRANOUS Muscle (See 6.18)

Styloid Process of FIBULA

Surface for POPLITEAL Muscle (See 6.21)

Head of FIBULA

Popliteal Line

④→   ←⑤

Posterior Surface

Posterior Surface

Color Guidelines:
2 = yellow; 3 = flesh
4 = orange; 5 = red.

Lateral Surface becomes posterior

Groove for TIBIALIS POSTERIOR Muscle (See 6.21)

Medial Malleolus

Lateral Malleolus

Articular Surface

Posterior View

Similar to the upper extremities, the lower extremities contain sixty bones, with thirty bones in each extremity (twenty-six of which are in the ANKLE and FOOT). The epicondyles serve for ligament and tendon attachment.

The projected trochanters (greater and lesser) serve as points of attachment for some of the thigh and buttock muscles.

The TIBIA is the larger of the two bones of the leg. At the proximal end, this medial bone bears the major portion of the weight of the leg, where it articulates proximally at the knee joint. The MEDIAL and LATERAL CONDYLES of the TIBIA articulate with the FEMORAL CONDYLES. The lateral condyle of the tibia carries a FACET for articulation with the HEAD OF THE FIBULA. The shaft of the tibia is triangular in cross-section (with the triangular base posterior) and contains three surfaces (posterior, medial and lateral) and three borders (the anterior, medial and interosseous). On the proximal, anterior portion of the shaft lies the TIBIAL TUBEROSITY for attachment of the patellar ligament. The distal end of the tibia

bears an INFERIOR ARTICULAR SURFACE, with a medially oriented MEDIAL MALLEOLUS for articulation with the TALUS (an ankle bone). On the distal-lateral end, there is a FIBULAR NOTCH for articulation with the FIBULA.

The FIBULA is a long, narrow bone parallel and lateral to the tibia, and considerably smaller. Its fragility lends little to weight-bearing or support. It extends the area for muscle attachment (along with the INTEROSSEOUS MEMBRANE that connects it to the tibia). The medial surface of the shaft is obviously grooved to serve this function. The distal end of the bone bears a pointed, prominent knob called the LATERAL MALLEOLUS, which bears a facet for the talus on its medial aspect.

The tibia and fibula together form a cuplike articulating surface for the ankle. The medial malleolus (tibia) and lateral malleolus (fibula) are positioned on either side of the talus bone to offer stabilization to the ankle joint (see next page).

TWO ARCHES OF THE FOOT

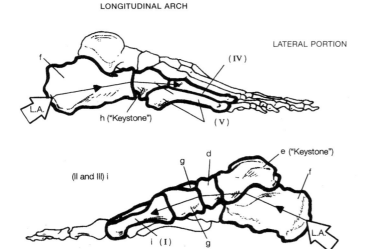

LONGITUDINAL ARCH

LATERAL PORTION

f

(IV)

h ("Keystone")

(V)

L.A.

(II and III) i

g   d

e ("Keystone")

f

i (I)

g

L.A.

MEDIAL PORTION (LARGER)

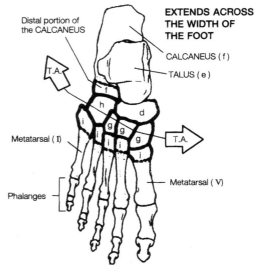

TRANSVERSE ARCH

EXTENDS ACROSS THE WIDTH OF THE FOOT

Distal portion of the CALCANEUS

CALCANEUS (f)

TALUS (e)

T.A.

f

h

d

i

g g

T.A.

Metatarsal (I)

i g g

i i g

Phalanges

j

Metatarsal (V)

Superior (Dorsal) View

THE ANKLE (TARSUS) — SEVEN BONES:
PROXIMAL ROW

d | Navicular | Scaphoid

e | Talus | Astragalus

f | Calcaneus | "Heelbone"
     (Os Calcis)

DISTAL ROW

g | 3 Cuneiforms | 1st, 2nd, 3rd

h | Cuboid | Os Cuboideum

THE FOOT (PES):
BONES OF THE SOLE — THE PLANTAR SURFACE OF THE FOOT
Metatarsus

i | 5 Metatarsal Bones | (I – V)

Toes (digits)

j | 14 Phalanges | (singular phalanx)

ARCHES OF THE FOOT

L.A. | Longitudinal Arch
     Medial Part
     Proximal Part

T.A. | Transverse Arch

The distal portion of the lower extremity consists of the ANKLE and the FOOT, which contain twenty-six bones (seven TARSAL [ankle] bones and nineteen bones of the foot). Of the nineteen foot bones there are five METATARSAL (SOLE) BONES and fourteen PHALANGES. Both ankles and feet therefore contain fifty-two bones: fourteen tarsals, ten metatarsals and twenty-eight phalanges.

The ankle is made up of the seven tarsal bones, of which only the TALUS bone is involved in forming the ANKLE JOINT. The talus contains large superior and lateral ARTICULAR SURFACES that connect to the tibia and fibula. The CALCANEUS is the largest of the tarsal bones and forms the heel of the foot. The tuberosity of the calcaneus is a roughened posterior surface for attachment of calf muscles. The NAVICULAR is boat-shaped and is anterior to the talus. The four DISTAL BONES OF THE ANKLE are a series of bones that articulate with the METATARSALS. From the medial to the lateral side they are the first, second and third CUNEIFORMS (wedge-shaped) and the cuboid (which is anterior to the calcaneus).

The FOOT has five METATARSAL bones comprising the SOLE, numbered I—V from the medial side (big toe side) to the lateral side. Metatarsal I is the largest and serves most of the weight-bearing function of the meta-tarsals. The ball of the foot is formed by the heads of metatarsals I and II. Metatarsals II—V are slender and function like an outrigger, providing stability. The proximal bases of metatarsals I-III articulate proximally with the three cuneiforms. The distal heads of all the metatarsals articulate distally with the PROXIMAL PHALANGES. The fourteen phalanges of the toes are arranged like those of hands, with proximal phalanges being the longest. Middle and distal phalanges are very short.

The two arches of the foot are formed by the structure and arrangement of the bones and held in place by ligaments and tendons, but are not rigid. Arches yield when force (weight) is applied, and spring back to normal when the force is lifted. They support the weight of the body and provide leverage when walking.

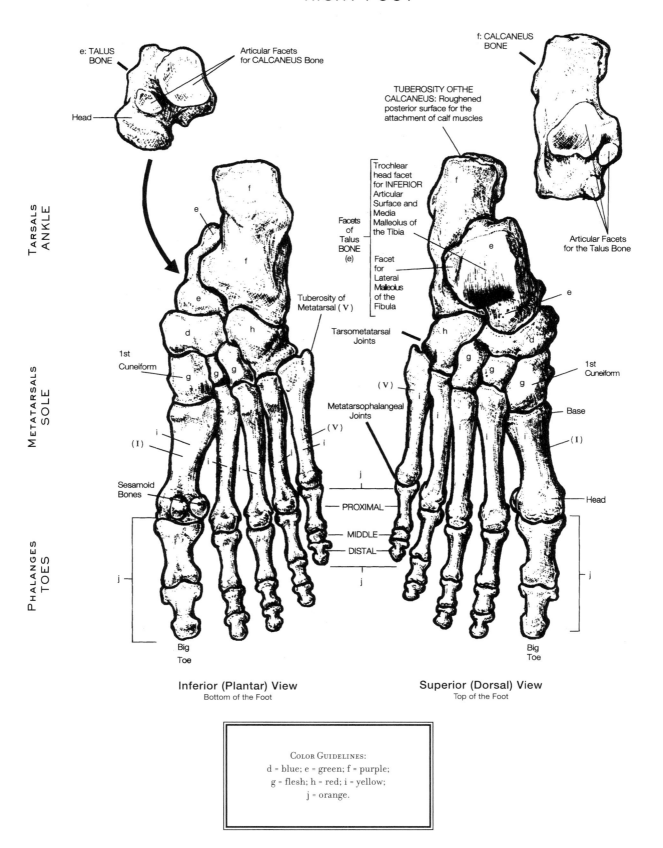

e: TALUS BONE

Head

Articular Facets for CALCANEUS Bone

f: CALCANEUS BONE

TUBEROSITY OFTHE CALCANEUS: Roughened posterior surface for the attachment of calf muscles

Articular Facets for the Talus Bone

TARSALS **A**NKLE

METATARSALS **S**OLE

PHALANGES **T**OES

Tuberosity of Metatarsal (V)

1st Cuneiform

Sesamoid Bones

Big Toe

Trochlear head facet for INFERIOR Articular Surface and Media Malleolus of the Tibia

Facets of Talus BONE (e)

Facet for Lateral Malleolus of the Fibula

Tarsometatarsal Joints

Metatarsophalangeal Joints

PROXIMAL

MIDDLE

DISTAL

1st Cuneiform

Base

Head

Big Toe

Inferior (Plantar) View
Bottom of the Foot

Superior (Dorsal) View
Top of the Foot

COLOR GUIDELINES:
d = blue; e = green; f = purple;
g = flesh; h = red; i = yellow;
j = orange.

CHAPTER 6: ARTICULAR SYSTEM

JOINTS, or ARTICULATIONS, do not comprise a system as such, but in this book we will treat them in total as the ARTICULAR SYSTEM. However, they are directly related to the SKELETAL SYSTEM and specifically to one of the functions of the skeletal system: permitting body movement. The articulations between the bones allow movement of body parts, and the range of movement permitted is determined by the structure of a joint. Articulations are necessary for such activities as walking, running and lifting, but also for many activities in which we are apt to take the role of the muscles and joints for granted, such as talking, eating and writing—even smiling requires coordinated effort of muscles, bones and joints.

Articulations may be classified according to the structure of the joint—the presence or absence of a cavity and the kind of supportive connective tissue associated with the joint; but also according to function—the degree of movement permitted within the joint. For each, there are three basic categories.

THE BIOMECHANICS OF ARTICULATION

The physics of articulation is simply the application of the principle of the lever to anatomy, though it is not always easy to determine how the body is applying those principles. A lever has four basic components: a rigid bar or structure (called simply "the lever"); a "fulcrum" or pivot which remains fixed; a weight or resistance that is set into motion by the lever; and a force or effort that moves the lever and thereby moves the weight or resistance. Depending on the configuration of its components (see illustrations below), there are three kinds of levers—called

CONTENTS

COLORING NOTES:
Choose a color for each symbol,
and color the illustrations.
Lever = bone;
pivot (fulcrum = joint);
effort (force = muscle);
resistance = object moved.

COLOR GUIDELINES:
Immovable joints = light blue;
Slightly movable joints = light purple;
Freely movable joints = light yellow;
Color all 6 diarthrotic (freely
movable—synovial) joints in warm
colors (oranges, reds, pinks, etc.).

SYSTEM COMPONENTS
Joints and their associated ligaments

SYSTEM FUNCTION
Flexible fibrous connective tissue at points of contact between bones or cartilage and bones.

PRINCIPLE OF LEVERAGE IN RELATION TO BONES AND JOINTS

The rigid lever (BONE) turns about a FULCRUM (JOINT) when force is applied (MUSCLE) to move the resisting object.

THREE CLASSES OF LEVERS

1   First-Class Lever

The fulcrum (joint) is positioned between the FORCE (muscle) and the resistance

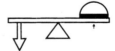

2   Second-Class Lever

The resistance is positioned between the force (muscle) and the fulcrum (joint)

3   Third-Class Lever

The force (muscle) is positioned between the fulcrum (joint) and the resistance

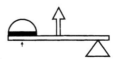

Choose a color for each symbol, and color the illustrations above.

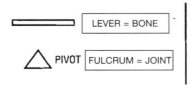

LEVER = BONE

PIVOT   FULCRUM = JOINT

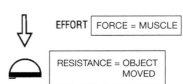

EFFORT   FORCE = MUSCLE

RESISTANCE = OBJECT MOVED

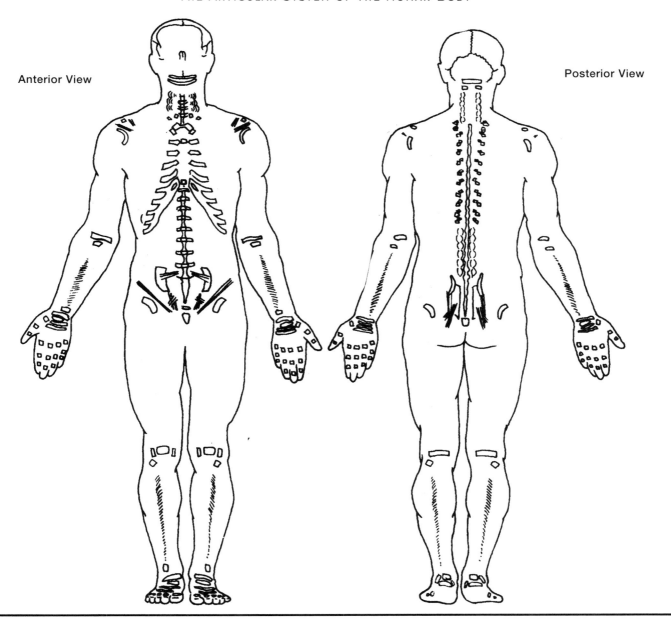

Anterior View

Posterior View

"first-class," "second-class" and "third-class"—and they are in descending order with respect to efficiency. In the human body, the most common type of lever configuration is (somewhat ironically) the third-class lever. The body has often had to trade speed for power, and range of movement for mechanical efficiency—the environment often dictated how the articulation developed.

An example of a first-class lever would be a pair of scissors. In the body, the head at the atlantooccipital joint is an example: the posterior neck muscles pull down on the skull around the joint as a fulcrum and maintain the weight of the front part of the head.

An example of a second-class lever is a wheelbarrow; and in the body, contracting the calf muscles to stand on one's toes (with the ball of the foot acting as the fulcrum).

An example of a third-class lever is a pair of tongs or forceps. In the body, the flexing of the biceps muscle in the arm moves the forearm as it hinges on the elbow in the rear—the effort (muscle tension) lies between the fulcrum (elbow joint) and the weight (the forearm).

ARTHROLOGY is the science concerned with the study of articulations (joints) and all aspects of joint classifications. KINESIOLOGY is an applied, practical, dynamic science concerned with the biomechanics of movements involving certain joints.

COMPOSITION AND CLASSIFICATION OF JOINTS

Joints are composed of FIBROUS CONNECTIVE TISSUE, CARTILAGE, or LIGAMENTS. Not all joints are flexible in permitting movement, as some joints remain rigid for body stabilization and balance. Joints may also require specific structures to serve as cavities, requiring their own support structure, such as protective membranes, cushioning tissue and lubricating fluid.

CLASSIFICATION OF JOINTS
FUNCTIONAL AND STRUCTURAL CATEGORIES

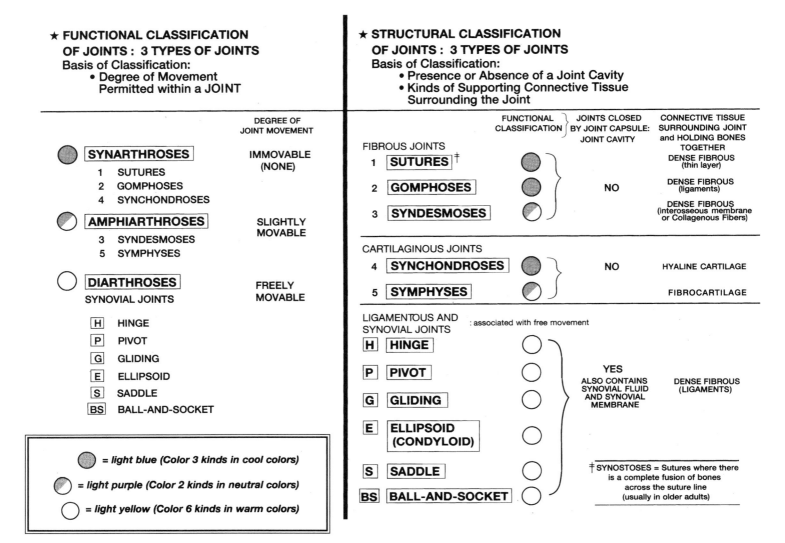

★ **FUNCTIONAL CLASSIFICATION OF JOINTS : 3 TYPES OF JOINTS**
Basis of Classification:
- Degree of Movement Permitted within a JOINT

★ **STRUCTURAL CLASSIFICATION OF JOINTS : 3 TYPES OF JOINTS**
Basis of Classification:
- Presence or Absence of a Joint Cavity
- Kinds of Supporting Connective Tissue Surrounding the Joint

DEGREE OF JOINT MOVEMENT

SYNARTHROSES — IMMOVABLE (NONE)
1 SUTURES
2 GOMPHOSES
4 SYNCHONDROSES

AMPHIARTHROSES — SLIGHTLY MOVABLE
3 SYNDESMOSES
5 SYMPHYSES

DIARTHROSES — FREELY MOVABLE
SYNOVIAL JOINTS

H HINGE
P PIVOT
G GLIDING
E ELLIPSOID
S SADDLE
BS BALL-AND-SOCKET

= light blue (Color 3 kinds in cool colors)

= light purple (Color 2 kinds in neutral colors)

= light yellow (Color 6 kinds in warm colors)

FUNCTIONAL CLASSIFICATION	JOINTS CLOSED BY JOINT CAPSULE: JOINT CAVITY	CONNECTIVE TISSUE SURROUNDING JOINT and HOLDING BONES TOGETHER

FIBROUS JOINTS
1 **SUTURES** † — DENSE FIBROUS (thin layer)
2 **GOMPHOSES** — NO — DENSE FIBROUS (ligaments)
3 **SYNDESMOSES** — DENSE FIBROUS (interosseous membrane or Collagenous Fibers)

CARTILAGINOUS JOINTS
4 **SYNCHONDROSES** — NO — HYALINE CARTILAGE
5 **SYMPHYSES** — FIBROCARTILAGE

LIGAMENTOUS AND SYNOVIAL JOINTS   : associated with free movement
H **HINGE**
P **PIVOT**
G **GLIDING** — YES ALSO CONTAINS SYNOVIAL FLUID AND SYNOVIAL MEMBRANE — DENSE FIBROUS (LIGAMENTS)
E **ELLIPSOID (CONDYLOID)**
S **SADDLE**
BS **BALL-AND-SOCKET**

† SYNOSTOSES = Sutures where there is a complete fusion of bones across the suture line (usually in older adults)

Classifying joints structurally yields three basic categories of joint (see the charts, above and on next page): FIBROUS JOINTS lack a SYNOVIAL cavity (i.e., one containing joint tissue and lubricating fluid), and the articulating bones are held together by fibrous connective tissue. Such joints permit little or no movement. The three types of fibrous joints are: SUTURES, found only in the skull, and consisting of thin layers of connective tissue that connect bones which were previously separated by soft tissue in infancy; GOMPHOSES, which are formed by a cone-shaped peg fitting into a socket, but, again, permitting no movement; and SYNDESMOSES, in which more fibrous tissue between bones allows for limited movement. (In a *functional* categorization scheme, both sutures and gomphoses would be classified as SYNARTHROSES—joints in which no movement occurs, but syndesmoses would be classified as AMPHIARTROSES —joints in which some movement occurs.)

A second structural category is CARTILAGINOUS JOINTS, in which the bones are held together with cartilage, which also highly restricts movement. There are two varieties: SYNCHONDROSES, where the cartilage is hyaline cartilage that ossifies in adulthood; and SYMPHYSES, where the bones are connected by a flat disc of fibrocartilage. This joint allows for some movement—an example are the discs between vertebrae in the spinal column (which permit limited bending of the torso). *Functionally,* symphyses are categorized with syndesmoses as amphiatroses.

The third *structural* category of joint is the SYNOVIAL or LIGA-MENTOUS JOINTS, in which a space exists between the articulating bones (called a SYNOVIAL or JOINT CAVITY) and the bones are connected by hyaline cartilage and are permitted to move (nearly) freely. The six kinds of synovial joints are categorized functionally as DIARTHROSES because of their ability to move freely (i.e., they are "diarthrotic").

Synovial joints are surrounded by a covering, the ARTICULAR CAPSULE, which encloses the joint and holds the bones together. An outer layer of the articular capsule (the FIBROUS CAPSULE)

FUNCTIONAL AND STRUCTURAL CLASSIFICATION OF ARTICULATIONS

STRUCTURAL		FUNCTIONAL
FIBROUS JOINTS (3 KINDS)		-ARTHROSES
SUTURES	• SERRATE ➞ Sagittal suture between Parietal Bones of the Skull • LAP (SQUAMOUS) ➞ Squamous suture beteen Temporal and Parietal Bones • PLANE (BUTT) ➞ Maxillary suture = the 2 Maxillary Bones form HARD PALATE †. SYNOSTOSIS ➞ Union of Right and Left Portions of the Frontal Bone	**SYN-**
1		
GOMPHOSES	• Styloid process in Temporal Bone inserted into a socket • **Teeth roots embedded in the Alveoli of the Maxillae and Mandible**	**SYN-**
2		
SYNDESMOSES	• **Distal end articulation of the Tibia and Fibula** • Articulations between shafts of the Ulna and Radius	**AMPHI-**
3		

STRUCTURAL		FUNCTIONAL
CARTILAGINOUS JOINTS (2 KINDS)		-ARTHROSES
SYNCHONDROSES	• Temporary joints forming the growth lines or EPIPHYSEAL PLATES in the long bones in children • Occipital-sphenoid joint in children • Joints ossify when bone growth is complete. Hyaline cartilage of the temporary joint is replaced by bone. • **Joint changes from a SYNCHONDROSES to a SYNOSTOSES (See p. 82)**	**SYN-**
4	• 1st rib STERNUM • Sacroiliac joint.	
SYMPHYSES	• Symphysis Pubis (between the anterior surfaces of the Coxal Bones). • Intervertebral discs between the vertebrae	**AMTHI-**
5		

STRUCTURAL		FUNCTIONAL
SYNOVIAL JOINTS (6 KINDS) - DIARTHROTIC		-ARTHROSES
H	• Bending in only one plane (like a door) • 1 bone surface is always concave, and the other is always convex _Most common type of DIARTHROSES :_ • KNEE (Tibiofemoral) JOINT • ELBOW (Humeroulnar) JOINT • Joints of the PHALANGES • TEMPOROMANDIBULAR JOINT	**DI-**
P	• Rotation about a central axis • ATLAS and AXIS JOINT (rotational movement of the head) • Proximal articulation of the RADIUS and ULNA (rotational movement of the forearm)	**DI-**
G	• Smooth sliding • Simplest kind of movement • COSTOVERTEBRAL JOINTS (between ribs, and vertebrae and vertebral processes) • Joints between the carpals • Joints between the tarsals • STERNOCLAVICULAR JOINT	**DI-**
E	• Angular movement in 2 directions (Biaxial): up-and-down and side-to-side • Oval, convex surface-elliptical, concave depression • RADIOCARPAL JOINT • ATLANTO-OCCIPITAL JOINT (between the ATLAS BONE and the Occipital Bone of the skull)	**DI-**
S	• Modified CONDYLOID (ELLIPTICAL) JOINT • Associated only with the thumb • Concave surface in one direction and convex surface in another direction • The articulation of the TRAPEZIUM BONE IN THE CARPUS (WRIST) with the FIRST METACARPAL BONE • CARPOMETACARPAL JOINT	**DI-**
BS	• Greatest range of movement of all DIARTHROSES (all planes and rotation) • Rounded convex surface and cuplike cavity • HIP JOINTS • SHOULDER JOINTS	**DI-**

consists of dense collagenous tissue that sometimes bundles to form LIGAMENTS that add strength to the joint; the inner layer is softer tissue, the SYNOVIAL MEMBRANE, that secretes SYNOVIAL FLUID that maintains lubrication in the joint.

An interesting feature of some synovial joints—particularly those with more congruent surfaces, such as those between the phalanges and metacarpals of the hand—may be observed when the bones on either side of the joint are pulled apart. This causes negative pressure on the synovial fluid, which releases carbon dioxide, which collects as a bubble. When the bone surfaces continue to separate, the pressure in the joint exceeds that of the bubble, and the bubble collapses, producing a cracking sound as the gas is reabsorbed by the synovial fluid.

The six types of synovial joints are described in the chart above. Many synovial joints contain additional tissues and structures—LIGAMENTS, fibrous cartilage and discs of fibrous cartilage that further cush-ion the bones and the structures of the joint.

These pads, called ARTICULAR DISCS or MENISCI, permit bones that would ordinarily not fit to remain connected and articulate with one another. They also direct synovial fluid to areas where friction is greatest.

Bodily movement creates friction throughout the body, necessitating a mechanism for reducing friction. This function is served by sac-like structures situated in body tissue called BURSAE. These structures resemble synovial capsules—they have walls comprised of connective tissue that is lined by a synovial membrane and they produce and distribute a fluid very much like synovial fluid. Bursae cushion the parts of the body that are in contact with bone and with each other. Inflammation of a bursa is called BURSITIS.

The drawing of the anatomy and structures of the knee joint (see pages 76–77) provides an excellent example of the various elements of the joint and joint cavity, including the bursae, ligaments and menisci.

Freely Movable (Synovial) Joints
Typical Synovial Joint Structure: The Knee Joint

Color Guidelines:
a and b = warm colors;
c = light brown; d = cream;
e = light yellow; f = light gray;
synovial cavity =light blue.

See
WWW.MCMURTRIESANATOMY.COM
for examples of major synovial
joints (of the skull, upper limbs
and lower limbs) and a review
of the articular system.

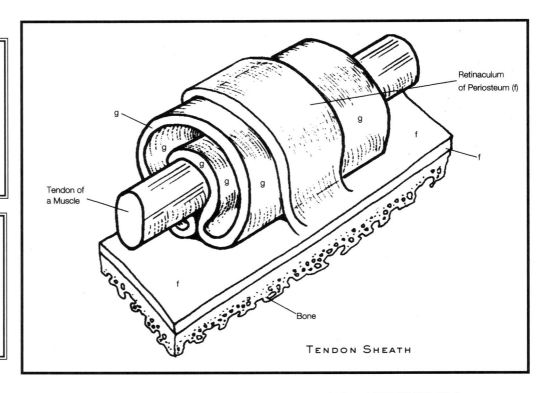

Retinaculum
of Periosteum (f)

Tendon of
a Muscle

Bone

TENDON SHEATH

The flexibility of the FIBROUS JOINT CAP-SULE permits movement at a DIARTHROTIC JOINT. Its great tensile strength resists dislocation. (Fibers arranged in parallel bundles adapted to resist recurrent strain are called LIGA-MENTS.) The fibrous joint capsule is attached to the PERIOSTEUM of the ARTICU-LATING BONES (at a distance away from the ARTICULAR CARTILAGE—which is smooth and caps the ends of the articulating bones, and resides within the synovial cavity filled with SYNOVIAL FLUID). There are three factors limiting the movement of a diarthrotic-synovial joint: the structure of the bone (sometimes a process will limit range and "lock" articulation), the strength and tautness of associated structures (ligaments, tendons, and joint capsules) and the size, arrange-ment, and action of associated muscles spanning the joint.

The cushioned BURSAE near diarthrotic-synovial joints are flattened, pouchlike synovial membrane sacs filled with syn-ovial fluid. They are located between mus-cles or within areas where tendon passes over bone; they cushion the muscles and facilitate movement of muscles or tendons over surfaces of bones or ligaments.

The TENDON SHEATH facilitates the glid-ing of a tendon as it traverses a tunnel (bony or fibrous). It is a modified synovial bursa (closed sac) where one layer of the synovial membrane folds over the surface of the tendon, and the other layer lines the formed tunnel. It is usually associated with important, strong FLEXOR MUSCLES. For example, it lubricates tendons of mus-cles that cross the wrist and ankle joints.

All DIARTHROSES are synovial joints. They are enclosed by a JOINT CAPSULE made of dense fibroelastic tissue, also called an ARTICULAR CAPSULE. A JOINT CAVITY within the capsule, also called a synovial cavity, creates a space between articulating bones. Synovial fluid (synovium), which fills the synovial cavity, is secreted by a thin synovial membrane that lines the inside of the joint capsule. The fluid con-tains the lubricant HYALURONIC ACID, and interstitial fluid from BLOOD PLASMA. The

TYPICAL STRUCTURE OF A DIARTHROTIC-SYNOVIAL JOINT

 Synovial cavity with synovial fluid/synovium

a | **Synovial Membrane**
Inner layer of the joint capsule

b | **Joint Capsule**
Fibrous articular capsule

c | **Articular Cartilage**

d | **Articulating Bones**
Never come in actual contact with one another

e | **Articular Disk**

f | **Periosteum of Bones**

g | **Tendon Sheath**
Modified synovial bursa (closed sac)

synovial membrane lubricates the joint, and contains phagocytic cells that remove microbes and debris. It also provides nourishment for the ARTICULAR CARTILAGE, which caps both articulating bones. Bones never come in contact with each other.

Coronal View

Coronal View (with Articulating Disk)
Example: Temporomandibular Disk

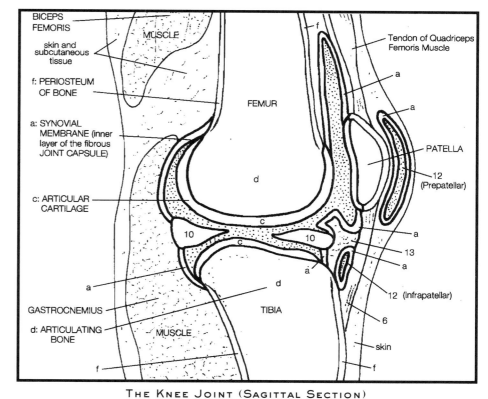

STRUCTURE OF THE KNEE JOINT

1	Fibula Collateral Ligament
2	Tibial Collateral Ligament
3	Anterior Cruciate Ligament
4	Posterior Cruciate Ligament
6	Patellar Ligament
7	Oblique Popliteal Ligament
8	Arcuate Popliteal Ligament
9	Ligament of Wrisberg
10	Lateral Meniscus
11	Medial Meniscus
12	Synovial Bursa
13	Subpatellar Fat
	Synovial Cavity (filled with Synovial Fluid)

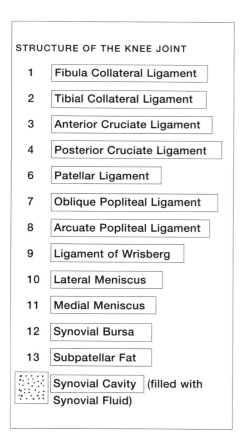

THE KNEE JOINT (SAGITTAL SECTION)

An example of a temporomandibular joint is the encapsulated DOUBLE SYNOVIAL JOINT between the condyles of the MANDIBLE and the TEMPORAL BONES of the SKULL. It is the only DIARTHROTIC JOINT in the skull with a combination of a HINGE JOINT and a GLIDING JOINT. The ARTICULAR DISK separates the synovial cavity into two separate compartments. It is clinically important due to TEMPOROMANDIBULAR JOINT (TMJ) syndrome.

The limiting factors of movement are: apposition of soft parts, structure of participating bones, tension of associated ligaments and tension, size and arrangement of associated muscles. There are many movements which work in opposition to one another. When one surface moves back-and-forth and side-to-side over another surface without angular or circular motion, the movement is called gliding.

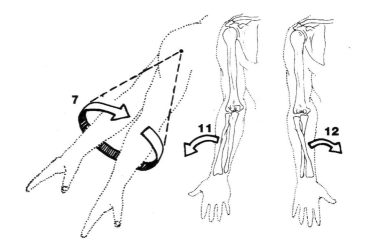

> COLOR GUIDELINES:
> Color in opposite colors for opposite movements.
> (e.g., yellow-purple, orange-blue, red-green, etc.)

ARTICULATIONS OF THE BODY AND MOVEMENT OF THE JOINTS

1 Gliding
Back-and-forth and side-to-side (one surface over another surface)

ANGULAR
Increase or decrease the joint angle produced by the articulating bones

2 Flexion
Movement that decreases the joint angle on an anterior-posterior plane (e.g.: bending the head forward, flexing arm upward

8 Dorsiflexion
Upward bending of the foot backward at the angle

3 Extension
Movement that increases the joint angle on an anterior-posterior plane (e.g.: bending the head backward, extending arm downward)

9 Plantar Flexion
Downward bending (extension) of the foot at the ankle joint (forepart depressed in respect to the ankle)

10 Hyperextension
Extension of a joint beyond the anatomical position

4 Abduction
Movement away from the midline of the body

5 Adduction
Movement toward the midline of the body

CIRCULAR
One bone with a rounded or oval surface articulates with a bone with a corresponding cup or depression, permitting circular movement

6 Rotation
Movement of a bone around its own axis

11 Supination
Assuming the supine position (e.g.:turning wrists so that the palms are upward)

12 Pronation
Assuming the prone position; turning wrists so that the palms are downward (Both supination and pronation pertain to position of body as a whole)

7 Circumduction
Movement at a synovial joint describing a cone-shaped figure, where the distal end of a bone moves in a circle, while proximal end remains stationary

SPECIALIZED MOVEMENTS

13 Inversion
Turning bottom of the feet (soles) inward

14 Eversion
Turning bottom of the feet (soles) outward

15 Protraction
To draw forward (e.g.: jutting of the lower jaw forward)

16 Retraction
To draw backward (e.g.: pulling the lower jaw backward behind upper teeth)

17 Elevation
An upward movement (e.g.: lifting the shoulders [scapulae] as in a shrug)

18 Depression
A downward movement (e.g.: moving shoulders [scapulae] in a downward direction)

*MONAXIAL MOVEMENT

H Hinge Flexion-Extension

P Pivot Rotation

*BIAXIAL MOVEMENT
(Flexion-Extension and Abduction-Adduction)

E Ellipsoidal

S Saddle

*TRIAXIAL MOVEMENT

G Gliding

BS Ball-and-Socket
Flexion-Extension
Abduction-Adduction
Rotation, Circumduction

*Refer to pages 74–75.

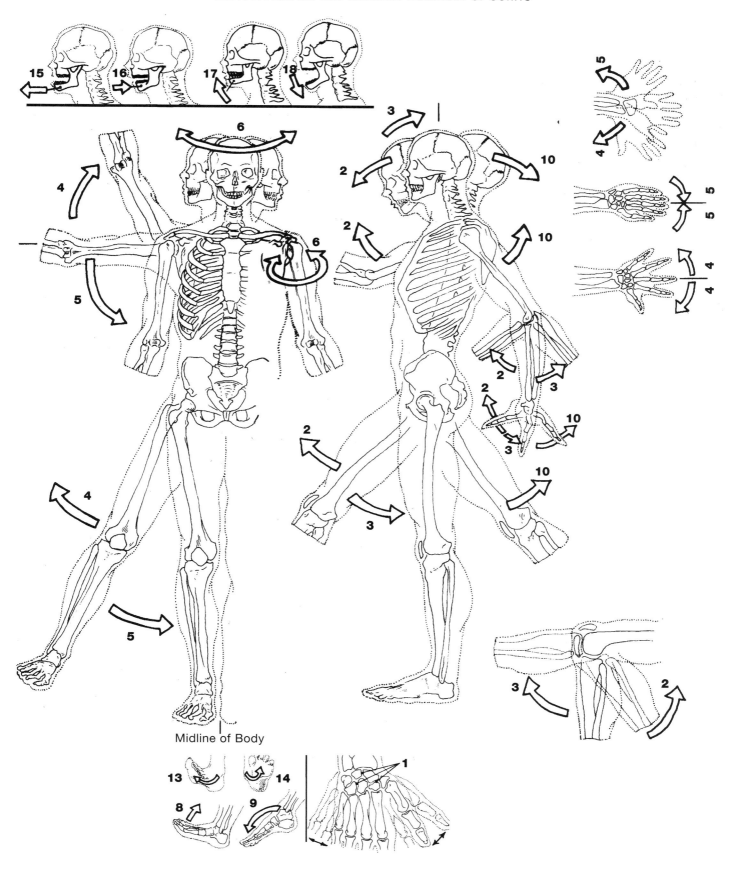

Midline of Body

CHAPTER 7: MUSCULAR SYSTEM

Although there are three types of contractile muscle tissue in the body—SKELETAL, CARDIAC, and SMOOTH—the MUSCULAR SYSTEM in this manual includes mainly a study of the skeletal muscles, which give form and stability to the body.

Cardiac muscle is found only in the heart, which is part of the CARDIOVASCULAR SYSTEM; it is involuntary and striated. SMOOTH MUSCLE is generally associated with other (usually hollow) internal organs and glands of involuntary body systems—CARDIOVASCULAR (walls of blood vessels), DIGESTIVE, GENITOURINARY, AND RESPIRATORY. It is involuntary and nonstriated.

SKELETAL MUSCLE TISSUE is highly specialized (voluntary and striated), and specifically adapted to contract and relax in order to carry out three general functions: BODY MOTION—whole body movements produced by functional integration with bones and joints, which provide the leverage and formative framework of the body; HEAT PRODUCTION and MAINTENANCE OF BODY TEMPERATURE (muscle tissue produces 40–50% of heat from cell respiration); and POSTURE and BODY SUPPORT around the flexible joints.

Muscle tissue comprises 40–50% of total body weight. Although the three types of muscle tissue differ in structure and function, they all possess four striking characteristics — CONTRACTIBILITY, EXCITABILITY (irritability), EXTENSIBILITY and ELASTICITY. The outstanding characteristic of muscle tissue is its ability to shorten, or contract. Muscle tissue possesses little intercellular material, so its cells or fibers lie close together.

The skeletal muscle system may be divided into two main groups associated with the AXIAL and APPENDICULAR SKELETONS. Muscles of the axial skeleton involve: FACIAL EXPRESSION; MASTICATION; EYE, TONGUE, and NECK MOVEMENT; VERTEBRAL COLUMN and TRUNK MOVEMENT; and RESPIRATION. They relate to the ABDOMINAL WALL and PELVIC OUTLET. Muscles of the appendicular skeleton involve muscles that act on the girdles and those that move appendage segments, such as the upper limb (pectoral girdle, arm, forearm, wrist, hand and fingers) and the lower limb (pelvic girdle, thigh, leg, ankle, foot and toes).

COLORING NOTES:
The optimal situation for coloring the muscles would be to use as many warm and fleshy colors as possible (reds, oranges, etc.), but in some situations that is not possible due to the number of muscles on a page. Use complementary (opposing) color schemes (red-green, orange-blue, yellow-purple) (warms vs. cools) when dealing with: opposing movements (flexion-extension), contrasting movements (actin-myosin), different layers of muscle, or different groups (extrinsic-intrinsic; suprahyoid-infrahyoid). Tendons should be similar in color to their associated muscle, but lighter or clearer in tone to distinguish them.

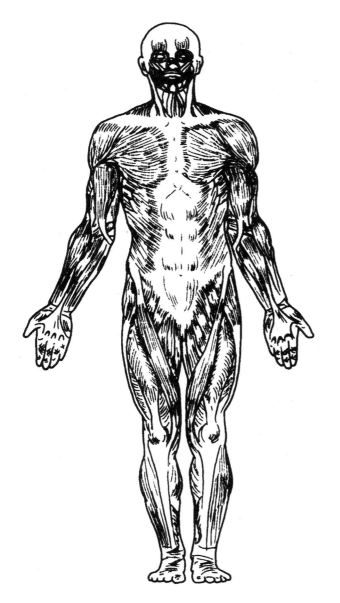

Anterior View

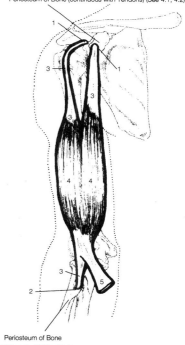

STRUCTURE OF A SKELETAL MUSCLE

Periosteum of Bone (continuous with Tendons) (See 4.1, 4.2)

Periosteum of Bone
(continuous with Tendons)

SYSTEM COMPONENTS

*The muscle tissue of the body (including skeletal,
cardiac and smooth contractile tissue).*

SYSTEM FUNCTION

*Movement
Heat production
Maintenance of body posture
Maintenance of body temperature*

STRUCTURE OF A SKELETAL MUSCLE

ATTACHMENTS: ORIGIN AND INSERTION

1 | Origin | Stationary Attachment

2 | Insertion | Movable Attachment

3 | Tendons |

4 | Belly (Gaster) |

Muscle Bulk

5 | Aponeurosis |

Sheetlike layer of connective tissue joining a
muscle to the part that it moves

BASIS OF NAMING SKELETAL MUSCLES ACCORDING TO PRIMARY FUNCTIONS*

1 Flexors

2 Extensors

3 Abductors

4 Adductors

5 Rotators

Lateral Rotators
Medial Rotators

6 Scapula Stabilizers

* Reference numbers in parentheses (1–6) on the illustrations indicate an additional and/or secondary function. Color them as dots of color on top of the primary colors.

DEFINITIONS OF PRIMARY FUNCTIONS AND ACTIONS IN NAMING SKELETAL MUSCLES

1 Flexors

Bend a part in proximal direction
Decrease anterior angle at a joint (generally)
Bend a limb at a joint

2 Extensors

Decrease anterior angle at a joint (generally)
Straighten a limb at a joint

3 Abductors

Move a bone away from the midline

4 Adductors

Move a bone closer to midline

5 Rotators

Move a bone around its longitudinal axis

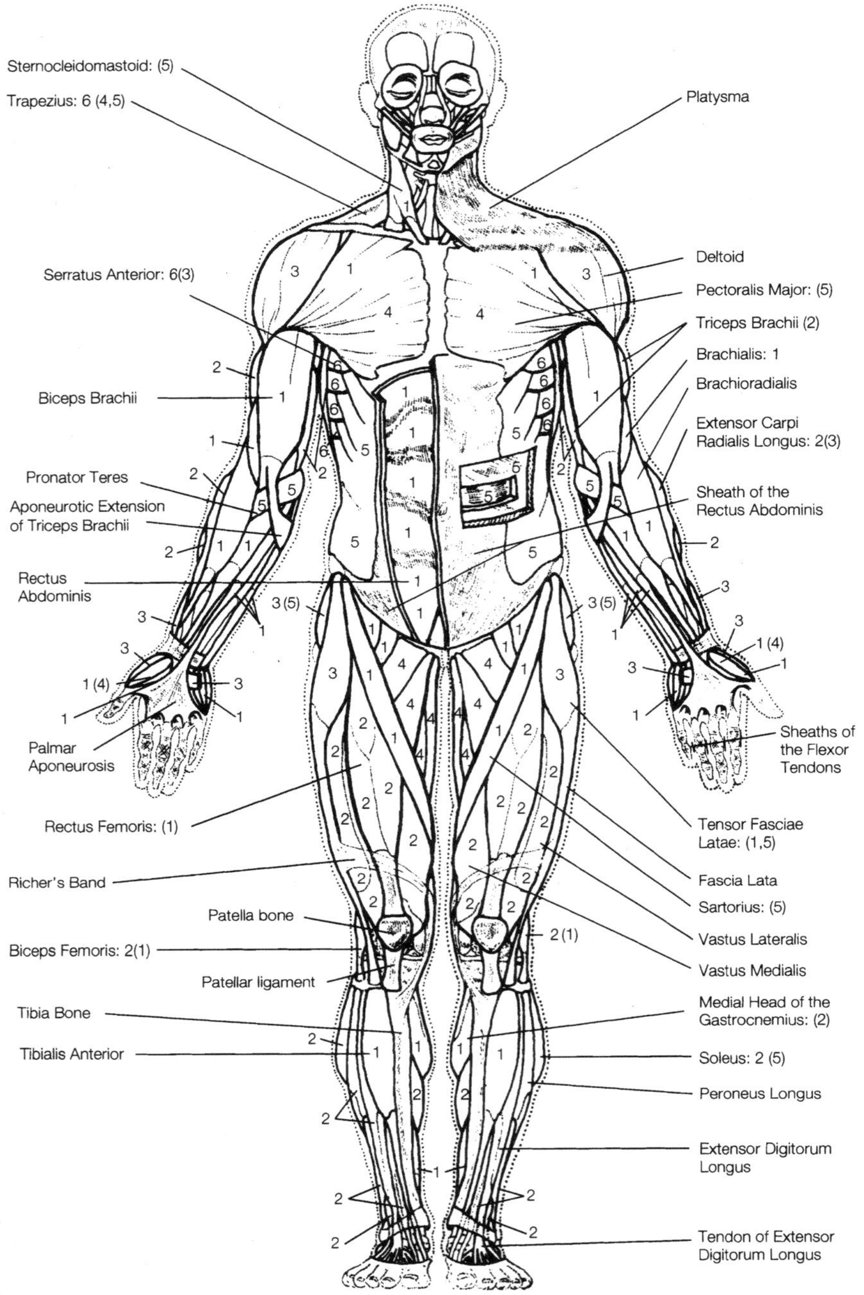

Sternocleidomastoid: (5)

Trapezius: 6 (4,5)

Platysma

Serratus Anterior: 6(3)

Deltoid

Pectoralis Major: (5)

Triceps Brachii (2)

Brachialis: 1

Brachioradialis

Extensor Carpi Radialis Longus: 2(3)

Biceps Brachii

Pronator Teres

Aponeurotic Extension of Triceps Brachii

Sheath of the Rectus Abdominis

Rectus Abdominis

Palmar Aponeurosis

Sheaths of the Flexor Tendons

Rectus Femoris: (1)

Richer's Band

Tensor Fasciae Latae: (1,5)

Fascia Lata

Sartorius: (5)

Vastus Lateralis

Vastus Medialis

Patella bone

Biceps Femoris: 2(1)

Patellar ligament

Medial Head of the Gastrocnemius: (2)

Tibia Bone

Soleus: 2 (5)

Tibialis Anterior

Peroneus Longus

Extensor Digitorum Longus

Tendon of Extensor Digitorum Longus

Anterior View

There are over 650 skeletal muscles in the body. Most of the muscles occur in pairs, in mirror images, on either side of a midsagittal symmetrical plane through the upright plane through the upright body (the right side is a mirror image of the left side). There are only five single (unpaired) muscles.

Muscles generally contract as FUNCTIONAL GROUPS rather than individually. SYNERGISTIC MUSCLES are muscles that function in cooperation and coordination with one another to achieve a particular movement(s).

FLEXOR-EXTENSOR ACTION and ABDUCTOR-ADDUCTOR ACTION are examples of ANTAGONISTIC MUSCLE GROUPS; antagonistic muscles are muscles that function in opposition to one

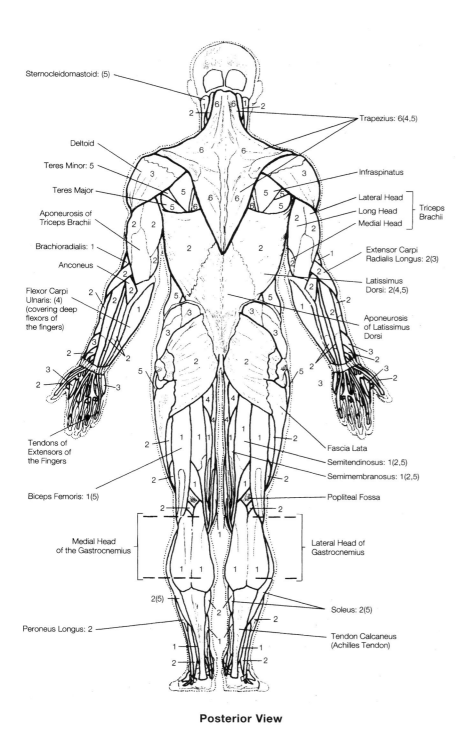

Sternocleidomastoid: (5)

Deltoid

Teres Minor: 5

Teres Major

Aponeurosis of
Triceps Brachii

Brachioradialis: 1

Anconeus

Flexor Carpi
Ulnaris: (4)
(covering deep
flexors of
the fingers)

Tendons of
Extensors of
the Fingers

Biceps Femoris: 1(5)

Medial Head
of the Gastrocnemius

Peroneus Longus: 2

Trapezius: 6(4,5)

Infraspinatus

Lateral Head ⎫
Long Head ⎬ Triceps
Medial Head ⎭ Brachii

Extensor Carpi
Radialis Longus: 2(3)

Latissimus
Dorsi: 2(4,5)

Aponeurosis
of Latissimus
Dorsi

Fascia Lata

Semitendinosus: 1(2,5)

Semimembranosus: 1(2,5)

Popliteal Fossa

Lateral Head of
Gastrocnemius

Soleus: 2(5)

Tendon Calcaneus
(Achilles Tendon)

Posterior View

**DEFINITIONS OF OTHER FUNCTIONS AND
ACTIONS IN NAMING SKELETAL MUSCLES**

SPECIALIZED FLEXORS

Dorsiflexors

Flex the foot (upward) at the ankle joint

SPECIALIZED EXTENSORS

Plantar Flexors

Extend the foot (downward) at the ankle joint

SPECIALIZED ROTATORS

Supinators

Turn the palm upward or anteriorly (forward)

Pronators

Turn the palm downward or posteriorly
(backward)

Levators

Elevation, or lifting, produce an upward movement

Depressors

Produce a downward movement

Invertors

Turn the sole of the foot inward

Evertors

Turn the sole of the foot outward

Sphincters

Decrease the size of an opening

Tensors

Make a body part more rigid to increase tension

---

another to achieve a particular movement. They are generally
located on opposite sides of a limb.

The names of skeletal muscles according to primary functions
are FLEXORS and EXTENSORS, ABDUCTORS, ADDUCTORS, AND ROTATORS
(LATERAL ROTATORS and MEDIAL ROTATORS), all of which have opposite
action (ANTAGONISTIC) and SCAPULA STABILIZERS.

The eight logical derivations for the naming of skeletal muscles
(with examples given) are: function (flexor, extensor, abductor,

adductor, pronator, levator); shape (rhomboideus, trapezius); size
(maximus, medius, minimus, longus, brevis); location (pectoralis,
femoral, intercostal, abdominal, brachii); relative position (lateral,
medial, internal, external); fiber orientation (rectus [straight],
oblique, transverse); attachment—origin and insertion (zygomati-
cus, temporalis, nasalis, sternocleidomastoid, stylohyoid); and
number of origins (triceps, biceps).

MUSCULAR SCHEME
SKELETAL MUSCLE: TISSUE STRUCTURE AND CONNECTIVE TISSUE SHEATHS

STRUCTURE OF SKELETAL TISSUE

1 **Belly (Gaster)**

2 **Fascicle (Bundle)** Muscle Fasciculus

3 **Skeletal Muscle Fiber (Cell)**

COMPONENTS OF MUSCLE CELL FIBER (CELL)

4 **Sarcolemma** Plasma Membrane "Muscle Husk"

5 **Sarcoplasm Cytoplasm**

  a **Sarcoplasmic Reticulum**

  b **Transverse (T-) Tubules**

  c **Myofibrils**
    Composed of myofilaments

  d **Mitochondria**

6 Many Peripheral Nuclei

CONTINUOUS THICK COVERINGS OF ASSOCIATED MUSCLE ELEMENTS (MUSCLES TENDONS AND THE PERIOSTEUM COVERING OF ATTACHED BONES)

7 **Epimysium** Perimysium externum
   Invests the entire belly, or gaster

8 **Periysium** Perimysium internum
   Invests each bundle, or fascicle

9 **Endomysium** invests each muscle cell (or muscle fiber)

COMPONENTS OF A MYOFIBRIL
PROTEIN STRANDS

Myofilaments

A **Actin** (F-Actin), Thin Filaments (50–60 Å)

M **Myosin** Thick Filaments (110 Å)

C-B **Cross-Bridges** (Myosin "Heads")
   Link with the binding sites on the actin during contraction

MYOFIBRIL SHOWING COMPARTMENTALIZATION OF SARCOMERES

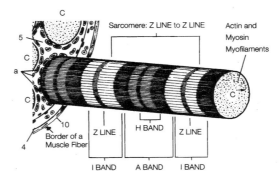

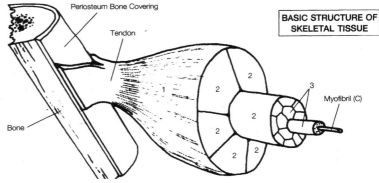

BASIC STRUCTURE OF SKELETAL TISSUE

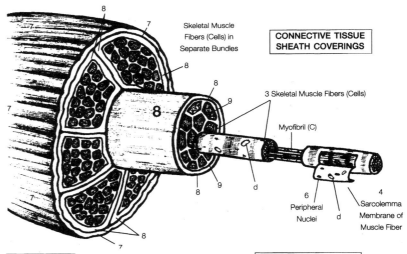

CONNECTIVE TISSUE SHEATH COVERINGS

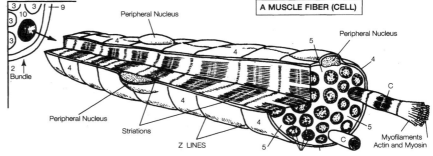

A MUSCLE FIBER (CELL)

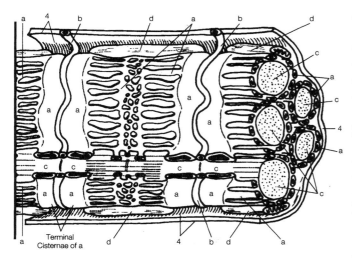

Terminal
Cisternae of a

A MUSCLE SHORTENS WHEN EACH SARCOMERE UNIT SHORTENS BETWEEN THE Z LINES, CAUSED BY SLIDING ACTION OF ACTIN FILAMENTS

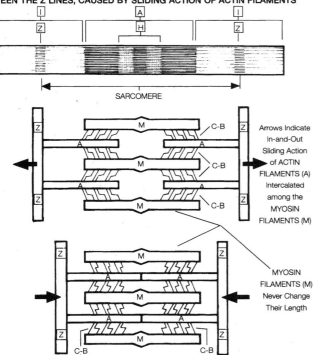

SARCOMERE

Arrows Indicate In-and-Out Sliding Action of ACTIN FILAMENTS (A) Intercalated among the MYOSIN FILAMENTS (M)

MYOSIN FILAMENTS (M) Never Change Their Length

The main, fleshy part of a skeletal muscle is the BELLY (GASTER). The entire muscle is wrapped by a large, thick connective tissue sheath—the EPIMYSIUM, an extension of the deep fascia.

Invaginations of the epimysium that divide the muscle into bundles or FASCICLES are called the PERIMYSIUM, which is also an extension of the deep fascia. Invaginations of the perimysium divide the bundles into each separate MUSCLE CELL, or FIBER, and are called the ENDOMYSIUM, also an extension of the deep fascia.

Histologically, a large skeletal muscle is composed of many elongated, parallel cylindrical MULTINUCLEATED CELLS called muscle fibers.

Each muscle fiber (cell) contains small, numerous parallel longitudinally arranged bundles called myofibrils. In turn, each myofibril is composed of MYOFILAMENTS (which contain the contractile ACTIN and MYOSIN). Each muscle fiber (cell) is surrounded by a plasma membrane called a SARCOLEMMA.

The CYTOPLASMIC interior of the cell is called SARCOPLASM.

SARCOPLASMIC RETICULUM is a network of membrane-enclosed tubules (channels) that form a sleeve around each MYOFIBRIL. The reticulum tubules are essential to the metabolic functions of the cell and are involved with protein synthesis.

Perpendicular to the sarcoplasmic reticulum, and running transversely through the muscle fiber, are T-TUBULES, which are internal extensions of the sarcolemma. They penetrate deep into the cell's interior, and open to the outside of the cell (carrying extracellular fluid in their lumina).

T-tubules pass through adjacent expanded chambers of the sarcoplasmic reticulum, called TERMINAL CISTERNAE. A T-tubule and its two adjacent cisternae comprise a MUSCLE TRIAD. The t-tubules, along with the reticulum, are involved in the transmission of nerve impulses to the muscle fiber.

Skeletal muscle fibers only contract when stimulated by SOMATIC MOTOR NEURONS (see pages 128–29).

A MOTOR UNIT consists of a single MOTOR NEURON and the aggregation of muscle fibers innervated by the motor neuron. As the muscle is penetrated, it splays into a number of branching neuron processes called AXONS. The terminal ends of the axons (telodendria) contact the sarcolemma of the individual muscle fibers by means of MOTOR END PLATES, forming the NEUROMUSCULAR JUNCTION (see pages 130–31).

A nerve impulse causes the release of a NEUROTRANSMITTER CHEMICAL across the junction. As it contacts the sarcolemma, physiological activity within the fiber creates contraction.

Skeletal muscle contracts according to an ALL-OR-NONE PRINCIPLE. When a nerve impulse travels through a motor unit, all the muscle fibers served by it simultaneously contract to their maximum capacity, but only when exposed to a stimulus that reaches a specific threshold strength. (Otherwise, there is no contraction at all.)

MYOFILAMENTS (ACTIN and MYOSIN) do not extend the entire length of a MUSCLE FIBER. They are stacked in compartments called SARCOMERES separated by thin, narrow zones of dark material called Z LINES (which are located under the T-tubules). This accounts for the striations of skeletal muscle; nonstriated smooth muscle does not contain sarcomeres.

Contraction of muscle tissue results from a sliding movement—of actin filaments in the myofibrils causing a reduction in the length of the sarcomeres.

DEEP MUSCLES OF MASTICATION

(Coronal Section of Skull, Immediately Behind the Jaw, Posterior View)

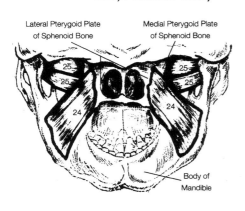

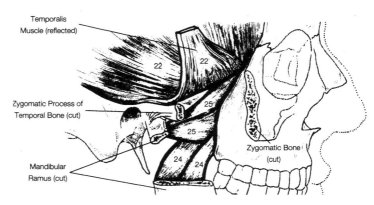

MUSCLES OF FACIAL EXPRESSION:

THE SCALP MUSCLES

Epicranius (Occipitofrontalis)

1 | Frontalis | Frontalis Bellies

2 | Occipitalis |

THE FACIAL MUSCLES
  The Eyelids

4 | Corrugator | Corrugator Supercilii

5 | Orbicularis Oculi | Orbicularis Palpebrarum (Pars Palpebralis)

  The Nose

6 | Procerus | Pyramidalis Nasi

Nasalis

7 | Transverse Nasalis |

8 | Alar Nasalis |

9 | Depressor Septi | Depressor Alae Nasi

The Mouth

10 | Levator Labii Superioris* |

11 | Levator Labii Superioris Alaeque Nasi |

12 | Levator Anguli Oris | Caninus

13 | Zygomaticus Major* |

14 | Zygomaticus Minor* |

15 | Risorius* |
"Laughing Muscle": Partly a continuation of the platysma

16 | Depressor Labii Inferioris | Quadratus Labii Inferioris (Quadratus Menti)

17 | Depressor Anguli Oris | Triangularis

18 | Mentalis | Levator Labii Inferioris (Levator Menti)

19 | Transversus Menti |

20 | Orbicularis Oris | "Kissing Muscle"

21 | Buccinator |

The Ears: Extrinsic Ear Muscles

26 | Auricularis Anterior | Attrahens Aurem

27 | Auricularis Posterior | Retrahens Aurem

28 | Auricularis Superior | Attolens Aurem

MUSCLES OF MASTICATION (FOUR PAIRS)

22 | Temporalis | Temporal

23 | Masseter |

24 | Medial Pterygoid | (Pterygoideus Medialis)

25 | Lateral Pterygoid | (Pterygoideus Lateralis)

* 10, 13, 14, and 15 insert at different locations on orbicularis oris (20) and cause the expressions of smiling and laughing.

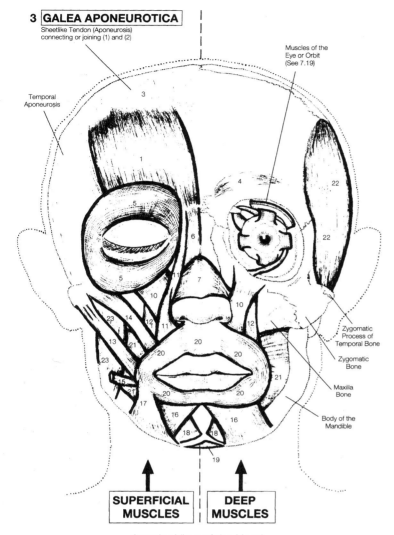

3 GALEA APONEUROTICA
Sheetlike Tendon (Aponeurosis)
connecting or joining (1) and (2)

Temporal
Aponeurosis

Muscles of the
Eye or Orbit
(See 7.19)

Zygomatic
Process of
Temporal Bone

Zygomatic
Bone

Maxilla
Bone

Body of the
Mandible

SUPERFICIAL MUSCLES | **DEEP MUSCLES**

Anterior View of the Head

M USCLES OF FACIAL EXPRESSION originate on the
different facial bones and insert into the
dermis (second major layer of skin). When these
muscles contract, surface features are affected,
and various emotions are expressed. An incredible
array of complex facial expressions are very important in social communication, especially nonverbal
communication.

Muscles of the EYES (ORBITS) are discussed on
pages 162–63, but their origin, insertion, action,
and innervation will be found in the charts found at
www.mcmurtriesanatomy.com.

The ORBICULARIS OCULI is a sphincter muscle that
encircles the eye. It permits blinking, winking, and
squinting. It also compresses the lacrimal gland, which
keeps the eyeball continuously moistened.

COLORING NOTES:
Use warm colors for superficial muscles
and muscles of mastication. Use cool
colors for deeper muscles.

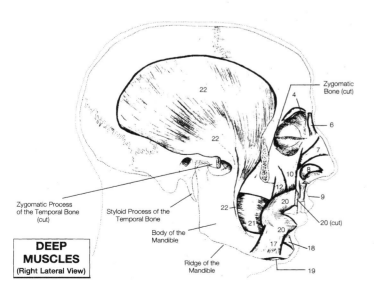

Zygomatic
Bone (cut)

Zygomatic Process
of the Temporal Bone
(cut)

Styloid Process of the
Temporal Bone

Body of the
Mandible

Ridge of the
Mandible

DEEP MUSCLES
(Right Lateral View)

20 (cut)

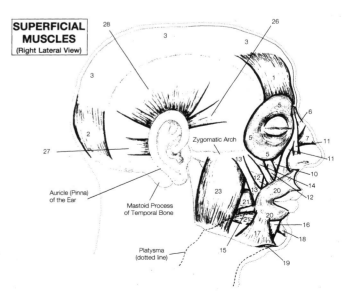

SUPERFICIAL MUSCLES
(Right Lateral View)

Zygomatic Arch

Auricle (Pinna)
of the Ear

Mastoid Process
of Temporal Bone

Platysma
(dotted line)

POSTERIOR VIEW OF PHARYNX

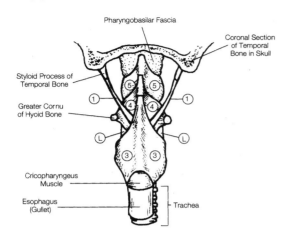

Pharyngobasilar Fascia

Coronal Section of Temporal Bone in Skull

Styloid Process of Temporal Bone

Greater Cornu of Hyoid Bone

Cricopharyngeus Muscle

Esophagus (Gullet)

Trachea

Muscles of the pharyngeal and laryngeal regions include those of the PHARYNX, LARYNX, TONGUE, and the SOFT PALATE.

The pharynx and larynx are located in the anterior region of the neck, or cervix. The supporting walls of the pharynx are composed of skeletal muscles, the CONSTRICTORS. The STYLOPHA-RYNGEUS connects the styloid process of the temporal bone to the pharynx. This muscle elevates and dilates the pharynx. The SALP-INGOPHARYNGEUS raises the upper third division of the pharynx, the NASOPHARYNX (see pages 236–37 and 260–61).

The pharynx is a passageway common to the respiratory and digestive systems. It is a conduit for air from the nasal and oral cavities to the larynx, and food from the oral cavity to the ESOPHAGUS.

The larynx, or voice box, is a musculocartilagenous structure comprising the enlarged upper end of the TRACHEA (WINDPIPE), just below the root of the tongue.

The intrinsic muscles of the larynx are shown here. Extrinsic muscles associated with the larynx include the INFRAHYOID MUS-CLES, located below the hyoid bone (see pages 88–89).

The tongue is a thick, highly specialized muscular organ covered with a mucous membrane. The INTRINSIC TONGUE MUSCLES are the muscles that comprise the tongue. They change the shape and size of the tongue for swallowing, and give the tongue motility. The EXTRINSIC TONGUE MUSCLES are muscles that insert on the tongue, but originate on other structures. They enable gross tongue movement (side-to-side, in-and-out) and maneuver food for chewing; they shape food into a rounded bolus and force it to the back of the mouth.

P | Pharynx | Throat, or Gullet

L | Larynx | Voice Box

Musculocartilaginous structure comprising the enlarged upper end of the TRACHEA (WINDPIPE), just below the root of the TONGUE. The organ of the voice.

ASSOCIATED MUSCLES OF THE PHARYNX

1 | Stylopharyngeus

2 | Salpingopharyngeus

SKELETAL SUPPORTING WALL OF THE PHARYNX

3 | Inferior Constrictor

Laryngopharyngeus (constrictor pharyngis inferior)

4 | Medial Constrictor

Hypopharyngeus (constrictor pharyngis medius)

5 | Superior Constrictor

Cephalopharyngeus (constrictor pharyngis superior)

INSTRINSIC MUSCLES

6 | Aryepiglottic | Aryepiglotticus

7 | Thyroepiglottic | Thyreoepiglotticus (Thyroepiglotticus)

8 | Arytenoid | (Arytenoideus)

9 | Thyroarytenoid | (Thyreoarytenoideus)

10 | Cricoarytenoid | (Cricoarytenoideus)

11 | Cricothyroid | (Cricothyroideus)

EXTRINSIC TONGUE MUSCLES

12 | Genioglossus

Depresses tongue; thrusts it forward

13 | Hyoglossus

Depresses sides of tongue

14 | Styloglossus

Antagonist to genioglossus; retracts and elevates tongue

15 | Stylohyoid

Does not attach directly to the tongue. Has an indirect effect on the tongue as it elevates the hyoid bone, where it inserts

CORORING NOTES:
Pharynx = warm colors; Larynx = cool colors.

THE SOFT PALATE (VELUM PALATINUM)

16 | Palatoglossus
Glossapalatinus (to tongue)

17 | Palatopharyngeus
Pharyngopalatinus

18 | Tensor Veli Palatini | Tensor Palati

19 | Levator Veli Palatini
Levator Palati

20 | Musculus Uvulae | Uvula

LEFT LATERAL VIEW OF LARYNX

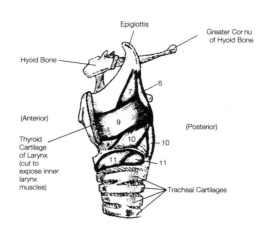

Epiglottis
Hyoid Bone
Greater Cornu of Hyoid Bone
(Anterior)
Thyroid Cartilage of Larynx (cut to expose inner larynx muscles)
(Posterior)
Tracheal Cartilages
6 7 9 10 10 11 11

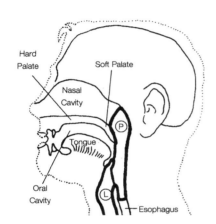

Hard Palate
Soft Palate
Nasal Cavity
Tongue
Oral Cavity
P
L
Esophagus

MUSCLES OF THE PHARYNX, LARYNX, AND SOFT PALATE
(Constrictor Muscle Pulled Back to Expose Interior)

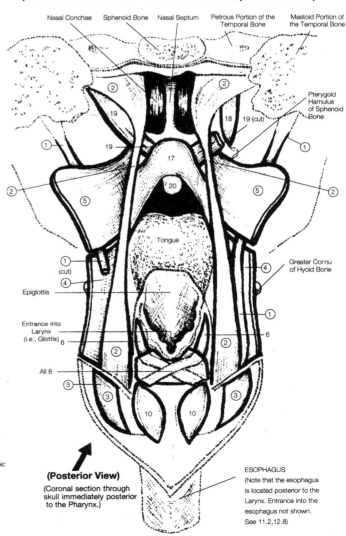

Nasal Conchae   Sphenoid Bone   Nasal Septum   Petrous Portion of the Temporal Bone   Mastoid Portion of the Temporal Bone
Pterygoid Hamulus of Sphenoid Bone
Tongue
Greater Cornu of Hyoid Bone
Epiglottis
Entrance into Larynx (i.e., Glottis)
All 8

(Posterior View)
(Coronal section through skull immediately posterior to the Pharynx.)

ESOPHAGUS
(Note that the esophagus is located posterior to the Larynx. Entrance into the esophagus not shown. See 11.2,12.8)

MUSCLES ASSOCIATED WITH THE TONGUE

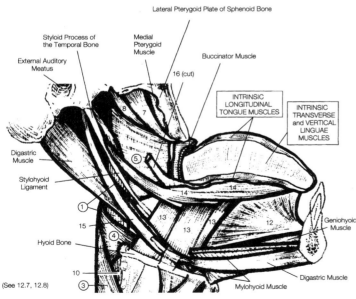

Lateral Pterygoid Plate of Sphenoid Bone
Styloid Process of the Temporal Bone
Medial Pterygoid Muscle
Buccinator Muscle
External Auditory Meatus
16 (cut)

INTRINSIC LONGITUDINAL TONGUE MUSCLES

INTRINSIC TRANSVERSE and VERTICAL LINGUAE MUSCLES

Digastric Muscle
Stylohyoid Ligament
Hyoid Bone
Geniohyoic Muscle
Digastric Muscle
Mylohyoid Muscle
(See 12.7, 12.8)

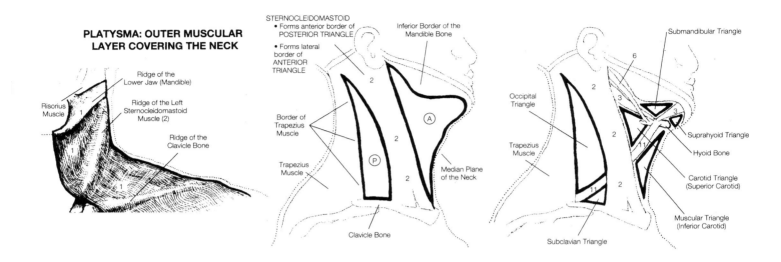

PLATYSMA: OUTER MUSCULAR LAYER COVERING THE NECK

Risorius Muscle

Ridge of the Lower Jaw (Mandible)

Ridge of the Left Sternocleidomastoid Muscle (2)

Ridge of the Clavicle Bone

STERNOCLEIDOMASTOID
• Forms anterior border of POSTERIOR TRIANGLE
• Forms lateral border of ANTERIOR TRIANGLE

Inferior Border of the Mandible Bone

Border of Trapezius Muscle

Trapezius Muscle

Median Plane of the Neck

Clavicle Bone

Occipital Triangle

Trapezius Muscle

Subclavian Triangle

Submandibular Triangle

Suprahyoid Triangle

Hyoid Bone

Carotid Triangle (Superior Carotid)

Muscular Triangle (Inferior Carotid)

CERVICALS

SUPERFICIAL CERVICAL

1 | Platysma | Platysma Myoides

ANTEROLATERAL CERVICALS

2 | Sternocleidomastoid | Sternocleidomastoideus

HYOIDS

SUPRAHYOIDS:

3 | Digastric | Digastricus

Anterior Belly/Posterior Belly

4 | Mylohyoid | Mylohyoideus

5 | Geniohyoid | Geniohyoideus

(not shown)

6 | Stylohyoid | Stylohyoideus

7 | Hyoglossus

INFRAHYOIDS

8 | Sternothyroid | Sternothyreoideus (not shown)

9 | Sternohyoid | Sternohyoideus

10 | Thyrohyoid | Thyrohyoideus

11 | Omohyoid | Omohyoideus

Prevertebrals: muscles of the neck region located immediately in front of the vertebral column.

PREVERTEBRALS

ANTERIOR PREVERTEBRALS

12 | Rectus Capitis | Anterior Rectus Capitis and Anticus Minor

13 | Rectus Capitis Lateralis

14 | Longus Capitis | Rectus Capitis and Anticus Major

15 | Longus Colli

Three parts: superior oblique, inferior oblique, vertical

LATERAL PREVERTEBRALS

16 | Scalenus Anterior | Scalenus Anticus

17 | Scalenus Medius

18 | Scalenus Posterior | Scalenus Posticus

2 LARGE TRIANGULAR NECK DIVISIONS

(A) | Anterior Triangle

Four lesser triangles, salivary glands, larynx, trachea, thyroid glands, various nerves and vessels

(P) | Posterior Triangle

Two lesser triangles, various nerves and vessels

6 LESSER TRIANGULAR NECK SUBDIVISIONS

POSTERIOR TRIANGLE

Occipital Triangle:

Cervical nerve plexus, accessory nerve

Subclavian Triangle:

Brachial nerve plexus, subclavian artery.

ANTERIOR TRIANGLE

Submandibular Triangle

Salivary glands

Suprahyoid Triangle

Muscles on the floor of mouth, salivary glands and ducts

Carotid Triangle

Carotid arteries, internal jugular vein, vagus nerve

Muscular Triangle

Larynx, trachea, thyroid glands, carotid sheath

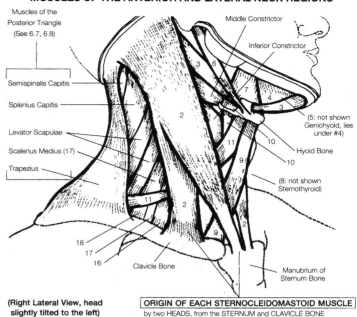

MUSCLES OF THE ANTERIOR AND LATERAL NECK REGIONS

Muscles of the Posterior Triangle (See 6.7, 6.8)

Middle Constrictor

Inferior Constrictor

Semispinalis Capitis

Splenius Capitis

Levator Scapulae

Scalenus Medius (17)

Trapezius

(5: not shown Geniohyoid, lies under #4)

Hyoid Bone

(8: not shown Sternothyroid)

Clavicle Bone

Manubrium of Sternum Bone

(Right Lateral View, head slightly tilted to the left)

ORIGIN OF EACH STERNOCLEIDOMASTOID MUSCLE
by two HEADS, from the STERNUM and CLAVICLE BONE

M uscles of the NECK either support and move the head or are associated with structures in the neck such as the larynx, pharynx and hyoid bone.

HYOIDS are neck muscles associated with the hyoid bone, the only bone in the body that does not articulate with any other bone. Seven paired muscles and one unpaired muscle (the MYLOHYOID) attach to it.

SUPRAHYOIDS are located above the hyoid bone. The DIGASTRIC is a two-bellied muscle of double origin. It can open the jaw or elevate the hyoid bone. The MYLOHYOID forms the floor of the mouth and elevates it to aid in swallowing; it also elevates the hyoid bone. The GENIOHYOID helps to depress the jaw; it also elevates and advances the hyoid bone. Other suprahyoids include the SYLOHYOID and HYOGLOSSUS (see page 87).

INFRAHYOIDS are located below the hyoid bone. The STERNOTHYROID pulls down the larynx (depresses thyroid cartilage). The STERNOHYOID and the OMOHYOID depress the hyoid. The THYROHYOID elevates the larynx (thyroid cartilage) and depresses the hyoid.

There are two types of cervical neck muscles—the ANTEROLATERAL CERVICALS and SUPERFICIAL CERVICALS. The two anterolateral cervicals are STERNOCLEIDOMASTOIDS. Contraction of one sternocelidomastid muscle rotates the head and tilts the chin upward to the opposite side (turns the head sideways). Contraction of both muscles pulls the head forward and down (flexes the neck on the chest) and obliquely transects the neck, dividing it into two triangles, an ANTERIOR TRIANGLE and a POSTERIOR TRIANGLE.

The two superficial cervicals are the two PLATYSMA muscles. They are broad, thin sheets of muscle beginning on the ANTERIOR CHEST and sweeping up each side of the neck over the large sternocleidomastoids. They terminate at the body of the mandible and the corners of the mouth, enabling screaming or expressions of sadness or horror when contracted.

The PREVERTEBRALS (ANTERIOR and LATERAL) are muscles of the neck region located immediately in front of the vertebral column. The ANTERIOR PREVERTEBRALS flex and support the head and cervical vertebrae. The RECTUS CAPITIS ANTERIOR turns and inclines the head. The RECTUS CAPITIS LATERALIS supports the head and inclines it laterally. The LONGUS CAPITIS flexes the head, and the LONGUS COLLI twists and bends the neck forward.

The lateral prevertebrals laterally bend (flex) the neck and elevate the first two ribs. The SCALENUS ANTERIOR/SCALENUS MEDIUS flexes the neck and elevates the first rib; the SCALENUS POSTERIOR flexes the neck and elevates the second rib.

ANTERIOR AND LATERAL PREVERTEBRAL NECK MUSCLES

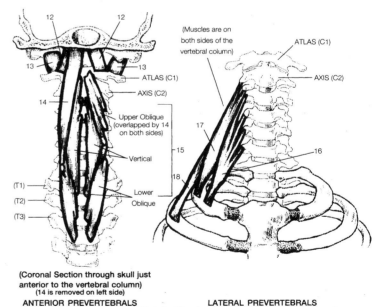

(Muscles are on both sides of the vertebral column)

ATLAS (C1)

AXIS (C2)

ATLAS (C1)

AXIS (C2)

Upper Oblique (overlapped by 14 on both sides)

Vertical

Lower Oblique

(T1)
(T2)
(T3)

(Coronal Section through skull just anterior to the vertebral column)
(14 is removed on left side)
ANTERIOR PREVERTEBRALS

LATERAL PREVERTEBRALS

(Anterior Views)

COLOR GUIDELINES:
Color only the muscles labeled with reference numbers.
A = yellow; P = flesh; 1 = light brown; 2 = red-orange;
Suprahyoids = pinks, oranges, and reds; Infrahyoids = blues and greens;
Prevertebrals = grays, browns and neutrals.

MUSCLES OF THE AXIAL SKELETON
TRUNK (TORSO) REGION: DEEP MUSCLES OF THE POSTERIOR NECK AND BACK

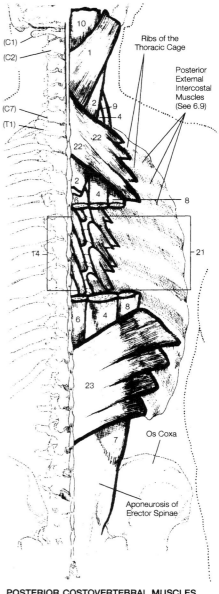

The superficial group of erector spinae (sacrospinalis) muscle = 3–9

DEEP POSTVERTEBAL MUSCLES OF THE VERTEBRAL COLUMN

SUPERFICIAL GROUP – THE COLUMNAR MUSCLES: PRINCIPAL MOVERS OF THE HEAD AND BACK:

Deep Neck Muscles:

"BANDAGE MUSCLES"

1 | Splenitus Capitis |

2 | Splenitus Cervicis | Splenius Colli

Longissimus (Intermediate Column): 3, 4, 5/Erector Spinae

3 | Longissimus Capitis |
Trachelomastoid

DEEP BACK MUSCLES:

4 | Longissimus Thoracis |
Longissimus Dorsi

5 | Longissimus Cervicis |
Transversalis Colli

SPINALIS (MEDIAL COLUMN)*

6 | Spinalis Thoracis | Spinalis Dorsi

ILIOCOSTALIS (LATERAL COLUMN)

7 | Iliocostalis Lumborum |
Sacrolumbalis

8 | Iliocostalis Thoracis | Iliocostalis
Dorsi (Accessorius)

9 | Iliocostalis Cervicis | Cervicalis
Ascendens

* Spinalis Capitis and Spinalis Cervicis are not shown. They are inseparable from one another.

DEEPER STRATUM – THE TRANSVERSOSPINAL MUSCLES: ROTATORS OF THE VERTEBRAL COLUMN

Semispinalis: 10, 11, 12

Deep Neck Muscles:

10 | Semispinalis Capitis |
Complexus

Deep Back Muscles:

11 | Semispinalis Thoracis |
Semispinalis Dorsi

12 | Semispinalis Cervicis |
Semispinalis Colli

13 | Multifidus | Multifidus Spinae

14 | Rotatores | Rotators Spinae

DEEPEST STRATUM – CONTIGUOUS VERTEBRAE MUSCLES:

(SUBOCCIPITAL MUSCLES ARE NOT PRIME MOVERS, BUT POSTURAL MUSCLES)
Suboccipital Muscles

Deep Neck Muscles:

15 | Rectus Capitis Posterior Major |
R.C. Posticus Major

16 | Rectus Capitis Posterior Minor |
R.C. Posticus Minor

17 | Obliquus Capitis Inferior |

18 | Obliquus Capitis Superior |

Deep Back Muscles:

19 | Interspinales | (A Series)

20 | Intertransversarii |
Intertransversales

THORACIC POSTERIOR COSTOVERTEBRAL MUSCLES

21 | Levatores Costarum |

22 | Serratus Posterior Superior |

23 | Serratus Posterior Inferior |

POSTERIOR COSTOVERTEBRAL MUSCLES OF THE THORAX (RIB CAGE) OVERLAPPING THE EXTENSORS OF THE ERECTOR SPINAE
(#2 and ERECTOR SPINAE MUSCLES #4, 6, and 8 have been cut to expose the deeper 14 and 21 muscles)

COLOR GUIDELINES:
Erector = warm colors;
Remaining muscles = cool and neutral colors.

DEEPER AND DEEPEST STRATUM OF DEEP BACK MUSCLES

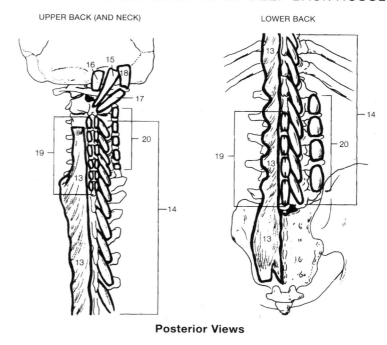

Posterior Views

POSTVERTEBRAL DEEP MUSCLES OF THE NECK

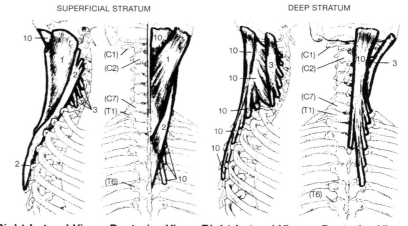

Right Lateral View   Posterior View   Right Lateral View   Posterior View

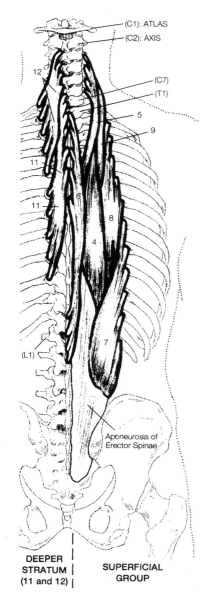

SUPERFICIAL GROUP AND DEEPER STRATUM OF DEEP BACK MUSCLES

The EXTENSOR MUSCLES, located on the posterior side of the vertebral column, are much stronger and more complex than the muscles that flex the vertebral column anterior to it (the RECTUS ABDOMINUS—long, straplike muscles of the anterior abdominal wall). These long extensors have to be stronger than the flexors because extension (e.g., lifting a weight) moves in opposition to the force of gravity (see pages 80–81).

These deep muscles on the posterior side of the vertebral column, which extend, rotate, and abduct the spine, can be divided into two main groups: the SUPERFICIAL GROUP and the DEEP GROUP (divided into DEEPER and DEEPEST STRATUM). The superficial group of the SPINE EXTENSORS is a massive group called the ERECTOR SPINAE or SACROSPINALIS, which extends vertically from the skull to the sacrum.

The erector spinae is made up of three main groups, each of which consists of overlapping strips of muscle: the LONGISSIMUS (intermediate columns); the SPINALIS (medial columns), which are in contact with the spinal processes of the vertebrae; and the ILIOCOSTALIS (lateral columns).

The POSTERIOR COSTOVERTEBRAL MUSCLES of the THORAX overlap the erector spinae, originating on the vertebral column and inserting on the ribs. Their primary action is on the ribcage (the LEVATORES COSTARUM also flexes the vertebral column). The other two muscles in this group are the SERRATUS POSTERIOR SUPERIOR and SERRATUS POSTERIOR INFERIOR.

SUPERFICIAL MUSCLES OF THE BACK AND NECK

EXTENSORS OF THE SHOULDER JOINT/SUPERFICIAL MUSCLES OF THE BACK

POSTERIOR/DORSAL (VERTEBRAL COLUMN)

1 | Trapezius |

2 | Latissimus Dorsi |

3 | Rhomboideus Major |

4 | Rhomboideus Minor |

5 | Levator Scapulae | Levator Anguli Scapulae

FLEXORS OF THE SHOULDER JOINT
VENTRAL (ANTEROLATERAL THORACIC)

6 | Pectoralis Major |

7 | Pectoralis Minor |

8 | Subclavius |
Acts on the clavides

9 | Serratus Anterior | Serratus magnus

Coracobrachialis is also flexor of the shoulder joint. (See pages 100–103.)

COLOR GUIDELINES:
Posterior/Dorsal Muscles = warm colors;
Ventral Muscles = cool colors.

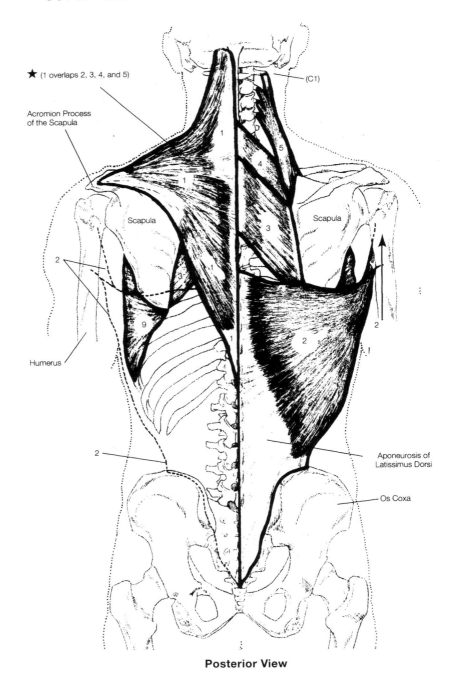

★ (1 overlaps 2, 3, 4, and 5)

(C1)

Acromion Process of the Scapula

Scapula

Scapula

Humerus

Aponeurosis of Latissimus Dorsi

Os Coxa

Posterior View

There are seven paired muscles and two superficial triangular back muscles that originate on the axial skeleton and connect to the shoulder. Most of these muscles act on the PECTORAL GIRDLE (the paired SCAPULAE and CLAVICLES), but some of them help to move the brachium (upper arm).

Among the axial muscles that help move the brachium are the large superficial LATISSIMUS DORSI BACK MUSCLES and the paired PECTORALIS MAJOR MUSCLES. They have their origins on the axial skeleton, but are not actually muscles of the axial skeleton. They develop in the forelimb, and extend to the trunk secondarily. The

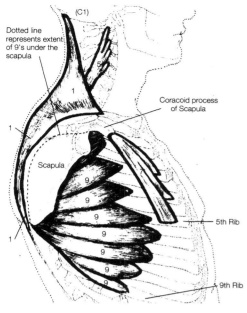

Right Lateral View

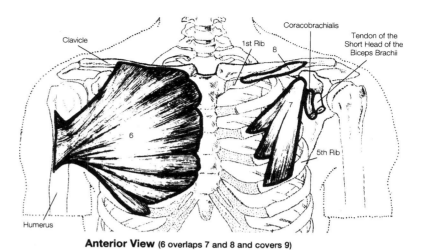

Anterior View (6 overlaps 7 and 8 and covers 9)

latissimus dorsi, along with eight paired muscles, span the shoulder joint to insert on the humerus bone.

The latissimus dorsi and pectoralis majors are the only muscles of the group that help move the brachium that do not originate on the scapula. Among the axial muscles that act on the pectoral girdle are the muscles that move the brachium, most of which originate on the scapula. During brachial movement the scapula has to be held fixed.

The axial muscles of scapular stabilization assist the pectoral girdle in holding the shoulder joint and scapulae firm. The pectoral girdle is an incomplete girdle, because it only has an anterior attachment to the axial skeleton at the STERNOCLAVICULAR JOINT (where the clavicles join the manubrium of the sternum, at the clavicular notches). Therefore, the pectoral girdle (in particular the scapulae) requires support from strong straplike muscles. Five paired muscles and one superficial triangular back muscle (the TRAPEZIUS) span the shoulder joint from the axial skeleton, insert on the two scapulae, and hold them firm. (For the muscles that originate on the scapulae to help move the brachium, see pages 100–101.)

The axial muscles that act on the clavicle are the paired SUBCLAVIUS MUSCLES that depress the clavicles (thus drawing the shoulder downward).

The trapezius muscle is a large, superficial triangular muscle of the back. A portion of the trapezius muscle extends over the posterior neck region, but it is primarily a superficial muscle of the back. It can adduct the scapula, elevate the scapula (e.g., shrugging the shoulder), and hyperextend the head.

The latissimus dorsi muscle is a large, flat triangular posterior muscle, often referred to as "the swimmer's muscle," and it covers the lower half of the thoracic region of the back. A powerful adductor of the arm, it draws it downward and backward while rotating the arm medially. It also extends the arm and retracts the shoulder.

There are a number of axial muscles that act on the pectoral girdle. Pectoral girdle bones (the scapulae and clavicles) serve as attachments for origin muscles that insert on the brachium and help move the brachium and forearm. These muscles originate on the scapulae, except for the latissimus dorsi and PECTORALIS MAJOR.

Most muscles that act on and stabilize the pectoral girdle itself originate on the axial skeleton and insert on the scapulae. They are divided into POSTERIOR and ANTERIOR groups (all but one insert on the scapulae). These are the muscles of scapular stabilization. The posterior group is composed of the trapezius, the MAJOR AND MINOR RHOMBOIDS, and the the LEVATOR SCAPULAE. The anterior group is composed of the PECTORALIS MINOR and the SERRATUS ANTERIOR.

The SUBCLAVIUS originates on the axial skeleton and inserts on the clavicle. Regarding the origin and insertion of the important trapezius, LATISIMUSS, dorsi and pecoralis major trunk connectors to the shoulder, note the following, the trapezius has its origin on the superior curved line of occipital bone, the nuchal ligament, the spinous processes of seventh cervical and all thoracic vertebrae. Its insertion is at the clavicle, the acromion of the scapula and the base of the spine of scapula.

The latissimus dorsi has its origin in the broad lumbar aponeurosis that is attached to the iliac crest and to the spines of the sacral, lumbar, and lower thoracic vertebrae. It inserts at the floor of the bicipital (intertubercular) groove of the humerus.

The pectoralis major originates on the clavicle, the sternum, the six upper costal cartilages and the aponeurosis of the external oblique muscle of the abdomen. Its insertion is at the crest of the great tubercle of the humerus.

THE MUSCLES OF RESPIRATION:

INTERCOSTALS (BETWEEN THE RIBS)

1 **External Intercostals**
Intercostales Externus

2 **Internal Intercostals**
Intercostales Internus

3 **Subcostals** Subcostales
(Infracostales)

STERNOCOSTALS (INSIDE ANTERIOR)

4 **Transversus Thoracis**
Triangularis Sterni

Floor of the Thoracic Cavity

5 **The Diaphragm**

* (Origin = Anteriorly: Xiphoid process of sternum, costal cartilages
of last 6 ribs (and intercostal spaces). Posteriorly: Lumbar vertebrae
7th–12th ribs
* (Insertion = Central tendon)
* The right half of the diaphragm rises higher than the left half

OPENINGS OF THE DIAPHRAGM:

a **Aortic Hiatus**

b **Esophageal Hiatus**

c **Vena Caval Foramen**

TENDONS OF THE DIAPHRAGM:

d **Central Tendon**

e **Right Crus**

f **Left Crus**

LIGAMENTS OF THE DIAPHRAGM:

g **Medial Arcuate Ligament**

h **Lateral Arcuate Ligament**

i **Median Arcuate Ligament**

MUSCLES OF THE POSTRERIOR
ABDOMINAL WALL (WORK
SYNERGISTICALLY)

6 **Quadratus Lumborum**

ILIOPSOAS (PREVERTEBRAL)

7 **Psoas Minor** Psoas Parvus

8 **Psoas Major** Psoas Magnus

9 **Iliacus**

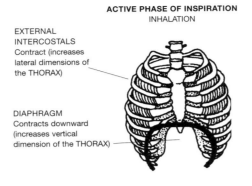

ACTIVE PHASE OF INSPIRATION
INHALATION

EXTERNAL INTERCOSTALS
Contract (increases lateral dimensions of the THORAX)

DIAPHRAGM
Contracts downward (increases vertical dimension of the THORAX)

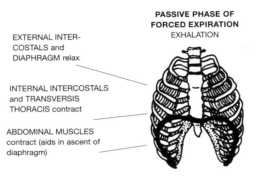

PASSIVE PHASE OF FORCED EXPIRATION
EXHALATION

EXTERNAL INTER-COSTALS and DIAPHRAGM relax

INTERNAL INTERCOSTALS and TRANSVERSIS THORACIS contract

ABDOMINAL MUSCLES contract (aids in ascent of diaphragm)

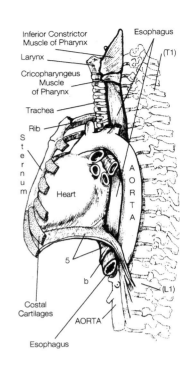

Inferior Constrictor Muscle of Pharynx
Esophagus
Larynx
(T1)
Cricopharyngeus Muscle of Pharynx
Trachea
Rib
Sternum
Heart
AORTA
5
b
Costal Cartilages
(L1)
AORTA
Esophagus

The THORAX (chest) is the closed cavity of the body between the base of the neck and muscular DIAPHRAGM inferiorly, which contains the HEART, GREAT VESSELS, portions of the ESOPHAGUS and TRACHEA, and the LUNGS.

The muscles associated with the thorax—mainly the diaphragm and the INTERNAL and EXTERNAL INTERCOSTALS—cause the continuous rhythmical (and usually involuntary) contractions associated with the changes in the capacity (air volume) of the thorax during breathing.

There are two phases to the process of breathing —INSPIRATION (inhalation) and EXPIRATION (exhalation). During the active phase (inspiration), contraction of the diaphragm and external intercostal muscles is prominent. During the active phase (expiration), contraction of the internal intercostals and TRANSVERSIS THORACIS (and some abdominal muscles) is prominent.

The diaphragm is a dome-shaped musculomembranous wall of skeletal muscle separating the THORACIC CAVITY from the ABDOMINAL CAVITY, with its convexity upward.

During the active phase (inspiration), the diaphragm contracts and flattens out downward, permitting the descent of the bases of the lungs and an increase in the capacity of the thorax. This condition creates an increased "suction pull" on the lung tissue as the lungs expand to fill the thoracic cavity. Because the air pressure within the lungs is less than the atmospheric pressure, air is sucked passively into the lungs (air pressure always tends to equalize). During the passive phase (forced expiration), the diaphragm relaxes, elevating it and restoring its inverted dome shape. The capacity of the thorax decreases, there is less "suction pull" on the lung tissue, and lungs recoil. Now the air pressure within the lungs is greater than the atmospheric pressure, so air is pushed out of the lungs into the atmosphere.

Thus, muscular action is the foundation of the breathing process, with the intake or expulsion of air following passively.

Structures that enclose and bound the thorax are above (upper ribs, neck tissues), at the sides (ribs, intercostal muscles), at the back (ribs, vertebral column, intercostal muscles) in front (ribs, costal cartilages, sternum, intercostal muscles) and below (diaphragm).

COLOR GUIDELINES:
1 = warm color; 2 = opposing cool color;
3 = yellow; 4 = flesh; 5 = light brown.

Anterior View

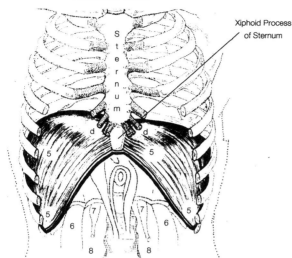

Xiphoid Process
of Sternum

Ribs cut (5–10) to expose DIAPHRAGM (5)

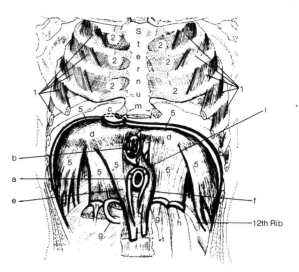

12th Rib

Coronal section midway through the DIAPHRAGM (5) to expose the
back wall and inside of DIAPHRAGM

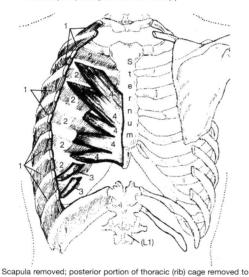

(L1)

Scapula removed; posterior portion of thoracic (rib) cage removed to
expose muscles on the inner side of the anterior portion of rib cage

Posterior View

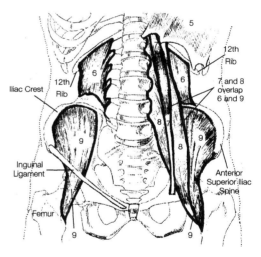

12th
Rib

7 and 8
overlap
6 and 9

Anterior
Superior Iliac
Spine

Iliac Crest

12th
Rib

Inguinal
Ligament

Femur

Anterior View

The intercostal muscles of respiration are those located between the ribs. The external intercostals are an outer layer of muscles between the ribs. They originate in the lower margin of each rib (excluding the twelfth) and they insert in the upper margin of the next rib. Their fibers are directed downward and forward.

Internal intercostals lie beneath the externals, and the fibers are directed downward and backward.

Muscles of the posterior abdominal wall include the QUADRATUS LUMBORUM, which is also considered a muscle of the vertebral column (see pages 90–91). It is an extensor of the lumbar region of the vertebral column when the right and left quadratus lumborum muscles contract together, and a lateral flexor and abductor of the vertebral column when these muscles contract separately.

The PSOAS MAJOR and MINOR and the ILIACUS MUSCLES of the posterior abdominal wall are frequently referred to as a single muscle—the ILIOPSOAS. These muscles cooperate in a joint synergistic action that independent action could not achieve. They are flexors and rotators of the thigh and flexors of the vertebral column. Iliopsoas are sometimes included as muscles of the hip (iliac-pelvic) region (see 110–111).

The QUADRATUS LUMBORUM is also considered a muscle of the VERTEBRAL COLUMN, an EXTENSOR of the LUMBAR REGION (when the muscles contract together) and LATERAL FLEXORS and ABDUCTORS (when they contract separately).

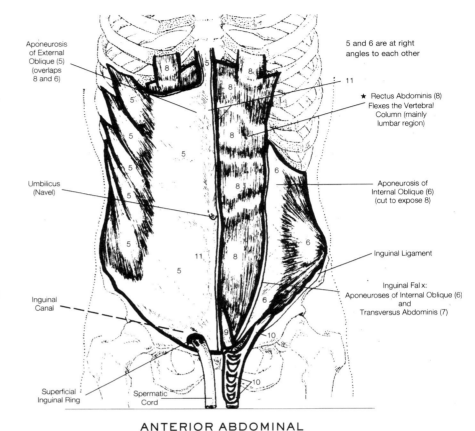

Aponeurosis of External Oblique (5) (overlaps 8 and 6)

5 and 6 are at right angles to each other

★ Rectus Abdominis (8) Flexes the Vertebral Column (mainly lumbar region)

Aponeurosis of Internal Oblique (6) (cut to expose 8)

Umbilicus (Navel)

Inguinal Ligament

Inguinal Falx: Aponeuroses of Internal Oblique (6) and Transversus Abdominis (7)

Inguinal Canal

Superficial Inguinal Ring

Spermatic Cord

ANTERIOR ABDOMINAL WALL OF THE TRUNK
(Anterior View)

ANTEROLATERAL ABDOMINAL WALL

FOUR PAIRS OF FLAT, SHEETLIKE MUSCLES (5–8):

Three-Layered Lateral Abdominal Wall

5 | External Oblique | Obliquus Externus Abdominis (Superficial)

6 | Internal Oblique | Obliquus Internus Abdominis (Middle)

7 | Transversus Abdominis | Transversalis Abdominis (Deep)

One-Layered Anterior Abdominal Wall

8 | Rectus Abdominis | (Deep)

9 | Pyramidalis

Inguinal Region (Lower Medial)

10 | Cremaster | (of the internal oblique)

11 | Linea Alba | "White Line"

12 | Inguinal Ligament

COLORING NOTES:
Superficial to Deep Layers =
warm to cool colors.

The sheath covering the RECTUS ABDOMINIS is composed of APONEUROSES of the three muscle layers of the LATERAL ABDOMINAL WALL: the EXTERNAL OBLIQUE, THE INTERNAL OBLIQUE, and the TRANSVERSUS ABDOMINIS.

Aponeuroses converge on the anterior midline of the abdomen to form the LINEA ALBA, or "white line." Also in the INGUINAL region, it forms the INGUINAL LIGAMENT.

The external oblique contracts the abdomen and viscera. The internal oblique compresses viscera and flexes the thorax forward. The transverus abdominis compresses the abdomen and flexes the thorax. The rectus abdominis compresses the abdomen and flexes the lumbar region of the vertebral column. The PYRAMIDALIS tightens the linea alba; and the CREMASTER raises the testicle.

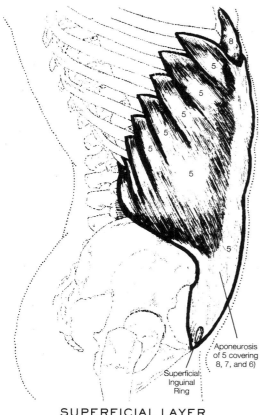

SUPERFICIAL LAYER

5
5
5
5
5
5
5
5
8

Aponeurosis of 5 covering 8, 7, and 6)

Superficial Inguinal Ring

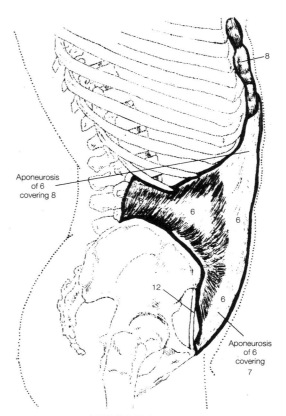

MIDDLE LAYER
(Right Lateral View)

8

Aponeurosis of 6 covering 8

6
6

12

6

Aponeurosis of 6 covering 7

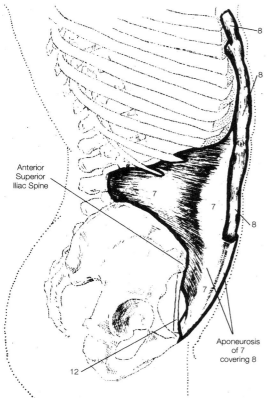

DEEP LAYER

8

8

8

Anterior Superior Iliac Spine

7

7

7

12

Aponeurosis of 7 covering 8

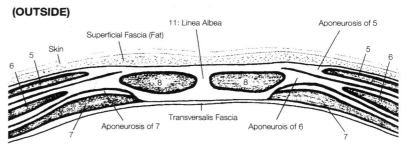

(OUTSIDE)

11: Linea Albea

Superficial Fascia (Fat)

Aponeurosis of 5

Skin

5

6

5
6

6

8

8

7

Aponeurosis of 7

Transversalis Fascia

Aponeurois of 6

7

ANTERIOR ABDOMEN (ABOVE UMBILICUS)
(Transverse Section, Superior View)

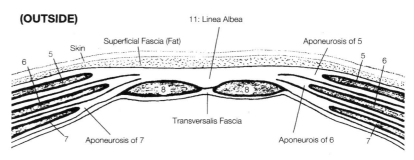

(OUTSIDE)

11: Linea Albea

Superficial Fascia (Fat)

Aponeurosis of 5

Skin

5

6

5
6

6

8

8

7

Aponeurosis of 7

Transversalis Fascia

Aponeurois of 6

7

ANTERIOR ABDOMEN
(LEVEL OF ILIAC CREST, JUST BELOW UMBILICUS)
(Transverse Section, Superior View)

The muscles of the pelvic outlet comprise the pelvic floor (of the perineum). These ten muscles are composed of two groups of muscular sheets, the POSTERIOR and ANTERIOR group (and their associated perineal fascia). The PERINEUM consists of the structures occupying the PELVIC OUTLET and comprising the PELVIC FLOOR. It is the region containing the external GENITALIA (sexual organs) and the ANAL OPENING. The perineum is composed of SKIN, PERINEAL FASCIAE, and MUSCLE.

In the male, the perineum comprises the external region between the SCROTUM and ANUS. In the female, the perineum comprises the external region between the VULVA and ANUS (see pages 306–7).

The posterior group, known as the PELVIC DIAPHRAGM, consists of three muscles. They are similar in the male and female. The anterior group, known as the urogenital diaphragm, consists of seven muscles associated with the genitalia. They differ greatly in the male and female.

The pelvic diaphragm (posterior group) consists of the LEVATOR ANI and COCCYGEUS. The two levator ani muscles (PUBOCOCCYGEUS and ILIOCOCCYGEUS) help support the pelvic viscera,

and help constrict and pull forward the lower part of the rectum (thus aiding in defecation). The coccygeus muscle aids the levator ani in its functions and pulls the coccyx forward following defecation. The pelvic diaphragm as a whole resists increased intra-abdominal pressure.

The UROGENITAL DIAPHRAGM (anterior group) consists of three superficial muscles, two deep muscles, and two muscles of the anal region. The OBTURATOR INTERNUS and PRIFORMIS muscles of the PELVIC WALL (DEEP LATERAL ROTATORS of the GLUTEAL REGION) are sometimes considered part of the muscles of the pelvic outlet (see pages 110–11).

COLOR GUIDELINES:
Pelvic Diaphragm = warm colors;
Urogenital Diaphragm = cool colors.

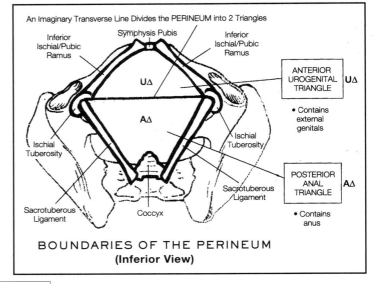

An Imaginary Transverse Line Divides the PERINEUM into 2 Triangles

Inferior Ischial/Pubic Ramus — Symphysis Pubis — Inferior Ischial/Pubic Ramus

UΔ

AΔ

ANTERIOR UROGENITAL TRIANGLE UΔ
• Contains external genitals

POSTERIOR ANAL TRIANGLE AΔ
• Contains anus

Ischial Tuberosity — Ischial Tuberosity

Sacrotuberous Ligament

Sacrotuberous Ligament — Coccyx

BOUNDARIES OF THE PERINEUM
(Inferior View)

MUSCLES OF THE PELVIC OUTLET
POSTERIOR GROUP: PELVIC DIAPHRAGM

1 **Levator Ani**

1a **Pubococcygeus** Anterior portion of levator ani

1b **Iliococcygeus** Posterior portion of levator ani

2 **Coccygeus** Ischiococcygeus

ANTERIOR GROUP: UROGENITAL DIAPHRAGM
(Associated with the genitalia)

SUPERFICIAL MUSCLES

3 **Superficial Transversus Perinei**
T. Perinei Superficialis

Action in urogenital diaphragm in both sexes: helps to stabilize the central tendon of perineum (in both the male and the female) and assists in supporting pelvic viscera (in both the male and the female).

4 **Bulbospongiosus**

Action in urogenital diaphragm in both sexes: in the male (called the bulbocavernosus), it constricts the bulbous urethral canal, assists in emptying urethra (of urine or semen), and assists in erection of penis. In the female (called the sphinctervaginae), it constricts the vaginal orifice and assists erection of clitoris.

5 **Ischiocavernosus**

Action in urogenital diaphragm in both sexes: in male (called the erector penis)—maintains erection of penis. In female (called the rector clitoridus)—maintains erection of clitoritis. In male = Erector penis. In female = Erector clitoris.

DEEP MUSCLES

In male, helps eject last drops of urine or semen. In female, helps eject last drops of urine.

6 **Profundus Transversus Perinei**

7 **Sphincter Urethrae** Compressor Urethrae, Constrictor Urethrae.

ANAL REGION

Keeps anal canal (both male and female) and orifice closed

8 **Sphincter Ani Externus**
External anal sphincter

9 **Sphincter Ani Internus**
(not shown)

MUSCLES OF THE PERINEUM, PELVIC OUTLET, AND PELVIC FLOOR

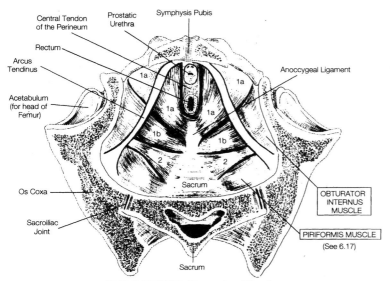

MALE PELVIS (Superior View)
(Transverse-Coronal Section)

Central Tendon of the Perineum
Prostatic Urethra
Symphysis Pubis
Rectum
Arcus Tendinus
Acetabulum (for head of Femur)
Os Coxa
Sacroiliac Joint
Sacrum
Anococcygeal Ligament
OBTURATOR INTERNUS MUSCLE
PIRIFORMIS MUSCLE (See 6.17)

ANTERIOR GROUP
UROGENITAL DIAPHRAGM

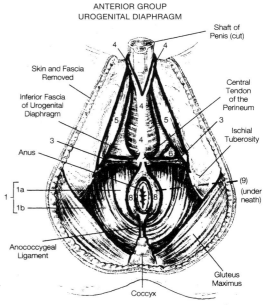

MALE PERINEUM (Inferior View)

Shaft of Penis (cut)
Skin and Fascia Removed
Inferior Fascia of Urogenital Diaphragm
Anus
Anococcygeal Ligament
Coccyx
Central Tendon of the Perineum
Ischial Tuberosity
(9) (under neath)
Gluteus Maximus

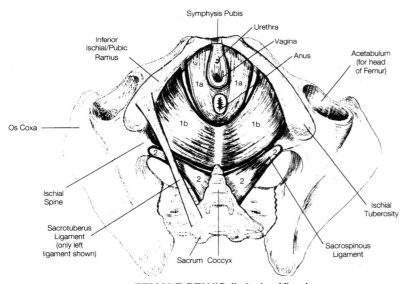

FEMALE PELVIS (Inferior View)
(Anterior Group removed)

Symphysis Pubis
Urethra
Vagina
Anus
Inferior Ischial/Pubic Ramus
Acetabulum (for head of Femur)
Os Coxa
Ischial Spine
Sacrotuberus Ligament (only left ligament shown)
Sacrum Coccyx
Ischial Tuberosity
Sacrospinous Ligament

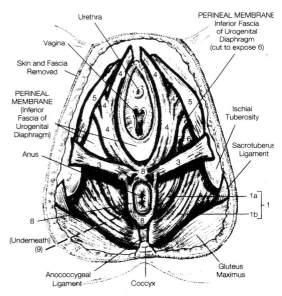

FEMALE PERINEUM (Inferior View)

Urethra
Vagina
Skin and Fascia Removed
PERINEAL MEMBRANE (Inferior Fascia of Urogenital Diaphragm)
Anus
(Underneath) (9)
Anococcygeal Ligament
Coccyx
PERINEAL MEMBRANE Inferior Fascia of Urogenital Diaphragm (cut to expose 6)
Ischial Tuberosity
Sacrotuberus Ligament
Gluteus Maximus

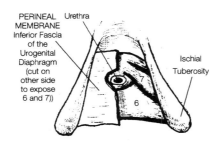

DEEP FEMALE PERINEUM

PERINEAL MEMBRANE Inferior Fascia of the Urogenital Diaphragm (cut on other side to expose 6 and 7))
Urethra
Ischial Tuberosity

Anterior View

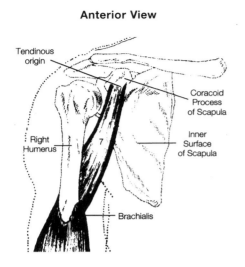

Tendinous origin

Coracoid Process of Scapula

Right Humerus

Inner Surface of Scapula

7

Brachialis

Anterior View

PECTORALIS MAJOR (cut) See 6.8

2

2

4

1

6

4

6

LATISSIMUS DORSI (cut) (arises from the back) See 6.8

Tendon of Long Head of Biceps Brachii (cut) See 6.13

Inner Surface of Scapula

Humerus (Left)

Rib cage removed (sternum and clavicle still intact) to expose the inside surface of each scapula. The deltoid muscle (1) is removed to reveal the right scapula.

MUSCLES THAT MOVE THE BRACHIUM (HUMERUS):

SEVEN SCAPULAR MUSCLES OF THE SHOULDER
(Originate on the scapulae and insert on the humerus)

Abductors of the Shoulder Joint

1 | Deltoid | Deltoideus

2 | Supraspinatus

Rotators of the Shoulder Joint

3 | Infraspinatus

4 | Subscapularis

5 | Teres Minor

6 | Teres Major

Flexor/Adductor of Shoulder Joint

7 | Coracobrachialis

FOUR MUSCLES OF THE MUSCULOTENDINOUS (ROTATOR) CUFF: (2, 3, 4, 5)

2 | Supraspinatus Abductor

3 | Infraspinatus/5 Teres Minor Lateral Rotators

4 | Subscapularis Medial Rotator

Refer back to page 92. Recall that there are five paired muscles and one large, triangular superficial back muscle (the TRAPEZIUS) that specifically function as SCAPULA STABILIZERS. These muscles originate on the axial skeleton and insert on the SCAPULA, holding the scapula firm and stable for the muscles that help move the brachium (arm).

Recall that the muscles of scapular stabilization fall into two groups. The posterior group consists of the trapezius, the RHOMBOIDEUS MAJOR and MINOR, and the LEVATOR SCAPULAE; the anterior group consists of the PECTORALIS MINOR and the SERRATUS ANTERIOR.

There are eight paired muscles and one large, triangular superficial back muscle (the LATISSIMUS DORSI) that span the shoulder joint, insert on the HUMERUS BONE, and help move the brachium. The latissimus dorsi and the paired PECTORALIS MAJOR muscles originate on the axial skeleton and are termed axial muscles. The remaining seven paired muscles originate on the scapula and insert on the humerus. They are the major muscles involved in the tremendous complexity of movement offered by the arm. (The scapular security offered by the muscles of scapular stabilization offers almost unlimited motility for the arm.)

The seven paired scapular muscles of the shoulder are divided into three groups: shoulder abductors (DELTOID and SUPRASPINATUS), the shoulder rotators (INFRASPINATUS, SUBSCAPULARIS, TERES MINOR, TERES MAJOR) and the shoulder flexor/adductor (CORACOBRACHIALIS).

The deltoid muscle is a thick, powerful muscle involved in a variety of arm movements. It raises the arm, rotates it, and abducts it; and it extends or flexes the humerus. It also caps the shoulder joint. Its origin is at the acromion process of the scapula, the clavicle, and the spine of the scapula. It inserts at the deltoid tuberosity of the humerus. Individual muscle groups within the deltoid can act separately for different movements.

There are four muscles of the MUSCULOTENDINOUS (ROTATOR) CUFF that originate on the scapulae, span the shoulder (glenohumeral) joint, and insert about the head of the humerus. These four muscles and their tendons reinforce the shoulder joint, whose socket is too flat to offer security to the head of the humerus on its own. The muscles of scapular stabilization secure the scapula for the stable base of origin for these four muscles. In general, the musculotendinous rotator cuff muscles correspond to the action of the deep lateral rotators of the

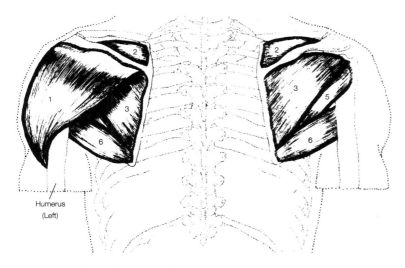

Posterior View

Humerus
(Left)

COLOR GUIDELINES:
Coracobrachialis = yellow;
Abductors = warm colors;
Rotators = cool colors.

The deltoid muscle (1) is removed from the right arm to expose the scapular muscles.
(Note that the deltoid caps the shoulder joint in both views.)

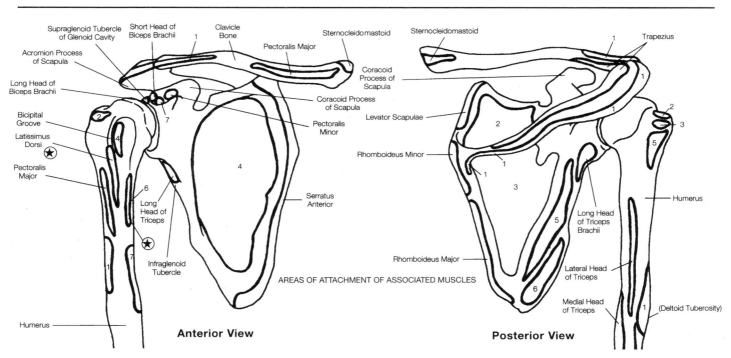

Supraglenoid Tubercle
of Glenoid Cavity

Short Head of
Biceps Brachii

Clavicle
Bone

Sternocleidomastoid

Sternocleidomastoid

Trapezius

Acromion Process
of Scapula

Pectoralis Major

Coracoid
Process of
Scapula

Long Head of
Biceps Brachii

Levator Scapulae

Coracoid Process
of Scapula

Bicipital
Groove

Pectoralis
Minor

Rhomboideus Minor

Latissimus
Dorsi

Pectoralis
Major

Serratus
Anterior

Humerus

Long
Head of
Triceps

Long Head
of Triceps
Brachii

Infraglenoid
Tubercle

AREAS OF ATTACHMENT OF ASSOCIATED MUSCLES

Rhomboideus Major

Lateral Head
of Triceps

Humerus

Medial Head
of Triceps

(Deltoid Tuberosity)

Anterior View

Posterior View

RIGHT CLAVICLE, SCAPULA, AND HUMERUS

(See pages 102–3 for biceps and triceps brachii; see pages 60–61 for details of skeletal structure.)

iliac-pelvic region. The rotator cuff of the shoulder is commonly a problem site for baseball pitchers.

The movements possible for the arm are ABDUCTION (supraspinatus, deltoid), LATERAL ROTATION (infraspinatus, teres minor), and MEDIAL ROTATION (deltoid [anterior fibers], subscapularis, teres major, latissimus dorsi, and pectoralis major).

Although both the teres major and the latissimus dorsi arise from the back, they cross from the back to the anterior side of the humerus, where their insertions to the bicipital groove of the humerus anteriorly cause medial arm rotation.

COLOR GUIDELINES:
Coracobrachialis = yellow;
Biceps Brachii = orange;
Triceps Brachii = blue (opposing color).

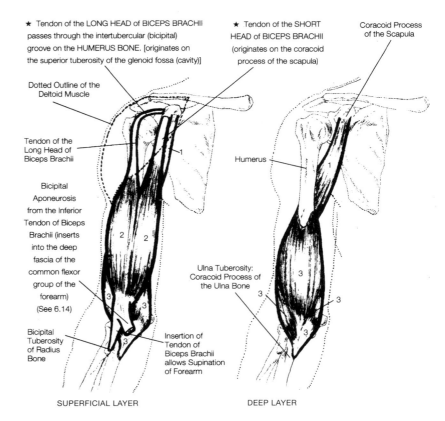

★ Tendon of the LONG HEAD of BICEPS BRACHII passes through the intertubercular (bicipital) groove on the HUMERUS BONE. [originates on the superior tuberosity of the glenoid fossa (cavity)]

★ Tendon of the SHORT HEAD of BICEPS BRACHII (originates on the coracoid process of the scapula)

Coracoid Process of the Scapula

Dotted Outline of the Deltoid Muscle

Tendon of the Long Head of Biceps Brachii

Bicipital Aponeurosis from the Inferior Tendon of Biceps Brachii (inserts into the deep fascia of the common flexor group of the forearm) (See 6.14)

Bicipital Tuberosity of Radius Bone

Insertion of Tendon of Biceps Brachii allows Supination of Forearm

Humerus

Ulna Tuberosity: Coracoid Process of the Ulna Bone

SUPERFICIAL LAYER

DEEP LAYER

Anterior View (Right Arm)

MUSCLES OF THE BRACHIUM (HUMERUS): THE ARM

FLEXOR/ADDUCTOR OF SHOULDER JOINT

1   Coracobrachialis

BRACHIAL MUSCLES THAT ACT ON THE FOREARM:

2   Biceps Brachii (2 Heads/Origins)

3   Brachialis

4   Brachioradialis   Supinator Longus

5   Triceps Brachii (3 Heads/Origins)

MUSCLES ACTING ON THE FOREARM
Four Flexors of the Elbow Joint

2   Biceps Brachii (Synergistic action of principal flexors with 3 brachialis)

3   Brachialis (Synergistic action of principal flexors with 2 brachialis)

4   Brachioradialis*

Pronator Teres*

Two Extensors of Forearm

5   Triceps Brachii   Principal Extensor

6   Anconeus*

Three Supinators of Hand (and Forearm)

2   Biceps Brachii (synergistic action with 4 and 7)

4   Brachioradialis (synergistic action with 2 and 7)*

7   Supinator (synergistic action with 2 and 4)

TWO PRONATORS OF HAND (ELBOW AND FOREARM)

8   Pronator Teres*

9   Pronator Quadratus*

* These are muscles of the forearm (see pages 104–7): the pronator teres (anterior superficial), pronator quadratus (anterior deep), brachioradialis and aconeus (posterior superficial) and supinator (posterior deep).

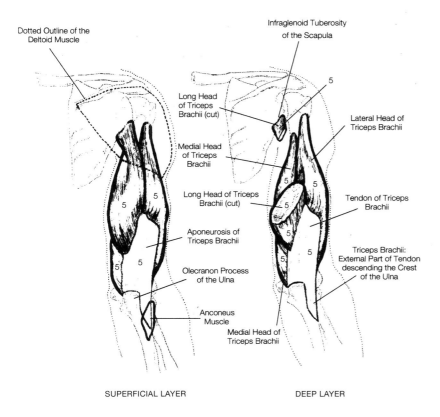

Dotted Outline of the
Deltoid Muscle

Infraglenoid Tuberosity
of the Scapula

Long Head
of Triceps
Brachii (cut)

Lateral Head of
Triceps Brachii

Medial Head
of Triceps
Brachii

Long Head of Triceps
Brachii (cut)

Tendon of Triceps
Brachii

Aponeurosis of
Triceps Brachii

Olecranon Process
of the Ulna

Triceps Brachii:
External Part of Tendon
descending the Crest
of the Ulna

Anconeus
Muscle

Medial Head of
Triceps Brachii

SUPERFICIAL LAYER

DEEP LAYER

Posterior View (Right Arm)

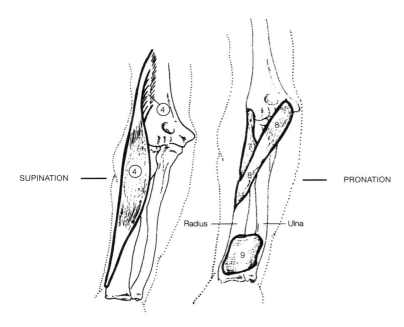

SUPINATION

PRONATION

Radius

Ulna

Anterior View (Right Arm)

The UPPER ARM, more commonly called the arm, is the proximal portion of the upper limb or upper extremity. The muscles associated with the upper arm (HUMERUS, or BRACHIUM) are very powerful. They act on the forearm at the elbow joint and the radioulnar joint (assisted by smaller muscles from the forearm proper).

The BICEPS BRACHII is the most prominent muscle (structurally and functionally) of the humerus bone, but it has no attachments on the humerus. It lies on the anterior surface of the humerus and flexes the forearm. It has a dual origin; the short head originates from the CORACOID PROCESS of the scapula, and the long head originates on the superior tuberosity of the GLENOID FOSSA. Both heads insert on the RADIAL TUBEROSITY.

The TRICEPS BRACHII is located on the posterior surface of the humerus; it is antagonistic to the biceps brachii, and it extends the forearm. It has a triple origin. The LONG HEAD originates on the INFRAGLENOID TUBEROSITY of the scapula. The LATERAL HEAD's origin is on the posterior surface of the humerus below the GREAT TUBERCLE and The origin of the MEDIAL HEAD is the humerus below the RADIAL GROOVE. The common tendinous insertion is the olecranon process of the ulna.

The BRACHIORADIALIS—apart from its flexion and SUPINATION action on the forearm—also plays a part in rapid extension (countering the centrifugal force produced by extension).

Supination refers to the rotation of the forearm and hand so that the palm faces upward (directed forward or anteriorly). This movement, analogous to tightening a screw with a screwdriver, is aided by the wrapping form of the SUPINATOR MUSCLE.

PRONATION—the opposite of supination—refers to the rotation of the forearm and hand so that the palm faces downward (directed backward or posteriorly). This movement, analagous to loosening a screw with a screwdriver, is aided by the PRONATOR MUSCLES, which cross the RADIUS on the anterior side of the forearm (since the radius bone rotates about the ULNA).

ANTERIOR (PALMAR) MUSCLES OF THE FOREARM

SUPERFICIAL GROUP [LATERAL (1) to MEDIAL (4)]

FLEXORS OF THE WRIST AND HAND

1 | Pronator Teres | Humeral Head and Ulnar Head

2 | Flexor Carpi Radialis | Radiocarpus

3 | Palmaris Longus |

4 | Flexor Carpi Ulnaris | Humeral Head and Ulnar Head

5 | Flexor Digitorum Superficialis | Flexor Digitorum Sublimis

ANTERIOR MUSCLES ACTING ON WRIST JOINT, HANDS AND FINGERS

DEEP GROUP

6 | Flexor Digitorum Profundus |

7 | Flexor Pollicis Longus |

8 | Pronator Quadratus |

ABDUCTOR OF THE WRIST AND HAND:

2 | Flexor Carpi Radialis |

ADDUCTOR OF THE WRIST AND HAND:

4 | Flexor Carpi Ulnaris |

PRONATORS OF THE WRIST, HAND, ELBOWS AND FOREARM:

1 | Pronator Teres |

8 | Pronator Quadratus |

FLEXORS OF THE FINGERS (DIGITS):

5 | Flexor Digitorum Superficialis |

6 | Flexor Digitorum Profundus |

FLEXOR OF THE THUMB:

7 | Flexor Pollicis Longus |

ROTATORS OF THE FOREARM:

1 | Pronator Teres |

8 | Pronator Quadratus |

9 | Supinator |

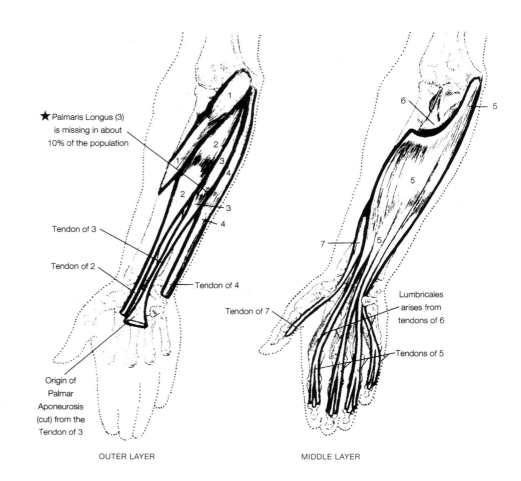

★ Palmaris Longus (3) is missing in about 10% of the population

Tendon of 3

Tendon of 2

Tendon of 4

Origin of Palmar Aponeurosis (cut) from the Tendon of 3

OUTER LAYER

Tendon of 7

Lumbricales arises from tendons of 6

Tendons of 5

MIDDLE LAYER

SUPERFICIAL FLEXORS

ANTERIOR (PALMAR) MUSCLES OF THE FOREARM
Anterior Views (Right Arm)

COLORING NOTES:
Color flexors of the forearm in warm colors, and the other anterior forearm muscles in light warm colors.

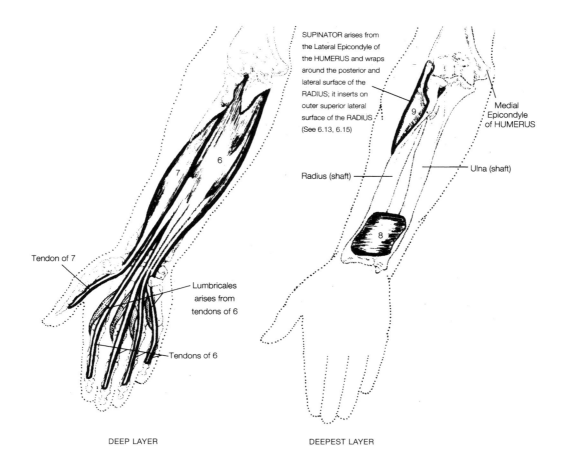

SUPINATOR arises from the Lateral Epicondyle of the HUMERUS and wraps around the posterior and lateral surface of the RADIUS; it inserts on outer superior lateral surface of the RADIUS (See 6.13, 6.15)

Medial Epicondyle of HUMERUS

Radius (shaft)

Ulna (shaft)

Tendon of 7

Lumbricales arises from tendons of 6

Tendons of 6

DEEP LAYER

DEEPEST LAYER

DEEP FLEXORS

ANTERIOR (PALMAR) MUSCLES OF THE FOREARM
Anterior Views (Right Arm)

Most of the ANTERIOR (or PALMAR) PORTION of the FOREARM is made up of FLEXOR MUSCLES of the WRIST AND FINGERS (DIGITS). As a group, these flexors originate from the medial epicondyle of the humerus, upper radius and ulna, and the interosseus membrane of the forearm.

The FLEXOR CARPI MUSCLES, as they cross the wrist, insert on the distal carpal bones, or metacarpals.

The two flexors for each finger (digit) insert on the MIDDLE and DISTAL PHALANGES; when the fingers are flexed, they close over the palm as in a fist.

The PALMARIS LONGUS inserts, or merges with (originates on) the PALMAR APONEUROSIS.

The FLEXOR DIGITORUM SUPERFICIALIS (of the superficial group) consists of three heads—the HUMERAL, ULNAR, AND RADIAL.

The LUMBRICAL MUSCLES of the hand (see page 110) arise from tendons of the FLEXOR DIGITORUM PROFUNDUS.

The anterior forearm muscles acting on the wrist joint, hands, and fingers include the flexors of the wrist and hand, consisting of all of the anterior forearm muscles except pronators (mainly the superficial group), one abductor of the wrist and hand, one adductor of the wrist and hand, two pronators of the wrist and hand (and forearm), two flexors of the fingers, one flexor of the thumb and three rotators of the forearm.

Most of the POSTERIOR (DORSAL/BACK) portion of the forearm is made up of EXTENSOR MUSCLES OF THE WRIST AND FINGERS (DIGITS), though they are less massive than those of the flexor side.

The ANCONEUS is a small superficial forearm muscle that helps extend the forearm itself. The BRACHIORADIALIS muscle originates dorsally on the supracondylar ridge of the humerus bone, but curves most of its mass to the anterior (palmar) side of the forearm. It helps to flex and supinate the forearm, although it has been shown to play a part in rapid extension (countering the centrifugal force produced by flexion).

As a group, these extensors originate from the lateral epicondyle of the humerus, upper radius and ulna, and the interosseus membrane of the forearm.

The EXTENSOR CARPI MUSCLES insert on the distal carpal bones, or metacarpals.

The three finger (digit) extensors insert (form a tendon expansion) on the middle and distal phalanges, and the small intrinsic muscles of the hand (see pages 108–9) insert on these expanded tendons. If a fist were made over the palm by flexors, the finger extensors would pull the fingers back out into the anatomical position.

The muscles that extend the hand are located on the posterior side of the forearm, and their tendons extend along the dorsal surface of the hand, mainly over the DORSAL INTEROSSEI (see page 108) and onto the phalanges. Note the complex division of the tendon of the EXTENSOR DIGITORUM COMMUNIS into the four phalanges (fingers).

The ANATOMICAL "SNUFFBOX" is a small depression in the skin at the base of the thumb. It is created by the boundaries of the two thumb extensors and the thumb abductor (and their tendons).

POSTERIOR (DORSAL/ BACK) MUSCLES OF THE FOREARM
SUPERFICIAL GROUP

[Lateral (1) to Medial (7)]

1 | Brachioradialis | Supinator Longus

2 | Extensor Carpi Radialis Longus

3 | Extensor Carpi Radialis Brevis

4 | Extensor Digitorum Communis

5 | Extensor Digiti Minimi | Extensor Digiti Quinti Proprius

6 | Extensor Carpi Ulnaris

7 | Anconeus

DEEP GROUP

8 | Supinator | Supinator Radii Brevis

9 | Abductor Pollicis Longus Extensor Ossis Metacarpi Pollicis

10 | Extensor Pollicis Brevis Extensor Secondii Internodii Pollicis

11 | Extensor Pollicis Longus Extensor Secondii Internodii Pollicis

12 | Extensor Indicis | Extensor Indicis Proprius

POSTERIOR FOREARM MUSCLES ACTING ON THE WRIST JOINT, HANDS, AND FINGERS
Extensors of the Hand and Wrist – All posterior forearm muscles of the superficial group except brachioradialis and anconeus

ABDUCTORS OF THE WRIST AND HAND

2 | Extensor Carpi Radialis Longus

3 | Extensor Carpi Radialis Brevis

6 | Extensor Carpi Ulnaris

ADDUCTORS OF THE WRIST AND HAND

10 | Extensor Pollicis Brevis

11 | Extensor Pollicis Longus

EXTENSORS OF THE FINGERS

4 | Extensor Digitorum Brevis

5 | Extensor Digiti Minimi

12 | Extensor Indicis

EXTENSORS OF THE THUMB

10 | Extensor Pollicis Brevis

11 | Extensor Pollicis Longus

ABDUCTOR OF THE THUMB

9 | Abductor Pollicis Longus

ROTATOR OF THE FOREARM

8 | Supinator

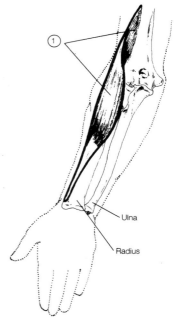

ANTERIOR (PALMAR) VIEW
OF RIGHT ARM

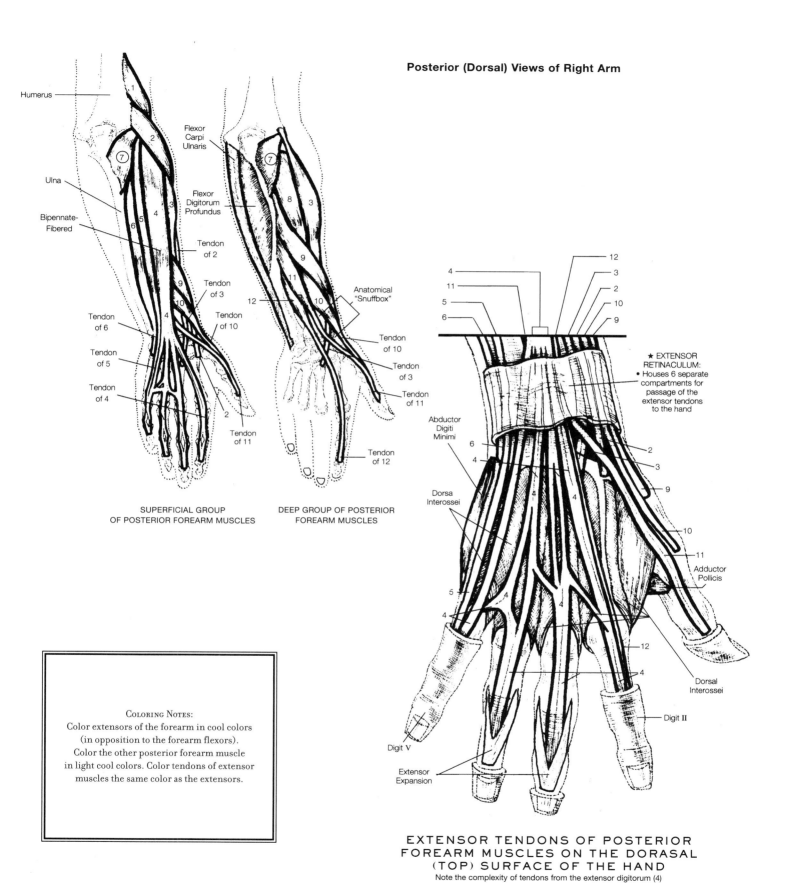

Posterior (Dorsal) Views of Right Arm

Humerus

Flexor Carpi Ulnaris

Ulna

Flexor Digitorum Profundus

Bipennate-Fibered

Tendon of 2

Tendon of 3

Tendon of 6

Tendon of 10

Tendon of 5

Tendon of 4

Tendon of 10

Tendon of 3

Tendon of 11

Anatomical "Snuffbox"

Tendon of 11

Tendon of 12

SUPERFICIAL GROUP
OF POSTERIOR FOREARM MUSCLES

DEEP GROUP OF POSTERIOR
FOREARM MUSCLES

COLORING NOTES:
Color extensors of the forearm in cool colors
(in opposition to the forearm flexors).
Color the other posterior forearm muscle
in light cool colors. Color tendons of extensor
muscles the same color as the extensors.

★ EXTENSOR
RETINACULUM:
• Houses 6 separate
compartments for
passage of the
extensor tendons
to the hand

Abductor Digiti Minimi

Dorsa Interossei

Adductor Pollicis

Dorsal Interossei

Digit II

Digit V

Extensor Expansion

EXTENSOR TENDONS OF POSTERIOR
FOREARM MUSCLES ON THE DORASAL
(TOP) SURFACE OF THE HAND
Note the complexity of tendons from the extensor digitorum (4)

The complex functions of the hand are accomplished by several sets of muscles. The flexion and extension of hands and fingers is accomplished with the forearm muscles; precise finger movements (flexion and extension coordinated with abduction and adduction) are accomplished with the small intrinsic muscles of the hand. The extensors of the thumb are the EXTENSOR POLLICIS BREVIS and the EXTENSOR POLLICIS LONGUS (see page 108).

Circumduction of the thumb involves the cyclic activity of three muscles, the OPPONENS POLLICIS to the FLEXOR POLLICIS BREVIS to the ABDUCTOR POLLICIS BREVIS and back to the OPPONENS

BREVIS. The act of grasping requires oppositional activity back and forth between the opponens pollicis and the OPPONENS DIGITI MINIMI. The INTEROSSEI and LUMBRICALS of the INTERMEDIATE THUMB GROUP insert into the EXTENSOR EXPANSION (along with certain tendons of arm muscles), which aids in complex functions required for extending and flexing the fingers. The lumbricals arise from tendons of the FLEXOR DIGITORUM PROFUNDUS.

Even when relaxed, the hand is contracted inward slightly toward the palmar surface. This is because the flexor muscles are naturally stronger (and more massive as a whole) than the extensor muscles.

INTRINSIC MUSCLES OF THE HAND (MANUS)
THENAR EMINENCE OF THE THUMB

1 Abductor Pollicis Brevis

2 Opponens Pollicis

3 Flexor Pollicis Brevis

HYPOTHENAR EMINENCE OF THE LITTLE FINGER

4 Abductor Digiti Minimi
Abductor Digiti Quinti

5 Opponens Digiti Minimi
Opponens Digit Quinti

6 Flexor Digiti Minimi Brevis
Flexor Digit Quinti Brevis

INTERMEDIATE GROUP OF THE PALM
(Deep Muscles between Metacarpal Bones)

7 Adductor Pollicis Oblique and Transverse Heads

8 Palmar Interossei (3) Interossei Volares

8 Dorsal Interossei (4) Interossei Dorsales Manus

10 Lumbricales (4) Lumbricales Manus

SUBCUTANEOUS MUSCLE

11 Palmaris Brevis

ACTION OF INTRINSIC HAND MUSCLES
ABDUCTOR OF THE THUMB

1 Abductor Pollicis Brevis

ADDUCTOR OF THE THUMB

7 Adductor Pollicis

FLEXORS OF THE THUMB

2 Opponens Pollicis

3 Flexor Pollicis Brevis

ABDUCTORS OF THE FINGERS

4 Abductor Digiti Minimi

8 Palmar Interossei (3)

9 Dorsal Interossei (4)

ADDUCTORS OF THE FINGERS

8 Palmar Interossei (3)

9 Dorsal Interossei (4)

FLEXORS OF THE FINGERS

6 Flexor Digiti Minimi Brevis

10 Lumbricales (4)

MUSCLES THAT CHANGE THE SHAPE OF THE PALM

5 Opponens Digiti Minimi

11 Palmar Brevis

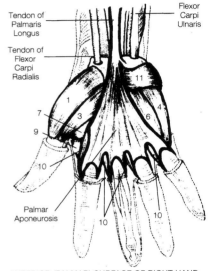

ANTERIOR (PALMAR) SURFACE OF RIGHT HAND
(SUPERFICIAL LAYER)

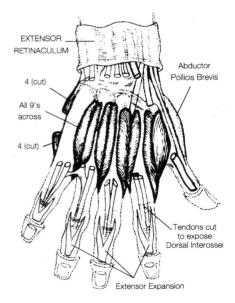

POSTERIOR (DORSAL) SURFACE OF RIGHT HAND

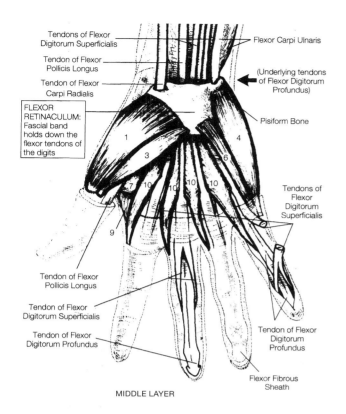

Tendons of Flexor Digitorum Superficialis

Tendon of Flexor Pollicis Longus

Tendon of Flexor Carpi Radialis

FLEXOR RETINACULUM: Fascial band holds down the flexor tendons of the digits

Flexor Carpi Ulnaris

(Underlying tendons of Flexor Digitorum Profundus)

Pisiform Bone

Tendons of Flexor Digitorum Superficialis

Tendon of Flexor Pollicis Longus

Tendon of Flexor Digitorum Superficialis

Tendon of Flexor Digitorum Profundus

Tendon of Flexor Digitorum Profundus

Flexor Fibrous Sheath

MIDDLE LAYER

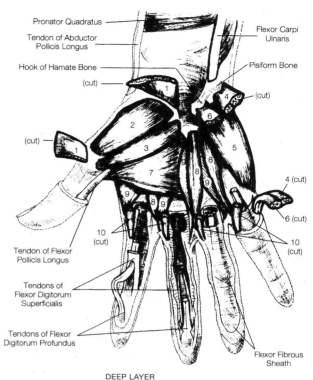

Pronator Quadratus

Tendon of Abductor Pollicis Longus

Hook of Hamate Bone (cut)

Flexor Carpi Ulnaris

Pisiform Bone

(cut)

(cut)

4 (cut)

6 (cut)

10 (cut)

10 (cut)

Tendon of Flexor Pollicis Longus

Tendons of Flexor Digitorum Superficialis

Tendons of Flexor Digitorum Profundus

Flexor Fibrous Sheath

DEEP LAYER
(Lumbricales, tendons, and superficial muscles cut to expose deep muscles)

PALMAR SURFACES of RIGHT HAND

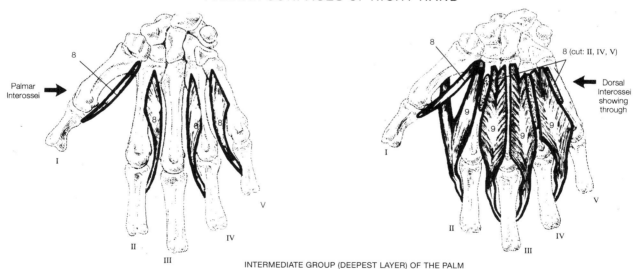

Palmar Interossei

I

II

III

IV

V

Dorsal Interossei showing through

I

II

III

IV

V

INTERMEDIATE GROUP (DEEPEST LAYER) OF THE PALM

COLOR GUIDELINES:
Thenar eminence = reds and oranges;
Hypothenar eminence = pinks and purples;
Intermediate group = blues and greens;
Palmar brevis = yellow.

SUPERFICIAL POSTERIOR BUTTOCK MUSCLES

1 Gluteus Maximus

2 Gluteus Medius (Synergistic Action with 3)

3 Gluteus Minimus (Synergistic Action with 2)

4 Tensor Fasciae Latae (Tensor Fasciae Femoris)

SIX DEEP LATERAL BUTTOCK MUSCLES (ROTATORS)

5 Piriformis*

6 Obturator Internus*

7 Gemellus Superior

8 Gemellus Inferior

9 Quadratus Femoris

10 Obturator Externus

Buttock muscles act on the hip joint and extend, abduct, and rotate the THIGH (FEMUR).

The GLUTEUS MAXIMUS is a powerful thigh extensor (one to two inches thick) that contributes to bipedal stance and locomotion, running and climbing. Superficial fascia (fat) adds form. Its origins are at the SUPERIOR CURVED ILIAC LINE and CREST, the SACRUM, COCCYX, and the APONEUROSIS OF THE BACK. Its insertion is on the GLUTEAL TUBEROSITY of the femur (just below greater TROCHANTER), and the ILIOTIBIAL TRACT.

The GLUTEUS MEDIUS is a hip stabilizer that keeps hips level during walking and running. It originates on the lateral surface of the lower part of the ilium, and inserts at the greater trochanter of the femur.

The GLUTEUS MINIMUS is the smallest and deepest of the gluteals. It originates at the lateral surface of the lower part of the ilium and inserts on the greater trochanter.

The TENSOR FASCIA LATAE is actually part of the thigh region (being a thigh flexor), but is also attached to the buttocks. Its origin is on the iliac crest and its inserttion is at the TIBIA, by way of the iliotibial tract and fascia lata.

The ILIOTIBIAL TRACT is a tendinous band that extends down the thigh. The FASCIA LATA is a broad, wide lateral fascia encasing the thigh muscles. It is continuous with the iliotibial tract. The iliotibial tract and the fascia lata (near the junction of the gluteus maximus and tensor fascia latae) are collectively referred to as the FEMORAL APONEUROSIS.

In general, the six deep LATERAL ROTATORS correspond to the action of the musculo-tendinous cuff of the shoulder joint, located directly over the posterior aspect of the hip joint (see pages 100–1).

There are two groups of deep lateral buttock muscles that act on the hip joint and the thigh (femur); they are the ILIOPSOAS (the compound ILIACUS and PSOAS MAJOR muscle) and the lateral rotators (all six deep lateral buttock muscles plus the gluteus maximus). The single synergistic ILIOPSOAS not only laterally rotates the thigh but also flexes the thigh and the vertebral column.

The PIRIFORMIS AND OBTURATOR INTERNUS (sometimes included as muscles of the pelvic wall) laterally rotate the thigh as well as abduct it.

The superficial posterior buttock muscles that act on the hip joint and the thigh (femur) are the gluteus maximus (extensor of thigh), the tensor fascia latae (flexor of thigh), all the superficial posterior buttock muscles (abductors of the thigh), and all the superficial posterior buttock muscles except gluteus maximus (rotators of thigh and medial rotators).

The deep lateral buttock muscles that act on the hip joint and the thigh (femur) are all six deep laterals plus the gluteus maximus as well as the iliopsoas muscles (see pages 94–95), the synergistic compund ILIACUS and PSOAS MAJOR MUSCLE.

COLOR GUIDELINES:
1 = yellow; 2 = orange; 3 = red;
4 = light pink; deep laterals = cool colors.

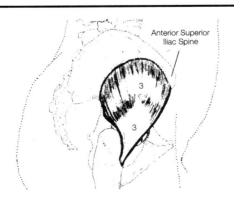

Anterior Superior Iliac Spine

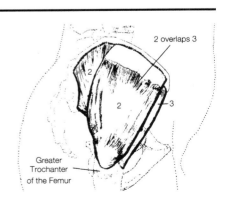

2 overlaps 3

Greater Trochanter of the Femur

Right Lateral View

GLUTEAL MUSCLES OF THE ILIAC-PELVIC (GIRDLE) REGION

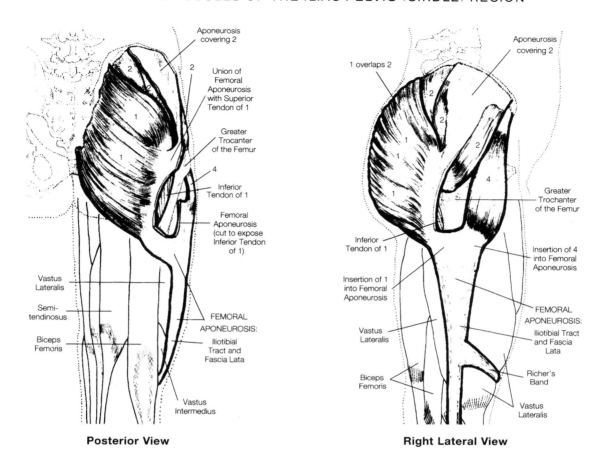

Aponeurosis covering 2

Union of Femoral Aponeurosis with Superior Tendon of 1

Greater Trocanter of the Femur

Inferior Tendon of 1

Femoral Aponeurosis (cut to expose Inferior Tendon of 1)

FEMORAL APONEUROSIS:

Iliotibial Tract and Fascia Lata

Vastus Lateralis

Semi-tendinosus

Biceps Femoris

Vastus Intermedius

Posterior View

Aponeurosis covering 2

1 overlaps 2

Greater Trochanter of the Femur

Insertion of 4 into Femoral Aponeurosis

Inferior Tendon of 1

Insertion of 1 into Femoral Aponeurosis

FEMORAL APONEUROSIS:

Iliotibial Tract and Fascia Lata

Vastus Lateralis

Richer's Band

Biceps Femoris

Vastus Lateralis

Right Lateral View

DEEP LATERAL ROTATOR BUTTOCK MUSCLES

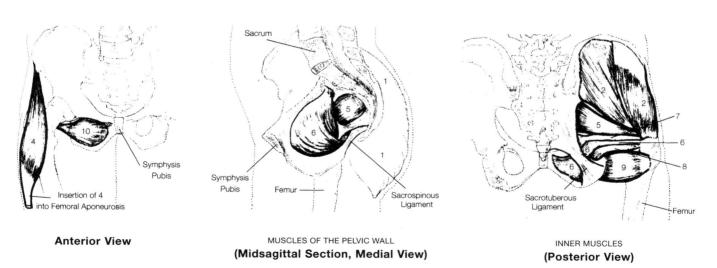

Insertion of 4 into Femoral Aponeurosis

Symphysis Pubis

Anterior View

Sacrum

Symphysis Pubis

Femur

Sacrospinous Ligament

MUSCLES OF THE PELVIC WALL
(Midsagittal Section, Medial View)

Sacrotuberous Ligament

Femur

INNER MUSCLES
(Posterior View)

MEDIAL FEMORAL (ADDUCTOR) THIGH MUSCLES

1 | Gracilis |

2 | Pectineus |

3 | Adductor Longus | (Synergistic Action with 4, 5)

4 | Adductor Brevis | (Synergistic Action with 3, 5)

5 | Adductor Magnus | (Synergistic Action with 3, 4)

The medial thigh muscles that act on the hip joint (and thigh) are all the medial femoral thigh muscles (adductors of thigh), the flexors of the thigh (pectineus, adductor longus, adductor brevis), the lateral rotators of the thigh (adductor longus, adductor brevis, adductor magnus) and medial thigh muscles that act on the knee joint (and lower leg)—flexor of the leg (gracilis) and extensor of the leg (adductor magnus [superficial]).

POSTERIOR FEMORAL (FLEXOR) THIGH MUSCLES—THE THREE "HAMSTRINGS":

6 | Biceps Femoris |

7 | Semitendinosus |

8 | Semimembranosus |

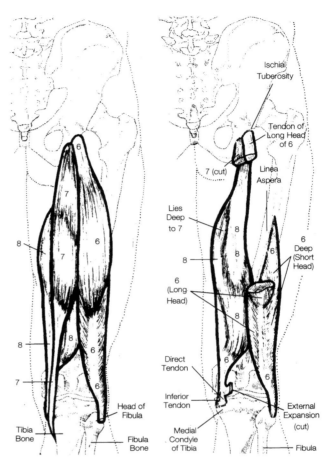

Posterior Views of the Three "Hamstrings"

The THIGH is the proximal portion of the lower extremity, lying between the hip joint and the knee.

There are five primary MEDIAL FEMORAL THIGH MUSCLES. They are basically ADDUCTORS, and make up most of the thigh mass (recall that an adductor muscle draws toward the medial line of the body). The long, thin GRACILIS is the most superficial of the medial thigh muscles, and is a two-joint muscle which adducts the thigh and flexes the leg.

The PECTINEUS is flat and quadrangular, and is the uppermost of the medial thigh muscles. The ADDUCTOR LONGUS, BREVIS, AND MAGNUS muscles are basically synergistic in adducting, flexing and rotating the thigh.

Note the hiatus (opening) in the adductor magnus for the passing of the femoral artery and vein from anterior thigh to posterior thigh.

The muscles of the posterior thigh, or the "hamstrings," are the BICEPS FEMORIS, the SEMITENDINOSUS, and the SEMIMEMBRANOSUS. They act as extensors of the hip joint (more efficient when the KNEE JOINT is extended), and they also act as flexors of the knee joint (more efficient when the hip joint is extended). They act as flexors of the knee joint (more efficient with the hip joint extended). When the leg is raised (as in kicking a ball), the posterior thigh muscles restrict knee extension. The three hamstrings are antagonistic to the QUADRICEPS FEMORIS when the leg is flexed (see pages 114–5).

The biceps femoris has a superficial long head (which originates at the ischial tuberosity), and a deep, short head (which originates at the linea aspera of the femur). It inserts at the head of the FIBULA and the LATERAL CONDYLE of the TIBIA. Its action is complicated, working over both the hip and knee joints.

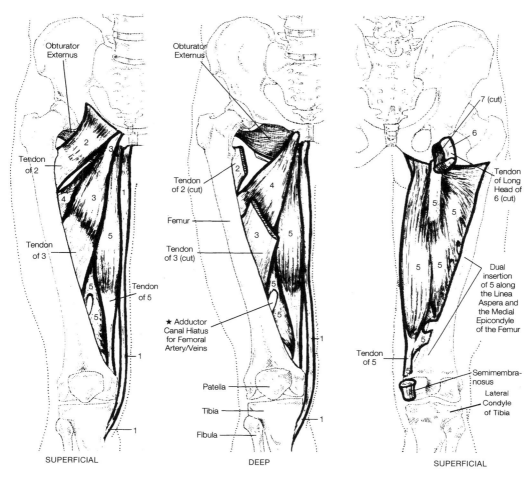

Obturator
Externus

Tendon
of 2

Tendon
of 3

Tendon
of 5

SUPERFICIAL

Obturator
Externus

Tendon
of 2 (cut)

Femur

Tendon
of 3 (cut)

★ Adductor
Canal Hiatus
for Femoral
Artery/Veins

Patella

Tibia

Fibula

DEEP

Anterior Views

7 (cut)

6

Tendon
of Long
Head of
6 (cut)

Dual
insertion
of 5 along
the Linea
Aspera and
the Medial
Epicondyle
of the Femur

Tendon
of 5

Semimembra-
nosus

Lateral
Condyle
of Tibia

SUPERFICIAL

Posterior View

The semitendinosus is a fusiform muscle whose origin is the ISCHIAL TUBEROSITY from a common tendon with the biceps femoris. It is a long slender tapering tendon, and it inserts at the proximal lateral portion of the tibia. Like the biceps femoris, its action is complicated and it works over both the knee and hip joints.

The semimembranosus—whose action is similar to that of the semitendinosus—is a flat muscle which also originates at the ischial tuberosity, and inserts at the MEDIAL CONDYLE of the tibia.

The POSTERIOR THIGH MUSCLES that act on the hip joint and thigh are all the POSTERIOR FEMORAL THIGH MUSCLES (the extensors of the thigh). The posterior thigh muscles that act on the knee joint and lower leg are the biceps femoris (the lateral rotator of the leg and thigh).

The medial rotator of the leg and thigh is the semitendinosus; the flexors of the leg are all the posterior femoral thigh muscles.

COLOR GUIDELINES:
1 = yellow; 2 = yellow-orange; 3, 4, 5 = light cool colors (blues, greens); 6 = oranges; 7 = red; 8 = pink.

The QUADRICEPS FEMORIS arises from four distinct heads (four separate origins): RECTUS FEMORIS (whose origin is at the anterior inferior spine of the ilium and the lip of the acetabulum), VASTUS LATERALIS (whose origin is at the greater trochanter and linea aspera of femur), VASTUS MEDIALIS (whose origin is at the medial surface of the femur), and VASTUS INTERMEDIUS (whose origin is at the anterior and lateral surfaces of femur). All four muscle heads of the quadriceps femoris insert on its COMMON TENDON.

The common tendon inserts on the superior aspects of the patella KNEECAP itself, along with a THICK PATELLAR TENDON, which attaches to the tibial tuberosity. The patellar tendon is techni-cally a LIGAMENT because the patella is a sesamoid bone that develops inside of the tendon.

The quadriceps femoris are the only extensors of the knee joint.

THE rectus femoris is the main flexor of the hip joint (aided by the iliopsoas) and sartorius (weak). It is the only one of the four quadriceps that contracts over both hip and knee joints.

The sartorius can act on both the hip and knee joint. It flexes and rotates the thigh, and flexes the leg and knee. The sartorious is the longest MUSCLE in the body; it is a long, ribbon-shaped thigh muscle that obliquely crosses the anterior aspect of the thigh (it originates at the ANTERIOR SUPERIOR ILIAC SPINE and inserts at the TIBIAL TUBEROSITY).

ANTERIOR FEMORAL (EXTENSOR) THIGH MUSCLES

1 | Sartorius | The "Tailor's Muscle"

Quadriceps Femoris: Quadriceps, Extensor Femoris

COMPOSITE OF 4 DISTINCT MUSCLES (HEADS)

2 | Rectus Femoris

3 | Vastus Lateralis | Vastus Externus (Synergistic Action of Extension with 4, 5)

4 | Vastus Medialis | Vastus Internus (Synergistic Action of Extension with 3, 5)

5 | Vastus Intermedius | Crureus (Synergistic Action of Extension with 3, 4)

ANTERIOR THIGH MUSCLES THAT ACT ON THE THIGH (AND HIP JOINT)
ABDUCTOR OF THIGH (AND HIP)

1 | Sartorius

LATERAL ROTATOR OF THIGH (AND HIP)

1 | Sartorius

FLEXORS OF THIGH (AND HIP)

1 | Sartorius (weak)

2 | Rectus Femoris

ANTERIOR THIGH MUSCLES THAT ACT ON THE KNEE JOINT (AND LOWER LEG)
FLEXOR OF THE KNEE JOINT/MEDIAL ROTATOR OF LEG

1 | Sartorius

EXTENSORS OF LEG AT KNEE (KNEE JOINT)

2–5 | Quadriceps Femoris | All 4 Heads

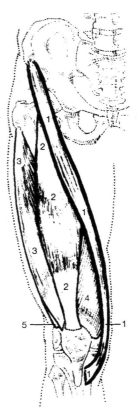

Anterior View
Quadriceps Femoris = 2, 3, 4 and 5

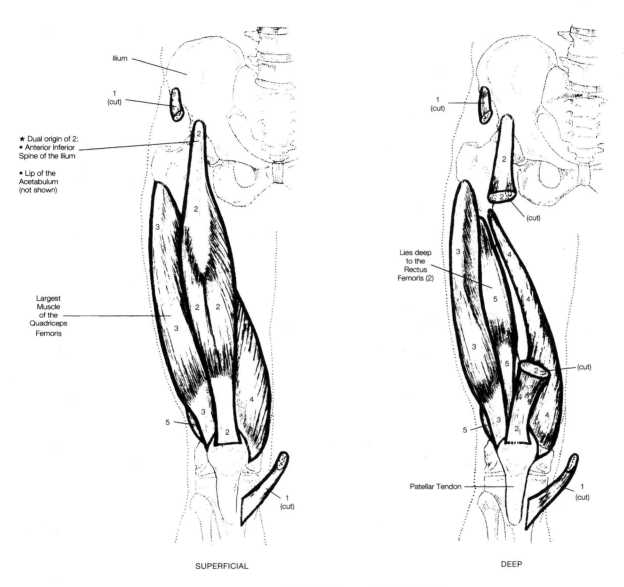

Ilium

1
(cut)

★ Dual origin of 2:
• Anterior Inferior
Spine of the Ilium

• Lip of the
Acetabulum
(not shown)

Largest
Muscle
of the
Quadriceps
Femoris

1
(cut)

1
(cut)

(cut)

Lies deep
to the
Rectus
Femoris (2)

(cut)

Patellar Tendon

1
(cut)

SUPERFICIAL

DEEP

Anterior Views of Right Thigh

COLOR GUIDELINES:
1 = flesh; 2 = yellow;
3 = green; 4 = blue; 5 = purple.

The CRURAL LEG MUSCLES move the foot. They include the ANTERIOR LEG MUSCLES, which cross several joints—the EXTENSORS OF THE TOES and the DORSIFLEXORS OF THE ANKLE.

The TIBIALIS ANTERIOR is an INVERTOR OF THE FOOT as well as an ankle dorsiflexor, and it is found in the ANTEROLATERAL PORTION of the leg. The tibialis anterior has its origin at the upper tibia (under the lateral condyle), interosseus membrane, and intermuscular septum. It inserts at the INTERNAL CUNEIFORM and the first METATARSAL.

The EXTENSOR DIGITORUM LONGUS originates under the lateral condyle of the tibia and body of the fibula. It inserts at the second to the fifth phalanges of the toes.

The EXTENSOR HALLUCIS LONGUS originates in the front of the fibula shaft and interosseus membrane; its insertion is at the terminal phalanx of the great toe.

The PERONEUS TERTIUS is actually an extension of the extensor digitorum longus. Its eversion qualities are suspect.

The LATERAL (PERONEAL) LEG MUSCLES have their tendons passing laterally and under the foot. They are EVERTORS OF THE FEET.

The PERONEUS LONGUS originates at the head of the fibula, and the lateral condyle of the tibia and inserts by tendon under the internal cuneiform and first metatarsal.

The PERONEUS BREVIS originates at the midportion of the front of the fibula shaft and inserts at the fifth metatarsal.

There is no muscle attachment along the ANTEROMEDIAL ASPECT of the leg along the TIBIAL SHAFT.

The anterior and lateral muscles of the lower leg that act on the ankle joint, foot, and toes include all anterior crural muscles (dorsiflexion of the ankle and foot); the TIBIALIS ANTERIOR (inversion of the foot); all peroneal muscles—tertius, longus, and brevis (plantar flexion of the foot); the PERONEUS LONGUS (adduction of the foot); the EXTENSOR DIGITORUM LONGUS and the EXTENSOR HALLUCIS LONGUS (extension of the toes).

MUSCLES OF THE LEG: CRURAL MUSCLES

ANTERIOR CRURAL MUSCLES: DORSIFLEXORS

1 | Tibialis Anterior | Tibalis Anticus

2 | Extensor Digitorum Longus |

3 | Extensor Hallucis Longus | Great Toe

4 | Peroneus Tertius |

LATERAL CRURAL MUSCLES: EVERTORS

5 | Peroneus Longus | (Synergistic Action with 6)

6 | Peroneus Brevis | (Synergistic Action with 5)

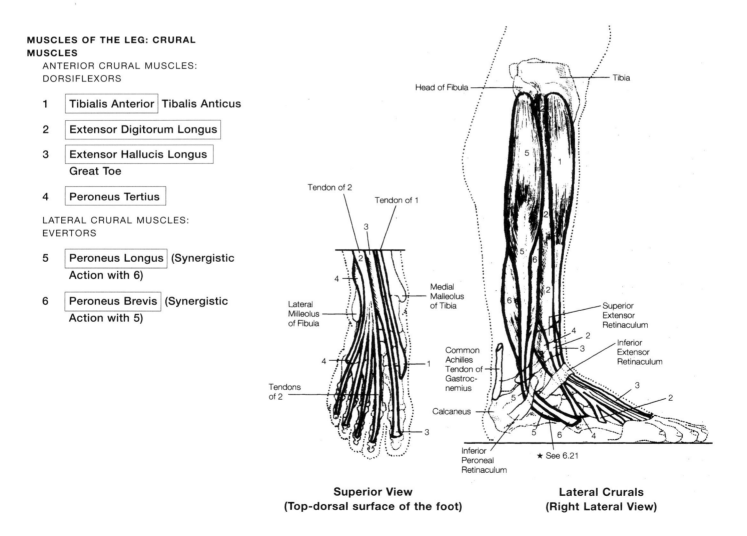

Superior View
(Top-dorsal surface of the foot)

Lateral Crurals
(Right Lateral View)

MUSCLES OF THE LEG:
(ANTERIOR AND LATERAL): RIGHT CRURALS

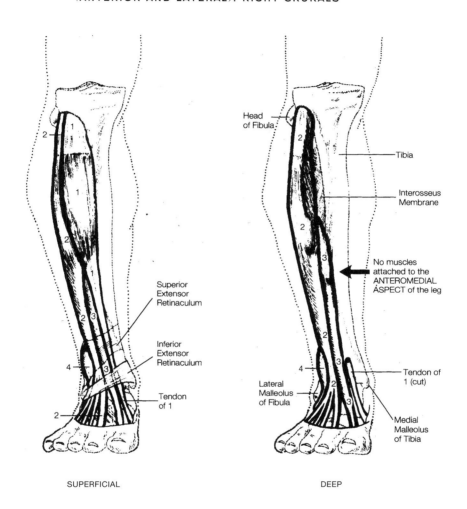

SUPERFICIAL

DEEP

Head of Fibula

Tibia

Interosseus Membrane

No muscles attached to the ANTEROMEDIAL ASPECT of the leg

Superior Extensor Retinaculum

Inferior Extensor Retinaculum

Tendon of 1

Lateral Malleolus of Fibula

Tendon of 1 (cut)

Medial Malleolus of Tibia

COLORING NOTES:
Dorsiflexors = warm colors;
Evertors = cool colors.

There are seven POSTERIOR CRURAL LEG MUSCLES. The three superficial muscles, known as the TRICEPS SURAE, are the GASTROCNEMIUS, SOLEUS, and PLANTARIS. The gastrocnemius covers the major portion of the calf muscles, and has two distinct heads and origins. The origin of the LATERAL HEAD is at the posterior surface of the lateral condyle of the femur. The origin of the MEDIAL HEAD is at the posterior surface of the medial condyle of the femur.

The SOLEUS has its origin at the head and upper shaft of the fibula and the oblique line of the tibia.

A TRICIPITAL MUSCLE is formed from the two heads of the gastrocnemius and the soleus. It inserts at the common TENDON OF ACHILLES, or TENDON CALCANEUS, which inserts on the CALCANEUS BONE of the foot (os calcis). It can lift the entire body onto the heads of the metatarsal bones in a special movement (PLANTAR FLEXION). This action involves only one joint, the ankle.

The PLANTARIS has its origin at the external supracondyloid ridge of the femur. Insertion is at the inner border of Achilles tendon.

Note how the tendons of the FLEXOR HALLUCIS LONGUS and the FLEXOR DIGITORUM LONGUS pass under the sole of the foot to insert on the terminal phalanges of the toes.

MUSCLES OF THE LEG: CRURAL MUSCLES
POSTERIOR CRURAL MUSCLES
SUPERFICIAL GROUP: TRICEPS SURAE

1 | Gastrocnemius | 2 heads (lateral & medial)

2 | Soleus |

3 | Plantaris |

DEEP GROUP:

4 | Popliteus |

5 | Flexor Hallucis Longus |

6 | Flexor Digitorum Longus |

7 | Tibialis Posterior | Tibialis Posticus (Also supports arches)

POSTERIOR MUSCLES OF THE LEG THAT ACT ON THE LEG:
FLEXOR OF THE KNEE JOINT:

1 | Gastrocnemius |

FLEXORS OF THE LEG:

1 | Gastrocnemius |

3 | Plantaris |

MEDIAL ROATATOR OF LEG

4 | Popliteus |

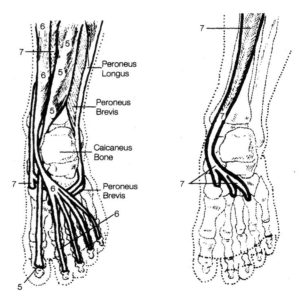

Peroneus Longus

Peroneus Brevis

Calcaneus Bone

Peroneus Brevis

Inferior Views

COLORING NOTES:
Superficial group = warm colors;
Deep group = cool colors.

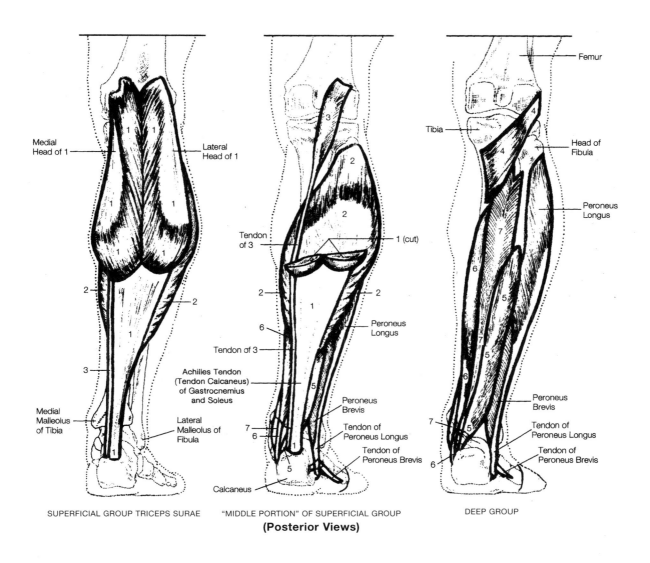

SUPERFICIAL GROUP TRICEPS SURAE   "MIDDLE PORTION" OF SUPERFICIAL GROUP   DEEP GROUP

(Posterior Views)

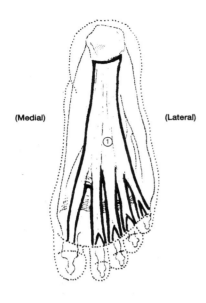

**MOST SUPERFICIAL
BOTTOM OF THE SOLE
(Right Foot)**

SUPERFICIAL SHEATH OF THE SOLE

1	Plantar Aponeurosis

MUSCLES OF THE SOLE

Plantar (Bottom) Surface: Sole
of the Foot

SUPERFICIAL LAYER

2	Abductor Hallucis
3	Abductor Digiti Minimi
4	Flexor Digitorum Brevis

MIDDLE LAYER

5	Quadratus Plantae	Pronator

Pedis, Flexor Accessorius

6	Lumbricales	(4)

DEEP LAYER

7	Flexor Hallucis Brevis
9	Flexor Digiti Minimi Brevis
11	Plantar Interossei
12	Dorsal Interrosei

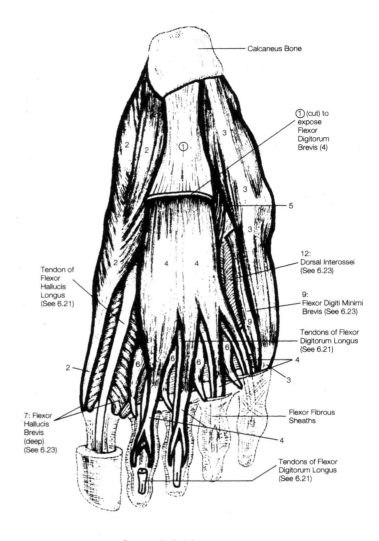

Superficial Layers

SUPERFICIAL AND MIDDLE PLANTAR LAYERS

THE MUSCLES OF THE FOOT are arbitrarily divided into three PLANTAR (BOTTOM) LAYERS (superficial, middle, and deep) and the DORSAL SURFACE MUSCLES. Here we will concentrate on the outermost plantar layers of the sole. The plantar surface is the bottom surface of the foot, as opposed to the dorsal surface (top of the foot) we readily see with our eyes.

The muscles of the foot are similar in name and number to those of the hand (with the exception of one short intrinsic muscle, the EXTENSOR DIGITORUM BREVIS). Functions of the foot and hand are different, however: the foot provides support and bears heavy body weight, while the hand is a complex instrument for precise movements, such as grasping.

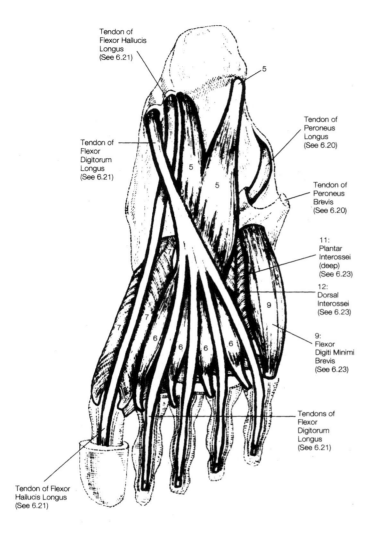

Tendon of
Flexor Hallucis
Longus
(See 6.21)

5

Tendon of
Flexor
Digitorum
Longus
(See 6.21)

Tendon of
Peroneus
Longus
(See 6.20)

Tendon of
Peroneus
Brevis
(See 6.20)

11:
Plantar
Interossei
(deep)
(See 6.23)

12:
Dorsal
Interossei
(See 6.23)

9:
Flexor
Digiti Minimi
Brevis
(See 6.23)

Tendons of
Flexor
Digitorum
Longus
(See 6.21)

Tendon of Flexor
Hallucis Longus
(See 6.21)

Middle Layer
(1, 2, 3, and 4 removed)

COLOR GUIDELINES:
1 = yellow. Superficial layer = oranges and reds;
middle layer = pinks and purples;
tendons = light or neutral colors

Topographically, the foot muscles are arranged into three layers, although they are not distinct (even in dissection), and different authorities have differing opinions. Action of the foot muscles provides movement of the toes and support for arches of the foot (by contracting).

The four LUMBRICALES MUSCLES arise from the TENDONS of the FLEXOR DIGITORUM LONGUS (see pages 118–9).

To increase overall understanding refer to pages 116–7 and review the tendons of peroneus longus and brevis. Then refer to pages 118–9 and note the tendons of flexor hallicus longus and flexor digitorum longus. In the next pages, we will discuss the deep layer plantar muscles in more detail.

MUSCLES OF THE FOOT
PLANTAR (BOTTOM) SURFACE

Sole of the Foot
DEEP LAYERS

7 | Flexor Hallucis Brevis |

8 | Adductor Hallucis | (Transversus or Obliquus) Hallucis

9 | Flexor Digiti Minimi Brevis |

10 | Opponens Digiti Minimi |
Opponens Digiti Quinti

11 | Plantar Interossei | (3)
Interosseus Plantaris

DORSAL (TOP) SURFACE

12 | Dorsal Interossei | (4)
Interosseus Dorsalis Pedis

13 | Extensor Digitorum Brevis |

14 | Extensor Hallucis Brevis |

FASCIA RETINACULUM

15 | Superior Extensor Retinaculum |

16 | Inferior Extensor Retinaculum |

MUSCLES OF THE FOOT THAT ACT ON THE FOOT AND TOES
FLEXORS OF TOES

2 | Flexor Hallucis Brevis |
(Great Toe)

3 | Adductor Hallucis | (Great Toe)

8 | Abductor Hallucis | (Great Toe)

4 | Flexor Digiti Minimi Brevis |
(Little Toe)

5 | Opponens Digiti Minimi |
(Little Toe)

11, 12 | Interossei Plantar and Dorsal |

10 | Flexor Digitorum Brevis |

6 | Quadratus Plantae |

FLEXORS OF METATARSOPHALANGEAL JOINTS

7 | Lumbricales |

Adductor of Toes

3 | Adductor Hallucis | (Great Toe)

5 | Opponens Digiti Minimi | (Little Toe)

Abductor of Toes

8 | Abductor Hallucis | (Great Toe)

9 | Abductor Digiti Minimi | (Little Toe)

11, 12 | Interossei Plantar and Dorsal |

Extensors of Toes

13 | Extensor Digitorum Brevis |

14 | Extensor Hallucis Brevis |

7 | Lumbricales | (Distal Phalanges)

DEEP PLANTAR (BOTTOM) LAYERS OF THE RIGHT FOOT (SOLE)

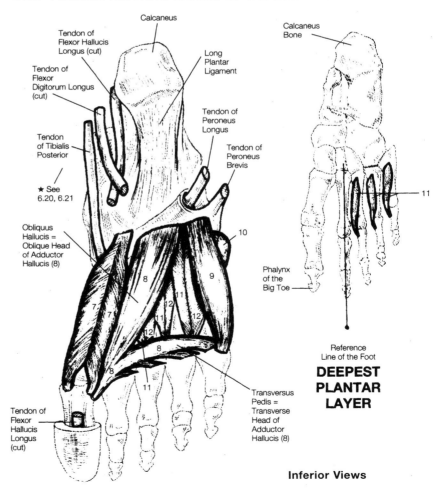

Calcaneus

Tendon of Flexor Hallucis Longus (cut)

Tendon of Flexor Digitorum Longus (cut)

Tendon of Tibialis Posterior

★ See 6.20, 6.21

Obliquus Hallucis = Oblique Head of Adductor Hallucis (8)

Tendon of Flexor Hallucis Longus (cut)

Long Plantar Ligament

Tendon of Peroneus Longus

Tendon of Peroneus Brevis

Transversus Pedis = Transverse Head of Adductor Hallucis (8)

Calcaneus Bone

Phalynx of the Big Toe

Reference Line of the Foot

DEEPEST PLANTAR LAYER

Inferior Views

In the previous section, we studied the SUPERFICIAL and MIDDLE LAYERS of the BOTTOM (OR SOLE) of the foot. Here, we look at the DEEP LAYER of the sole, directly underneath the muscles of the DORSAL (or TOP) surface of the foot, as well as those top dorsal muscles.

Note the SUPERIOR and INFERIOR EXTENSOR RETINACULUM, the thickenings of the deep fascia in the distal portion of the lower legs that hold tendons in position when muscles contract.

The superior extensor retinaculum crosses the extensor tendons of the foot and is attached to the lower portions of the TIBIA and FIBULA.

The inferior extensor retinaculum consists of two limbs. They have a common origin on the lateral surface of the CALCANEUS. The upper limb is attached to the medial malleolus and the lower limb curves around the instep; it is attached to the fascia of the ABDUCTOR HALLUCIS on the medial side of the foot.

The TRANSVERSUS PEDIS is the transverse head of ADDUCTOR HALLUCIS. The OBLIQUUS HALLUCIS is the oblique head of adductor hallucis. The EXTENSOR HALLUCIS BREVIS is that portion of EXTENSOR DIGITORUM BREVIS that inserts on the great toe. Shown here are the tendons of many of the anterior and lateral crural muscles seen on pages 116–7, and the tendons of the posterior crural muscles seen on pages 118–9.

DORSAL (TOP) SURFACE OF THE RIGHT FOOT

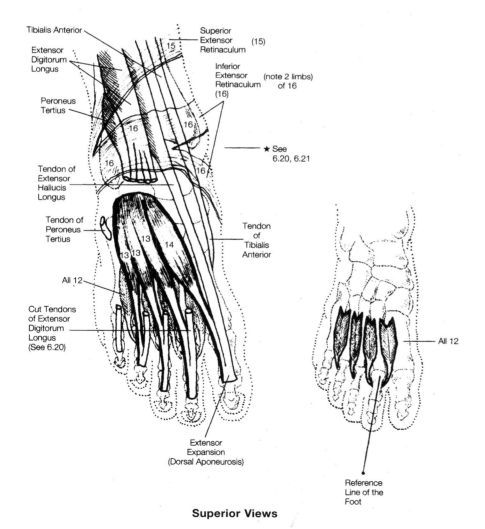

Tibialis Anterior

Extensor Digitorum Longus

Peroneus Tertius

Tendon of Extensor Hallucis Longus

Tendon of Peroneus Tertius

All 12

Cut Tendons of Extensor Digitorum Longus (See 6.20)

Superior Extensor Retinaculum (15)

Inferior Extensor Retinaculum (16) (note 2 limbs) of 16

★ See 6.20, 6.21

Tendon of Tibialis Anterior

All 12

Extensor Expansion (Dorsal Aponeurosis)

Reference Line of the Foot

Superior Views

COLORING NOTES:
Deep layer = blues and greens,
Dorsal surfaces = oranges and reds

CHAPTER 8: NERVOUS SYSTEM

CINING NOTES:
You may want to choose your color palette by comparing
colored illustrations from your main textbook.
Check each individual page for coloring suggestions,
but the general scheme followed is:
Gray matter = light gray; white matter and myelin = cream;
spinal cord = flesh-tones; brain = yellow.

The NERVOUS SYSTEM is concerned with the integration, control, regulation and coordination of all bodily functions through a vast communications network. It is comprised of billions of extremely delicate NERVE CELLS, interlaced with each other in an elaborate system that maintains the vital function of reception and RESPONSE TO STIMULI. Through EXCITABILITY, it is able to receive and bring about responses by which the body adjusts to changes from EXTERNAL and INTERNAL ENVIRONMENTS. These changes constitute STIMULI that initiate electrochemical NERVE IMPULSES in RECEPTORS or SENSE ORGANS. Through CONDUCTION, the nervous system is able to transmit these message-impulses to and from COORDINATING CENTERS.

The nerve impulses that are carried through the body by nerve tissue are *electrochemical* in the strictest sense: an amazing combined effect of chemical reactions and electrical activity pass from cell to cell and from nerve to organ, creating a complex series of processes that effectuate activity and allow the organism to survive. In an information age, we are apt to regard all such processes as electrical, as if the brain were a large computer and nerves were wiring. The reality is much more complex—miraculously so—as electrical and chemical processes interact to stimulate a response on the part of the organism that will enhance its survival in an ever-changing environment.

Two broad functions of the human nervous system are: the STIMULATION OF MOVEMENT (until stimulated by a NERVE IMPULSE, all skeletal muscle and most smooth muscle cells cannot contract, nor can exocrine glands—salivary, gastric, and sweat glands—release their secretions); and SHARING MAINTENANCE OF BODY HOMEOSTASIS by harmoniously interlocking with the endocrine system to coordinate activities of the other body systems. The nervous system can inhibit or stimulate the release of hormones from endocrine glands while the endocrine system can inhibit or stimulate the flow of nerve impulses.

The nervous system is divided into the CENTRAL NERVOUS SYSTEM (CNS) and the PERIPHERAL NERVOUS SYSTEM (PNS). The CNS consists of the BRAIN and SPINAL CORD, two organs that are surrounded by the bony structures of the skull and vertebral column, protected by a MENINGES MEMBRANE and suspended in CEREBROSPINAL FLUID. These areas form central regions of control for integration, interpretation, thought, and assimilation of experiences required in memory, learning, speech, language, interpretation of language, sensory awareness and movement, and intelligence and transmission of messages to and from the periphery. The PNS consists of all the other neural elements lying outside the CNS: twelve pairs of CRANIAL NERVES and their branches arising from the brain, thirty-one pairs of SPINAL NERVES and their branches arising from the spinal cord, their associated AUTONOMIC GANGLIA (peripheral nerve cell bodies), GANGLIONATED TRUNKS, NERVES and NERVE PLEXUSES, specialized sensory receptors within the sensory organs and specialized endings on muscles.

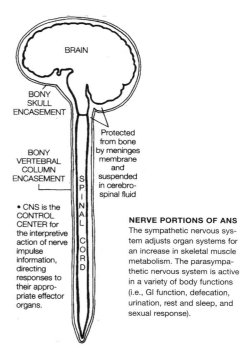

PERIPHERAL NERVOUS SYSTEM (PNS)
All the nerve processes that connect the brain and spinal cord with receptors, muscles and glands make up the PNS.

AFFERENT SYSTEM OF THE PNS
Sensory nerve receptors in the periphery of the body pick up informational changes (stimuli) from the external and internal environment. The stimuli are converted into nerve impulses that travel along nerve processes conveying the information to the CNS.

EFFERENT SYSTEM OF THE PNS
Motor nerve cells convey interpreted information from the CNS along their nerve processes to muscles and glands in the body periphery.

SOMATIC NERVOUS SYSTEM (SNS)
The efferent motor nerve cells conduct impulses from the cns to the skeletal muscle tissue. The SNS is under conscious control and voluntary.

Anterior View

AUTONOMIC NERVOUS SYSTEM (ANS)
The efferent motor nerve cells conduct impulses from the CNS to smooth muscle tissue (of viscera and blood vessels), cardiac muscle tissue (of the heart), and glands (salivary, gastric, sweat). The ANS operates without conscious control below and conscious level. It is involuntary.

BRAIN

BONY SKULL ENCASEMENT

BONY VERTEBRAL COLUMN ENCASEMENT

Protected from bone by meninges membrane and suspended in cerebro-spinal fluid

SPINAL CORD

• CNS is the CONTROL CENTER for the interpretive action of nerve impulse information, directing responses to their appropriate effector organs.

NERVE PORTIONS OF ANS
The sympathetic nervous system adjusts organ systems for an increase in skeletal muscle metabolism. The parasympathetic nervous system is active in a variety of body functions (i.e., GI function, defecation, urination, rest and sleep, and sexual response).

Functionally speaking, the PNS can be subdivided into an AFFERENT SYSTEM and an EFFERENT SYSTEM. The afferent system consists of nerve cells called AFFERENT (SENSORY) NEURONS that convey impulses toward the CNS. These nerve cells are the first cells to pick up informational changes (stimuli) from the external and internal environment. The efferent system consists of nerve cells called EFFERENT (MOTOR) NEURONS that convey information from the CNS to muscles and glands.

Note the functional subdivision of the EFFERENT SYSTEM of the PNS into two systems: a SOMATIC NERVE SYSTEM (SNS) conducting impulses from the CNS to SKELETAL MUSCLE TISSUE and an AUTONOMIC NERVOUS SYSTEM (ANS) conducting impulses from the CNS to SMOOTH MUSCLE TISSUE, CARDIAC MUSCLE TISSUE, and glands. (The controlling centers of the SNS and the ANS, which are the nerve cell bodies, are considered part of and within the CNS.)

The peripheral nerve processes of the ANS cell bodies that travel outside the CNS are subdivided into the SYMPATHETIC (STIMULATING) and PARASYMPATHETIC (INHIBITING) SYSTEMS. These complementary divisions carry opposite, or antagonistic, messages to the viscera of the body systems, regulating their activities. Their impulses are adapted and balanced by the CNS BRAIN CENTERS to integrate the actions of the body systems toward the vital maintenance of body homeostasis.

Despite the elaborate complexity and high specialization of the nervous system, it contains only two principal types of cells: NEURONS and NEUROGLIA.

Neuroglia offer structural and functional support and aid for the neurons, are five times more abundant than neurons and account for half of all brain cells. Additionally, they have limited mitotic activity and are neither RECEPTIVE nor CONDUCTING.

Neurons are the structural and functional units of the nervous system. They are specialized in EXCITABILITY (response to physical and chemical stimuli), CONDUCTION (impulse conduction from one part of the body to another) and INTEGRATION (thinking, storage of memory, controlling muscle activity, regulating organs and glands). The number of neurons in a human being are established shortly after birth. There-after, they are incapable of mitosis, dividing only in rare cases.

The NEURON CELL BODY controls metabolism of the neuron. It contains NEUROFIBRILS, which are delicate thread-like strands of protein, and NISSL BODIES, which are specialized layers of ROUGH ENDOPLASMIC RETICULUM involved in protein synthesis of neurofibrils and MICROTUBULES, and the metabolism and transportation of cellular material.

The cell body has two types of CYTOPLASMIC EXTENSIONS: DENDRITES and the AXON. Dendrites are short ramified branched processes providing a large surface area (DENDRITIC ZONE) to receive a stimulus and conduct impulses to the cell body. The AXON is a single, typically long cylindrical process that conducts nerve impulses away from the cell body to another neuron or tissue.

The AXON HILLOCK is a conical tapering region where the axon originates from the cell body. It contains no nissl bodies and is not enclosed by any sheath coverings.

THE TWO PRINCIPAL KINDS OF CELLS IN THE NERVOUS SYSTEM:
NEURONS

THE STRUCTURAL AND FUNCTIONAL UNITS NEURONS

PRINCIPAL COMPONENTS OF A NEURON

1. **1 Cell body** (perikaryon) – Soma
2. **Ramified dendrites** (Afferent processes)
3. **1 Axon (Axis Cylinder)** Efferent Process

COMPONENTS OF THE CELL BODY

a. **Nucleus**
b. **Nucleolus**
c. **Cytoplasm**
c1. **Nissl bodies**
c2. **Neurofibrils**

Extend into the axon and dendrites for support

COMPONENTS OF THE AXON

4. **Axon hillock**
5. **Axoplasm**
6. **Axolemma**
7. **Telodendria**
8. **Synaptic knobs** (end feet)

AXON SHEATH COVERINGS – PNS NEUROLOGICAL COVERING: SCHWANN

9. **Sheath of Schwann/outer neurilemma and cytoplasm**
   Only in fibers of PNS
10. **Nucleus of Schwann cell**
11. **Inner myelin sheath**
    Gives color to white matter
12. **Nodes of Ranvier**
    Gaps between segments of myelin sheath

NEUROGLIA

THE SUPPORTING AND PROTECTING UNITS: NEUROGLIA "NERVE GLUE" PNS: NEUROGLIA

SCHWANN CELLS
- Wrap around PERIPHERAL AXONS of the PNS, forming an inner MYELIN SHEATH and an outer NEURILEMMA SHEATH. These Schwann cell coverings plus the axon form the PERIPHERAL NERVE FIBER.
- Flattened cells arranged in series

SATELLITE CELLS
- Also called CAPSULE CELLS, form supportive capsulelike coverings around the neuron cell bodies within the PERIPHERAL GANGLIA of the PNS. May transfer nutrients to neurons.
- Small, flattened cells

■ CNS NEUROGLIA:

in GRAY MATTER / in WHITE MATTER

ASTROCYTES
- Star-shaped with numerous processes, which surround most of the outer surface of BRAIN CAPILLARIES. Contributes to BLOOD-BRAIN BARRIER. Provide structural support by attaching neurons to their blood vessels.
- Twine around nerve cells of CNS in transversely oriented supporting network Protoplasmic types are in CNS GRAY MATTER Fibrous types are in CNS WHITE MATTER

OLIGODENDROCYTES
- Resemble ASTROCYTES with fewer and shorter processes. Look like "plump pillows."
- Produce the MYELIN SHEATHS around axons of CNS fibers
- Form semi-rigid longitudinal supporting rows between neurons of BRAIN and SPINAL CORD.
- Guide development of CNS neurons

EPENDYMA
- Wedge-shaped, ciliated columnar cells that line the cavities (VENTRICLES and CENTRAL CANAL) of CNS. Produce small amounts of CEREBROSPINAL FLUID and help circulate it with ciliary motion.

MICROGLIA
- Also called MESOGLIA, the only NEUROGLIA of MESODERMAL (MESENCHYMAL) origin. (All others are of ECTODERMAL ORIGIN as is all neuron tissue.) Invade the CNS and respond to injury or infection. They become ameboid and phagocytic, engulfing and destroying microbes and cellular debris.
- Small, flattened, studded with spinelike GEMMULES.

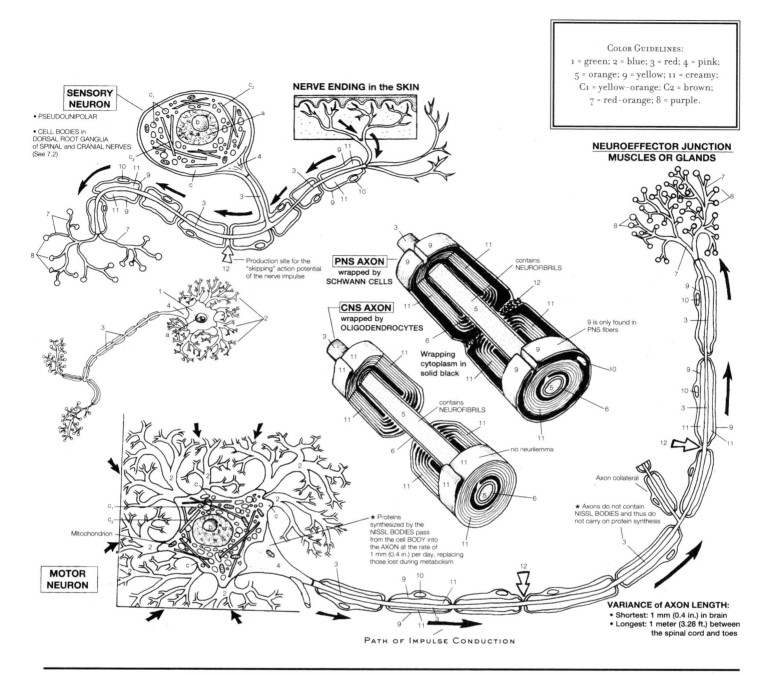

COLOR GUIDELINES:
1 = green; 2 = blue; 3 = red; 4 = pink;
5 = orange; 9 = yellow; 11 = creamy;
C1 = yellow-orange; C2 = brown;
7 = red-orange; 8 = purple.

SENSORY NEURON

• PSEUDOUNIPOLAR

• CELL BODIES in DORSAL ROOT GANGLIA of SPINAL and CRANIAL NERVES (See 7.2)

Production site for the "skipping" action potential of the nerve impulse

NERVE ENDING in the SKIN

NEUROEFFECTOR JUNCTION MUSCLES OR GLANDS

PNS AXON wrapped by SCHWANN CELLS

CNS AXON wrapped by OLIGODENDROCYTES

contains NEUROFIBRILS

Wrapping cytoplasm in solid black

9 is only found in PNS fibers

contains NEUROFIBRILS

no neurilemma

★ Proteins synthesized by the NISSL BODIES pass from the cell BODY into the AXON at the rate of 1 mm (0.4 in.) per day, replacing those lost during metabolism

Axon collateral

★ Axons do not contain NISSL BODIES and thus do not carry on protein synthesis

Mitochondrion

MOTOR NEURON

VARIANCE of AXON LENGTH:
• Shortest: 1 mm (0.4 in.) in brain
• Longest: 1 meter (3.28 ft.) between the spinal cord and toes

PATH OF IMPULSE CONDUCTION

AXOPLASM (cytoplasm of the axon) contains many MITOCHONDRIA, microtubules and neurofibrils. The newly synthesized proteins pass into the axoplasm from the cell body at the rate of 1 mm (0.4 inch) a day. They replace proteins lost during metabolism, and are involved in neuron growth and regeneration of severed portions of peripheral nerve fibers.

THE AXOLEMMA is the plasma membrane surrounding the axoplasm (continuation of cell membrane of the cell body).TELODENDRIA are fine terminal branching filaments of the axon. SYNAPTIC KNOBS are bulblike distal endings of telodendria, important in conduction of the nerve impulse at the synapse. The NERVE FIBER refers to an axon and its sheath coverings.

Large PERIPHERAL AXONS are surrounded by the myelin sheath, a multilayered, white, phospholipid segmented covering. The myelin sheath insulates the axon and increases the speed of the nerve impulse conduction. Axons covered by a myelin sheath are called MYELINATED; uncovered, UNMYELINATED.

In the PNS, the myelin sheath is formed by flattened neuroglial SCHWANN cells, which encircle and wind around the axon. Much like toothpaste is rolled from the bottom of a tube, the cytoplasm and nucleus of the Schwann cell are squeezed into another layer called the NEURILEMMA (SHEATH OF SCHWANN). The inner, fatty myelin sheath consists of many wrapped layers of the Schwann cell membrane.

Unmyelinated axons are enclosed by Schwann cells forming a living neurilemma sheath, but without the multiple wrappings characteristic of the myelin sheath. Neurilemma aids the regeneration of peripheral nerve fibers.

Neurons may be classified according to structure and function. Structural classification of neurons is based on the number of POLES or processes associated with each cell body. UNIPOLAR means one process leaves the cell body. These neurons originate embryologically as BIPOLAR NEURONS. The axon and dendrite fuse to form a single process, thus the term PSEUDOUNIPOLAR. They are all sensory (afferent) in function and found in the PNS; they transmit sensory information to the CNS (e.g.: posterior sensory root ganglia of the spinal and cranial nerves).

Bipolar neurons have two processes at either end of an elongated cell body, with one process highly modified to respond to particular forms of energy. They are all sensory (afferent) in function and are found in organs of special sense (exclusively in the retina of an eye, auditory nerve in the inner ear, taste buds, and olfactory epithelium).

MULTIPOLAR NEURONS have many cell processes that extend from the cell body (many dendrites, one axon). They are mostly motor (EFFERENT) or ASSOCIATION in function and are concerned

STRUCTURAL CLASSIFICATION BY NUMBER OF PROCESSES:

a   Unipolar or Pseudounipolar

b   Bipolar

c   Multipolar

TYPES OF PROCESSES
(2 AND 4 ARE IN A AND B)

2   Peripheral Process functions like a dendrite (afferent impulses to the cell body), structured like an axon

4   Central Process function and structure of an axon (efferent impulses from cell body to CNS)

d   Dendrite

8   Axon

FUNCTIONAL CLASSIFICATION BY DIRECTION OF IMPULSE CONDUCTION: NEURONS OF THE PERIPHERAL NERVOUS SYSTEM (PNS)

THE AFFERENT SYSTEM OF THE PNS: UNIPOLAR NEURONS
SENSORY NEURON (AFFERENT): (BOTH SOMATIC AFFERENT AND VISCERAL AFFERENT)

1   Receptor

2   Peripheral Process (Dendritic "Axon")

3   Cell Body in Dorsal Root Ganglia of Cranial or Spinal Nerves

4   Central Process (Axon)

5   Synapses /Junctions between neurons

6   Central Nervous System (CNS)

THE EFFERENT SYSTEM OF THE PNS: MULTIPOLAR NEURONS
SOMATIC NERVOUS SYSTEM (SNS): SOMATIC MOTOR NEURON (SOMATIC EFFERENT)

7   Cell Body in CNS

8   Axon

9   Motor End Plate (Synapse)

10   Skeletal Muscle (Effector)

AUTONOMIC NERVOUS SYSTEM (ANS)
AUTONOMIC MOTOR NEURON (VISCERAL EFFERENT)

11   Preganglionic Neuron (Cell Body in CNS)

5   Synapse

12   Postganglionic Neuron (Cell Body outside CNS)

13   Smooth Muscle, Cardiac Muscle, or Glands

NEURONS OF THE CENTRAL NERVOUS SYSTEM (CNS)
CONNECTING INTERNEURONS

14   Association Neuron/ Internuncial Neuron

Association neurons are found mostly in the CNS, making up the bulk of the neurons of the brain and spinal cord. Association neurons in the spinal cord can be directly related to incoming sensory (afferent) impulses and others to outgoing (efferent) impulses. Association neurons linked with higher brain centers integrate sensory input to effect an appropriate motor response (output). Cell bodies of these neurons make up the majority of the GRAY MATTER of the CNS, and their myelinated axons make up most of the WHITE MATTER nerve tracts of the CNS.

EXAMPLES OF ASSOCIATION NEURON LINKUPS

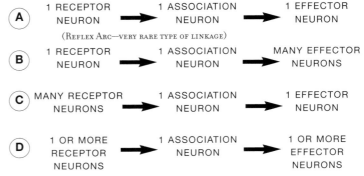

A   1 RECEPTOR NEURON → 1 ASSOCIATION NEURON → 1 EFFECTOR NEURON
(REFLEX ARC—VERY RARE TYPE OF LINKAGE)

B   1 RECEPTOR NEURON → 1 ASSOCIATION NEURON → MANY EFFECTOR NEURONS

C   MANY RECEPTOR NEURONS → 1 ASSOCIATION NEURON → 1 EFFECTOR NEURON

D   1 OR MORE RECEPTOR NEURONS → 1 ASSOCIATION NEURON → 1 OR MORE EFFECTOR NEURONS

Other association neurons synapsing with the effector neuron(s) allow complex linkups with CNS centers at higher and lower levels of the brain and spinal cord. These functional linkups make the establishment of CONDITIONED INVOLUNTARY REFLEXES possible (e.g., the swallowing reflex). These reflexes probably form the basis of all training. Thus, a fine line exists between the end of involuntary reflex behavior and the beginning of voluntary behavior.

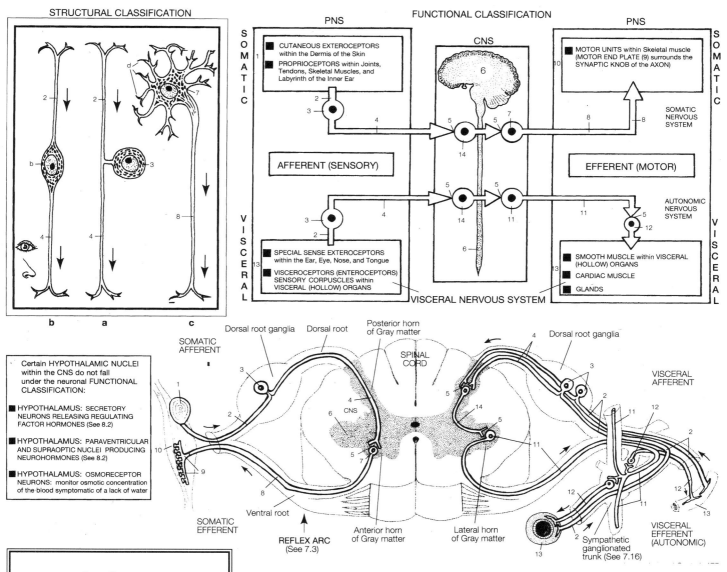

STRUCTURAL CLASSIFICATION

FUNCTIONAL CLASSIFICATION

SOMATIC

PNS

- CUTANEOUS EXTEROCEPTORS within the Dermis of the Skin
- PROPRIOCEPTORS within Joints, Tendons, Skeletal Muscles, and Labyrinth of the Inner Ear

CNS

6

PNS

- MOTOR UNITS within Skeletal muscle (MOTOR END PLATE (9) surrounds the SYNAPTIC KNOB of the AXON)

SOMATIC NERVOUS SYSTEM

SOMATIC

AFFERENT (SENSORY)

EFFERENT (MOTOR)

VISCERAL

- SPECIAL SENSE EXTEROCEPTORS within the Ear, Eye, Nose, and Tongue
- VISCEROCEPTORS (ENTEROCEPTORS) SENSORY CORPUSCLES within VISCERAL (HOLLOW) ORGANS

AUTONOMIC NERVOUS SYSTEM

- SMOOTH MUSCLE within VISCERAL (HOLLOW) ORGANS
- CARDIAC MUSCLE
- GLANDS

VISCERAL NERVOUS SYSTEM

VISCERAL

b   a   **c**

Certain HYPOTHALAMIC NUCLEI within the CNS do not fall under the neuronal FUNCTIONAL CLASSIFICATION:

- HYPOTHALAMUS: SECRETORY NEURONS RELEASING REGULATING FACTOR HORMONES (See 8.2)

- HYPOTHALAMUS: PARAVENTRICULAR AND SUPRAOPTIC NUCLEI PRODUCING NEUROHORMONES (See 8.2)

- HYPOTHALAMUS: OSMORECEPTOR NEURONS: monitor osmotic concentration of the blood symptomatic of a lack of water

SOMATIC AFFERENT

Dorsal root ganglia   Dorsal root   Posterior horn of Gray matter

SPINAL CORD

CNS

Dorsal root ganglia

VISCERAL AFFERENT

Ventral root

SOMATIC EFFERENT

REFLEX ARC (See 7.3)

Anterior horn of Gray matter

Lateral horn of Gray matter

Sympathetic ganglionated trunk (See 7.16)

VISCERAL EFFERENT (AUTONOMIC)

COLOR GUIDELINES:
d = blue; 2 = yellow-orange; 3 = green;
4 = red-orange; 5 = gray dots; 6 = yellow;
7 = yellow-green; 8 = red; 9 = light brown;
10 = flesh; 11 = pink; 12 = light blue.

with processing or transmitting motor impulses that lead to muscular contraction or gland secretion.

The functional classification of neurons is based on three main modes of direction of impulse conduction. Sensory (afferent) neurons transmit impulses from the body periphery (skin, sense organs, muscles and viscera) to the CNS (brain and spinal cord); they are unipolar.

Motor (efferent) neurons convey impulses from the CNS to body periphery

(to effector organs, muscles (skeletal or smooth) and glands; they are multipolar.

Association (INTERNUNCIAL) neurons form a network of interconnecting neurons carrying impulses from sensory neurons to motor neurons. They are located in the CNS (brain and spinal cord).

The processes (nerve fibers) of afferent and efferent neurons are arranged into bundles called ganglia, which lie outside the CNS, belonging to the PNS. Neurons and their nerve fibers within nerves may be grouped according to a general scheme.

Somatic afferent impulses travel from sensory receptors in the skin, sense organs, joints and skeletal muscle conducted to CNS for interpretation (unipolar).

VISCERAL AFFERENT impulses are from sensory receptors in structures with hollow cavities (visceral organs and blood vessels). and are conducted to the CNS for interpretation (unipolar).

SOMATIC EFFERENT impulses from the CNS are conducted to motor end plates or skeletal muscle, causing contraction of skeletal muscle (effector organ) (multipolar).

VISCERAL (AUTONOMIC) EFFERENT, which belong to the ANS, are impulses from the CNS conducted to smooth and cardiac muscle, causing contraction, and to glands, causing secretion. Two neurons are involved through GANGLIONATED TRUNKS outside the CNS.

Simply put, a NERVE IMPULSE is a wave of electrical negativity that travels along the surface of the membrane of a neuron, reversing the electrical state (positive charge outside-to-inside, and negative charge inside-to-outside) thereby depolarizing the membrane. The ability of a neuron to respond to a stimulus and convert it into an impulse is known as IRRITABILITY (or EXCITABILITY). Neurons have CONDUCTIVITY, or the ability to transmit an impulse to another neuron or to another tissue (a muscle or a gland).

The junction where one neuron connects to another is a SYNAPSE. The PRESYNAPTIC NEURON carries an impulse to the synapse; the POST-SYNAPTIC NEURON carries an impulse away from the synapse. There is no direct PROTOPLASMIC UNION between connecting neurons at the synapse (i.e. no physical contact). The two parts of the synapse are separated by a space called the SYNAPTIC CLEFT. Synapses may occur between the AXON TERMINATIONS of the presynaptic neuron (TELO-DENDRIA) and three possible areas of the postsynaptic neuron: the AXON (AXO-AXONIC SYNAPSE), the CELL BODY (AXO-SOMATIC SYNAPSE) and the DENDRITES (AXO-DENDRITIC SYNAPSE).

Impulse conduction at a synapse is always one-way (from the presynaptic neuron to the postsynaptic neuron). Impulses cannot back up into another presynaptic neuron. This prevents impulse conduction along unsuitable and incorrect pathways ("short-circuits"). One neuron usually connects with many other neurons widely scattered throughout the CNS. Thus, intricate complex pathways are built up in the CNS for the integration, coordination, association and modification of incoming information and memory storage into desired motor responses (muscle contraction). There are approximately 1×10^{13} synapses in the CNS, and some neurons are known to have over five thousand synapses per nerve cells.

A NEUROEFFECTOR JUNCTION is when an impulse is transmitted between a neuron and muscle cell (fiber). A NEUROGLANDULAR JUNC-TION is when an impulse is transmitted between a neuron and a glandular cell. Impulse conduction across a synapse or a neuroeffector junction is dependent on chemical activity by a NEUROTRANSMITTER SUBSTANCE (produced in the neuron by amino acids, energized for rapid production by mitochondria). After production and transportation to the synaptic knob, it is stored in thousands of tiny membrane-enclosed sacs called SYNAPTIC VESICLES. An approaching nerve impulse causes the synaptic vesicles to move and bind to the PRESYNAPTIC MEM-BRANE, releasing the neurotransmitter, which diffuses across the synaptic cleft and binds to a specific receptor on the POSTSYNAPTIC MEMBRANE. This binding forms the stimulus to cause depolarization of the postsynaptic membrane, and the impulse is conducted to the next neuron, muscle fiber, or glandular cell. The neurotransmitter substance is rapidly renewed in the presynaptic neuron.

Reflex Arc: Involuntary Functional Unit of the Nervous System

The neuron is the anatomical or structural unit of the nervous system. The nervous reflex is the physiological or functional unit. A reflex or REFLEX ACTION is an involuntary action, automatic and unconscious, caused by the stimulation of a SENSORY (AFFERENT) RECEPTOR whose res-ponse is always muscular contraction or glandular secretion. Important bodily functions such as digestion, respiration, heart beat and postural adjustments are all controlled through reflexes. Reflexes form the basis of all CNS activity, occurring at all levels of the brain and spinal cord. The structural basis for all reflex action is the REFLEX ARC, the simplest type of CONDUCTION NERVE PATHWAY. The arc leads by a short route from sensory to motor neurons and includes only two or three neurons. Reflex arcs involving only two neurons are termed MONOSYNAPTIC ARCS (only one synapse between the afferent and efferent neurons). In most reflex

---

TYPES OF SYNAPSES

a Axo-Axonic (a–b)

b Axo-Somatic (a–c)

c Axo-Dendritic (a–d)

SYNAPSE: INTERNEURONAL FUNCTIONAL JUNCTION

1 Presynaptic Axon (Telodendrion)

2 Synaptic Knob (end Foot)

3 Mitochondria

4 Synaptic Vesicles with neurotransmitter substance

5 Presynaptic Membrane

NONANATOMIC CONTINUITY JUNCTION:

6 Synaptic Cleft 150–200 Å wide

7 Postsynaptic Membrane with receptor sites for neurotransmitter

8 Axon, Cell Body, or Dendrite

SOMATIC NERVOUS SYSTEM (SNS) OF THE PNS: NEUROMUSCULAR JUNCTION

1a Telodendrion of Axon Branch

2 Synaptic Knob (End Foot) (axon terminal)

3 Mitochondria

4 Synaptic Vesicles with neurotransmitter substance

5 Presynaptic Membrane

NONANATOMIC CONTINUITY JUNCTION

6a Neuromuscular (Subneural) Cleft

7a Motor End Plate (Folded Sarcolemma)

8a Muscle Fiber (Cytoplasm)

MONOSYNAPTIC ATC (2 NEURONS AND SYNAPSE)

e Stretch Receptor
   Includes peripheral process "axon" dendrite of sensory neuron

f Afferent (Sensory) Axon
   Central process of sensory neuron

g Sensory Neuron Cell Body in Dorsal (Posterior) Root Ganglia

h Synapse

i Motor Neuron Cell Body in CNS

j Efferent (Motor) Axon (Axon Branch)

7a Motor End Plate

8a Skeletal Muscle (Effector)

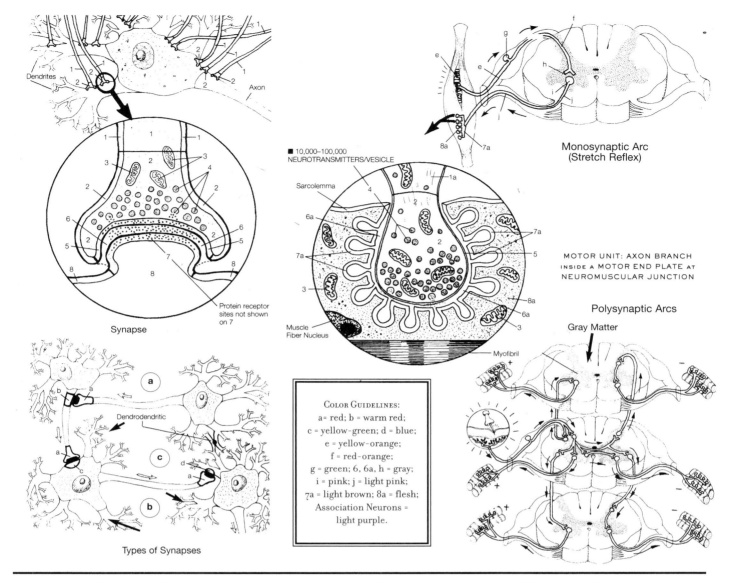

Synapse

■ 10,000–100,000
NEUROTRANSMITTERS/VESICLE

Sarcolemma

Muscle
Fiber Nucleus

Myofibril

Monosynaptic Arc
(Stretch Reflex)

MOTOR UNIT: AXON BRANCH
INSIDE A MOTOR END PLATE AT
NEUROMUSCULAR JUNCTION

Polysynaptic Arcs

Gray Matter

Dendrites

Axon

Protein receptor
sites not shown
on 7

Dendrodendritic

Types of Synapses

COLOR GUIDELINES:
a= red; b = warm red;
c = yellow-green; d = blue;
e = yellow-orange;
f = red-orange;
g = green; 6, 6a, h = gray;
i = pink; j = light pink;
7a = light brown; 8a = flesh;
Association Neurons =
light purple.

arcs in humans afferent and efferent neurons are linked by at least one association neuron, forming the simplest of the POLYSYNAPTIC ARCS (which is composed of three neurons and two synapses). In the majority of polysynaptic arcs, a chain of many association neurons forms complex linkups with various levels of the brain and spinal cord. Every sensory receptor neuron is potentially linked through the CNS with a large number of effector organs all over the body. Every effector organ is in similar CNS communication with a large number of receptors all over the body. Reflex action in the spinal cord can thus be modified by centers in the brain and brainstem, which can send inhibiting or facilitating impulses along nerve tracts to cells in the spinal cord. The FIVE COMPONENTS OF A REFLEX ARC are the receptor (includes the dendritic axon of sensory neuron), sensory neuron (unipolar: axon and cell body in posterior root ganglia), reflex center (within CNS: arc, synapses and linkups), motor neuron and effector organ. The simplest form of a reflex arc shows the structural basis of all reflex arcs: monosynaptic arc (two neurons and one synapse). An example is a stretch reflex (very rare), when the receptor and effector are on the same side of the spinal cord (ipsilateral).

Impulse conduction at a synapse is one-way, always moving forward from the presynaptic neuron to the postsynaptic neuron.

MONOSYNAPTIC ARC (STRETCH REFLEX)

Synaptic transmission (0.5–1 msec) is slower than that along a nerve fiber. A certain amount of control may be established on the impulse's further progress at the synaptic point. After the neuro-transmitter has exerted its effect—whether enhanced (facilitation) or reduced (inhibition) transmission—a chemical match to the neurotransmitter is always present to destroy it. This prevents continued stimulation and allows more messages to cross the synapse.

POLYSYNATPTIC ARCS (MULTINEURONAL)

Much more common than the rare stretch reflexes are the POLYSYNAPTIC REFLEXES, which involve many spinal cord segments or centers in the brain. They usually involve at least three neurons and association or INTERNUNCIAL NEURONS in the gray matter inserted between sensory input and motor output.

A pain receptor in the skin responds to a painful stimulus and the flexor muscles are contracted (excited) to remove the body part from harm; this is an example of a FLEXOR REFLEX. A CROSSED EXTENSOR REFLEX combines a flexor reflex on one side of the body with muscular extension on the other side of the body to maintain posture. It involves the transmission of information across the cord.

THE CENTRAL NERVOUS SYSTEM: BRAIN AND SPINAL CORD

The CENTRAL NERVOUS SYSTEM develops in the dorsal (posterior) half of the embryo as a simple, hollow PRIMITIVE NEURAL TUBE. The cells lining the tube become the NERVOUS TISSUE of the brain and spinal cord. The inner cavity or NEURAL CANAL undergoes many changes in shape during development of the brain and ultimately forms the VENTRICLES of the brain and the CENTRAL CANAL of the SPINAL CORD.

Rapid differentiation and growth begin in the head (cephalic) portion of the neural tube, and between the end of the third week and middle of the fourth week of embryological development, three distinct swellings are apparent: the PROSENCEPHALON (FOREBRAIN), MESENCEPHALON (MIDBRAIN) and RHOMBENCEPHALON (HINDBRAIN). Further development of these three swellings form five specific regions by the end of the fifth week. A massive TELENCEPHALON and a central DIENCEPHALON derive from the forebrain. The mesencephalon remains unchanged in its general tubular shape. An upper METENCEPHALON with a large, dorsal outpocketing and a lower, thin, narrow MYELENCEPHALON both derive from the hindbrain.

Rapid differentiation occurs in each of these five regions shortly after their formation, resulting in all of the major structures recognizable later in the adult. The rapid swelling of the telencephalon forms the two expansive CEREBRAL HEMISPHERES, which cover the diencephalon, midbrain and portions of the hindbrain. Within the diencephalon, in the lateral walls of the third ventricle, three swellings form the EPITHALAMUS, THALAMUS, and HYPOTHALAMUS. (The thalamus, a large ovoid mass of gray matter, comprises 4/5 of the diencephalon.) From the roof midline of the third ventricle the PINEAL GLAND is formed. From the ventral portion of the diencephalon, the posterior lobe of the PITUITARY GLAND is formed.

The mesencephalon undergoes few changes, the most notable being four aggregations of neurons, the paired superior and inferior COLLICULI and two neural fiber tracts, the CEREBRAL PEDUNCLES. The walls of the metencephalon expand into two dorsal midline swellings that join to form the CEREBELLUM and a smaller ventral swelling of nerve fiber bands called the PONS. The adult mylencephalon forms from an aggregation of nerve fibers, and is called the MEDULLA OBLONGATA. The end (caudal) portion of the medulla is continuous with, and structurally similar to, the spinal cord.

All of the nervous tissue of the CNS (brain and spinal cord) that develops from the walls of the primitive neural tube specializes into either gray or white matter.

FUNCTION

The CEREBRUM controls most sensory and motor activities— reasoning, memory, intelligence, limbic (emotional) functions and instinct. The THALAMUS is the relay center. All sensory impulses (except olfactory) are destined for the cerebrum synapse here. The HYPOTHALAMUS regulates certain metabolic activities: body temperature, sugar and fat metabolism, maintenance of water balance, regulation of urine formation, hunger, heartbeat, and the subcortical integration of ANS activities. The secretion of regulating and inhibiting hormones in the anterior lobe of the pituitary gland is

---

DEVELOPMENTAL SEQUENCE AND PRINCIPLE DIVISIONS OF THE BRAIN

THREE EMBRYOLOGICAL VESICLES (4TH WEEK)

- F (Forebrain) Prosencephalon
- M (Midbrain) Mesencephalon
- H (Hindbrain) Rhombencephalon

FIVE EMBRYOLOGICAL VESICLES (5TH WEEK)

- 1 Telencephalon "End Brain" (from F)
- 2 Diencephalon "In-Between Brain" (from M)
- 3 Mesencephalon (Midbrain) (from M)
- 4 Metencephalon "Change Brain"
- 5 Mylencephalon "Spinal Brain" (from H)

ADULT DERIVATIVES OF VESICLE WALLS

- a Cerebrum (2 Cerebral Hemispheres (from 1, F)
- b Thalamus (from 2, F)
- c Hypothalamus (from 2, F)
- d Epithalamus and Pineal Gland (from 2, F)
- e Pituitary Gland (from 2, F)
- f Corpora Quadrigemina (from 3, M)

(2 Superior Colliculi; 2 Inferior Colliculi)

- g 2 Cerebral Peduncles (2 Fiber Tracts) (from 3, M)
- h Cerebellum (from 4, H)
- i Pons (from 4, H)
- j Medulla Oblongata (from 5, H)

ADULT DERIVATIVES OF VESICLE CAVITIES

FROM FOREBRAIN

- 1V, 2V 1st and 2nd Ventricles (Lateral Ventricles) One in each cerebral hemisphere
- 3V 3rd Ventricle (thin slit)

FROM MIDBRAIN

- CA Cerebral Aqueduct of Sylvius

FROM HINDBRAIN

- 4V 4th Ventricle (Upper Portion)
- 4V 4th Ventricle (Lower Portion) (Central Cavity of the Spinal Cord)
- S Spinal Cord

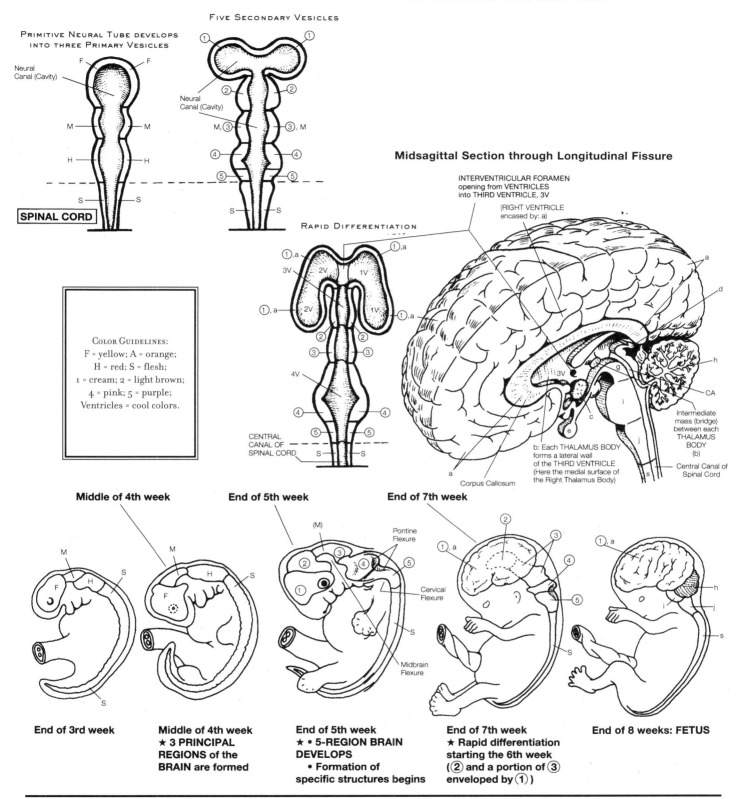

FIVE SECONDARY VESICLES

PRIMITIVE NEURAL TUBE develops INTO THREE PRIMARY VESICLES

Neural Canal (Cavity)

F F

M M

H H

S S

SPINAL CORD

Neural Canal (Cavity)

M, ③ ③, M

④ ④

⑤ ⑤

S S S S

RAPID DIFFERENTIATION

Midsagittal Section through Longitudinal Fissure

INTERVENTRICULAR FORAMEN opening from VENTRICLES into THIRD VENTRICLE, 3V

(RIGHT VENTRICLE encased by: a)

①,a ①,a
3V 2V 1V'
①,a 2V 1V\ ①,a
② ②
③ ③
4V
④ ④
⑤ ⑤
CENTRAL CANAL OF SPINAL CORD
S S

COLOR GUIDELINES:
F = yellow; A = orange;
H = red; S = flesh;
1 = cream; 2 = light brown;
4 = pink; 5 = purple;
Ventricles = cool colors.

3V

a g h
CA
i
b: Each THALAMUS BODY forms a lateral wall of the THIRD VENTRICLE (Here the medial surface of the Right Thalamus Body)
c
e j s

d

Intermediate mass (bridge) between each THALAMUS BODY (b)

Central Canal of Spinal Cord

Corpus Callosum

Middle of 4th week **End of 5th week** **End of 7th week**

M M S
F H S F H S
S

(M)
② ③ ④ ⑤
①
Pontine Flexure
Cervical Flexure
S
Midbrain Flexure

①,a ②
③
④
⑤
S

①,a
④
⑤
h
i j
S

End of 3rd week

Middle of 4th week
★ 3 PRINCIPAL REGIONS of the BRAIN are formed

End of 5th week
★ • 5-REGION BRAIN DEVELOPS
• Formation of specific structures begins

End of 7th week
★ Rapid differentiation starting the 6th week (② and a portion of ③ enveloped by ①)

End of 8 weeks: FETUS

done by the hypothalamus, as well as limbic (emotional) functions. The pituitary gland regulates other endocrine glands. The two superior colliculi deal with visual reflexes, the two inferior colliculi manage auditory reflexes. The two cerebral peduncles coordinate reflexes and are composed of many motor fibers. The cerebellum is the cen- ter for balance, motor coordination (along with the PONS) and the relay center (along with PONTINE NUCLEI). The medulla oblongata is a relay center with many nuclei. It is the center for visceral autonomic activity (respiration, heart rate, vasoconstriction).

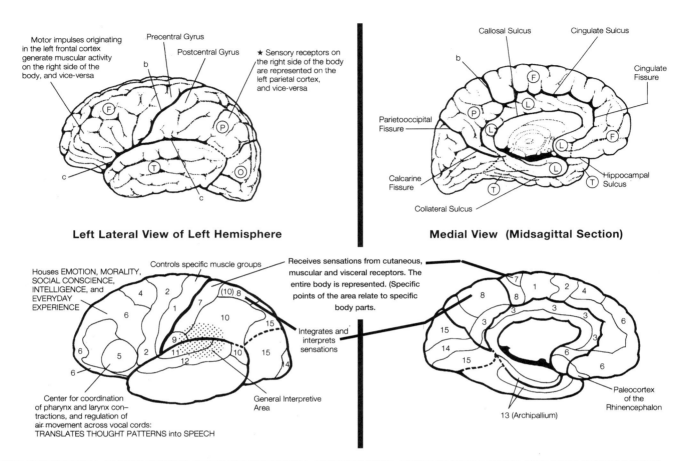

Motor impulses originating in the left frontal cortex generate muscular activity on the right side of the body, and vice-versa

Precentral Gyrus

Postcentral Gyrus

★ Sensory receptors on the right side of the body are represented on the left parietal cortex, and vice-versa

Left Lateral View of Left Hemisphere

Callosal Sulcus

Cingulate Sulcus

Cingulate Fissure

Parietooccipital Fissure

Calcarine Fissure

Collateral Sulcus

Hippocampal Sulcus

Medial View (Midsagittal Section)

Houses EMOTION, MORALITY, SOCIAL CONSCIENCE, INTELLIGENCE, and EVERYDAY EXPERIENCE

Controls specific muscle groups

Receives sensations from cutaneous, muscular and visceral receptors. The entire body is represented. (Specific points of the area relate to specific body parts.

Integrates and interprets sensations

Center for coordination of pharynx and larynx contractions, and regulation of air movement across vocal cords: TRANSLATES THOUGHT PATTERNS into SPEECH

General Interpretive Area

Paleocortex of the Rhinencephalon

13 (Archipallium)

THE FUNCTIONAL SUBDIVISIONS OF EACH CEREBRAL HEMISPHERE: FOUR OUTER LOBES AND TWO INNER LOBES

FOUR OUTER LOBES NAMED IN RELATION TO CRANIAL BONES THAT COVER THEM

- (F) Frontal Lobe
- (P) Parietal Lobe
- (T) Temporal Lobe
- (O) Occipital Lobe

MAJOR FUNCTIONAL AREAS OF EACH OUTER LOBE

- F Front Lobe
- 1 Primary Motor: (Precentral Gyrus)
- 2 Premotor Association
- 3 Supplementary Motor
- 4 Frontal Eyefield
- 5 Broca's Speech (Left Lobe Only)

- 6 Prefrontal Cortex
- 7 Primary Sensory: (Postcentral Gyrus)
- 8 Somatesthetic Association
- 9 Primary Gustatory
- 10 Gnostic
- 11 Primary Auditory
- 12 Auditory Association
- 13 Primary Olfactory (Archipallium)
- 14 Primary Visual
- 15 Visual Association

TWO INNER LOBES OF EACH HEMISPHERE

- I Insula: Island of Reil
- L Limbic Lobe: Cingulate Gyrus and Hippocampal Gyrus

MAJOR DEPRESSED GROOVES
Fissures (Deep Grooves)

- A Longitudinal Fissure
- B Central Fissure of Rolando (Central Sulcus)
- C Lateral Fissure of Sylvius (Lateral Sulcus)

Sulci (Shallow Grooves, Furrows, indicated on illustrations)

CEREBRAL CORTEX: (PALLIUM)
Outer gray matter (nerve cell bodies)
NEOCORTEX (NEOPALLIUM) = 90% of the cerebral cortex: 6 Types of cells in the 6-layered Neocortex:

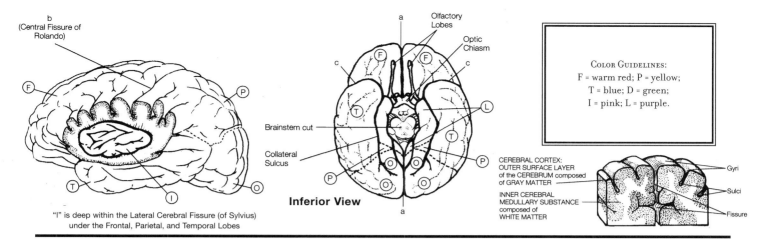

b
(Central Fissure of Rolando)

F

P

T

I

O

"I" is deep within the Lateral Cerebral Fissure (of Sylvius) under the Frontal, Parietal, and Temporal Lobes

a

Olfactory Lobes

Optic Chiasm

c   c

F   F

T   L

Brainstem cut

Collateral Sulcus

T

P

O   O

O

P

Inferior View

a

COLOR GUIDELINES:
F = warm red; P = yellow;
T = blue; D = green;
I = pink; L = purple.

CEREBRAL CORTEX:
OUTER SURFACE LAYER
of the CEREBRUM composed
of GRAY MATTER

INNER CEREBRAL
MEDULLARY SUBSTANCE
composed of
WHITE MATTER

Gyri

Sulci

Fissure

The CEREBRUM, derived from the TELENCEPHALON, is the largest portion of the brain, accounting for eighty percent of the brain mass. It is functionally subdivided into six paired lobes comprising two large, convoluted CEREBRAL HEMISPHERES. The inferior, medial portions of these hemispheres are connected internally by the CORPUS CALLOSUM, a large tract of white matter (and to a lesser extent by two other white matter tracts, the ANTERIOR and POSTERIOR HIPPOCAMPAL COMMISSSURES (see pages 138–39)). Where the right and left hemispheres are not connected by these three tracts they are incompletely separated by a large LONGITUDINAL FISSURE. Extending into this midline fissure is the FALX CEREBRI MEMBRANE from the cranial dura mater (see pages 146–47). There are four general areas of the two cerebral hemispheres: an outer surface layer, called the CEREBRAL CORTEX (PALLIUM), and three subcortical areas consisting of a CEREBRAL MEDULLARY SUBSTANCE (WHITE MATTER), the BASAL GANGLIA, and the LATERAL VENTRICLES. (see pages, 138–39 and 148–49)

The CEREBRAL CORTEX is 2 to 4 mm (0.08–0.16 in.) thick. It is composed of GRAY MATTER, or nerve cell bodies and their DENDRITES (and unmyelinated axons and neuroglia). The CONVOLUTED SURFACE AREA is characterized by elevated folds called GYRI (sing. GYRUS), and as these folds tuck away into the brain they form depressed grooves; shallow grooves are called SULCI (sing. SULCUS) and deeper, longer grooves are called FISSURES. These convolutions are important as they increase the surface area of the gray matter and thus increase the total number of nerve cell bodies. The six types of cortical gray matter neurons are arranged in six interconnecting layers to form an intricate three-dimensional network (For more on the neocortex, see WWW.MCMURTRIESANATOMY.COM).

The cerebral cortex is the most highly evolved area of the brain, and is concerned with the higher human functions: perception of sensory impulses, instigation of voluntary movement by skeletal muscles, and the mental processes of conscious thought, intelligence, memory, judgment, imagination and creativity.

Each cerebral hemisphere is functionally divided into four outer lobes and two inner lobes. Structurally, the outer external lobes of each cerebral hemisphere are mirror images of each other, but functionally they are quite different, as one hemisphere dominates the other in specific higher functions. Long tracts to and from the cortex (from both sensory and motor neurons) at some point cross to opposite sides of the CNS.

The FRONTAL LOBE manages intellectual processes (concentration, abstract thinking and decision-making). It has voluntary motor control of skeletal muscles and it translates thought patterns into verbal communication (speech and language). The frontal lobe deals with personality, specifically aggressive and sexual behavior. It also deals with olfaction (smell).

The PARIETAL LOBE is the somatesthetic interpreter of cutaneous and muscular sensations. It deals with the use of symbols in understanding, verbal articulation of thoughts and emotions, ideation and appropriate motor responses for carrying ideas through, the recollection and recognition of specific objects and the interpretation of textures and shapes of handled objects.

The TEMPORAL LOBE interprets auditory sensations through sensory fibers of the cochlea of the ear. It handles memory storage of both auditory and visual experiences. It also deals with olfaction, language, and emotional behavior associated with the preservation of the species and of the self (anger, hostility, sexuality).

The OCCIPITAL LOBE integrates eye movements by directing and focusing the eye. It deals with conscious vision and manages visual association (correlation of visual images with previous visual experiences). The GNOSTIC is the common integrative area. It integrates various sensory interpretations to form a common thought, and it transmits its integrated signal to other parts of the brain to cause the appropriate response.

The INSULA, made up of the ISLAND OF REIL, is the central deep lobe, a triangular area of the cerebral cortex, lying in the floor of the lateral fissure; it cannot be viewed on the surface. It is involved in the integration of cerebral activities and memory.

The limbic lobe is composed of the CINGULATE GYRUS and the HIPPOCAMPAL GYRUS. The limbic lobe is a marginal section of each cerebral hemisphere on the medial aspect, sometimes called the GYRUS FORNICATUS. It is the part of the limbic system involved in basic emotional activity. The other ten percent of the cerebral cortex is made up of the RHINENCEPHALON (phylogenetically, the oldest portion of the brain; it is represented by only three layers).

The PALEOCORTEX (PALEOPALLIUM) is restricted to olfactory regions of the frontal lobe (it includes the olfactory bulbs, tract and striae, and intermediate olfactory area). The ARCHICORTEX (ARCHIPALLIUM) is the oldest part of the rhinencephalon, and makes up large portions of the old olfactory cortex of the temporal lobe associated with the limbic system (including the fornix and the hippocampus).

There are three main subcortical areas of the cerebrum, those being the large cerebral medullary substance (body), the basal ganglia and the lateral ventricles. The layer underlying the OUTER CEREBRAL CORTEX of gray matter consists mainly of white matter (MYELINATED AXONS) arranged into FIBER TRACTS. These tracts offer lines of communication between the cerebral hemispheres involving the cerebral cortex in three principal directions: ASSOCIATION TRACTS, COMMISSURAL TRACTS and PROJECTION TRACTS. Association tracts are confined to a given hemisphere. They connect and transmit impulses between the anterior and posterior cortical areas of GYRI in various lobes. The commissural tracts connect the cortical areas of gyri of one hemisphere to the corresponding gyri in the opposite hemisphere. The CORPUS CALLOSUM is the largest commissural tract

(the GREAT COMMISSURE), composed of a thick layer of transverse fibers arching above the lateral ventricles. It is the major connector between the two cerebral hemispheres; the minor connectors are the ANTERIOR and posterior commissures.

The fornix is a fibrous, vaulted band connecting the cerebral lobes (see pages 140–141). The HIPPOCAMPAL COMMISSURE is a thin sheet of transverse fibers under the posterior portion of the corpus callosum, connecting the medial margins of the CRURA of the fornix. Projection tracts consist of fibers ascending (entering) and descending (leaving) the hemispheres as sensory and motor pathways, respectively. They enable the brain to keep in touch with the peripheral body area. The CORONA RADIATA is a radiating mass of ascending and descending fibers that, above the corpus callosum, extends in all directions to the cerebral cortex. Many of the fibers

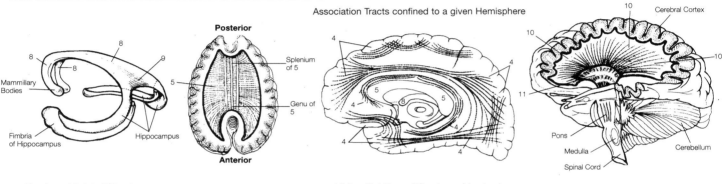

Association Tracts confined to a given Hemisphere

Fornix and Related Structures
Left Lateroposterior View

Transverse Section of Cerebrum through the Corpus Callosum

Midsagittal View of Cerebrum (viewing inner aspect of Right Cerebral Hemisphere)

Projection Tracts

THREE MAJOR SUBCORTICAL AREAS:

(1) Cerebral Medullary Substance (Body)

(2) Basal Ganglia

(3) Lateral Ventricles

ONE CEREBRAL MEDULLARY BODY: INNER WHITE MATTER BUNDLES OF DENDRITES AND MYELINATED AXONS: COMMUNICATING DIRECTIONAL FIBER TRACTS:
IN THE SAME HEMISPHERE

4 Association Tracts

BETWEEN THE HEMISPHERES

Commissural Tracts

5 Corpus Callosum

6 Anterior Commissure

7 Posterior Commissure
Connects the superior colliculi

8 Fornix

9 Hippocampal Commissure of Fornix

ASCENDING AND DESCENDING TRACTS CONNECTING THE CORTEX WITH OTHER STRUCTURES OF THE BRAIN AND SPINAL CORD
PROJECTION TRACTS

10 Corona Radiata

11 Internal capsule
Considered part of the corpus striatum and the thalamus

(2) Basal Ganglia (Cerebral Nuclei): paired masses of "central gray matter" within white matter

CORPUS STRIATUM: CAUDATE NUCLEUS AND THE LENTIFORM NUCLEUS

12 Caudate Nucleus (Upper Mass)

LENTIFORM NUCLEUS (LOWER MASS)

a Putamen (Lateral Portion)

b Globus Pallidus (Medial Portion)

13 Claustrum

14 Amygdaloid Nucleus   (part of limbic system)

15 Septum Pellucidum

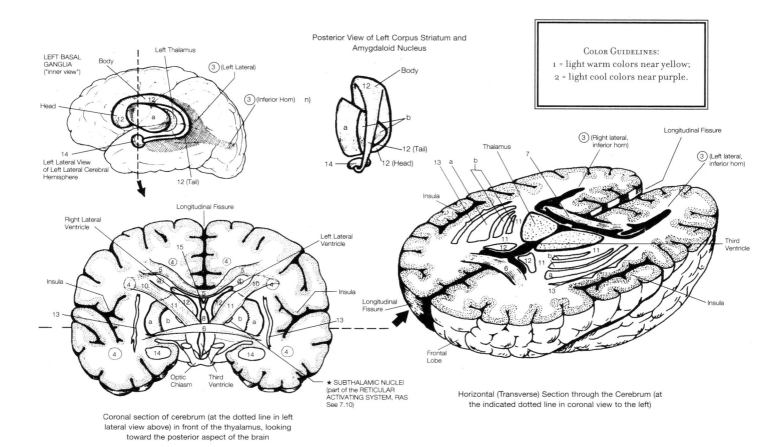

Posterior View of Left Corpus Striatum and
Amygdaloid Nucleus

LEFT BASAL
GANGLIA
("inner view")

Left Thalamus

Body

③ (Left Lateral)

③ (Inferior Horn) n)

Head

14

Left Lateral View
of Left Lateral Cerebral
Hemisphere

12 (Tail)

Body

12

a

b

14

12 (Tail)

12 (Head)

Thalamus

③ (Right lateral,
inferior horn)

Longitudinal Fissure

③ (Left lateral,
inferior horn)

Longitudinal Fissure

Right Lateral
Ventricle

15

Left Lateral
Ventricle

Insula

13 a b

Insula

Insula

Third
Ventricle

Insula

13

Insula

Longitudinal
Fissure

13

Frontal
Lobe

13

★ SUBTHALAMIC NUCLEI
(part of the RETICULAR
ACTIVATING SYSTEM, RAS
See 7.10)

Optic
Chiasm

Third
Ventricle

Coronal section of cerebrum (at the dotted line in left
lateral view above) in front of the thyalamus, looking
toward the posterior aspect of the brain

Horizontal (Transverse) Section through the Cerebrum (at
the indicated dotted line in coronal view to the left)

arise in the relay center of the thalamus (see pages 140–41). The corona radiata is continuous with the INTERNAL CAPSULE, which is a compact area composed mainly of efferent motor fibers leaving the cerebrum, connecting with the brainstem and spinal cord, and passing through the two BASAL GANGLIA.

The basal ganglia are paired discrete masses of gray matter at the base of the hemispheres on either side of the diencephalon deep within the white matter. They are interconnected by many fibers to the cerebral cortex, internal capsule, thalamus, hypothalamus, and gray matter masses of the midbrain. They facilitate or inhibit motor impulses from motor areas of the frontal lobe (except the precentral gyrus). The CAUDATE NUCLEUS and PUTAMEN assist in the control of large subconscious movements of the skeletal muscles (e.g.: swinging of arms while walking). The GLOBUS PALLIDUS regulates muscle tone necessary for specific intentional body movements. (Some authorities also include SUBSTANTIA NIGRA, SUBTHALAMIC NUCLEI, and the RED NUCLEUS [all of the midbrain] as part of the basal ganglia.) (See pages 142–43.)

Each cerebral hemisphere contains a large cavity called the LATERAL VENTRICLE inferior to (below) the CORPUS CALLOSUM. They develop as large cavities of the upper forebrain (telencephalon) in the embryo. Most of the CSF (cerebrospinal fluid) is secreted within the CHOROID PLEXUSES (specialized capillaries) intertwined with the ciliated

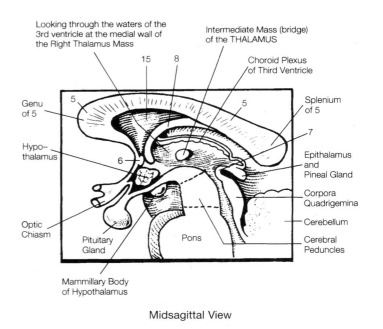

Looking through the waters of the
3rd ventricle at the medial wall of
the Right Thalamus Mass

Intermediate Mass (bridge)
of the THALAMUS

15 8

Choroid Plexus
of Third Ventricle

Genu
of 5

5

5

Splenium
of 5

Hypo—
thalamus

7

6

Epithalamus
and
Pineal Gland

Corpora
Quadrigemina

Optic
Chiasm

Cerebellum

Pituitary
Gland

Pons

Cerebral
Peduncles

Mammillary Body
of Hypothalamus

Midsagittal View

ependymal cells (EPENDYMAL VENTRICULORUM CEREBRAL) lining the lateral ventricles (see 148–49).

The SEPTUM PELLUCIDUM is the medial wall of the lateral ventricles. It is a thin, triangular sheet of nervous tissue. The two laminae are attached to the corpus callosum above, and the fornix below.

THE FOUR MAJOR AREAS OF THE DIENCEPHALON PLUS THE 3RD VENTRICLE:

(E) Epithalamus

(N) Neurohypophysis (Posterior Lobe) of the Pituitary Gland*

(T) Thalamus

(H) Hypothalamus

3V Third Ventricle

(E) EPITHALAMUS

1 Pineal Body (Gland)

2 Pineal Stalk (Habenula)

3 Habenular Commissure and Habenular Nuclei

4 Pia Mater Capsule

* For more information, see pages 178–81 and 204–05

> COLORING NOTES:
> Color all the hypothalamic nuclei,
> H, and 12 one color.
> Carry colors over from pages 134–35.
> Choose your own colors for the rest.

(T) THALAMUS: RELAY CENTER

Commissure Connecting The 2 Oval Thalamic Masses:

5 Intermediate Mass (Bridge)

White Matter Dividing The Thalamic Gray Matter Into 3 Groups:

6 Internal Medullary Lamina

3 Nuclear Groups of Gray Matter Masses: Anterior Group

7 Anterior Nucleus

Medial Group

8 Medial Nucleus (part of the Reticular Activating System)

Lateral Group

9 Sensory Relay Centers

10 Interpretive Sensory Centers

11 Motor Centers

14 Fornix

(H) HYPOTHALAMUS

12 Hypothalamic Nuclei Masses

13 Mammillary Bodies

The Limbic System ("Emotional Brain")

15 Hippocampus

16 Cerebral

LIMBIC LOBE

17 Amygdaloid Nuclei of Basal Ganglia

13 Mammillary Bodies of Hypothalamus

7 Anterior Nuclei of Thalamus

3 Habenular Nuclei of Epithalamus

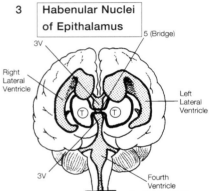

3V
5 (Bridge)
Right Lateral Ventricle
Left Lateral Ventricle
T   T
3V
Fourth Ventricle
Anterior View

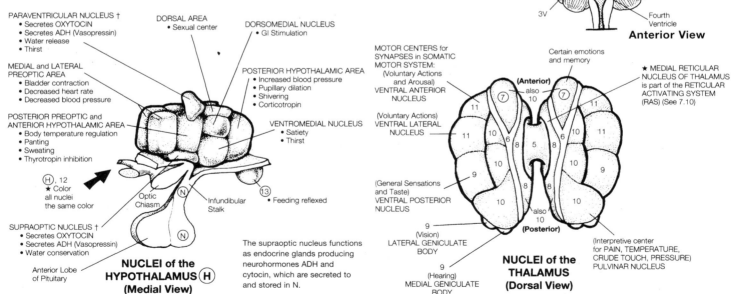

PARAVENTRICULAR NUCLEUS †
• Secretes OXYTOCIN
• Secretes ADH (Vasopressin)
• Water release
• Thirst

DORSAL AREA
• Sexual center

DORSOMEDIAL NUCLEUS
• GI Stimulation

MEDIAL and LATERAL PREOPTIC AREA
• Bladder contraction
• Decreased heart rate
• Decreased blood pressure

POSTERIOR HYPOTHALAMIC AREA
• Increased blood pressure
• Pupillary dilation
• Shivering
• Corticotropin

POSTERIOR PREOPTIC and ANTERIOR HYPOTHALAMIC AREA
• Body temperature regulation
• Panting
• Sweating
• Thyrotropin inhibition

VENTROMEDIAL NUCLEUS
• Satiety
• Thirst

(H), 12
★ Color all nuclei the same color

Optic Chiasm
(N)
Infundibular Stalk
(13)
• Feeding reflexed

SUPRAOPTIC NUCLEUS †
• Secretes OXYTOCIN
• Secretes ADH (Vasopressin)
• Water conservation

Anterior Lobe of Pituitary
(N)
NUCLEI of the HYPOTHALAMUS (H) (Medial View)

The supraoptic nucleus functions as endocrine glands producing neurohormones ADH and cytocin, which are secreted to and stored in N.

MOTOR CENTERS for SYNAPSES in SOMATIC MOTOR SYSTEM:
(Voluntary Actions and Arousal)
VENTRAL ANTERIOR NUCLEUS

(Voluntary Actions)
VENTRAL LATERAL NUCLEUS

(General Sensations and Taste)
VENTRAL POSTERIOR NUCLEUS

Certain emotions and memory

(Anterior)
also 10
7   7
11   11
10   10
6   6
11   11
8   5   8
10   10
9   10   10   9
8   8
also 10
(Posterior)

★ MEDIAL RETICULAR NUCLEUS OF THALAMUS is part of the RETICULAR ACTIVATING SYSTEM (RAS) (See 7.10)

(Interpretive center for PAIN, TEMPERATURE, CRUDE TOUCH, PRESSURE) PULVINAR NUCLEUS

9 (Vision) LATERAL GENICULATE BODY

9 (Hearing) MEDIAL GENICULATE BODY

NUCLEI of the THALAMUS (Dorsal View)

The DIENCEPHALON is the second subdivision of the "OLD" FOREBRAIN that gave rise to the cerebrum. It is a MAJOR AUTONOMIC REGION, positioned atop the BRAINSTEM between the inferior portions of the two cerebral hemispheres. It is almost completely engulfed and surrounded by the hemispheres. The third ventricle forms a thin, slitlike midline cavity within the diencephalon. The chief components of the diencephalon are the EPITHALAMUS, THALAMUS, HYPOTHALAMUS, and PITUITARY GLAND (POSTERIOR LOBE).

The Limbic System ("Emotional Brain")

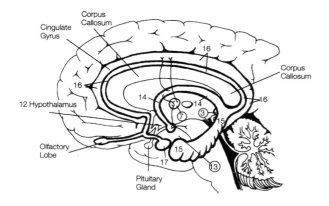

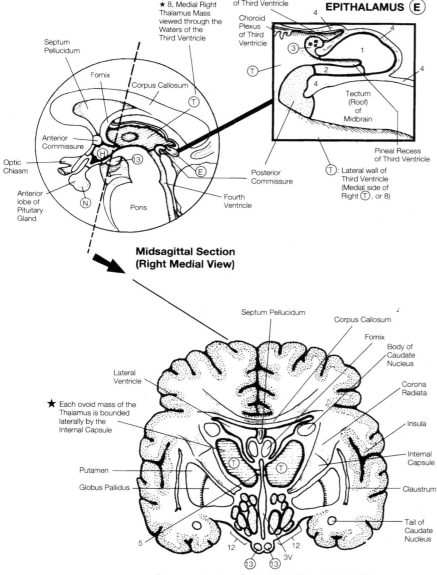

**Midsagittal Section
(Right Medial View)**

★ 8, Medial Right Thalamus Mass viewed through the Waters of the Third Ventricle

EPITHALAMUS (E)

Suprapineal recess of Third Ventricle

Choroid Plexus of Third Ventricle

Tectum (Roof) of Midbrain

Pineal Recess of Third Ventricle

(T): Lateral wall of Third Ventricle (Medial side of Right (T), or 8)

Posterior Commissure

Fourth Ventricle

Septum Pellucidum

Fornix

Corpus Callosum

Anterior Commissure

Optic Chiasm

Anterior lobe of Pituitary Gland

Pons

★ Each ovoid mass of the Thalamus is bounded laterally by the Internal Capsule

Septum Pellucidum

Corpus Callosum

Fornix

Body of Caudate Nucleus

Corona Radiata

Insula

Internal Capsule

Claustrum

Tail of Caudate Nucleus

Lateral Ventricle

Putamen

Globus Pallidus

**Coronal Section through DIENCEPHALON
(at the indicated dotted line in above circle)**

The EPITHALAMUS is the uppermost portion of the diencephalon. It consists of the PINEAL GLAND, the HABENULAR NUCLEI, (PINEAL STALK) and HABENULAR COMMISSURE. It includes a thin roof over the third ventricle, whose inside lining forms a CHOROID PLEXUS that produces cerebrospinal fluid. The pineal gland extends outward from the posterior end of the epithalamus. Its functions, which are obscure, may be concerned with the onset of puberty and establishment of circadian rhythms. The habenular nuclei receives impulses from the olfactory system and projects them to the limbic system. It is also called the "emotional brain." Thus, the nuclei establish a connection between the sense of smell and the sex drive, rage reactions, and associated visceral changes.

The thalamus is the largest portion of the diencephalon, comprising four-fifths of its mass. It consists of many groups of cell bodies or nuclei formed into two paired ovoid masses. Each thalamic mass is positioned immediately below the lateral ventricle in its respective hemisphere, in one of the lateral walls of the third ventricle. The masses are connected by a commissure called the INTERMEDIATE MASS, which passes through a foramen of the third ventricle. It is a relay center for all sensory impulses (except smell) to the cerebral cortex. Sensory stimuli are associated, synthesized and then relayed through the CORONA RADIATA to specific cortical areas. Impulses are also received from the CORTEX, CORPUS STRIATUM and HYPOTHALAMUS and are relayed to visceral and somatic effectors. It is the center for appreciation of primitive, uncritical sensations of crude touch, temperature and pain, and assists in the initial autonomic response to intense pain (shock).

The hypothalamus is very small in size and consists of several masses of nuclei interconnected to other vital parts of the brain. It forms the floor and lower lateral walls of the third ventricle. The AUTONOMIC CENTER is concerned with HOMEOSTASIS (facilitation and inhibition) of several vital functions necessary for organism survival, most of which relate to regulation of visceral activities (but some relate to emotion [limbic] and instinct); regulation of heart rate, body temperature, water and electrolyte balance (OSMORECEPTOR and thirst center), hunger and gastrointestinal activity (feeding and satiety center), conscious alertness (sleep and wakefulness center) and sexual response (sexual center). The hypothalamus also produces POLYPEPTIDES that regulate the activity of the pituitary gland in the secretion or suppression of hormones. Hunger, rage and increased blood pressure are regulated by the lateral hypothalamic areas. There is no autonomic regulation of respiratory activity by the hypothalamus.

The BRAINSTEM is the stem-shaped part of the brain that connects the forebrain (cerebral hemispheres and diencephalon) with the spinal cord. It consists of the MIDBRAIN (MESENCEPHALON), the PONS (anterior portion of the metencephalon), and the MEDULLA OBLONGATA (MYELENCEPHALON).

The midbrain is a short (1.5 cm), wedge-shaped section of the brainstem between the diencephalon and the pons. The structures that compose it are the CORPORA QUADRIGEMINA (dorsal portion), CEREBRAL PEDUNCLES (ventral portion), SPECIALIZED NUCLEI and the CEREBRAL AQUEDUCT OF SYLVIUS (associated ventricle). The corpora quadrigemina consists of four rounded elevations on the dorsal portion of the midbrain. The two upper "balls," the SUPERIOR COLLICULI, receive fibers from the visual cortex and send fibers to eye and neck muscles for visual reflexes. The two lower "balls," the INFERIOR COLLICULI, receive fibers from the cochlea of the ear and send fibers to the MEDIAL GENICULATE BODY OF THE THALAMUS for auditory reflexes.

The cerebral peduncles are a pair of cylindrical ascending and descending PROJECTION FIBER TRACTS that support and connect the cerebrum to other regions of the brain (bulk of the midbrain). The specialized nuclei are considered by some authorities to be part of the basal ganglia. The cerebral aqueduct of sylvius connects the third and fourth ventricles.

The red nucleus connects the cerebral hemispheres and the cerebellum. It is the motor nucleus concerned with muscle tone, skilled motor coordination and posture; its efferent fibers give rise to the descending RUBROSPINAL TRACT. The SUBSTANTIA NIGRA is the midbrain motor nucleus concerned with suppression of purposeless movements.

The "old" HINDBRAIN (RHOMBENCEPHALON) gives rise to the UPPER METENCEPHALON (pons and cerebellum) and the LOWER MYELENCEPHALON (MEDULLA OBLONGATA).

The PONS VAROLII (ANTERIOR METENCEPHALON) is a rounded eminence of the ventral surface of the brainstem, about 2.5 cm long, lying between the cerebral peduncles and the medulla. A bulge on the VENTRAL PONS, called the BASAL PONS, consists of a broad band of surface transverse fiber tracts receiving motor fibers from one cerebral hemisphere and relaying them to the opposite cerebellar hemisphere via the middle cerebellar peduncles. (PONTINE NUCLEI, which send fibers to the cerebellum, are also found in the basal pons.) A deeper DORSAL PONS, called the TEGMENTUM, contains longitudinal motor and sensory tracts that connect the medulla with tracts of the midbrain (ascending sensory fibers pass to the thalamus and descending fibers pass through the CORTICOSPINAL and RUBROSPINAL TRACTS). (The nuclei of cranial nerves II–VII are also found in the tegmentum.) The CEREBRAL AQUEDUCT

THREE MAJOR AREAS OF THE BRAINSTEM PLUS THE FOURTH VENTRICLE

1 | Midbrain (Mesencephalon)

4 | 2 Cerebral Peduncles

CA | Cerebral Aqueduct of Sylvius

TECTUM (ROOF): CORPORA QUADRIGEMINA

5 | 2 Superior Colliculi

6 | 2 Inferior Colliculi

7 | 2 Superior Cerebellar Peduncles

III, IV (in cerebral peduncles)

2 | PONS (PONS VAROLII)

8 | Basal Pons

4V | 4th Ventricle (Upper Portion)

9 | Tegmentum (Floor of 4th Ventricle)

10 | 2 Middle Cerebellar Peduncles

V, VI, VII | Cochlear Divisions of VIII
Nuclei of cranial nerves (in tegmentum)

3 | MEDULLA OBLONGATA (MYELENCEPHALON

11 | 2 Pyramids (Cortex-To-Spinal Cord)

12 | Decussation Area of Pyramids

13 | 2 Olives (One On Each Lateral Surface)

4V | 4th Ventricle (Lower Portion)

14 | Peripheral White Matter Bundle Tracts
Up and Down Medulla-Spinal Cord Floor of 4th Ventricle

15 | 2 Inferior Cerebellar Peduncles

IX, X, XI, XII | (Vestibular Divisions of VIII)

Color Guidelines:
1 = orange; 2 = pink; 3 = purple;
4 = red; h = white; g = flesh.
Choose your own colors for the rest.

THE CEREBELLUM
Portions According to Phylogenetic Age:
Oldest Archicerebellum:
Flocculonodular Lobe

a | Flocculus

b | Nodulus (Nodule)

Paleocerebellum

c | Vermis (Sup. and Inf.)

Neocerebellum
2 Cerebellar Hemispheres:

d | Anterior Lobe

e | Posterior Lobe

Extensions of Cranial Dura Mater:
Separate the Cerebellum from Cerebrum:

f | Tentorium Cerebelli

Between the Hemispheres:
Falx Cerebelli (not shown)

INTERNAL ANATOMY
Outer Gray Matter

g | Cortex

Inner Tree of White Matter Tracts

h | Arbor Vitae

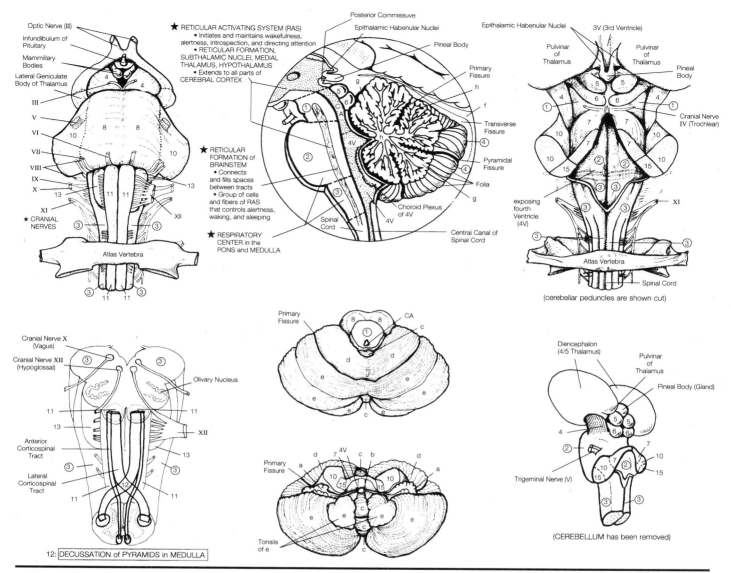

of the mesencephalon enlarges within the entire metencephalon to become the fourth ventricle.

The medulla oblongata (myelencephalon) is the lower 3 cm. of the brainstem, the enlarged portion of the spinal cord in the cranium after it enters the foramen magnum of the occipital bone. It consists mostly of ascending and descending white matter tracts that communicate between the spinal cord and various parts of the brain. Most of these fibers cross to the other side at the decussation area of two VENTRAL PYRAMIDS. The specialized nuclei are for cranial nerves VII-XII, sensory relay to the thalamus, motor relay from the cerebrum to the cerebellum, and autonomic control centers for cardiac, vasomotor, and respiratory activity. The NUCLEUS GRACILIS and NUCLEUS CUNEATUS receive sensory fibers from ascending tracts in the dorsal area of the spinal cord. Axons cross the medulla midline to the opposite side of the medulla. The information is then relayed to the thalamus, and then to the sensory areas of the cerebral cortex. The OLIVARY NUCLEI (INFERIOR) receive fibers from the basal ganglia and the cerebral cortex, and send the fibers to the cerebellum. They are part of the motor body system and form part of the reticular formation and the RAS.

The cerebellum (POSTERIOR METENCEPHALON) is the second largest structure of the brain. It occupies the inferior and posterior aspect of the cavity of the skull (cranial cavity), overhanging the medulla oblongata. It consists of two lateral cerebellar hemispheres and a narrow, medial portion, the VERMIS. It is composed of gray matter on the surface, and deeper "tree tracts" of white matter collectively called the ARBOR VITAE. The cerebellar surface is thrown into numerous parallel ridges called FOLIA that are separated by CEREBELLAR FISSURES. It functions totally at the subconscious (involuntary) level to coordinate body movements and to maintain balance. It is connected to the brainstem by three pairs of fiber bundles, the superior, middle and inferior CEREBELLAR PEDUNCLES. The incoming afferent impulses from propriocenters within muscles, tendons, joints, and two special sense organs communicate through the peduncles; they are then assimilated in the cerebrum. Efferent impulses are discharged through the peduncles to other neurological structures to coordinate voluntary skeletal-muscle contractions. The cerebellum does not serve as a reflex center, but may reinforce some reflexes and inhibit others.

GENERAL STRUCTURE OF THE SPINAL CORD
EXTERNAL ANATOMY TO REGIONS OF THE SPINAL CORD

C Cervical Enlargement (C3-T2)

T Thoracic Section (T3-T8)

L Lumbar Enlargement (T9-T12)

S Sacral Section (S1-S4)

a Conus Medullaris

b Filum Terminale

c Cauda Equina

SPINAL GROOVES:
ANTERIOR (VENTRAL) SURFACE

d Anterior Median Fissure

POSTERIOR (DORSAL) SURFACE

e Posterior Median Sulcus

THIRTY-ONE SPINAL SEGMENTS EACH GIVE RISE TO A PAIR OF

SN Spinal Nerves

INTERNAL ANATOMY

GM Gray Matter

Composed of nerve cell bodies, neuroglia and unmyelinated association neurons

WM White Matter

Tracts (bundles) of myelinated fibers of sensory and motor neurons

GRAY MATTER

1 2 Anterior (Ventral) Horns

Cell bodies of motor neurons (motor axons of bodies exit) and association neurons

2 2 Posterior (Dorsal) Horns

Where central processes of sensory neurons enter and consist of association neurons

3 2 Lateral Horns

Have autonomic motor neurons and are only in T1-L2, S2-4 levels

4 2 Gray Commissures

Have crossing of axons of various neurons and a central canal in the middle

5 Intermediate Zone

Mostly association neurons and some motor neurons

6 Reticular Formation

A diffuse gray column network of cell bodies and fibers that receives fibers from the sensory lemniscus tract and sends fibers to the cerebral cortex, cerebellum, and spinal portion of the RAS (reticular activating system)

WHITE MATTER
Six Funiculi (Columns)

7 2 Anterior (Ventral) Funiculi

8 2 Posterior (Dorsal) Funiculi

9 2 Lateral Funiculi

10 2 White Commissures

ASCENDING (SENSORY) WHITE MATTER TRACTS
PAIN/TEMPERATURE RECEPTOR PATHWAY:

11 Lateral Spinothalamic Tract

Thalamus (same side), to the thalamocortical tract (internal capsule and corona radiata) to postcentral gyrus of cerebral cortex

CRUDE TOUCH AND PRESSURE RECEPTOR PATHWAY

12 Anterior Spinothalamic Tract

FINE TOUCH, PROPRIOCEPTION AND VIBRATION RECEPTOR PATHWAY:

13 Fasciculus Gracilis And Fasciculus Cuneatus Tracts

Medulla (decussation) to medial lemnicus to the thalamus to the thalamocortical tract to the postcentral gyrus

MUSCLE/POSITION SENSE RECEPTOR PATHWAY:
Tendon stretch impulses

14 Posterior Spinocerebellar Tracts

Inferior cerebellar peduncle to cerebellum cortex

Position Sense Impulses

15 Anterior Spinocerebellar Tracts

Association neuron crosses to superior cerebellar peduncle to the cerebellum cortex

DESCENDING (MOTOR) WHITE MATTER TRACTS (ALL TERMINATE IN ANTERIOR (VENTRAL) GRAY HORNS)

(1) Corticospinal (Pyramidal)Tracts

Neuron cell bodies in precentral gyrus of frontal lobe which effect coordinated, precise voluntary movements of skeletal muscle

85% DECUSSATE IN PYRAMIDS OF MEDULLA:

16 Lateral Corticospinal Tract

25% UNCROSSED IN MEDULLA:

17 Anterior Corticospinal Tract

Descend directly (no synapses) to lower motor neurons in anterior gray horns

(2) Extrapyramidal Tracts and Their (Effector) Activity

Major extrapyramidal tracts originate in reticular formation

18 Reticulospinal Tracts

(muscle tone, sweat glands)

No direct descending tracts from the cerebellum—only indirectly through vestibular nuclei, red nuclei, superior colliculi origins, respectively:

19 Vestibulospinal Tracts

(muscle tone, equilibrium)

20 Rubrospinal Tracts

(muscle tone, posture)

21 Tectospinal Tracts

(head movements)

The spinal cord is an ovoid column of nervous tissue about 44 cm long, flattened anteroposteriorly. It is continuous with the medulla oblongata, beginning at the FORAMEN MAGNUM of the occipital bone, extending inferiorly through the neural canal of the vertebral column, and ending at the level of the second LUMBAR VERTEBRA, which terminates in the tapering CONUS MEDULLARIS. It has an average diameter of 1 cm and contains two prominent bulges in the lower cervical and lumbar regions (CERVICAL ENLARGEMENT and LUMBAR ENLARGEMENT), expressive of an increased number of neurons that serve the upper and lower extremities, respectively. In cross section, it does

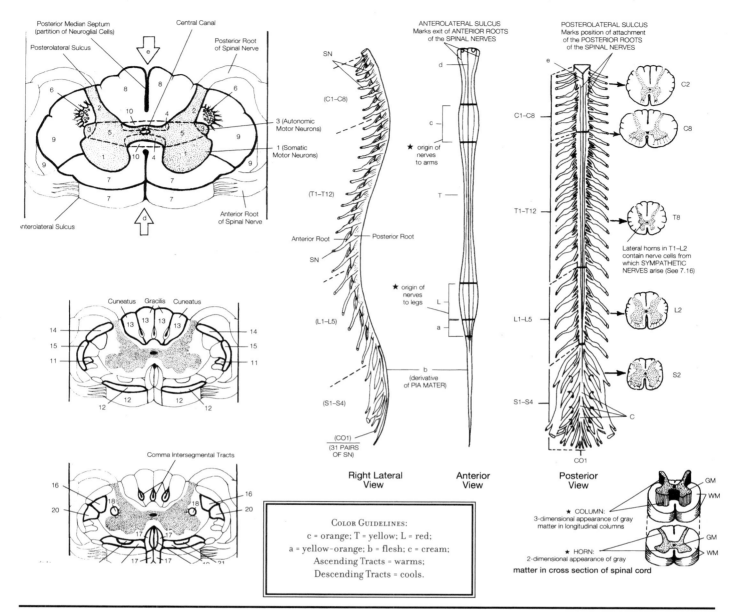

Posterior Median Septum
(partition of Neuroglial Cells)

Central Canal

Posterior Root
of Spinal Nerve

Posterolateral Sulcus

3 (Autonomic
Motor Neurons)

1 (Somatic
Motor Neurons)

Anterolateral Sulcus

Anterior Root
of Spinal Nerve

ANTEROLATERAL SULCUS
Marks exit of ANTERIOR ROOTS
of the SPINAL NERVES

POSTEROLATERAL SULCUS
Marks position of attachment
of the POSTERIOR ROOTS
of the SPINAL NERVES

SN

(C1–C8)

C1–C8

★ origin of
nerves
to arms

T

(T1–T12)

T1–T12

Anterior Root

Posterior Root

SN

★ origin of
nerves
to legs

L

(L1–L5)

a

L1–L5

b
(derivative
of PIA MATER)

(S1–S4)

S1–S4

(CO1)
(31 PAIRS
OF SN)

CO1

Right Lateral
View

Anterior
View

Posterior
View

C2

C8

T8

Lateral horns in T1–L2
contain nerve cells from
which SYMPATHETIC
NERVES arise (See 7.16)

L2

S2

c

Cuneatus Gracilis Cuneatus

Comma Intersegmental Tracts

COLOR GUIDELINES:
c = orange; T = yellow; L = red;
a = yellow-orange; b = flesh; c = cream;
Ascending Tracts = warms;
Descending Tracts = cools.

GM
WM
★ COLUMN:
3-dimensional appearance of gray
matter in longitudinal columns

GM
WM
★ HORN:
2-dimensional appearance of gray
matter in cross section of spinal cord

not fill the vertebral space. It is surrounded immediately by the PIA MATER, the cerebrospinal fluid in the SUBARACHNOID SPACE, the ARACHNOID, and the DURA MATER, which fuses with the periosteum of the inner vertebral surfaces. It floats in CSF (see pages 148–49).

Distal to the conus medullaris, the pia mater continues as a fibrous strand, the FILUM TERMINALE, that attaches to the coccyx and joins the extended DURAL SAC of the dura mater of the second sacral vertebra. This DURAL SAC is filled with CSF, as is the LUMBAR CISTERN. A collection of nerve fibers resembling a horse's tail called the CAUDA EQUINA radiate interiorly from the CONUS MEDULLARIS, floating within the lumbar cistern as they travel to the SACRAL INTERVERTEBRAL FORAMINA. The spinal cord develops as thirty-one segments, each segment giving rise to a pair of spinal nerves that exit posterolaterally from the cord through the intervertebral foramina. Two grooves (anterior median fissure and posterior median sulcus) extend the length of the spinal cord and partially divide the cord into right and left portions.

There are two principal functions of the spinal cord. It is a center for spinal reflexes (centrally located gray matter) and has specific nerve pathways for reflexive, involuntary movement. The gray matter approximates the shape of an H: the transverse crossbar of the H is the GRAY COMMISSURE; the upright projections are the ANTERIOR (VENTRAL) and POSTERIOR (DORSAL) HORNS. Located between these horns to the sides are the short LATERAL HORNS. The second function is the neural communication to and from the brain (peripherally located white matter tracts). This "gray matter H" divides the surrounding white matter into FUNICULI (COLUMNS). Each funiculus contains a number of specific TRACTS (FASCICULI). The SENSORY ASCENDING LONG TRACTS conduct impulses from peripheral sensory receptors of the body to the brain and the MOTOR DESCENDING LONG TRACTS conduct motor impulses from the brain to muscles and glands. SHORT INTERSEGMENTAL TRACTS are involved in various spinal reflexes operating on more than one cord level, jumping from segment to segment.

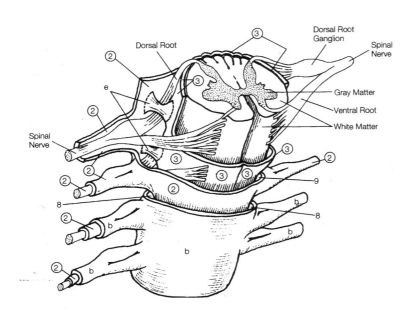

Dorsal Root

Dorsal Root Ganglion

Spinal Nerve

3

2

e

2

Gray Matter

Ventral Root

White Matter

Spinal Nerve

2

3

3

2

2

2

9

2

8

b

8

2

b

2

b

The delicate organs of the CNS are surrounded and protected by a bony encasement—the CRANIUM (SKULL) surrounds the brain, and the VERTEBRAL COLUMN surrounds the spinal cord. Three membranes, collectively referred to as the MENINGES (SINGULAR MENINX) form a protective barrier between the bone and the soft CNS tissue. The three connective tissue membranous coverings are the DURA MATTER, ARANCHNOID MEMBRANE and PIA MATER.

The dura mater (PACHYMENINX) is the outermost and thickest covering. It is composed of collagen fibers and fibroblasts for tough protection. CRANIAL DURA MATER consists of two layers; not only does it enclose the brain, but its outer thick layer also acts as the periosteum for the interior of the cranial cavity as it adheres firmly to the cranial bones of the skull. The thin inner layer follows the general contour of the brain.

These two layers are fused around most of the brain. In specific regions, they separate and enclose dural sinuses (whose function is to collect venous blood and drain it to the internal jugular veins of the neck). In four locations, the inner layer turns inward to form SEPTAL PARTITIONS between areas on the surface of the brain and BRAIN ANCHORS to the inside of the cranial case. The aforementioned dural sinuses are found in the margins of these infoldings of the inner meningeal layer of the cranial dura mater.

The SPINAL DURA MATER forms a loose, tubular sheath around the spinal cord. It is a single layer corresponding to the inner cranial dura layer; it connects to the cranial dura at the FORAMEN MAGNUM opening and continues down to the level of the second sacral vertebra, where it is firmly attached. An outer periosteal layer is not necessary since vertebrae have their own periosteum. As the spinal nerves exit from the vertebral canal, a short portion of their spinal roots are covered by sleeves of the spinal dura.

The arachnoid membrane is the intermediate membrane. It is a thin and delicate netlike membrane, an avascular spider web; it lines the inner surface of the dura, spreading over the entire CNS. In the brain, it also follows the FALX and TENTORIUM but does not extend into the sulci or fissures of the brain. From it, the delicate strands of connective tissue extend down through the subarachnoid space (see pages 148–49) and connect to the pia mater.

THREE PROTECTIVE MENINGES (FROM OUTSIDE IN)

① Dura Mater (Pachymenix)

② Arachnoid Membrane (Spider Membrane)

③ Pia Mater

ONE EXTERNAL MENINX: DURA MATER

a Double-Layered Cranial Dura Mater (Dura Mater Cerebri)

a1 Thick, Outer Periosteal Layer

a2 Thin, Inner Meningeal Layer

b Single-Layered Spinal Dural Mater (Dura Mater Spinalis) Thin, Meningeal Dural Sheath

INFOLDINGS FORMING SEPTA

4 Falx Cerebri

5 Tentorium Cerebelli

6 Falx Cerebelli

Diaphragm Sellae

TWO NETLIKE MIDDLE MENINX: ARACHNOID MEMBRANE CRANIAL ARACHNOID (ARACHNOIDEA ENCEPHALI)

c Arachnoid Granulations (Villi)

SPINAL ARACHNOID (ARACHNOIDEA SPINALIS)

d 3 Dorsal Arachnoid Septa

THREE THIN, INTERNAL MENINX: PIA MATER

Pia Mater (Extensions of the Spinal Cord)

LATER EXTENSIONS

e Ligamentum Denticulatum

INFERIOR EXTENSIONS

f Filum Terminale

SPACES INVOLVING THE MENINGES: SEPARATING VERTEBRAL CANAL WALL AND DURA MATER

7 Epidural Space (Potential)

SEPARATING DURA MATER AND ARACHNOID MEMBRANE

8 Subdural Space

SEPARATING ARACHNOID MEMBRANE AND PIA MATER

9 Subarachnoid Space

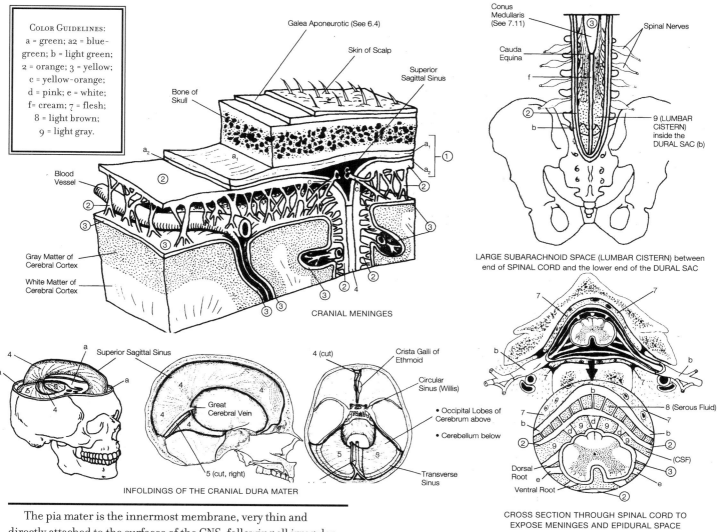

Galea Aponeurotic (See 6.4)

Skin of Scalp

Superior
Sagittal Sinus

Bone of
Skull

Blood
Vessel

Gray Matter of
Cerebral Cortex

White Matter of
Cerebral Cortex

CRANIAL MENINGES

Conus
Medullaris
(See 7.11)

Spinal Nerves

Cauda
Equina

9 (LUMBAR
CISTERN)
inside the
DURAL SAC (b)

LARGE SUBARACHNOID SPACE (LUMBAR CISTERN) between
end of SPINAL CORD and the lower end of the DURAL SAC

Superior Sagittal Sinus

Great
Cerebral Vein

5 (cut, right)

INFOLDINGS OF THE CRANIAL DURA MATER

Crista Galli of
Ethmoid

4 (cut)

Circular
Sinus (Willis)

• Occipital Lobes of
  Cerebrum above

• Cerebellum below

Transverse
Sinus

8 (Serous Fluid)

(CSF)

Dorsal
Root

Ventral Root

CROSS SECTION THROUGH SPINAL CORD TO
EXPOSE MENINGES AND EPIDURAL SPACE

The pia mater is the innermost membrane, very thin and directly attached to the surfaces of the CNS, following all irregular contours of the brain and spinal cord. It is highly specialized over the roofs of the ventricles to form the CHOROID PLEXUSES (along with the arachnoid). The pia mater is highly vascular, supporting blood vessels derived from the cerebral and cerebellar arteries that nourish the underlying cells of the brain and spinal cord.

Infoldings form septa from the thin, inner meningeal layer of the cranial dura mater; these septa are the FALX CEREBRI, TENTORIUM CEREBELLI, FALX CEREBELLI and DIAPHRAGM SELLAE. The falx cerebri extends downward to lie within the LONGITUDINAL FISSURE. It separates the two cerebral hemispheres. It is anchored anteriorly by the CRISTA GALLI of the ethmoid bone and posteriorly by the TENTORIUM. The tenotrium cerebelli supports the occipital lobes of the cerebrum and separates them from the cerebellum. It is anchored by the tentorium, petrous bones and occipital bone. The falx cerebelli forms a vertical partition between the two cerebellar hemispheres and is anchored by the occipital crest of the occipital bone. The diaphragm sellae forms the floor of the SELLA TURCICA, the concavity on the superior surface of the sphenoid bone that houses the pituitary gland.

SPACES INVOLVING THE MENINGES

Unlike the adherence of the outer layer of the dura mater to the cranial bones, there is no connection between the spinal dural sheath and the vertebrae forming the vertebral canal; there is a potential cavity called the EPIDURAL SPACE that forms a protective pad around the spinal cord (outside the dural sheath), consisting of areolar and adipose (fatty) connective tissue, and is highly vascular. It contains a venous plexus and branches of spinal nerves as they enter or exit the spinal cord. The epidural space separates the vertebral canal and the dura mater. The SUBDURAL SPACE is a very narrow space found in both portions of the CNS, in which lies a thin film of serous fluid. It separates the dura mater and arachnoid membrane. The SUBARACHNOID SPACE separates the arachnoid membrane from the pia mater. In most areas throughout the CNS, a considerable space is maintained by the weblike strands of the arachnoid membrane. The subarachnoid space contains cerebrospinal fluid. In some areas of the brain, this space is hardly recognizable, and the pia mater is almost adherent to the arachnoid (the PIA ARACHNOID). At the base of the brain, this space becomes very wide in areas, as in the SUBARACHNOID CISTERNAE. Another large subarachnoid space, the LUMBAR CISTERN, lies between the end point of the spinal cord and the end of the dural sac. This space is filled with spinal nerves (roots) as they enter or exit the spinal cord.

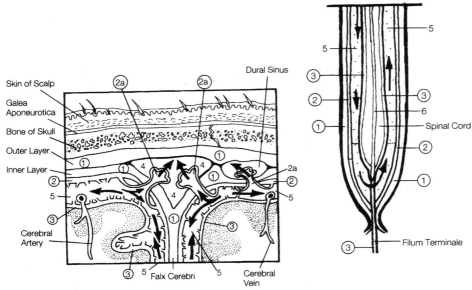

Circulation of Cerebrospinal Fluid

End of Spinal Cord

Embryologically, the CNS develops as a neural tube. Structural spaces develop within the brain and spinal cord as remnants of the original tube cavity: the brain contains several ventricles and the CEREBRAL AQUEDUCT. The spinal cord contains a small CENTRAL CANAL. All spaces are lined with EPENDYMAL CELLS.

Two large LATERAL VENTRICLES (spaces) lie within the cerebral hemispheres. Each ventricle has a body from which arise three horns. Lying within the body and part of the inferior horn (which extends into the temporal lobe) is a CHOROID PLEXUS, a network of capillaries intertwined with the ependymal lining of the ventricles. Two choroid plexuses in the lateral ventricles secrete the majority of the cerebrospinal fluid (CSF). CSF is also formed to a lesser extent by choroids plexuses in the roofs of the third and fourth ventricles (as well as medial walls of lateral ventricles). Ciliated ependymal lining covering the plexuses and the central canal is formed by the active transport and ultrafiltration of substances in blood plasma (800 ml produced each day).

CSF is a clear, colorless, watery, lymph-like alkaline fluid. It is similar to blood plasma but does not clot. There is 100 to 140 ml of it (23 ml in ventricles, 117ml in CNS subarachnoid spaces).

FUNCTIONS OF CSF

CSF forms a protective cushion around and within the CNS, protecting the brain and spinal cord from physical impact. It also buoys the CNS, so that the brain may function as a heavy organ. Brain weight is about 1500 grams, but buoyed in CSF, it weighs about 50 grams. The specific gravity of CSF is 1.007—a density close to that of brain tissue. Thus, the brain has a near neutral buoyancy floating in CSF. Shrinking or expanding of cranial contents (due to fluctuations in the amount of blood in the skull) is usually quickly balanced by increase of decrease of CSF, keeping cranial contents constant. CSF receives waste products from nerve cells.

STRUCTURES OF THE CNS THROUGH WHICH CEREBROSPINAL FLUID (CSF) CIRCULATES:

Ventricles of the Brain

5 | Subarachnoid Space of the entire CNS

6 | Central Canal of the Spinal Cord

FOUR VENTRICULAR CAVITIES OF THE BRAIN:

1V, 2V | 2 Lateral Ventricles (1st and 2nd) | in each cerebral hemisphere, inferior to the corpus callosum

3V | 3rd Ventricle | in the Diencephalon, between the Thalami

4V | 4th Ventricle | in the Brainstem, within the Pons, Medulla and Cerebellum

1 | Dura Mater

2 | Arachnoid Membrane

2a | Arachnoid Granulations (Villi)

3 | Pia Mater

4 | Superior Sagittal Sinus | (within upper margin of Falx Cerebri)

FORAMINA LINKING VENTRICLES:
OPENINGS LINKING LATER VENTRICLES TO 3RD VENTRICLE:

7 | Interventricular Foramen (Foramen of Monro)

OPENING LINKING 3RD VENTRICLE TO 4TH VENTRICLE:

8 | Cerebral Aqueduct (Aqueduct of Sylvius)

THREE FORAMINA LINKING THE 4TH VENTRICLE TO THE SUBARACHNOID SPACE (GREAT CISTERN):

9 | Median Aperture (Foramen) of Magendie

10 | 2 Lateral Apertures (Formina) of Luschka

SPECIALIZED CAPILLARIES IN THE BRAIN FOR THE MAIN PRODUCTION OF CSF:

11 | Choroid Plexuses

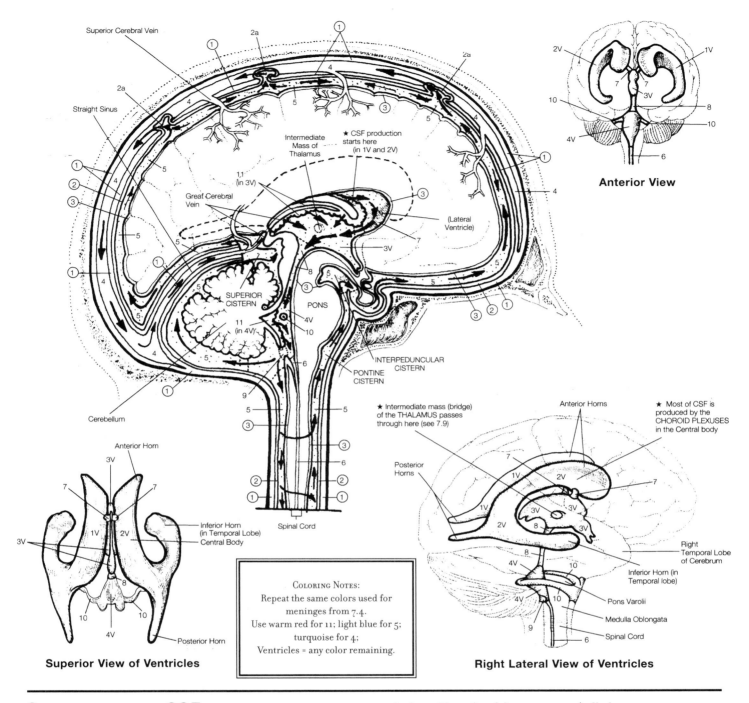

Anterior View

★ CSF production starts here (in 1V and 2V)

★ Intermediate mass (bridge) of the THALAMUS passes through here (see 7.9)

★ Most of CSF is produced by the CHOROID PLEXUSES in the Central body

Superior View of Ventricles

COLORING NOTES:
Repeat the same colors used for meninges from 7.4.
Use warm red for 11; light blue for 5; turquoise for 4;
Ventricles = any color remaining.

Right Lateral View of Ventricles

CIRCULATION OF CSF

The bulk of CSF formed in the lateral ventricles passes through the narrow oval openings of the interventricular foramina to the single, slitlike cavity of the third ventricle; it then follows down through the cerebral aqueduct to the fourth ventricle. CSF also seeps down into the central canal of the spinal cord. CSF formed by the ependymal cells of the central canal of the spinal cord drains upward into the fourth ventricle. Three openings in the roof of this ventricle permit it to pass from there into the subarachnoid space (just below the cerebellum) called the CISTERNA MAGNA (GREAT CISTERN). CSF circulates over the surface of the brain and spinal cord in the cisterna magna. It is then reabsorbed into the bloodstream from the cranial space through vas-cular berrylike tufts of the ARACHNOID (called ARACHNOID GRANULATIONS OF VILLI), which project into the upper cranial midline superior sagittal sinus.

The cisterna magna (GREAT CISTERN) is the first receiving station in the subarachnoid space of CSF from all ventricles (but directly from the fourth ventricle) via the MEDIA APERTURE OF MAGENDIE and LATERAL APERTURES OF LUSCHKA (see superior view).

CSF pools in certain locations of the subarachnoid space due to flexures and depressions of the brain structure. The five main cisterns are the great, superior, lumbar, pontine and interpeduncular. At the end of the spinal cord, the LUMBAR CISTERN is the safe site of CSF sampling for diagnostic purposes (at the level of L4 or L5). The spinal cord ends at L2.

PERIPHERAL NERVOUS SYSTEM (PNS)
CRANIAL NERVES

COMPOSITION OF IMPULSES (SENSORY AND MOTOR)

THE TWELVE PAIRS OF CRANIAL NERVES

I | Olfactory I |

II | Optic II |

III | Oculomotor III |

IV | Trochlear IV | (Smallest)

V | Trigeminal V | (Largest) two
Roots: Small Motor Root and
Large Sensory Root, which
enlarges into a swelling, from
which arises three branches:

(1) | The Ophthalmic Branch |

(2) | The Maxillary Branch |

(3) | The Mandibular Branch |

VI | Abducens VI | (Abducent)

VII | Facial VII |

VIII | Vestibulocochlear VIII |

IX | Glossopharyngeal IX |

X | Vagus X |

XI | Accessory XI | (Arises from both
Brain and Spinal Cord)

XII | Hypoglossal XII |

Posterior View of Brainstem

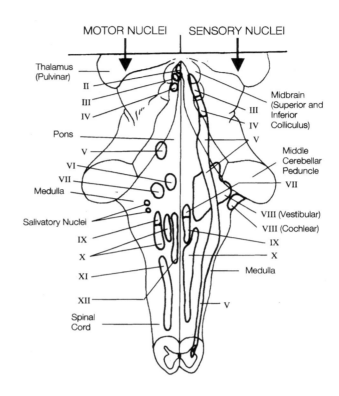

MOTOR NUCLEI   SENSORY NUCLEI

Thalamus (Pulvinar)
II
III
IV
Pons
V
VI
VII
Medulla
Salivatory Nuclei
IX
X
XI
XII
Spinal Cord

Midbrain (Superior and Inferior Colliculus)
III
IV
V
Middle Cerebellar Peduncle
VII
VIII (Vestibular)
VIII (Cochlear)
IX
X
Medulla
V

COLORING NOTES:
Use a color wheel palette, starting at olfactory (I) in yellow and then continuing through yellow-orange, orange, reds, reaching (VI) as purple, then continuing through the blues and greens, ending with hypoglossal (XII) as yellow-green.

There are twelve pairs of CRANIAL NERVES. Of these, two pairs arise from the FOREBRAIN (OLFACTORY I and OPTIC II, both of which contain sensory fibers exclusively). The other ten pairs arise from the MIDBRAIN and BRAINSTEM. All twelve pairs leave the skull through its foramina. They are designated with Roman numerals, indicating the order in which the nerves are positioned and arise from the front of the brain (I) to the back (XII). They are designated with names indicating distribution to innervated structures or function of the nerves.

Three cranial nerves contain SENSORY FIBERS only: OLFACTORY I, OPTIC II and VESTIBULOCOCHLEAR VIII; these are associated with a special sense perception. The TRIGEMINAL V is primarily sensory, except for a small motor function in the mandibular division. The cell bodies of these sensory fibers are located in ganglia outside the brain.

All the other cranial nerves contain both sensory and motor fibers (to varying degrees), and are termed MIXED NERVES. Recall that all spinal nerves are mixed nerves. Thus, all nerves in the body, both cranial and spinal, are mixed, except for cranials I, II, and VIII.

Although the SOMATIC NERVOUS SYSTEM has been defined as a conscious, voluntary system, some motor fibers control subconscious, involuntary movements. This is because the somatic fibers of the cranial nerves (which carry conscious functions) leave the brain bundled together with some fibers of the AUTONOMIC NERVOUS SYSTEM (which carry subconscious functions). (The same case may be made for the spinal nerves.)

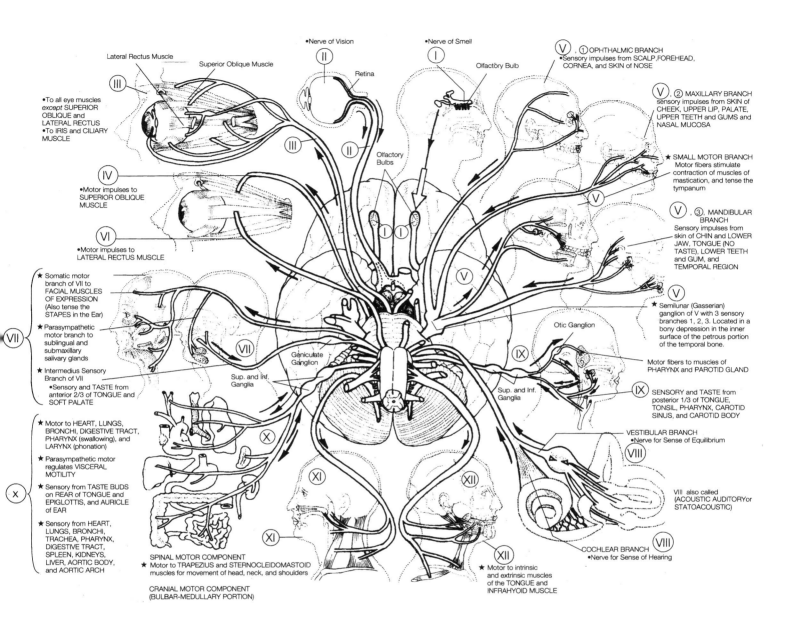

Nerve of Vision
II
Retina

Nerve of Smell
I
Olfactory Bulb

Lateral Rectus Muscle
Superior Oblique Muscle

III

Olfactory Bulbs

V, ① OPHTHALMIC BRANCH
•Sensory impulses from SCALP, FOREHEAD, CORNEA, and SKIN of NOSE

V, ② MAXILLARY BRANCH
sensory impulses from SKIN of CHEEK, UPPER LIP, PALATE, UPPER TEETH and GUMS and NASAL MUCOSA

•To all eye muscles *except* SUPERIOR OBLIQUE and LATERAL RECTUS
•To IRIS and CILIARY MUSCLE

III

II

★ SMALL MOTOR BRANCH
Motor fibers stimulate contraction of muscles of mastication, and tense the tympanum

V

IV

•Motor impulses to SUPERIOR OBLIQUE MUSCLE

V, ③, MANDIBULAR BRANCH
Sensory impulses from skin of CHIN and LOWER JAW, TONGUE (NO TASTE), LOWER TEETH and GUM, and TEMPORAL REGION

VI

•Motor impulses to LATERAL RECTUS MUSCLE

V

V

★ Semilunar (Gasserian) ganglion of V with 3 sensory branches 1, 2, 3. Located in a bony depression in the inner surface of the petrous portion of the temporal bone.

★ Somatic motor branch of VII to FACIAL MUSCLES OF EXPRESSION (Also tense the STAPES in the Ear)

VII

★ Parasympathetic motor branch to sublingual and submaxillary salivary glands

Otic Ganglion

Motor fibers to muscles of PHARYNX and PAROTID GLAND

★ Intermedius Sensory Branch of VII
•Sensory and TASTE from anterior 2/3 of TONGUE and SOFT PALATE

VII

Geniculate Ganglion

IX

Sup. and Inf. Ganglia

IX SENSORY and TASTE from posterior 1/3 of TONGUE, TONSIL, PHARYNX, CAROTID SINUS, and CAROTID BODY

Sup. and Inf. Ganglia

★ Motor to HEART, LUNGS, BRONCHI, DIGESTIVE TRACT, PHARYNX (swallowing), and LARYNX (phonation)

VESTIBULAR BRANCH
•Nerve for Sense of Equilibrium

VIII

★ Parasympathetic motor regulates VISCERAL MOTILITY

X

VIII also called (ACOUSTIC AUDITORY or STATOACOUSTIC)

★ Sensory from TASTE BUDS on REAR of TONGUE and EPIGLOTTIS, and AURICLE of EAR

XI

XII

★ Sensory from HEART, LUNGS, BRONCHI, TRACHEA, PHARYNX, DIGESTIVE TRACT, SPLEEN, KIDNEYS, LIVER, AORTIC BODY, and AORTIC ARCH

COCHLEAR BRANCH
•Nerve for Sense of Hearing

VIII

SPINAL MOTOR COMPONENT
★ Motor to TRAPEZIUS and STERNOCLEIDOMASTOID muscles for movement of head, neck, and shoulders

XI

XII

★ Motor to intrinsic and extrinsic muscles of the TONGUE and INFRAHYOID MUSCLE

CRANIAL MOTOR COMPONENT (BULBAR-MEDULLARY PORTION)

MAIN BRANCHINGS AND GANGLIA OF SPECIFIC CRANIAL NERVES

The TRIGEMINAL V has two roots, a small MOTOR ROOT and a large SENSORY ROOT. The large sensory root enlarges into a swelling called the SEMILUNAR (GASSERIAN) GANGLION, from which arise three large branches, the OPHTHALMIC, MAXILLARY and MANDIBULAR branches.

The FACIAL VII has two branches, an INTERMEDIUS SENSORY BRANCH with GENICULATE GANGLION and a SOMATIC MOTOR BRANCH. The VESTIBULOCOCHLEAR VIII also has two sensory branches, a VESTIBULAR BRANCH (with VESTIBULAR GANGLION) and a COCHLEAR (AUDITORY) BRANCH. The GLOSSOPHARYNGEAL IX has superior and inferior ganglia and OPTIC GANGLION. The

VAGUS has superior and inferior ganglia, and PETROSAL GANGLION. The HYPOGLOSSAL XII has two motor components: a SPINAL MOTOR COMPONENT and a CRANIAL MOTOR COMPONENT (BULBAR [MEDULLARY] PORTION).

The composition of cranial nerves is divided into motor and sensory. The three entirely sensory nerves are I, II, and VIII (except the Edinger-Westphal nucleus of II, which controls motor function of intrinsic eye muscles). None are entirely motor. V is primarily sensory. The primarily motor ones are III, IV, VI, XI, XII. Some are equally motor and sensory; they are VII, IX and X.

The PERIPHERAL NERVOUS SYSTEM (PNS) is the portion of the nervous system outside the limits of the CNS. The parts of the PNS convey impulses to and from the brain and spinal cord. The PNS consists of sensory receptors within sensory organs, NERVE FIBERS, PLEXUSES, GANGLIA, and SPECIALIZED MOTOR END ORGANS that are derived from the two main types of nerves of the PNS: CRANIAL NERVES (arising from the brain) and SPINAL NERVES (arising from the spinal cord).

The thirty-one pairs of spinal nerves are named and numbered according to the segment level of the spinal cord from which they emerge. All spinal nerves (except c1) leave the spinal cord and vertebral canal through the intervertebral foramina to ultimately innervate a DERMATOME. The first cervical nerve, c1, emerges between the atlas bone and the occipital bone, and does not innervate a dermatome for lack of a DORSAL ROOT. During fetal development, the spinal cord grows more slowly than the vertebral column. (Recall that the spinal cord terminates at the second LUMBAR VERTEBRA. Thus, the lower lumbar, sacral, and coccygeal nerves that constitute the cauda equina must descend to reach their foramina.) Spinal nerves are mixed collections of axons of sensory and motor neurons bound in an intricate array of fibrous connective tissue.

A spinal nerve attaches at two points to the spinal cord: the DORSAL (POSTERIOR) ROOT and the VENTRAL (ANTERIOR) ROOT. In the dorsal root, sensory axons (central processes) enter the POSTERIOR HORN of the cord, conveying sensory impulses into the cord. The DORSAL ROOT GANGLION is a swelling within the root composed of the unipolar cell bodies of the sensory neurons. In the ventral root, motor axons arise from multipolar cell bodies of motor neurons within the ANTERIOR HORN of the cord, conveying motor impulses away from the CNS. These two roots unite a short distance from the cord at the intervertebral foramina to form the spinal nerve.

THE THIRTY-ONE PAIRS OF SPINAL NERVES

C	8 Cervical	C1-C8
T	12 Thoracic	T1-T12
L	5 Lumbar	L1-L5
5	5 Sacral	S1-S5
Co	1 Coccygeal	Co1

GENERAL STRUCTURE OF A SPINAL NERVE:

SN   Spinal Nerve

POSTERIOR SENSORY ROOT

1   Dorsal Spinal Root

 1a   Dorsal-Root Ganglion

ANTERIOR MOTOR ROOT

2   Ventral Spinal Root

SPINAL NERVE BRANCHES: RAMI

3   Dorsal Ramus (Posterior Ramus)

4   Ventral Ramus (Anterior Ramus)

RAMI COMMUNICANTES OF THE AUTONOMIC NERVOUS SYSTEM

5   Gray Ramus Communicans

6   White Ramus Communicans

7   Sympathetic Chain Ganglion

8   Splanchnic Nerves

SMALL RETURNING BRANCH

9   Meningeal Branch

Re-enters spinal canal through the intervertebral foramen to supply the vertebrae, vertebral ligaments, meninges and blood vessels of the cord.

CUTANEOUS SENSORY AREAS

C   Dorsal Roots of Cervical Nerves (C2–C8)

L   Dorsal Roots of Lumbar Nerves (L1–L5)

T   Dorsal Roots of Thoracic Nerves (T1–T12)

S   Dorsal Roots of Sacral Nerves (S15)

PERIPHERAL NERVES: CRANIAL AND SPINAL NERVES

10   Fasciculus

Small cylindrical bundle of nerve fibers.

11   Endoneurium

Delicate connective tissue sheath—Henle's sheath—that surrounds each nerve fiber and its myelin and/or neurilemma covering within a fasciculus.

12   Perineurium

Thicker fibrous sheath investing each fasciculus.)

13   Epineurium

General outermost thick covering of fibrous connective tissue around the entire nerve —containing intrafascicular vessels and adipose cells; the spinal dura mater fuses with the epineurium for a short distance as the nerve exits from the vertebral canal. It is juxtaposed and continuous with the surrounding deep or surperficial fascia.

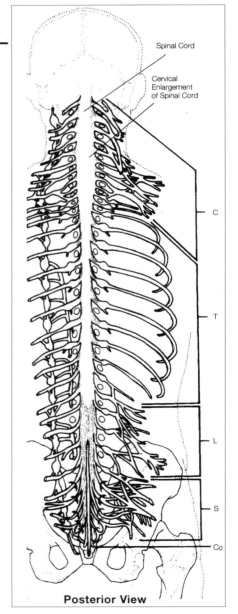

Spinal Cord

Cervical Enlargement of Spinal Cord

C

T

L

S

Co

Posterior View

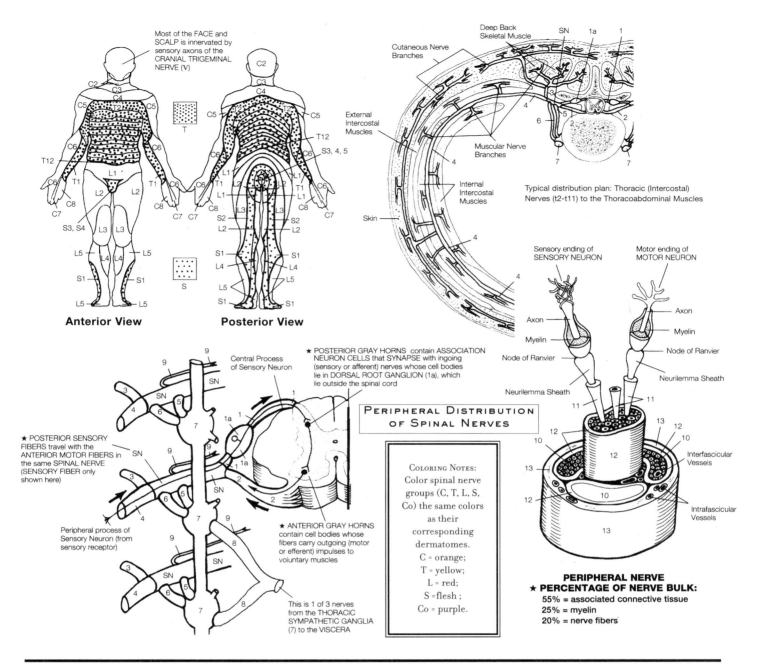

Anterior View

Posterior View

Most of the FACE and SCALP is innervated by sensory axons of the CRANIAL TRIGEMINAL NERVE (V)

T

S

Cutaneous Nerve Branches

Deep Back Skeletal Muscle

External Intercostal Muscles

Muscular Nerve Branches

Internal Intercostal Muscles

Skin

Typical distribution plan: Thoracic (Intercostal) Nerves (t2-t11) to the Thoracoabdominal Muscles

Sensory ending of SENSORY NEURON

Motor ending of MOTOR NEURON

Axon

Myelin

Node of Ranvier

Neurilemma Sheath

Axon

Myelin

Node of Ranvier

Neurilemma Sheath

Interfascicular Vessels

Intrafascicular Vessels

Central Process of Sensory Neuron

★ POSTERIOR GRAY HORNS contain ASSOCIATION NEURON CELLS that SYNAPSE with ingoing (sensory or afferent) nerves whose cell bodies lie in DORSAL ROOT GANGLION (1a), which lie outside the spinal cord

PERIPHERAL DISTRIBUTION OF SPINAL NERVES

★ POSTERIOR SENSORY FIBERS travel with the ANTERIOR MOTOR FIBERS in the same SPINAL NERVE (SENSORY FIBER only shown here)

Peripheral process of Sensory Neuron (from sensory receptor)

★ ANTERIOR GRAY HORNS contain cell bodies whose fibers carry outgoing (motor or efferent) impulses to voluntary muscles

This is 1 of 3 nerves from the THORACIC SYMPATHETIC GANGLIA (7) to the VISCERA

COLORING NOTES:
Color spinal nerve groups (C, T, L, S, Co) the same colors as their corresponding dermatomes.
C = orange;
T = yellow;
L = red;
S = flesh ;
Co = purple.

PERIPHERAL NERVE
★ **PERCENTAGE OF NERVE BULK:**
55% = associated connective tissue
25% = myelin
20% = nerve fibers

A typical spinal nerve distribution (thoracic) (T2-T12) divides immediately into several branches after emerging from the intervertebral foramina. The spinal nerve splits into two large branches, or RAMI, each of which has a number of smaller MUSCULAR and CUTANEOUS BRANCHES. The DORSAL RAMUS innervates the deep back skeletal muscles (and joints) along the vertebral column and the overlying fascia and skin. The VENTRAL RAMUS innervates the rest of the body (muscles, skin, and fascia) on the lateral and anterior side of the torso, coursing between the intercostal muscle layers of the rib cage and muscles of the abdominal wall). The spinal nerve also has two branches, the RAMI COMMUNICANTES, that connect to a SYMPATHETIC CHAIN GANGLION, part of the sympathetic division of the autonomic nervous system. The GRAY RAMUS (in all spinal nerves) contains UNMYELINATED POSTGANGLIONIC FIBERS. The WHITE RAMUS (only T1-L2) contains MYELINATED PREGANGLIONIC FIBERS.

THE HUMAN DERMATOMAL MAP

The spinal nerves innervate specific, constant segments of the skin (cutaneous areas) over the entire body. A skin segment whose sensory receptors and axons are supplied by a single dorsal (posterior) sensory root of a spinal nerve is called a DERMATOME. All spinal nerves supply branches to the skin, except the first cervical spinal nerve (C1). (C1 has no dermatome because it lacks a dorsal root.) In the neck and trunk, there are consecutive dermatome bands of skin. Especially in the trunk area, three or four dorsal roots may receive inputs from a single dermatome. Thus, overlapping occurs between adjacent dermatomes. The human dermatomal map is of clinical importance when determining which segment of the spinal cord or spinal nerve may be malfunctioning, or when anesthetizing a specific portion of the body. Note that C1 does not innervate the skin. Most of the face and scalp is innervated by sensory axons of the CRANIAL TRIGEMINAL NERVE (V).

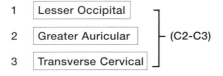

CERVICAL PLEXUS (C1–C4), and a portion of C5

BRACHIAL PLEXUS (C5–T1), and portions of CA and T2

Posterior View

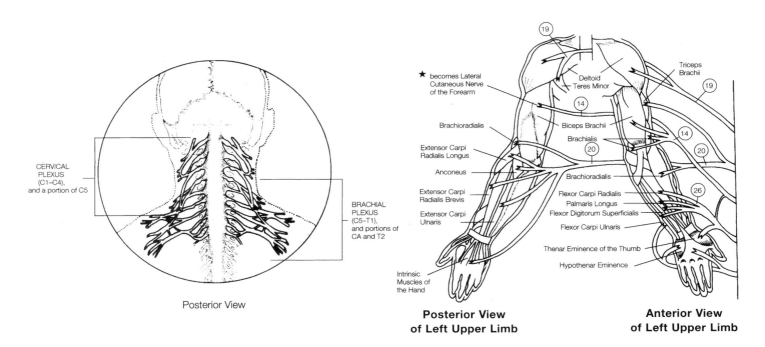

★ becomes Lateral Cutaneous Nerve of the Forearm

Deltoid
Teres Minor

Triceps Brachii

Brachioradialis
Extensor Carpi Radialis Longus
Anconeus
Extensor Carpi Radialis Brevis
Extensor Carpi Ulnaris
Intrinsic Muscles of the Hand

Biceps Brachii
Brachialis
Brachioradialis
Flexor Carpi Radialis
Palmaris Longus
Flexor Digitorum Superficialis
Flexor Carpi Ulnaris
Thenar Eminence of the Thumb
Hypothenar Eminence

Posterior View of Left Upper Limb

Anterior View of Left Upper Limb

SPINAL NERVES TO THE HEAD AND NECK: THE CERVICAL PLEXUS (C1-C4) (WITH CONTRIBUTIONS FROM C5)

| R | Roots | (C1 -C4) |

SUPERFICIAL CUTANEOUS BRANCHES

1	Lesser Occipital
2	Greater Auricular	(C2-C3)
3	Transverse Cervical
4	Supraclavicular	(C3, C4)

DEEP MOTOR BRANCHES

5	Ansa Cervicalis	(C1-C3)
5a	Superior Root	(C1, C2)
5b	Inferior Root	(C2, C3)
*6	Phrenic	(C3-C5)
7	Segmental Branches	(C1-C5)
XI	Accessory Cranial Nerve XI
XII	Hypoglossal Cranial Nerve XII

SPINAL NERVES TO THE UPPER LIMB: THE BRACHIAL PLEXUS (C5-T1) (WITH CONTRIBUTIONS FROM C4 AND T2)

R	5 Roots:	(C5-T1)
8	Dorsal Scapular Branch	(C5)
9	Long Thoracic Branch	(C5, C6, C7)
T	3 Trunks (Upper Middle Lower	(C5-C6)
10	Subclavius	(C5-C6)
11	Suprascapular
D	6 Divisions	(C5-C8)
12	3 Anterior Divisions	(C5-C8)
13	3 Posterior Divisions
C	3 Cords	(C5-T1)

LATERAL CORD: INNERVATES LATERAL ASPECT OF THE LIMB AND SOME SUPERFICIAL BACK MUSCLES

*14	Musculocutaneous
15	Median (Lateral Head)	(C5-C7)
16	Lateral Pectoral

POSTERIOR CORD: INNERVATES POSTERIOR ASPECT OF THE LIMB AND TWO SUPERFICIAL BACK MUSCLES

17	Subscapular
18	Thoracodorsal	(C5-C6)
*19	Axillary (Circumflex)
*20	Radial (C5-C8, T1)

MEDIAL CORD INNERVATES ANTERIOR ASPECT OF THE LIMB

21	Medial Pectoral
22	Medial Brachial Cutaneous	(C8, T1)
23	Medial Antebrachial Cutaneous
24	Medial (Medial Head)	(C5-C8, T1)
*25	Ulnar	(C8, T1)
*26	Median	(C6-C8, T1)

* indicates major nerves of the plexus.

COLOR GUIDELINES:
R = yellow ; T = orange; 12 = light blue;
13 = light green; C = red.

The branching distribution of the THORACIC SPINAL NERVE GROUP (T2–T12) is called the INTERCOSTAL NERVES. This is the only group of spinal nerves not termed a PLEXUS, as they go directly to the structures they innervate. The ventral rami of all other twenty pairs of spinal nerves (including T1) combine with adjacent nerves on either side of the body and then split again as a network of nerves referred to as a plexus, from which muscular and cutaneous branches emerge in a definite pattern. There are four plexuses of spinal nerves: CERVICAL, BRACHIAL, LUMBAR and SACRAL PLEXUSES. The names of the nerves indicate structures innervated or general courses taken (nerve numbers in parentheses indicate formation of the plexus from root origins of VENTRAL RAMI).

The roots (ventral rami) of the CERVICAL PLEXUS are C1–C4 and a portion of C5. They are positioned deep on each side of the neck, alongside the first four CERVICAL VERTEBRAE. They supply the muscles, fascia, and skin of the head, neck, and upper part of the shoulders. The branches connect with cranial nerves XI (ACCESSORY) and XII (HYPOGLOSSAL), supplying dual innervation to specific neck and pharyngeal muscles. The major nerve is the PHRENIC NERVE formed from fibers of C3, C4, and C5 supplying motor fibers to the dia-phragm. (These motor impulses cause contraction of the dia-phragm, inspiring air into the lungs, essential for normal breathing.)

The BRACHIAL PLEXUS consists of five roots (ventral rami): C5–C8, T1, and portions of C4 and T2.They are positioned on either side of the last four cervical and first THORACIC VERTEBRAE. The brachial plexus extends downward and laterally, passes over the first rib behind the clavicle bone and enters the axillary space below the shoulder. It constitutes the entire nerve supply for the upper limbs, the majority of the shoulder region, and some neck muscles. It is structurally divided into ROOTS, TRUNKS, DIVISIONS, and CORDS. Five roots unite to form three trunks: roots C5 and C6 converge to form the UPPER TRUNK, root C7 becomes the MIDDLE TRUNK

and roots C8 and T1 converge to form the LOWER TRUNK. The three trunks branch into six divisions (three of which make up an ANTERIOR DIVISION and three of which make up a POSTERIOR DIVISION). The six divisions converge into three cords: the anterior division of the upper and middle trunk converge to form the LATERAL CORD, the anterior division of the lower trunk continues to become the MEDIAL CORD (thus mostly contains C8 and T1 fibers) and the posterior divisions of all three trunks converge to form the POSTERIOR CORD (thus contains C5–C8 fibers). This posterior cord is deep to the other two cords. The three cords are formed in the axilla about the axillary artery. These cords give rise to the PERIPHERAL NERVES, which supply the musculoskeletal structures of the upper limbs.

There are five major nerves (TERMINAL NERVES) from the cords of the brachial plexus. From the lateral cord issues the MUSCULOCUTANEOUS NERVE which innervates the anterior upper arm muscles (including biceps brachii). From the posterior cord issues the AXILLARY NERVE wraps around the surgical neck of the humerus bone to innervate the deltoid and teres minor muscles. The RADIAL NERVE innervates triceps brachii, turns halfway about the humerus bone and innervates the posterior upper arm and forearm extensor muscles (including brachioradialis). The ULNAR NERVE innervates the anterior forearm flexor muscles and the hypothenar and intrinsic flexor muscles of the hand in the medial cord. The MEDIAN NERVE (LATERAL and MEDIAL HEADS) innervates the anterior forearm flexor muscles and the thenar muscles of the thumb.

Many injuries often arise from these five major brachial nerves at points where they pass close to the surface (e.g.: the ulnar nerve on the back of the elbow, known as the "funny bone") or around a bone.

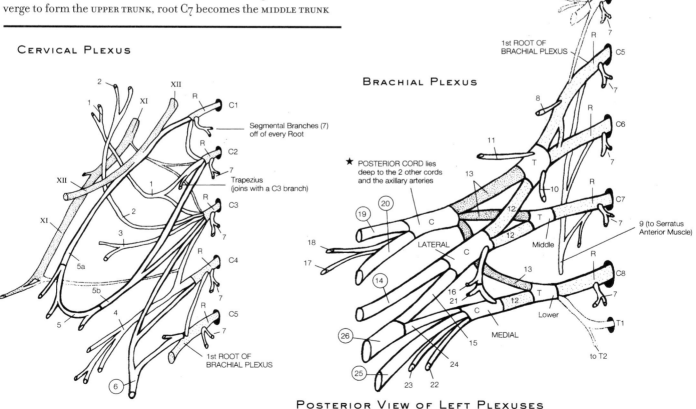

CERVICAL PLEXUS

Segmental Branches (7) off of every Root

Trapezius (joins with a C3 branch)

1st ROOT OF BRACHIAL PLEXUS

BRACHIAL PLEXUS

Last ROOT OF CELIAC PLEXUS

1st ROOT OF BRACHIAL PLEXUS

★ POSTERIOR CORD lies deep to the 2 other cords and the axillary arteries

9 (to Serratus Anterior Muscle)

LATERAL

Middle

MEDIAL

Lower

to T2

POSTERIOR VIEW OF LEFT PLEXUSES

The LUMBAR PLEXUS contains five ROOTS (VENTRAL RAMI) L1–L4, and a portion of T12. They are positioned on either side of the first and fourth lumbar vertebrae. The lumbar plexus innervates the lower abdominal wall, male scrotum (or female labia majora), and anterior and medial portions of the thigh. It is not as complex as the BRACHIAL PLEXUS, with no intricate interlacing of fibers. Structurally, the lumbar plexus has only roots and divisions (with no trunks or cords as in the brachial plexus). The five roots branch into two divisions: the ANTERIOR DIVISION superficial to the quadratus lumborum muscle, and the POSTERIOR DIVISION that passes obliquely outward behind the psoas major muscle. Those divisions give rise to its PERIPHERAL NERVES. The largest nerve of the lumbar plexus is the FEMORAL NERVE (posterior division) which innervates the anterior thigh muscles and extensor leg muscles. The OBTURATOR NERVE (anterior division) is a large nerve that innervates the adductor leg muscles.

The seven roots (ventral rami) of the SACRAL PLEXUS are L4–L5, S1–S4. It is positioned immediately caudal to the lumbar—largely in front of the sacrum—and innervates the lower back, pelvis, perineum, pudendal genitalia, buttocks, the posterior surface of the thigh and leg, and the dorsal and ventral surfaces of the foot. Like the lumbar plexus, it contains roots that branch into ANTERIOR and POSTERIOR DIVISIONS from which the peripheral nerves arise. The major nerve is the SCIATIC NERVE (L4–S3), the largest nerve in the body, supplying the entire musculature of the leg and foot. It is actually composed of two nerves (the COMMON PERONEAL and the TIBIAL NERVES, wrapped in a common SCIATIC SHEATH). It passes from the pelvis through the GREATER SCIATIC NOTCH of the os coxa and extends down the posterior aspect of the thigh among the hamstrings and adductor magnus. Just above and behind the knee, at the popliteal fossa, it divides separately into the TIBIAL and COMMON PERONEAL NERVES.

Another large nerve, the tibial nerve (L4–S3) passes down between the heads of the gastrocnemius muscle (innervating the posterior leg), then curves under the medial arch of the foot (innervating the sole of the foot through plantar branches). The large common peroneal (L4–S3) rounds the neck of the fibular bone and divides into two branches: the DEEP PERONEAL BRANCH innervates the anteriorlateral leg and the SUPERFICIAL PERONEAL BRANCH innervates the peroneal muscles of the lateral leg.

SPINAL NERVES TO THE GENITALS, ANTEROABDOMINAL WALL, AND LOWER LIMBS

THE LUMBAR PLEXUS (L1–L4) (With Contributions From T12)

R [5 Roots] (T12–L4)

(1) [Lumbosacral Trunk] (L4, L5)

DIVISIONS:
Anterior Division

2 [Ilioinguinal] (L1)

3 [Genitofemoral] (L1, L2)

(4) [Obturator] (L2–L4)

5 [Saphenous] (Accessory Obturator)

Posterior Division

6 [Iliohypogastric (T12, L1)]

7 [Lateral Femoral Cutaneous (L2, L3)]

(8) [Femoral (L2–L4)]

SPINAL NERVES TO THE BUTTOCKS, PERINEUM, AND LOWER LIMBS

THE SACRAL PLEXUS (L4–L5, S1–S4)

R [7 Roots:] (L4–S4)

[1 Lumbosacral Trunk] (L4–L5)

DIVISIONS:
Anterior Division

9 [Pudendal] (S2–S4)

10 [Nerve To Quadratus Femoris M.] (L4–L5, S1)

11 [Nerve To Obturator Internus M.] (L5–S2)

(12) [Tibial (Medial Popliteal)] (L4–S3)

  a [Medial Plantar]

  b [Lateral Plantar]

Posterior Division

(13) [Common Peroneal (Lateral Popliteal)] (L4–S3)

  c [Superficial Peroneal]

  d [Deep Peroneal]

Sciatic (bracket spanning 12–13)

14 [Superior Gluteal] (L4–L5, S1)

15 [Inferior Gluteal] (L5–S2)

16 [Nerve To Piriformis M.] (S1–S2)

17 [Posterior Cutaneous] (S1–S3)

(18) [Sciatic:] (Tibial And Common Peroneal) (L4–S3)
(largest nerve in the body)

Co1 [Coccygeal Nerves (Co1)]

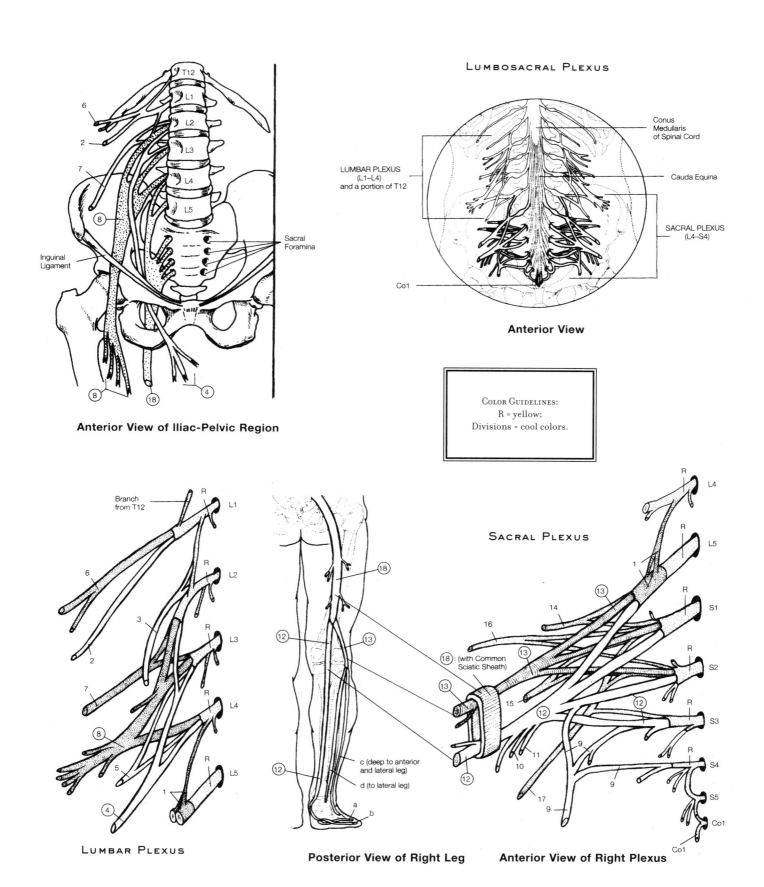

LUMBOSACRAL PLEXUS

Anterior View of Iliac-Pelvic Region

T12
L1
L2
L3
L4
L5

6
2
7
8
Inguinal Ligament
Sacral Foramina
8
18
4

Conus Medullaris of Spinal Cord

LUMBAR PLEXUS (L1–L4) and a portion of T12

Cauda Equina

SACRAL PLEXUS (L4–S4)

Co1

Anterior View

COLOR GUIDELINES:
R = yellow;
Divisions = cool colors.

LUMBAR PLEXUS

Branch from T12
R
L1
6
R
L2
3
2
R
L3
7
8
R
L4
5
1
R
L5
4

Posterior View of Right Leg

18
12
13
12
12
c (deep to anterior and lateral leg)
d (to lateral leg)
a
b

18: (with Common Sciatic Sheath)
13
13
12

SACRAL PLEXUS

R
L4
R
L5
1
13
14
16
13
15
11
10
9
17
9
9
R
S1
R
S2
12
R
S3
R
S4
R
S5
Co1
Co1

Anterior View of Right Plexus

Autonomic Nervous System (ANS)
Sympathetic Division

The AUTONOMIC NERVOUS SYSTEM (ANS) is the portion of the efferent system of the PNS that innervates and regulates smooth muscle (of visceral organs and blood vessels), cardiac muscle (of the heart) and glands (salivary, gastric, sweat, and the adrenal medulla).

The ANS is entirely motor, often referred to as the VISCERAL EFFERENT NERVOUS SYSTEM. All of its axons are visceral efferent fibers, conveying impulses from the CNS to visceral effectors (not to skeletal muscles, as in the SOMATIC NERVOUS SYSTEM [SNS])

The ANS is concerned with the control of involuntary bodily functions, but is not self-governing or independent of the CNS. The ANS is itself regulated by brain centers of the CNS, consti-

tuting the visceral nervous system: Sensory impulses from VISCERAL EFFECTORS pass over VISCERAL AFFERENT NEURONS (whose cell bodies are in the dorsal root ganglia of the spinal nerves). These afferent impulses influence mainly the hypothalamus, but also the medulla oblongata and spinal cord. These centers integrate the sensory visceral input with input from higher CNS centers (cerebral cortex and limbic system). Appropriate responses are then sent back to the visceral effector organs via the ANS.

The ANS consists of two principal divisions (organs that receive impulses from both divisions are said to have dual innervation). The PARASYMPATHETIC DIVISION is associated with decrease in organ activity, "vegetative reactions" (e.g., after a

VISCERAL EFFERENT (MOTOR) PATHWAYS OF THE ANS SUMPATHETIC DIVSION: EXPANDS BODY ENERGY

A | Spinal Cord

A1 | Lateral Gray Horns

PREGANGLIONIC NEURONS (CONVEY EFFERENT IMPULSES FROM CNS TO AUTONOMIC GANGLIA):

1 | Preganglionic cell bodies

2 | Preganglionic axons

TYPES OF AUTONOMIC GANGLIA:

3 | Sympathetic double chain of 22 paravertebral ganglia

4 | 3 Collateral (prevertebral) ganglia

POSTGANGLIONIC NEURONS (RELAY IMPULSES FROM AUTONOMIC GANGLIA TO VISCERAL EFFECTORS):

5 | Postganglionic cell bodies

Most of the postganglionic fibers leaving the sympathetic trunk join the spinal nerves through the GRAY (UNMYELINATED) RAMUS COMMUNICANTES before supplying visceral effectors (however, some pass directly from the trunk to visceral effectors).

POSTGANGLIONIC AXONS TO FIVE BASIC REGIONS:

6 | The head and neck

Postganglionic axons leave the sympathetic chain at the superior cervical ganglion (the uppermost of the three cervical ganglia in the cervical portion of the sympathetic trunk, located in the neck anterior to the prevertebral muscles). These axons follow the carotid arteries en route to the head and neck.

7 | The thoracic viscera

Postganglionic axons to the THORACIC VISCERA leave the upper portions of the sympathetic chain (the three cervical ganglia and T_1–T_4) and travel directly to the heart, bronchi, lungs and esophagus, forming cardiac, pulmonary, and esophageal plexuses along the way.

8 | The abdominal viscera

Postganglionic axons to the abdominal viscera have their cell bodies in the three sets of PREVERTEBRAL (COLLATERAL) GANGLIA arranged around the major branches of the abdominal aorta. Preganglionic axons (T_5–L_2) pass through the sympathetic trunk without terminating in the trunk, passing through it and forming SPLANCHNIC NERVES, which terminate in the CELIAC (SOLAR) PLEXUS, forming the three collateral ganglia. The GREATER SPLANCHIC NERVE passes to the large, upper CELIAC GANGLION.

9 | The pelvic/perineal viscera

Postganglionic axons to the pelvis and perineum leave the lower of the three COLLATERAL GANGLIA, the INFERIOR MESENTERIC GANGLIA, downward toward the HYPOGASTRIC PLEXUS, and from there to the lower intestine, and the pelvic and perineal organs.

10 | The skin

Postganglionic axons to the skin leave the sympathetic chain via the gray rami communicantes at each and every level of the spinal cord to reach blood vessels associated with skeletal muscle, arrector pili hair muscles, and sweat glands.

11 | White rami communicantes (only in T1–L2)

12 | Gray rami communicantes

13 | 3 Splanchnic nerves

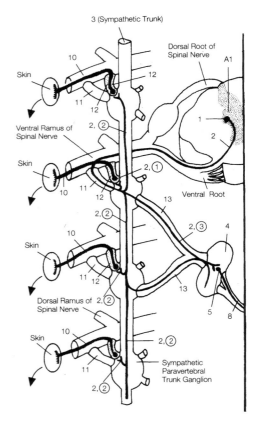

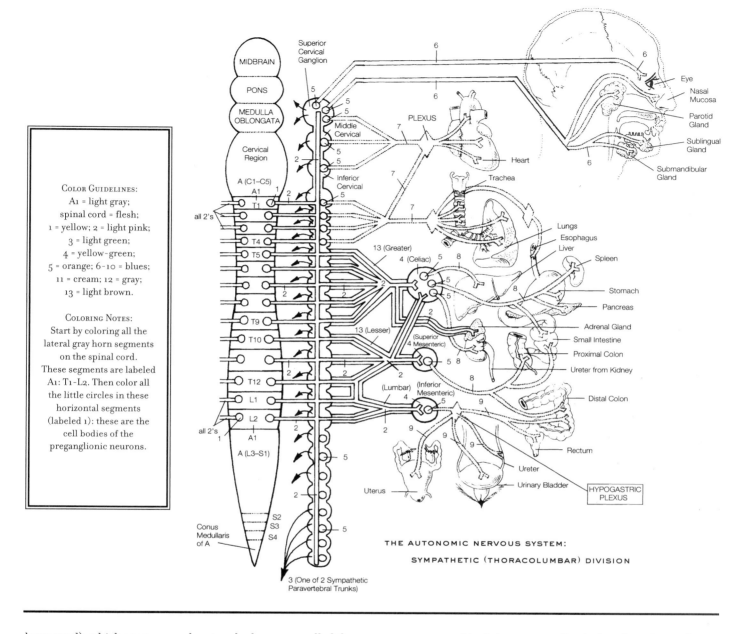

Color Guidelines:
A1 = light gray;
spinal cord = flesh;
1 = yellow; 2 = light pink;
3 = light green;
4 = yellow-green;
5 = orange; 6-10 = blues;
11 = cream; 12 = gray;
13 = light brown.

Coloring Notes:
Start by coloring all the
lateral gray horn segments
on the spinal cord.
These segments are labeled
A1: T1-L2. Then color all
the little circles in these
horizontal segments
(labeled 1): these are the
cell bodies of the
preganglionic neurons.

THE AUTONOMIC NERVOUS SYSTEM:

SYMPATHETIC (THORACOLUMBAR) DIVISION

large meal), which conserve and restore body energy, called the REST-REPOSE SYSTEM. The SYMPATHETIC DIVISION is associated with stimulation to start or increase organ activity, such as "fight or flight" responses, which are the expenditure of body energy in response to the need to flee or fight, due to being frightened.

Autonomic visceral efferent pathways always consist of two neurons. The first of the visceral efferent neurons is called a PRE-GANGLIONIC NEURON.

In the sympathetic division, the cell bodies of these neurons are exclusive to the lateral gray horns of the spinal cord in all the THO-RACIC (T1–T12) and the first two LUMBAR SEGMENTS (L1–L2). (Thus, the sympathetic division is also called the thoracolumbar division.)

Their lightly myelinated axons, the PREGANGLIONIC FIBERS, leave the spinal cord through the ventral roots along with the SOMATIC EFFERENT FIBERS of spinal nerves T1–L2. The pregan-glionic fibers branch off from the spinal nerve through the

WHITE RAMI, and in doing so are collectively referred to as the white rami communicantes.

In one of three routes, each preganglionic fiber courses to an AUTONOMIC GANGLION, where it synapses with the dendrites and cell body of the POSTGANGLIONIC NEURON, the second neuron of the visceral efferent pathway. The postganglionic neuron lies entirely outside the CNS. Its axon, the POSTGANGLIONIC FIBER, is unmyeli-nated and terminates in a visceral effector organ.

Each preganglionic fiber via the WHITE RAMI enters the SYMPA-THETIC TRUNK or VERTEBRAL CHAIN GANGLIA, a double series of ganglia that lie in a vertical row on either side of the vertebral column from the base of the skull to the coccyx. (Typically, there are twenty-two ganglia arranged more or less segmentally along the trunk chain: three CERVICAL, eleven THORACIC, four LUMBAR, and four SACRAL (most of the preganglionic fibers enter the eleven thoracic ganglia, just ventral to the necks of the corresponding ribs).

It is noteworthy that in the sympathetic division there are no preganglionic fibers that arise from the brainstem or sacral area of the spinal cord. However, this is what occurs in the PARASYMPATHETIC DIVISION of the ANS: the preganglionic cell bodies are found in nuclei of the BRAINSTEM (MIDBRAIN, PONS and MEDULLA OBLONGATA) and the lateral gray horns of the second through fourth sacral segments of the spinal cord. (Thus, the parasympathetic division is also referred to as the CRANIOSACRAL DIVISION of the ANS.)

The PREGANGLIONIC AXONS emerge as either a part of a cranial nerve (limited to III, VII, IX, X) or as part of the VENTRAL ROOT of a SPINAL SACRAL NERVE. These preganglionic axons synapse in GANGLIA that are located next to or within the visceral effector organs. These ganglia are referred to as TERMINAL GANGLIA and are exclusive to the parasympathetic division. Recall that sympathetic innervation of the skin originates from each and every level of the spinal cord. Note that, except for three spinal sacral nerves, parasympathetic fibers do not travel within spinal nerves. Thus, cutaneous effectors in the skin (blood vessels, arrector pili hair muscles, and sweat glands) do not receive parasympathetic postganglionic innervation and are not found in the long peripheral nerves as are sympathetic postganglionic fibers.

CRANIAL PARASYMPATHETIC OUTFLOW has five components: four pairs of HEAD (CRANIAL) GANGLIA, plus the PLEXUSES, associated with the VAGUS NERVE (X). Preganglionic axons of cranial nerves III, VII, and IX synapse in the CRANIAL GANGLIA. Short POSTGANGLIONIC AXONS arise from these ganglia to supply specific head structures. The OCULOMOTOR (III) nerve provides parasympathetic innervation to the CILIARY GANGLIA (located near the back of the orbit lateral to each optic nerve).

FACIAL (VII) innervates two ganglia: the PTERYGOPALATINE GANGLIA, which is lateral to a spenopalatine foramen in the lateral wall of the nasal cavity near the nasopharynx, and the SUBMANDIBULAR GANGLIA, which is deep to submandibular salivary gland under the mandible near its duct.

GLOSSOPHARYNGEAL (IX) innervates the otic ganglia just below the foramen ovale in the posterior margin of the great sphenoidal wing of the sphenoid bone.

Long preganglionic axons from the vagus nerve (X) pass down the neck through the thorax to the abdomen (to the level of the descending colon), branching off into many plexuses along the way. (Within the plexuses they are joined with POSTGANGLIONIC SYMPATHETIC AXONS destined for the same visceral effector.)

In the thorax, VAGAL FIBERS join with SYMPATHETIC FIBERS in the CARDIAC and PULMONARY PLEXUSES. From here, postganglionic axons travel to the heart, lungs, and trachea.

In the abdomen, continuing vagal fibers mix with sympathetic fibers around the three PREVERTEBRAL GANGLIA of the CELIAC (SOLAR) PLEXUS. From here, postganglionic axons follow arteries to the abdominal organs they supply.

The SACRAL PARASYMPATHETIC OUTFLOW begins with the PREGANGLIONIC FIBERS that arise from cell bodies in the lateral gray horns of S2-S4 and leave through the ventral roots, joining the spinal nerves. These fibers then bend quickly away toward the pelvic area collectively forming the PELVIC SPLANCHNIC NERVES. The nerves form a HYPOGASTRIC PLEXUS on the rectum with the inferior MESENTERIC GANGLIA. These axons then follow the arteries to the viscera they supply.

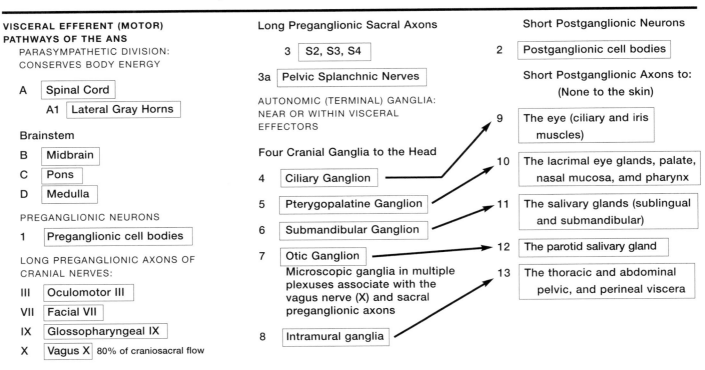

VISCERAL EFFERENT (MOTOR) PATHWAYS OF THE ANS
PARASYMPATHETIC DIVISION: CONSERVES BODY ENERGY

A Spinal Cord

   A1 Lateral Gray Horns

Brainstem

B Midbrain
C Pons
D Medulla

PREGANGLIONIC NEURONS

1 Preganglionic cell bodies

LONG PREGANGLIONIC AXONS OF CRANIAL NERVES:

III Oculomotor III
VII Facial VII
IX Glossopharyngeal IX
X Vagus X 80% of craniosacral flow

Long Preganglionic Sacral Axons

3 S2, S3, S4

3a Pelvic Splanchnic Nerves

AUTONOMIC (TERMINAL) GANGLIA: NEAR OR WITHIN VISCERAL EFFECTORS

Four Cranial Ganglia to the Head

4 Ciliary Ganglion
5 Pterygopalatine Ganglion
6 Submandibular Ganglion
7 Otic Ganglion

Microscopic ganglia in multiple plexuses associate with the vagus nerve (X) and sacral preganglionic axons

8 Intramural ganglia

Short Postganglionic Neurons

2 Postganglionic cell bodies

Short Postganglionic Axons to:
(None to the skin)

9 The eye (ciliary and iris muscles)

10 The lacrimal eye glands, palate, nasal mucosa, amd pharynx

11 The salivary glands (sublingual and submandibular)

12 The parotid salivary gland

13 The thoracic and abdominal pelvic, and perineal viscera

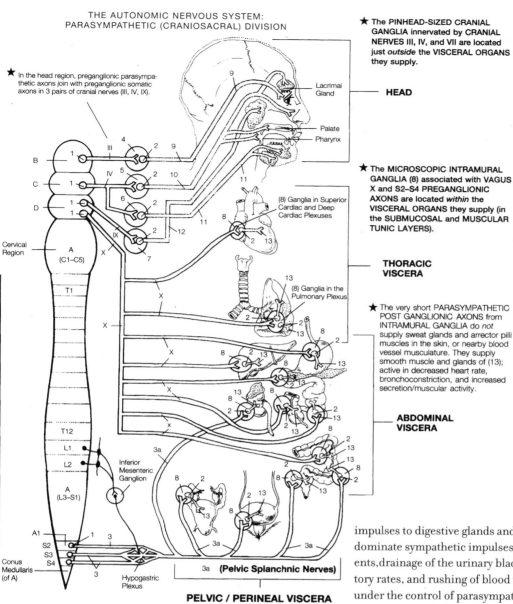

THE AUTONOMIC NERVOUS SYSTEM:
PARASYMPATHETIC (CRANIOSACRAL) DIVISION

★ In the head region, preganglionic parasympathetic axons join with preganglionic somatic
axons in 3 pairs of cranial nerves (III, IV, IX).

★ The PINHEAD-SIZED CRANIAL
GANGLIA innervated by CRANIAL
NERVES III, IV, and VII are located
just *outside* the VISCERAL ORGANS
they supply.

HEAD

Lacrimal
Gland

Palate
Pharynx

Cervical
Region

A
(C1–C5)

T1

T12

L1
L2

A
(L3–S1)

Inferior
Mesenteric
Ganglion

A1
S2
S3
S4

Conus
Medullaris
(of A)

Hypogastric
Plexus

(8) Ganglia in Superior
Cardiac and Deep
Cardiac Plexuses

(8) Ganglia in the
Pulmonary Plexus

★ The MICROSCOPIC INTRAMURAL
GANGLIA (8) associated with VAGUS
X and S2–S4 PREGANGLIONIC
AXONS are located *within* the
VISCERAL ORGANS they supply (in
the SUBMUCOSAL and MUSCULAR
TUNIC LAYERS).

**THORACIC
VISCERA**

★ The very short PARASYMPATHETIC
POST GANGLIONIC AXONS from
INTRAMURAL GANGLIA do *not*
supply sweat glands and arrector pilli
muscles in the skin, or nearby blood
vessel musculature. They supply
smooth muscle and glands of (13);
active in decreased heart rate,
bronchoconstriction, and increased
secretion/muscular activity.

**ABDOMINAL
VISCERA**

3a (**Pelvic Splanchnic Nerves**)

PELVIC / PERINEAL VISCERA

COLOR GUIDELINES:
A1 = light gray;
spinal cord = flesh;
1 = yellow; 2 = orange;
III, VII, IX, X = warm reds;
3 = light pink;
brainstem = light creams;
4-7 = greens;
9-13 = blues; 8 = light purple.

STRUCTURAL FEATURES OF SYMPATHETIC AND PARASYMPATHETIC DIVISIONS

The sympathetic division has a thoracolumbar nerve outflow (T1-L2). The ganglia are PARAVERTEBRAL (double chain sympathetic trunks) and PREVERTEBRAL (collateral). They are far from the visceral effector organs (close to the CNS). One PREGANGLIONIC FIBER synapses with twenty or more POSTGANGLIONIC NEURONS that pass to many VISCERAL EFFECTORS. POSTGANGLIONIC FIBERS are relatively long. The sympathetic division has a comprehensive distribution throughout the entire body, including the skin.

The parasympathetic division has a craniosacral nerve outflow (III, VII, IX, X S2-S4). The ganglia are terminal, near or within the visceral effector organs. One preganglionic fiber synapses with four to five postganglionic neurons that pass to one visceral effector. The postganglionic fibers are relatively short. The parasympathetic division has a limited distribution—primarily to the head, and to thoracic, abdominal, and pelvic viscera.

DUAL INNERVATION OF VISCERAL EFFECTORS

Most visceral effector organs receive fibers (axons) from both sympathetic and parasympathetic divisions of the ANS. Impulses from one division stimulate the organ's activities, and impulses from the other division inhibit the organ's activities. The stimulating division, however, is not always the sympathetic. In some instances (for example, in the abdominal viscera) the parasympathetic impulses increase digestive activities, while sympathetic impulses inhibit them. Under normal conditions, parasympathetic impulses to digestive glands and smooth muscle of digestive organs dominate sympathetic impulses. Digestion and absorption of nutrients, drainage of the urinary bladder, reduction in heart and respiratory rates, and rushing of blood to digestive viscera are all activities under the control of parasympathetic stimulation, which restores and conserves body energy in a REST-REPOSE SYSTEM.

Integration of the activities of the two ANS divisions helps to maintain homeostasis. During maintenance of homestais, the primary function of the sympathetic division is to counteract the parasympathetic activity just enough to carry out normal energy-requiring processes. During stressful situations, the sympathetic activity (which expends body energy) dominates the parasympathetic activity.

The classification of AUTONOMIC AXONS is based on production of a NEUROTRANSMITTER SUBSTANCE: the cholinergic axons release ACETYLCHOLINE (ACH); its effects are short-lived and local. The ADRENERGIC AXONS release NOREPINEPHRINE (NE), also called (NOR)ADRENALINE or SYMPATHIN, whose effects are long-lasting and widespread.

Preganglionic axons at synapses in the ganglia, in both the sympathetic and parasympathetic divisions, are CHOLOINGERGIC. The postganglionic axons at NEUROEFFECTOR JUNCTIONS in the sympathetic division are mostly ADRENERGIC (cholinergic exceptions): skin, sweat glands, external genitalia, adrenal medulla, and smooth muscles of blood vessels to skeletal muscle. In the parasympathetic division they are cholinergic (none to skin).

1 Bony orbit (Orbital cavity)

2 Eyebrows

EYELIDS AND EYELASHES

3 Eyelids (Palebrae)

4 Eyelashes

  a Sebaceous glands of Zeis

5 Tarsal plates

  b Tarsal meibomian glands

CONJUNCTIVA

6 Palpebral conjunctiva

7 Bulbar (ocular) conjunctiva

8 Conjunctival sac

9 Palebral fissure
  space between eyelids

COMMISSURES (CANTHI): ANGLES OF EYELIDS

10 Lateral canthus (narrow)

11 Medial canthus (broad)

12 Lacrimal caruncle
  small, reddish elevation in
  the medial canthus

FACIAL MUSCLES OF THE EYELIDS

13 Orbicularis oculi muscle

14 Levator palepebrae
  superioris muscle

EXTRINSIC OCULAR MUSCLES OF THE EYEBALL: EYE MOVEMENT

FOUR RECTUS MUSCLES

SR Superior rectus Muscle

LR Lateral rectus Muscle

MR Medial rectus Muscle

IR Inferior rectus Muscle

TWO OBLIQUE MUSCLES

SO Superior oblique Muscle

IO Inferior oblique Muscle

LACRIMAL APPARATUS

15 Lacrimal gland

16 Lacrimal ducts

17 Puncta lacrimalia

18 Lacrimal canals

19 Lacrimal sac

20 Nasolacrimal duct

COLOR GUIDELINES:
Muscles = warm colors;
Conjunctiva = light, cool colors.

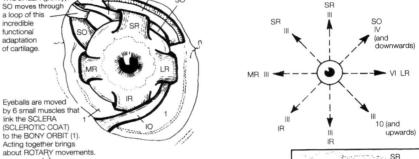

VIEWS of the LEFT EYE

★ TROCHLEA (pulley): SO moves through a loop of this incredible functional adaptation of cartilage.

★ Eyeballs are moved by 6 small muscles that link the SCLERA (SCLEROTIC COAT) to the BONY ORBIT (1). Acting together brings about ROTARY movements.

EXTRAOCULAR LEFT EYE MUSCLES
Movements and cranial nerve supply (iii, iv, vi)
(applies also to right eye as mirror image)

Trochlea

Nerve Fibers of OPTIC NERVE (II)

Optic Foramen

Annulus of Zinn

Left Lateral View

Left eye removed

LACRIMAL APPARATUS

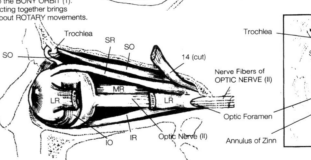

Pars Orbitalis (superior portion)

Pars Palpebralis (inferior portion)

PUPIL

IRIS

Papilla

Papilla

★ The ORBICULARIS OCULI MUSCLE (13) compresses LACRIMAL GLAND (15).

Drains tears into back of nose

Inferior Nasal Concha

Inferior Meatus

Nasal Cavity

Right Eye (Anterior View)

Eighty percent of all knowledge that the human being acquires is assimilated via the eyes.

The two eyes are the organs of vision that bend and focus incoming light waves onto SPECIALIZED PHOTORECEPTOR CELLS (RODS and CONES) at the back of each eye. The photoreceptor cell layer (RETINA) is encased within a fibrous globe (SCLERA), which becomes transparent in front (CORNEA). The eyeballs thus formed are subserved by many associated structures that either protect the eyeball or provide eye movement.

The eyeball is protected by the BONY ORBIT, FACIAL MUSCLES, LACRIMAL APPARATUS, EYEBROW, EYELASHES, EYELIDS, and CONJUNCTIVA. The bony orbit, the hidden posterior four-fifths of the eyeball, is encased in a bony socket, the ORBITAL CAVITY (ORBIT), in the skull. Seven bones of the skull form the orbit. A thick layer of AREOLAR and ADIPOSE TISSUE forms a compact cushion between the bone and the eyeball. The anterior fifth of the eyeball is exposed and protected from injury by the structures described below.

Eyebrows are short, thick hairs transversely oriented along the superior orbital ridges of the skull directed laterally. They shade the eye from the sun, and prevent perspiration and falling objects from entering it. Eyebrow movement is caused by the upper portion of the ORBICULARIS OCULI MUSCLE (and a portion of the CORRUGATOR MUSCLE).

Eyelids (PALPEBRAE) consist of upper and lower eyelids that shade eyes during sleep, protect the eyes from excessive light and foreign objects, and protect the eye from desiccation by reflexively blinking every seven seconds and moving fluid over the anterior eyeball surface. The upper eyelid is more moveable than the lower. Contraction of the ORBICULARIS OCULI MUSCLE closes the eyelid over the eye; contraction of the special LEVATOR PALEBRAE SUPERIORIS MUSCLE elevates the upper eyelid exclusively to expose the eye. The space that exposes the eyeball between the two eyelids is the PALPEBRAL FISSURE.

The eyelids develop as reinforced folds of skin. From superficial to deep, each eyelid consists of the EPIDERMIS, DERMIS, subcutaneous areolar connective tissue, orbicularis oculi muscle fibers, a TARSAL PLATE, TARSAL GLANDS, and a conjunctiva.

The tarsal plate is the thick fold of dense connective tissue that forms most of the inner wall of the eyelids, maintains their shape, and gives them support.

The TARSAL MEIBOMIAN GLANDS are specialized elongated glands embedded in grooves on the deep surface of each tarsal plate in a row. The ducts of these glands open onto the edge of the eyelids. Their oily secretions keep the eyelids from adhering to each other.

The conjunctiva is a thin, mucus-secreting epithelial membrane that lines the interior surface of each eyelid as the PALPEBRAL CONJUNCTIVA and reflects from the eyelids onto the eyeball (to the periphery of the cornea) as the BULBAR, or OCULAR, CONJUNCTIVA. When the eyelids are closed, a CONJUNCTIVAL SAC forms to protect the eyeball as a barrier against foreign objects (e.g., a contact lens).

Eyelashes are a row of short, thick hairs projecting from the border of each lid that protect the eye from airborne objects. GLANDS OF ZEIS pour a lubricating fluid into the hair follicles at their base.

The LACRIMAL APPARATUS refers to the group of structures that manufacture and secrete LACRIMAL FLUID (TEARS) and drain them away (through a series of ducts) into the nasal cavity.

The LACRIMAL GLAND, located in the superior lateral portion of each bony orbit, is a compound, tubuloacinar gland, the size and shape of an almond. It produces one milliliter of lacrimal fluid per gland each day.

There are six to twelve EXCRETORY DUCTS, or LACRIMAL DUCTSS, that empty tears into the conjunctival sac of the upper eyelid. Tears are an aqueous, mucus secretion that moisten and lubricate the conjunctival sac, and also protect the sac from infection with a bactericidal enzyme, LYSOZYME.

With each blink of the eyelids, the tears spread medially and downward over the bulbar conjunctiva to two small pore openings (on both sides of the LACRIMAL CARUNCLE) in each papilla of the eyelid, at the MEDIAL CANTHUS. These two pores are the PUNCTA LACRIMALIA. Located in the LACRIMAL GROOVES of the lacrimal bones of the orbit, LACRIMAL CANALS are the two ducts through which tears are drained. From the two lacrimal canals, the tears are conveyed into the LACRIMAL SAC, a superior expanded portion of the NASOLACRIMAL DUCT, which is the final canal that empties the tears into the inferior meatus of the nasal cavity. (This is the reason the nose runs during crying.)

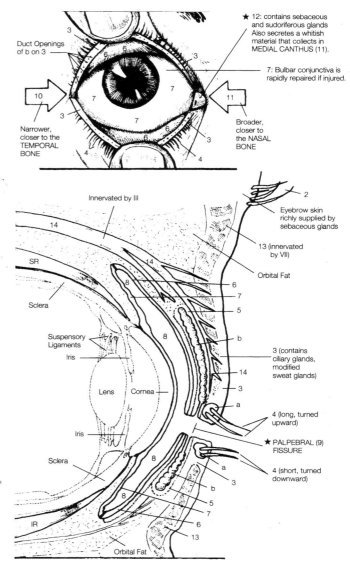

Midsagittal Section (Medial View) of Left Eye

The globelike structure of the eyeball is approximately 25 mm (one inch in diameter). Only the anterior fifth is exposed, the remainder being recessed and protected by the bony orbit in which it snugly fits. The eyeball is composed of three basic coats, or TUNICS, from the outside in: the OUTER FIBROUS TUNIC, MIDDLE VASCULAR and MUSCULAR TUNIC (UVEA), and the INNER NERVOUS TUNIC (RETINA). These three tunic layers enclose two liquid-filled cavities, basically separated by the lens, which itself is not part of any tunic.

The outer fibrous tunic has two parts, the SCLERA and the CORNEA. The sclera is a white outer coat of tough fibrous tissue, composed of tightly bound elastic and collagenous fibers. It is opaque and avascular, and located four-fifths along the posterior position. It gives shape to the eyeball and protects its inner parts. (Its posterior surface is pierced by the OPTIC NERVE [II]). The cornea is transparent (due to dense, tightly packed avascular connective tissue) and convex (permits passage and bending of incoming light waves). Its curvature is greater than the remainder of the bulb, enabling it to function as an important refractive medium. It is composed of five layers arranged in unusually regular patterns. The outer layer, the BULBAR (CORNEAL) EPITHELIUM, is continuous with the outer conjunctiva of the sclera.

The middle vascular tunic (uvea) is composed of three parts: the CHOROID, THE CILIARY BODY, and the IRIS.

The choroid is in the posterior position and is a thin, dark brown vascular membrane that lines most of the internal surface of the sclera. It contains numerous blood vessels united by connective tissue containing pigmented cells, and is composed of five layers. Its blood supply nourishes the inner nervous tunic, the RETINA. The deep choroid layer, the LAMINA VITREA, is placed next to the PIGMENTARY LAYER of the retina. Like the sclera, it is pierced by the optic nerve at the back of the eyeball, and extends from the optic nerve to the anterior, jagged margin of the retina,

THREE BASIC LAYERS OF THE EYEBALL

(1) Fibrous tunic

(2) Vascular tunic

(3) Internal tunic (retina) and lens
(not part of any tunic)

(1) **OUTER AVASCULAR FIBROUS TUNIC (TWO PARTS)**
POSTERIOR 4/5: OPAQUE REGION

4 Sclera "White of the eye"

a Venous Sinus/Canal of Schlemm
at the junction of the sclera and cornea

ANTERIOR 1/5: TRANSPARENT REGION

5 Cornea and muscular

(2) **MIDDLE VASCULAR TUNIC: UVEA (THREE PARTS)**
POSTERIOR PORTION

6 Choroid

ANTERIOR PORTION

7 Ciliary body

b 3 Ciliary smooth muscle planes

c Ciliary processes

8 Iris continuous with choroid through ciliary body

d Pupil black hole opening in center of iris

e Sphincter Pupillae circularly arranged

f Dilator pupillae radially arranged

(3) **INNERMOST NERVOUS TUNIC: RETINA**

9 Retina

10 Ora serrata

11 Macula lutea

12 Fovea centralis

13 Optic disc (blind spot)

14 Pars ciliaris retinae
Thin portion of the retina, situated in front of the ora serrata, which lines the inner surface of the ciliary body.

15 Pars iridica
Covers posterior surface of the iris

INTERIOR CAVITIES

ANTERIOR CAVITY

16 Anterior chamber

17 Posterior chamber

18 Fluid aqueous humor of the anterior cavity

POSTERIOR CAVITY

19 Vitreous chamber filled with vitreous humor

20 Hyaloid canal
Lymph channel in vitreous chamber (contains the hyaloid artery in the fetus)

FOCAL APPARATUS

21 Lens with surrounding lens capsule

22 Suspensory ligament (zonular fibers of zinn)

NERVE TRACT

(II) Optic cranial nerve II

BLOOD SUPPLY

23 Central retinal artery

24 Central vein

25 Ciliary arteries

26 Ciliary veins

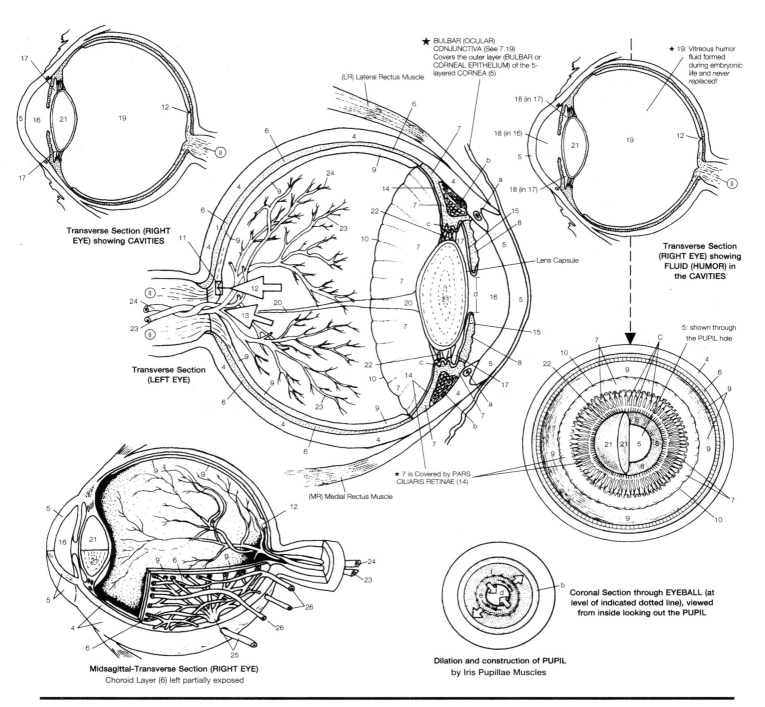

★ BULBAR (OCULAR) CONJUNCTIVA (See 7.19) Covers the outer layer (BULBAR or CORNEAL EPITHELIUM) of the 5-layered CORNEA (5)

(LR) Lateral Rectus Muscle

★ 19: Vitreous humor fluid formed during embryonic life and *never replaced!*

Transverse Section (RIGHT EYE) showing CAVITIES

Transverse Section (LEFT EYE)

Transverse Section (RIGHT EYE) showing FLUID (HUMOR) in the CAVITIES

Lens Capsule

5: shown through the PUPIL hole

★ 7 is Covered by PARS CILIARIS RETINAE (14)

(MR) Medial Rectus Muscle

Coronal Section through EYEBALL (at level of indicated dotted line), viewed from inside looking out the PUPIL

Dilation and construction of PUPIL by Iris Pupillae Muscles

Midsagittal-Transverse Section (RIGHT EYE)
Choroid Layer (6) left partially exposed

called the ORA SERRATA. The choroid absorbs light rays so they are not reflected back out of the eyeball.

Beyond the ora serrata, in the anterior portion of the vascular tunic, the choroid becomes the ciliary body, a very thick internal muscular ring composed of three distinct planes of smooth muscle fibers (CILIARY MUSCLES). Extensions of the ciliary body called CILIARY PROCESSES attach to the ZONULAR FIBERS OF ZINN, collectively called the SUSPENSORY LIGAMENT. This circular ligament attaches to the LENS (FOCAL APPARATUS) at the margins of the thin, clear LENS CAPSULE which encloses the lens. (The ciliary muscles alter the shape of the pliable lens for near or far vision.)

The iris is a contractile pigmented membrane consisting of circular and radial smooth muscle fibers arranged in the shape of a donut. The black donut hole is the PUPIL, through which light enters the eyeball. Contraction of the smooth muscle fibers regulates the diameter of the pupil. The iris is attached at its outer margin to the ciliary body, and is suspended between the lens and the cornea in the fluid-filled AQUEOUS HUMOR of the ANTERIOR CAVITY, separating this cavity into an ANTERIOR and POSTERIOR CHAMBER. These two chambers are connected through the pupil. (The larger, posterior cavity of the eye, the VITROUS CHAMBER, is filled with a transparent, jellylike substance called VITROUS HUMOR, and fills the inner space of the eyeball between the lens and the retina.)

COLOR GUIDELINES:
4 = cream; 5 = light gray; 6 = pink; 7 = orange; 8 = light brown; 9 = yellow; Chambers = light blues and greens; 23 = red; 24 = blue.

SPECIAL SENSORY ORGANS, CONTINUED
VISUAL SENSE: RETINAL STRUCTURE AND VISUAL PATHWAYS

The RETINA, the light-sensitive portion of the eye upon which light rays come to a focus, is the third and innermost tunic of the eye. It receives inverted light images through the REFRACTIVE MEDIA OF THE CORNEA (principal), INNER CAVITY FLUIDS (minimal), and the LENS (refining and altering of refraction). It is the immediate instrument of vision, whose primary function is image formation. It extends from the point of entrance of the optic nerve posteriorly towards the margin of the pupil anteriorly, completely lining the interior of the eye. The retina consists of three parts.

The PARS OPTICA is the nervous or sensory portion of the retina and is only in the posterior portion of the eye. It extends from the OPTIC DISC forward to the ORA SERRATA, the ending scalloped border or jagged margin of the retina, where the CHOROID LAYER ends and the CILIARY BODY begins. The ten-layered pars optica consists of an OUTER PIGMENTED LAYER (covering the choroid's bottom layer) and a nine-layered nervous tissue portion. This nervous tissue layer ends at the ora serrata. The PARS CILIARIS is the thin part of the retina lining the inner surface of the CILIARY PROCESS. The outer pigmented layer (along with the INTERNAL LIMITING MEMBRANE) extends anteriorly over the back of the ciliary body and processes. The PARS IRIDICA are the two layers that continue to form the posterior surface of the IRIS.

RODS and CONES (dendrites of the photoreceptor neurons) are visual receptors highly specialized for stimulation by light rays. Rods are specialized for vision in dim light and for black and white vision (no discrimination between colors). Rods are positioned on the peripheral (EXTRAFOVEAL) parts of the pars optica (over 100 million per eye), and provide poor visual acuity. Cones are specialized for color vision and visual acuity (sharpness of vision), stimulated only by bright light (thus we cannot see color by moonlight). There are an estimated seven million cones in each eye.

THE MACULA LUTEA is the special "yellow spot" in the exact center of the retina, approximately 2mm lateral to the exit of the optic nerves. It contains an abundance of cones and a pit, the FOVEA CENTRALIS, where the retina is reduced to a layer of closely-packed cones, the area of the most acute vision (central vision). When we look at an object, our eyes are directed so that the image will fall on the fovea of each eye. There are no rods here. The OPTIC DISC (PAPILLA), the "blind spot," is the area where nerve fibers from all parts of the retina converge to leave the eyeball as the optic nerve. There are no rods or cones present here, so the "blind spot" is not sensitive to light. (Retinal arteries and veins enter or leave the eyeball here.)

The VISUAL AXIS is a line drawn between the fovea and the apparent center of the pupil. It establishes the path of light rays when a person is looking at a distant object. The EQUATOR is a line drawn perpendicular to the visual axis through the greatest lateral expansion of the eyeball. The OPTICAL (ANATOMICAL) AXIS is a line drawn between the ANTERIOR POLE (center of the corner) and the POSTERIOR POLE (a point between the fovea and optic disc on the posterior retina at the greatest diameter in the anteroposterior dimension).

Note that when the eye receives and integrates centers of vision in the OCCIPITAL LOBES of the cerebral cortex, the image received is the reverse of that actually visualized. Light rays coming from a visual object in the nasal (medial) half of visual fields fall on the lateral (temporal) half of the retina; light rays coming from a visual object in the temporal half of the visual field fall on the nasal half of the retina.

Neurons from the right half of each eye project to the RIGHT LATERAL GENICULATE BODY of the THALAMUS, and neurons from the left half of each eye project to the LEFT LATERAL GENICULATE. Thus, neurons from the temporal portion of each eye (containing a nasal image) pass through the OPTIC CHIASMA without crossing, and neurons from the nasal portion of each eye (containing a temporal image) decussate (cross) through the optic chiasma. Thus, the RIGHT OCCIPITAL LOBE receives left lateral and right medial images, and the LEFT OCCIPITAL receives right lateral and left medial images.

THE TEN LAYERS OF THE RETINA / THREE PARTS OF THE RETINA:

A Pars optica (Pars optica retinae)

B Pars ciliaris (Pars ciliaris retinae)

C Pars Iridica (Pars iridica retinae)

SPECIAL POINTS ON THE PARS OPTICA RETINAE (A):

D Ora serrata

E Macula lutea ("Yellow spot")

F Fovea centralis

G Optic disc ("Blind spot") (Optic papilla)

Ten Layers of the retina are composed of three types of neurons:
PHOTORECEPTOR, BIPOLAR AND GANGLION

1 Pigmented epithelium

Next to the lamina vitrea of the choroid coat. (In bright light, PIGMENT GRANULES migrate into cell processes between rod and cones, preventing spread of light from one receptor to another.)

2 Layer of rods and cones (Bacillar layer)

a Rods

b Cones

The first receptive "neurons" dendrites of the photoreceptor neurons:

3 External limiting membrane

Separates rods and cones from their photoreceptor cell bodies (supports the rod and cone segments)

4 Outer nuclear layer

c Cell bodies and nuclei of photoreceptor neurons

5 Outer plexiform layer

Nerve processes and synapses between photoreceptor and bipolar neurons

6 Inner nuclear layer

d Cell bodies of second or bipolar neurons

7 Inner plexiform layer

Nerves processes and synapses between bipolar and ganglion neurons

8 Layer of ganglion cells

e Cell bodies of third, integrating ganglion neurons

9 Layer of optic nerve fibers

Axons of GANGLION NEURONS converge on the optic papilla to leave the eyeball as the optic nerve (II)

10 Internal limiting membrane

Glial membrane of the retina, which continues as the PARS CILIARIS RETINAE and the PARS IRIDICA

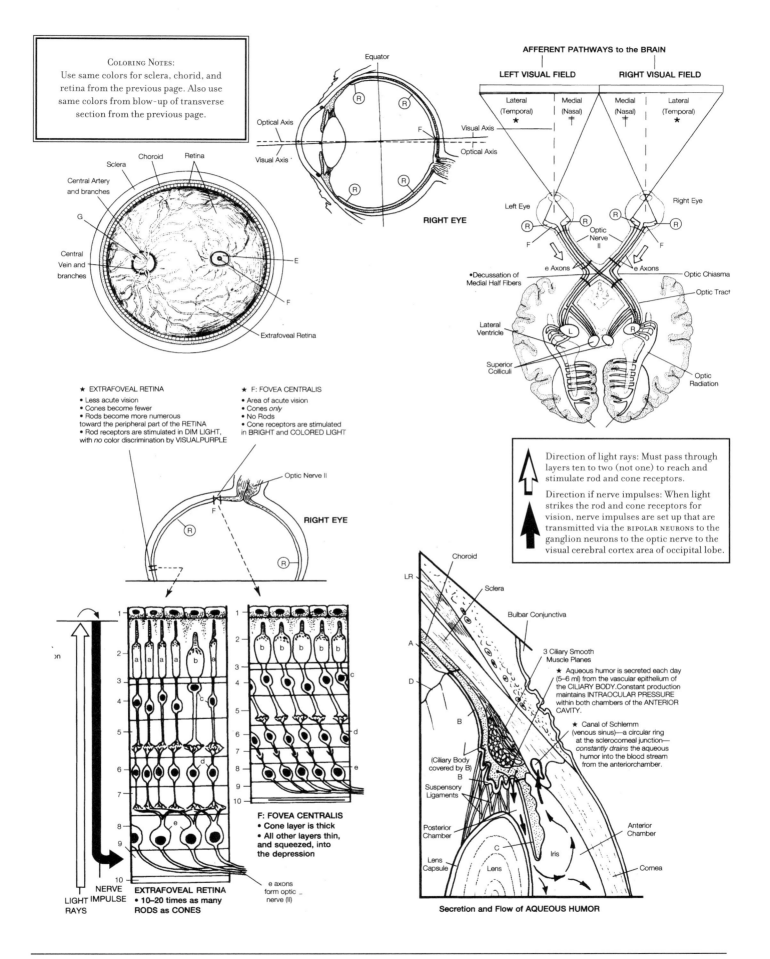

Sclera
Choroid
Retina
Central Artery and branches
G
Central Vein and branches
E
F
Extrafoveal Retina

Equator
Optical Axis
Visual Axis
Visual Axis
Optical Axis
R
R
F
R
R
RIGHT EYE

AFFERENT PATHWAYS to the BRAIN

LEFT VISUAL FIELD | RIGHT VISUAL FIELD

Lateral (Temporal) ★ | Medial (Nasal) ‡ | Medial (Nasal) ‡ | Lateral (Temporal) ★

Visual Axis

Left Eye
Right Eye
R | R
R | R
Optic Nerve II
F
F

e Axons | e Axons

•Decussation of Medial Half Fibers
Optic Chiasma
Optic Tract

Lateral Ventricle
L | R

Superior Colliculi
Optic Radiation

★ EXTRAFOVEAL RETINA
• Less acute vision
• Cones become fewer
• Rods become more numerous toward the peripheral part of the RETINA
• Rod receptors are stimulated in DIM LIGHT, with *no* color discrimination by VISUALPURPLE

★ F: FOVEA CENTRALIS
• Area of acute vision
• Cones *only*
• No Rods
• Cone receptors are stimulated in BRIGHT and COLORED LIGHT

Optic Nerve II
F
RIGHT EYE
R
R

Direction of light rays: Must pass through layers ten to two (not one) to reach and stimulate rod and cone receptors.

Direction if nerve impulses: When light strikes the rod and cone receptors for vision, nerve impulses are set up that are transmitted via the BIPOLAR NEURONS to the ganglion neurons to the optic nerve to the visual cerebral cortex area of occipital lobe.

1
2
a a a b a
3
4
c
5
6
d
7
8
e
9
10
LIGHT RAYS | NERVE IMPULSE | **EXTRAFOVEAL RETINA**
• 10–20 times as many RODS as CONES

1
2
b b b b b
3
4
c
5
6
d
7
8
e
9
10
F: FOVEA CENTRALIS
• Cone layer is thick
• All other layers thin, and squeezed, into the depression

e axons form optic nerve (II)

LR
A
D
B
B
B

Choroid
Sclera
Bulbar Conjunctiva

3 Ciliary Smooth Muscle Planes

★ Aqueous humor is secreted each day (5–6 ml) from the vascular epithelium of the CILIARY BODY. Constant production maintains INTRAOCULAR PRESSURE within both chambers of the ANTERIOR CAVITY.

★ Canal of Schlemm (venous sinus)—a circular ring at the sclerocorneal junction—*constantly drains* the aqueous humor into the blood stream from the anteriorchamber.

(Ciliary Body covered by B)
Suspensory Ligaments
Posterior Chamber
Lens Capsule
Lens
C
Iris
Anterior Chamber
Cornea

Secretion and Flow of AQUEOUS HUMOR

The ear is the organ of hearing and equilibrium. It contains receptors that convert sound waves into nerve impulses (for hearing) and receptors that respond to head movements (for equilibrium). The VESTIBULCOCHLEAR CRANIAL NERVE (VIII) transmits impulses from both types of receptors to the brain for interpretation.

Anatomically, the ear is divided into three principal regions: the EXTERNAL (OUTER) EAR, MIDDLE EAR, AND INNER EAR (LABYRINTH). The external or outer ear consists of three structures: the AURICLE (PINNA), EXTERNAL AUDITORY MEATUS, and TYMPANIC MEMBRANE (EARDRUM). A funnel-like, trumpet-shaped flap of elastic cartilage covered by thick skin, the pinna is attached to the skull by ligaments and poorly developed AURICULAR MUSCLES.

The external auditory meatus is a 2.5 cm (1 in) slightly S-shaped canal, running slightly upward, leading from the pinna to the eardrum. This small tube lies in the external auditory meatus of the temporal bone. The walls are lined with cartilage continuous with cartilage of the pinna, and are covered with thin, highly sensitive skin, containing fine hair and sebaceous glands at the entrance.

The eardrum is a thin, double-layered, semitransparent epithelial partition between the external auditory meatus and the middle ear. The external surface is concave and covered with skin. The internal surface is convex and covered with a mucous membrane. Sound waves set the eardrum in vibration.

The middle ear is a small air-filled chamber deep within the PETROUS PORTION of the temporal bone. Extending across this TYMPANIC CAVITY are three small bones, the AUDITORY OSSICLES, connected by SYNOVIAL JOINTS to form a small lever. The "handle" of the MALLEUS BONE is attached to the internal surface of the eardrum. Its head articulates with the INCUS, which is linked to the STAPES, which fits into the OVAL WINDOW, a small opening between the middle and inner ear. These ossicles are attached to the tympanic cavity by means of ligaments. Vibration of air moves

the lever of the ossicles. This amplifies and transmits vibrations across the middle ear so that the FOOTPLATE of the stapes moves backwards and forwards in the oval window. (There are two openings into the tympanic cavity.)

The inner ear (labyrinth) is a series of canals consisting of two main divisions, a BONY LABYRINTH and a MEMBRANOUS LABYRINTH that fits inside the bony labyrinth. The bony labyrinth is a series of bony canals in the petrous portion of the temporal bone, lined with PERIOSTEUM whose cells secrete a pale, limpid fluid, PERILYMPH, that fills the space between the two labyrinths.

The membranous labyrinth is a series of sacs and tubes lying inside the bony labyrinth and having the same basic shape as the bony labyrinth. The membranous labyrinth is lined with EPITHELIUM and contains an ENDOLYMPH fluid. These two fluids provide a liquid-conducting medium for the vibrations set up by the movement of the stapes in the oval window. These vibrational movements through fluid stimulate receptors in the organ of corti in the inner ear, where nerve impulses are set up in the AUDITORY NERVE (VIII) and pass to auditory centers in the brain.

Within the UTRICLE and SACCULE of the vestibule of the membranous labyrinth, the MACULA contains OTOLITHS (calcium carbonate crystals) that move back and forth in a GELATINOUS MATRIX with changes in head positions. These moving otoliths mechanically stimulate the hair processes of hair cells. Information is then carried in fibers of the vestibular branch of the auditory nerve to the brain. Within the AMPULLA of each semicircular canal, the CRISTA AMPULLARIS contains a gelatinous mass, the CUPULA, that stretches across each ampulla. The APEX swings free in reaction to the starting or stopping of rotary movements of the head, when the endolymph moves. The swinging of the CUPULA APEX causes the hair processes of the hair cells to bend, which stimulates them. This information is carried in fibers of the vestibular branch of the auditory nerve to centers in the brain.

STRUCTURES OF THE EAR

THREE BASIC REGIONS OF THE EAR:

1. **EXTERNAL (OUTER) EAR:**
   - 4 | Auricle (pinna)
   - 5 | External auditory meatus
   - 6 | Tympanic membrane (eardrum)

2. **MIDDLE EAR:**
   - 7 | Tympanic cavity
   - 8 | Eustachian (auditory) tube

   (Two openings into tympanic cavity)
   - 9 | Epitympanic recess

   (Two openings into tympanic cavity)

3. **INNER EAR: LABYRINTH** – TWO PARTS
   - BL | Outer bony labyrinth
   - ML | Inner membranous labyrinth

AUDITORY APPARATUS:

FLUIDS OF THE LABYRINTH:
- 10 | Perilymph | between the labyrinths
- 11 | Endolymph | within the membranous labyrinth

INNERVATION OF THE LABYRINTH:
- VII | Facial cranial nerve VII
- VIII | Vestibulocochlear cranial nerve (VIII): auditory nerve
- vb | Vestibular branch
- cb | Cochlear branch (not shown)

AUDITORY OSSICLES OF THE MIDDLE EAR
- 12 | Malleus | (hammer)
- 13 | Incus | (anvil)
- 14 | Stapes | (stirrup)

SKELETAL MUSCLES CONNECTED TO OSSICLES AND BONY WALL
- 15 | Tensor tympani | (attaches to malleus)
- 16 | Stapedius | (attaches to stapes)

Within the membranous labyrinth

MECHANISMS OF EQUILIBRIUM:

STATIC EQUILIBRIUM: MACALA
- MU | Macula utriculi
- MS | Macula sacculi
   - a | Gelatinous matrix
   - b | Otoliths | (protein-sugar layers)
   - c | Hair cells
   - d | Supporting cells

DYNAMIC EQUILIBRIUM:
- CA | Crista ampullaris
   - e | Cupula | (protein-sugar layers)
   - c | Hair cells
   - d | Supporting cells

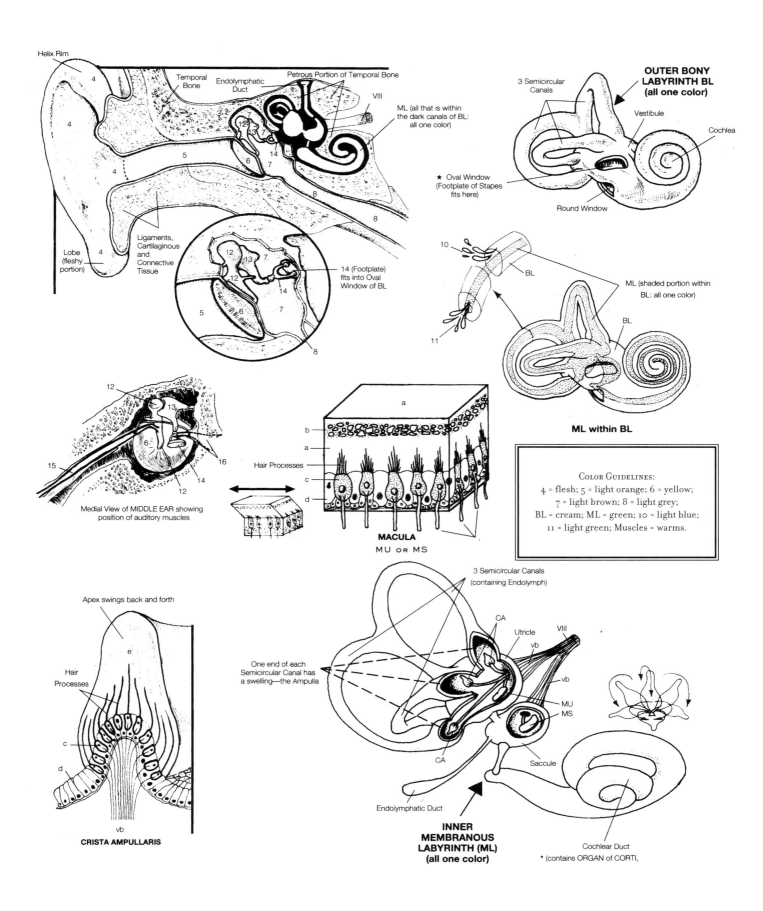

Helix Rim

Temporal Bone

Endolymphatic Duct

Petrous Portion of Temporal Bone

VIII

4

4

4

4

Lobe (fleshy portion)

4

5

4

Ligaments, Cartilaginous and Connective Tissue

6

7

14

7

7

8

8

8

ML (all that is within the dark canals of BL: all one color)

12 13 7

12

14

5 6 7

14 (Footplate) fits into Oval Window of BL

8

OUTER BONY LABYRINTH BL
(all one color)

3 Semicircular Canals

Vestibule

Cochlea

★ Oval Window (Footplate of Stapes fits here)

Round Window

10

BL

11

ML (shaded portion within BL: all one color)

BL

ML within BL

12

13

6

15

14

16

12

Medial View of MIDDLE EAR showing position of auditory muscles

a

b

a

Hair Processes

c

d

MACULA
MU or MS

COLOR GUIDELINES:
4 = flesh; 5 = light orange; 6 = yellow;
7 = light brown; 8 = light grey;
BL = cream; ML = green; 10 = light blue;
11 = light green; Muscles = warms.

Apex swings back and forth

Hair Processes

e

c

d

vb

CRISTA AMPULLARIS

3 Semicircular Canals (containing Endolymph)

CA

Utricle

VIII

vb

vb

MU
MS

One end of each Semicircular Canal has a swelling—the Ampulla

CA

Saccule

Endolymphatic Duct

INNER MEMBRANOUS LABYRINTH (ML)
(all one color)

Cochlear Duct
* (contains ORGAN of CORTI,

The INNER EAR (LABYRINTH) contains the organs of hearing and equilibrium. The bony labyrinth is structurally and functionally divided into three areas: the VESTIBULE, SEMICIRCULAR CANALS, and COCHLEA. The vestibule is the central oval portion of the bony labyrinth. It contains the OVAL WINDOW (into which the stapes fits) and the ROUND WINDOW, directly below it on the opposite end.

The MEMBRANOUS LABYRINTH within the vestibule contains two sacs connected by a small duct: a larger UTRICLE in the upper portion, and a smaller SACCULUS. These two sacs contain special proprioreceptors (MACULA UTRICULI and MACULA SACCULI, respectively) involved in the mechanism of static equilibrium.

The SEMICIRCULAR CANALS are three bony canals of each ear at right angles to one another, projecting upward and posteriorly from the vestibule; both ends of each canal open into the vestibule. (One end of each canal contains a swelling called the AMPULLA.) The SEMICIRCULAR DUCTS of the membranous labyrinth are almost identical in shape to their bony counterparts and communicate with the utricle. (Each of the three ducts has a MEMBRANOUS AMPULLA that connects with the upper back of the utricle.) The position of the three ducts permits detection of an imbalance in three planes. Special proprioreceptors called CRISTAE in each ampulla help to maintain dynamic equilibrium in the nonauditory part of the inner ear.

The bony cochlea lies in front of the vestibule, which spirals two and 3/4 times around a central bony axis (pillar) called the MODIOLUS. The view of a cross section shows that the spiral canal is divided into three separate tunnels resembling the letter Y on its side. The stem of the Y is a bony spiral lamina that projects part way into the interior of the SPIRAL CANAL of the cochlea and divided into two SCALAE CHAMBERS: SCALA VESTIBULI (which begins at the oval window and is continuous with the vestibule), and a lower SCALA TYMPANI (which terminates at the round window). The wings of the Y are composed of MEMBRANOUS LABYRINTH, and the third chamber produced between the wings is the SCALA MEDIA, or COCHLEAR DUCT. The upper wing of the Y, the root of the cochlear duct, which separates the cochlear duct from the scala vestibuli, is the VESTIBULAR MEMBRANE (OF REISSNER). The lower wing of the Y, the floor of the cochlear duct, which separates the cochlear duct from the scala tympani, is the BASILAR MEMBRANE. The basilar membrane joins the tip of the SPIRAL LAMINA and attaches to the outer bony wall by the EXTERNAL SPIRAL LIGAMENT.

Resting on the basilar membrane (within the cochlear duct) is the ORGAN OF CORTI, the functional unit of hearing. The tips of the hairlike processes of the hearing receptor cells of the ORGAN OF CORTI are embedded in a flexible gelatinous membrane called the TECTORIAL MEMBRANE. Impulses are passed on from the hair cells of corti to the branches of the COCHLEAR NERVE (VIII). The cell bodies of these first neurons lie within the bone of the spiral lamina.

The cochlear duct is filled with ENDOLYMPH (secreted by the STRIA VASCULARIS, pigmented granular cells with a profuse blood supply), and ends at the round window. The scala vestibuli and scala tympani contain PERILYMPH, and are completely separated except where the COCHLEAR SPIRAL ends at a narrow apex, the HELICOTREMA, where they are continuous. The perilymph of the scala vestibuli is continuous with that of the VESTIBULE.

AUDITORY STRUCTURES OF THE EAR

THE BONY LABYRINTH:

THREE STRUCTURAL AND FUNCTIONAL DIVISIONAL AREAS:

A   Vestibule

B   3 Semicircular Canals
(90-degree angles to each other)

C   Cochlea

CENTRAL PORTION: VESTIBULE:

1   Oval Window
(Fenestral Vestibuli)

2   Round Window
(Fenestral Cochlea)
Enclosed by the secondary tympanic membrane

SNAIL-SHAPED COCHLEA:

Central Bony Axis:

3   Modiolus (Central Pillar)

Projecting Bony Shelf:

4   Osseous Spiral Lamina

Two Chambers Formed by the Spiral Lamina:

5   Scala Vestibuli (Upper Chamber)

6   Scala Tympani (Lower Chamber)

INNERVATION OF INNER EAR—VESTIBULOCOCHLEAR NERVE (VIII):

cb   Cochlear Branch of VIII

vb   Vestibular Branch of VIII

THE MEMBRANOUS LABYRINTH:

WITHIN THE VESTIBULE:

7   Utriculus (Utricle)

8   Sacculus (Saccule)
(7 and 8 contain structures involved in equilibrium)

9   Endolymphatic Duct

10   Ductus Reuniens (Hensen's Canal)
Connects vestibule to cochlear duct

WITHIN THE THREE SEMICIRCULAR CANALS:

B1   3 Semicircular Ducts

B2   Membranous Ampullae of the Semicircular Ducts

WITHIN THE COCHLEA (THE ESSENTIAL ORGAN OF HEARING)

MIDDLE CHAMBER

C1   Cochlear Duct (Scala media)

11   Vestibular membrane (of Reissner)
(Roof of cochlear duct)

12   Basilar membrane
(Floor of cochlear duct)

OC   Organ of corti
(Organ of hearing)

13   Tectorial membrane

14   External spiral ligament

15   Stria vascularis

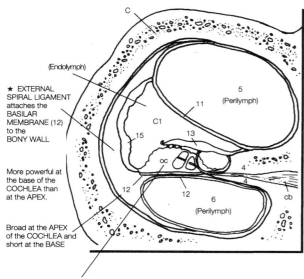

★ EXTERNAL SPIRAL LIGAMENT attaches the BASILAR MEMBRANE (12) to the BONY WALL

More powerful at the base of the COCHLEA than at the APEX.

Broad at the APEX of the COCHLEA and short at the BASE

(Endolymph)

5 (Perilymph)

11

C1

15

13

oc

4

4

cb

12

12

6 (Perilymph)

B1 (Posterior)

B1 (Superior)

B1 (Lateral)

B_2

B_2

B_2

7

7

vb

Ganglia

vb

Facial VII

9

10

8

cb

C_1

Spiral Ganglia of cb

C_1

C_1

Apex of the Cochlear Duct (C1) (point of the Helicotrema) ★

OC: The ORGAN of CORTI

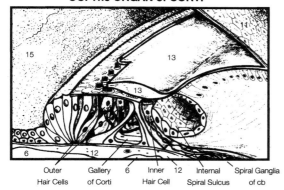

15

13

13

11

6

12

| Outer Hair Cells | Gallery of Corti | 6 | Inner Hair Cell | 12 Internal Spiral Sulcus | Spiral Ganglia of cb |

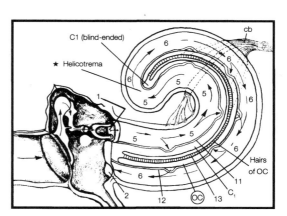

C1 (blind-ended)

cb

★ Helicotrema

6

6

6

5

5

5

1

5

5

6

5

5

6

6

Hairs of OC

2

12

OC

13

11

C_1

Idealized Schematic showing flow of sound waves through fluid of Labyrinth.

The BONY LABYRINTH

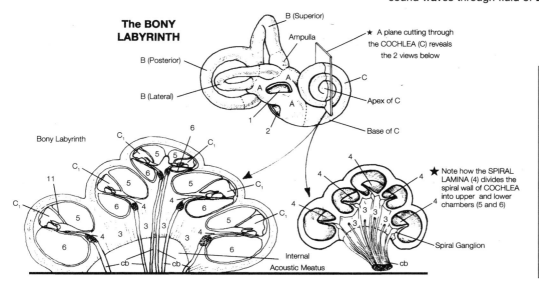

B (Superior)

Ampulla

B (Posterior)

B (Lateral)

A

A

A

1

2

★ A plane cutting through the COCHLEA (C) reveals the 2 views below

C

Apex of C

Base of C

Bony Labyrinth

C_1

6

C_1

C_1

5

5

11

C_1

5

6

5

C_1

6

4

C_1

5

4

3

3

3

C_1

5

4

6

4

6

cb

cb

Internal Acoustic Meatus

4

4

4

4

4

4

3

3

3

3

4

Spiral Ganglion

cb

★ Note how the SPIRAL LAMINA (4) divides the spiral wall of COCHLEA into upper and lower chambers (5 and 6)

COLOR GUIDELINES:
A = orange; B = purple;
C = green; 5 = light blue;
6 = turquoise;
7 = yellow-orange;
8 = red; B1 = light purple;
C1 = light green;
OC = yellow; 13 = orange;
14 = gray; 15 = pink.

GUSTATORY SENSE: TASTE

A — Tongue

   1 — Circumvallate papillae

   2 — Fungiform papillae

      1 and 2 contain the taste buds

B — Taste buds

   3 — Encapsulating supporting cells

   4 — Gustatory receptor cells

     a — Dendritic gustatory hairs

     b — Taste pore canal

C — Sensory nerve fibers

V — Branch of trigeminal nerve

(VII) — Facial cranial nerve (VII)

   Chorda tympani branch

(IX) — Glossopharyngeal cranial nerve (IX)

(X) — Vagus cranial nerve (X)

OLFACTORY SENSE: SMELL

D — Nasal cleft in nasal cavity

E — Cribiform plate of ethmoid bone

F — Olfactory nasal epithelium
Mucosa

   5 — Sustentacular/supporting cells

   6 — Bowman's glandular goblet cells

   7 — Olfactory receptor cells/ bipolar sensory neurons

     c — Dendritic olfactory hairs (cilia)

     d — Olfactory cell bodies

     e — Unmyelinated olfactory axons

(I) — Olfactory cranial nerve fibers (I)

G — Olfactory bulb

F — Mitral cell bodies

H — Olfactory tract
Axons of mitral neurons

OLFACTORY CENTER OF CEREBRAL CORTEX

   8 — Lateral and medial olfactory striae

   9 — Olfactory center

   10 — Thalamic centers

UMAMI

A fifth taste receptor has been long debated since first posited in 1907 by chemistry professor Kikunae Ikeda, creator of monosodium glutamate (MSG). In 2000, University of Miami researchers claimed to isolate an l-glutamate taste receptor, which they named "taste-mGluR4," given the popular name, *umami*, Japanese for "delicious essence."

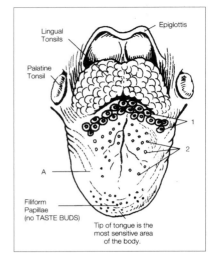

Lingual Tonsils

Palatine Tonsil

Epiglottis

Filiform Papillae (no TASTE BUDS)

Tip of tongue is the most sensitive area of the body.

TASTE

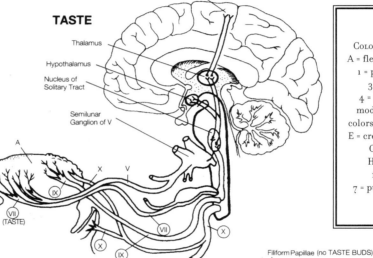

Sensory Cortex of Parietal Lobe

Thalamus

Hypothalamus

Nucleus of Solitary Tract

Semilunar Ganglion of V

VII (TASTE)

X (Motor innervation) and TASTE

COLOR GUIDELINES:
A = flesh;, B = yellow;
1 = pink; 2 = red;
3 = creamy;
4 = red-orange;
modalities = cool colors; D = light blue;
E = creamy; F = pink;
G = yellow;
H = orange;
f = green;
7 = purple; e = red.

---

4 MODALITIES of TASTE
■ (EVOKED BY SUBSTANCE in SOLUTION)

SWEET
■ Sugars
  Glycols
  Aldehydes

SOUR
■ Acids
  (Hydrogen ions, H+)

BITTER
■ Alkaloids

SALTY
■ Anions of ionizable salts

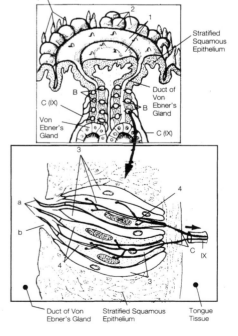

Filiform Papillae (no TASTE BUDS)

Stratified Squamous Epithelium

Duct of Von Ebner's Gland

Von Ebner's Gland

C (IX)

Duct of Von Ebner's Gland

Stratified Squamous Epithelium

Tongue Tissue

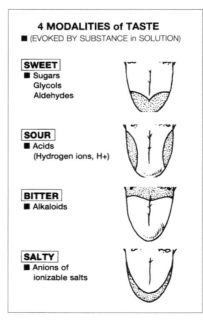

One way of classifying sensations is according to the simplicity or complexity of both the receptor and neutral pathways involved.

The GENERAL SENSES have receptors and neural pathways that are simple in structure and widespread throughout the body in the skin, subcutaneous tissue, muscles, tendons, joints, and viscera.

SPECIAL SENSES have receptors that are complex and the neural pathways are extensive in structure and localized in one or two specific areas of the body. Survival of the humanorganism depends on the ability to sense the environment and make the necessary homeostatic adjustments. A SENSATION is the arrival of a sensory impulse to the brain. A PERCEPTION is the interpretation (conscious registration) of a sensation occurring in the CEREBRAL CORTEX—thus you ultimately taste, smell, see, hear, and balance with your brain. For more information, visit www.mcmurtriesanatomy.com. The special senses of taste, smell, sight, hearing, and equilibrium will be discussed in the following six sections.

GUSTATORY SENSATIONS OF TASTE

Taste is a chemical sense (i.e., gustatory receptor cells respond to chemical stimuli, which must be in a solution, usually saliva). The essential organ of taste is the TONGUE. The specialized EPITHELIAL RECEPTOR CELLS are most numerous on the DORSUM of the tongue (but are also present on the PALATE, OROPHARYNX, EPIGLOTTIS, and LARYNX).

A TASTE BUD is a cluster of forty to sixty-eight GUSTATORY RECEPTOR CELLS encapsuled by supporting cells. The taste buds are found on the connective tissue projections on the upper surface of the TONGUE known as PAPILLAE.

Each GUSTATORY CELL contains a hairlike dendritic ending (GUSTATORY HAIR) that projects to the external surface through an opening in the bud called a TASTE PORE CANAL.

There are four types of taste buds, all structurally similar, but each responding more strongly to one of only four modalities of taste (sweet, sour, bitter, salt) on specific areas on the tongue.

Each gustatory cell is innervated by an afferent neuron. The afferent pathway to the brain (via the medulla and thalamus) involves mainly two cranial nerves, VII and IX. Impulses from one side of the tongue pass to the taste center of the POSTCENTRAL GYRUS in the parietal lobe on the opposite side of the cerebral cortex.

Taste discrimination is actually a combination of taste sensations, food texture, temperature and to a significant degree the sensation of smell.

OLFACTORY SENSATIONS OF SMELL

Smell is a chemical sense (i.e., OLFACTORY RECEPTOR CELLS respond to chemical stimuli in solution). However the substance of smell must originally be airborne (in a gaseous state), and then dissolve in solution.

The exclusive organ of smell is the NOSE (also serves as the main air passage to the respiratory system). The OLFACTORY RECEPTORS are located in the OLFACTORY NASAL EPITHELIUM within the roof of the nasal cavity in the nasal cleft, an upper space on both sides of the medial nasal septum (see pages 54–55).

The olfactory receptors are BIPOLAR SENSORY NEURONS whose cell bodies lie between supporting SUSTENTACULAR CELLS and regular PSEUDOSTRATIFIED COLUMNAR EPITHELIUM. The rounded tips of the RECEPTOR CELL BODIES project into the mucus of the nasal cavity. The mucus is secreted by BOWMAN'S GOBLET CELLS, and it lines the epithelium. Several dendritic hairs (CILIA) from each tip project into the cavity.

Gaseous substances drawn upwards dissolve into solution in the mucus, and, through the cilia, stimulate the OLFACTORY RECEPTOR CELLS. Unmyelinated OLFACTORY AXONS unite to form the OLFACTORY CRANIAL NERVES (I), which pass through the foramina of the CRIBRIFORM PLATE of the ethmoid bone, and synapse in the paired OLFACTORY BULBS with the dendrites of MITRAL NEURONS. Axons of these mitral neurons form the olfactory tract, which conveys impulses to the brain for odor translation. Unlike taste (divisible into only four modalities), smell is divisible into thousands of distinct odors. (The molecular bases of olfaction is not satisfactorily understood.) Only two to three percent of inhaled air comes in contact with the smell receptors, due to their location above the main airstream.

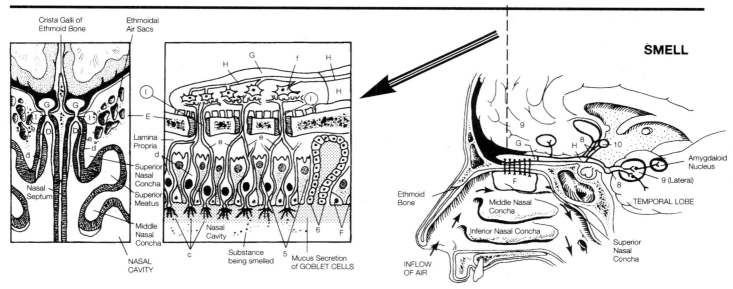

Frontol (Coronal) Section through the nose (at the level of the indicated dotted line in the illustration on the right)

CHAPTER 9: ENDOCRINE SYSTEM

The body could not function without the well-coordinated integration and control mechanisms of both the ENDOCRINE SYSTEM and the NERVOUS SYSTEM. Both systems work together harmoniously in an interlocking manner to maintain BODY HOMEOSTASIS by inhibiting or stimulating cell activity and organ function (a dynamic state of the body's internal environment maintained by feedback and regulation processes). The nervous system can inhibit or stimulate the release of hormones (biologically active chemicals-regulatory molecules) and the endocrine system can inhibit or stimulate the flow of nerve impulses. The nervous system controls and integrates body activities by the TRANSMISSION OF ELECTRICAL IMPULSES OVER NEURONS; the endocrine system does so by SECRETION OF HORMONES INTO THE BLOOD STREAM where they are transported throughout the body to all body tissues.

ENDOCRINE GLANDS secrete their hormones into the extracellular spaces around the secretory cells and directly into the blood via capillaries. (They do not secrete their products via ducts into body cavities or organ lumens, or onto a free surface, as do EXOCRINE GLANDS.) Thus, endocrine glands are also called DUCTLESS GLANDS. Hormones are delivered to every cell in the body, but each hormone only affects specific TARGET CELLS and ORGANS. (The effect a hormone has on a specific site varies with its concentration and the concentration of other hormones in the blood.)

A target cell must have specific RECEPTOR PROTEINS in order to respond to a specific hormone. Hormones can be functionally categorized into three groups based on the location of receptor proteins at their receptor protein site. For THYROID HORMONES, its receptor is within the nucleus of the target cells (THYROID GLAND). For STEROID HORMONES, it is within the cytoplasm of the target cells (ADRENAL CORTEX, GONADS). And for CATECHOLAMINES, POLYPEPTIDES, and GLYCOPROTEINS, it is in the outer surface of the target cell membrane (all endocrine glands except those above). Hormones do not generally accumulate in the blood because their half-life is very short (from under two minutes to two hours). They are removed quickly by the target organs (and by the liver, where they are converted to less active products by enzyme reactions).

CONTENTS

SYSTEM COMPONENTS

Glands of hormone production

SYSTEM FUNCTION

Integration, control and regulation of body activities through transportation of hormones by the blood vascular system.

COLORING NOTES:
In general, you may choose your own colors, but here is a sample guide:
Pancreas = flesh;
Adrenals = light brown;
Thymus = light purple;
Parathyroids = red;
Thyroid gland = pink;
Testes = orange;
Ovaries = light blue;
Pituitary gland = gray.

THE MAJOR ENDOCRINE GLANDS

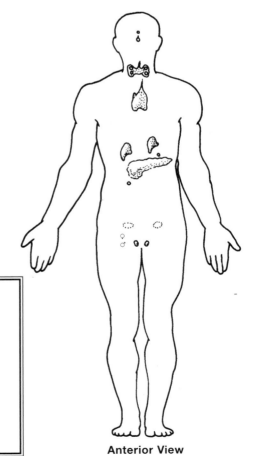

Anterior View

THE GONADS (HORMONES SECRETED)

THE OVARIES

a Ovarian Follicles with their Granulosa and Theca Interna

b Corpus Luteum

THE TESTES

c Interstitial Cells of Leydig

THE PLACENTA

COLORING NOTES:
Testes = warms;
Ovaries = cools.

1 Estrogen (Estradiol-17)

1 Estrogen (Estradiol-17)
2 Progesterone

3 Testosterone (Main Androgen)
4 Other Androgens

1 Estrogen (Estradiol-17)
2 Progesterone
5 (HCG) Human Chorionic Gonadotrophin
6 Somatomammotropin

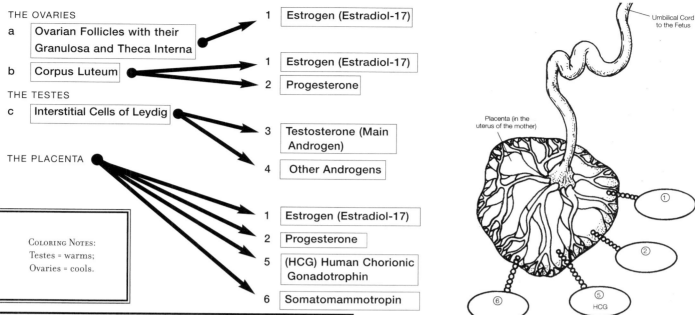

Umbilical Cord to the Fetus

Placenta (in the uterus of the mother)

T he MALE and FEMALE GONADS, found in the PELVIC-PERINEUM REGION of the body, are a prime example of organs that are not exclusively endocrine (serving other functions in addition to hormonal production and secretion) and whose endocrine glands are found in small islands within the organ. The male and female gonads—reproductive organs which produce sex cells—also secrete hormonal sex steroids. The OVARIAN secretions—ESTROGENS and PROGESTERONES—occur in the first half of the menstrual cycle.

During this time, the female gonads, or OVARIES, produce estrogens and progestrogens. The OVARIAN FOLLICLES (which contain the OOCYTE egg cell), along with their GRANULOSA CELLS and THECA INTERNA, secrete estrogen. In midcycle, one of the follicles grows very large, ruptures, and discharges its ovum from the ovary. The empty follicle is stimulated by LH (LUTEINIZING HORMONE) from the ANTERIOR PITUITARY, and is transformed into a different endocrine structure—the CORPUS LUTEUM or YELLOW BODY (see page 180). The corpus luteum secretes estrogen and progesterone. The TESTICULAR secretions are ANDROGENS produced by the LEYDIG CELLS (in "islands") which are found in the interstitial tissue between the SEMINIFEROUS TUBULES of the TESTES. They secrete TESTOSTERONE (the main androgen), other androgens, and even small amounts of estrogen (ESTRADIOL-17).

The hormones secreted by the gonads are both regulated and stimulated by other hormones (FSH, LH and ICSH), secreted themselves from the anterior lobe of the pituitary gland (see page 180 and the triangles in the illustrations).

In the female, the secretions of the gonads stimulate development and maintenance in the ovaries of estrogen (female secondary sex characteristics and cyclic changes in vaginal epithelium and uterine endothelium), estrogen (same as estrogen but in lesser amounts) and progesterone (changes in uterine endometrium in second [secretory] phase of menstrual cycle, placenta and mammary glands). In the testes, testosterone and other androgens stimulate development and maintenance of the male genitalia (penis and scrotum), male sex accessory organs (seminal vesicles, epididymides, ductus deferens and prostate gland) and male secondary sex characteristics.

OVARIES
The ovarian cycle showing cell changes and their hormone secretions

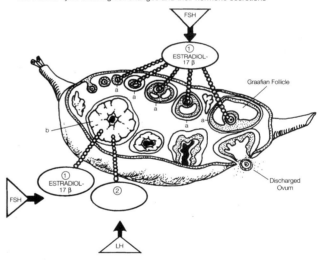

FSH

ESTRADIOL-17 β

Graafian Follicle

FSH

ESTRADIOL-17 β

Discharged Ovum

LH

TESTES

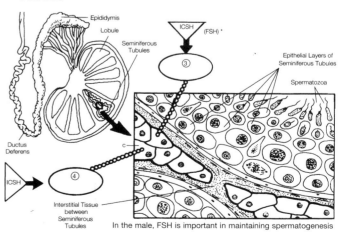

Epididymis

Lobule

Seminiferous Tubules

ICSH (FSH) *

Epithelial Layers of Seminiferous Tubules

Spermatozoa

Ductus Deferens

ICSH

Interstitial Tissue between Seminiferous Tubules

In the male, FSH is important in maintaining spermatogenesis

ENDOCRINE GLAND AND TISSUE LOCATIONS
EXCLUSIVE ORGANS AND NON-EXCLUSIVE ISLETS

There are two main types of ENDOCRINE (DUCTLESS) GLANDS: exclusively endocrine (located generally in the head and neck regions, close to the CNS control centers) and endocrine glands within organs that serve other functions—they occur in "islands" of endocrine tissue within the larger structures (located generally in the lower abdominal areas). The endocrine structures do not form a usual system where several organs work together to accomplish a single process. They are diverse organs that produce a wide variety of hormones that help regulate total body metabolism and many aspects of homeostasis, growth and reproduction. In many cases, the effects of certain hormones are vital to the survival of the organism and necessary for life.

The criteria for determining if a structure is an endocrine gland are if it is ductless, passing products directly into blood vessels within the gland itself; if it is extremely VASCULAR; if it consists of specific cells (circumscribed in groups) that are obviously different in both structure and function from other body cells (they produce specific hormones); or if its hormones have specific and well-defined effects on body function (their removal or injection cause clear-cut alterations in the body).

The organs that are exclusively endocrine (meaning their only function is the secretion of hormones) are the PITUITARY GLAND, PINEAL GLAND, THYROID GLAND, PARATHYROID GLANDS and ADRENAL GLANDS. Organs that are not exclusively endocrine (meaning they are serving other functions in addition to the production and secretion of hormones) are endocrine glands found in small islands: the THYMUS GLAND, PANCREAS, GASTROINTESTINAL TRACT (mostly STOMACH), DUODENUM and SMALL INTESTINE, OVARIES and TESTES. Also included are the SKIN, LIVER, KIDNEYS, HYPOTHALAMUS and the PLACENTA during pregnancy.

Glands regressive in the adult are the PINEAL and THYMUS glands. The pineal gland is very small (5 to 8 millimeters long, and 5 millimeters wide). It is cone-shaped (flattened) and is attached to the root of the third VENTRICLE, just above the tectum (root) of the MIDBRAIN. Located in a pocket near the splenium of the corpus callosum, it is covered by a capsule formed from the PIA MATER (part of the MENINGES covering the BRAIN). The pineal gland is larger in children, and begins to harden at puberty (calcification deposits form BRAIN SAND). Around the secretory cells (PINEALOCYTES and NEUROGLIAL CELLS) is a high level of innervation by the sympathetic nervous system from the SUPERIOR CERVICAL GANGLION. These nerves are connected to the RETINAL CELLS and appear to be an evolutionary remnant. The pineal gland has no direct connections to the rest of the brain. The physiology of the pineal gland is obscure, but its main activity appears to be the inhibition of the pituitary-gonad axis (specifically the ovaries, thus affecting MENSTRUATION). MELATONIN is the only chemical produced solely by the pineal gland.

REGIONS OF THE BODY

CRANIAL REGION

A | Pituitary Gland (Hypophysis)

B | Pineal Gland (Epiphysis Cerebri)

CERVICAL REGION

C | Thyroid Gland

D | Parathyroid Glands (4)
Behind and embedded in the thyroid gland

THORACIC REGION

E | Thymus Gland

ABDOMINAL REGION

F | Adrenal (Suprarenal) Glands (2)

G | Pancreas

H | Gastrointestinal Secretory Cells Of Stomach And Duodenum

PELVIC-PERINEAL REGION

I | Ovaries (female)

J | Testes (male)

GLANDS REGRESSIVE IN THE ADULT

B | Pineal Gland

E | Thymus Gland

Hormones that are secreted by pineal gland are melatonin (which is active in the onset of puberty, inhibits the pituitary-gonad axis and is involved with entraining circadian rhythms), ADRENOGLOMERULOTROPIN (which stimulates ADRENAL CORTEX to secrete ALDOSTERONE), SEROTONIN (which maintains normal brain physiology) and GROWTH-INHIBITING FACTOR (which inhibits growth).

The thymus gland is a bi-lobed lymphatic gland located in the anterior (and superior) mediastinum, posterior to the sternum in front of the aorta, between the lungs. Its size and structure varies with age. It is very large in infants and has a maximum size at puberty (about 40 grams). After puberty, the thymic tissue is replaced by fat and connective tissue, and at maturity, the

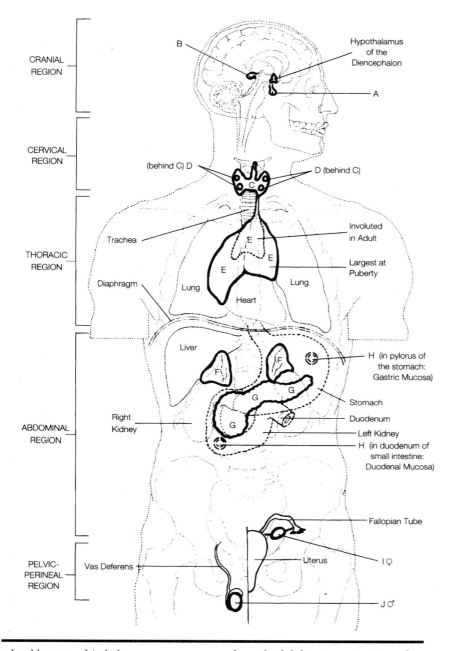

CRANIAL
REGION

CERVICAL
REGION

THORACIC
REGION

ABDOMINAL
REGION

PELVIC-
PERINEAL
REGION

B

Hypothalamus
of the
Diencephalon

A

(behind C) D

D (behind C)

C

Trachea

Involuted
in Adult

E

E   E

E

Largest at
Puberty

Diaphragm

Lung   Lung

Heart

Liver

F   F

F

H (in pylorus of
the stomach:
Gastric Mucosa)

G   G

Stomach

G

Duodenum

Right
Kidney

Left Kidney

H (in duodenum of
small intestine:
Duodenal Mucosa)

Fallopian Tube

Vas Deferens

Uterus

I ♀

J ♂

When the heart exhibits periods of weak
myocardial contraction, the two atria secrete
ANP (atrial natiriuetic peptide) which drives
increased excretion of water and sodium).

Multiple enzyme secretions and increased
intestinal motility are directly associated
with an array of endocrine factors secreted
by the cells of the gastrointestinal tract
(stomach, doudenum, and small intestine).

gland has atrophied. A FIBROUS CAPSULE envelopes both lobes; invaginations of the capsule separate the lobes into many LOBULES. The OUTER CORTEX of each lobule is the site for the production and maturation of densely packed THYMUS-DEPENDENT LYMPHOCYTES of the immune system, called THYMOCYTES or T-CELLS, which are packed in EPITHELIORETICULAR FIBROUS TISSUE. In the inner medulla of each lobule the T-cells are widely scattered amid large EPITHELIORETICULAR CELLS and large THYMIC (HASSALL'S) CORPUSCLES. A number of hormones secreted by the thymus gland (including THYMOSIN) may help regulate the immune system.

The main hormone secreted by the thymus gland is the THYMOSIN. It possibly influences B-CELL LYMPHOCYTES (which may be processed in the fetal liver and spleen) to develop into PLASMA CELLS, which produce ANTIBODIES against ANTIGENS (see chapter 12: Lymphoid Immune System).

COLOR GUIDELINES:
C = pink; D = red; E = light purple;
F = light brown; G = flesh;
I = light blue; J = orange.
A–J will be used throughout the chapter.

The PITUITARY GLAND (HYPOPHYSIS) is small and round, about 1/2 inch wide, and shaped like a pea. It is attached to the inferior aspect of the brain (region of the diencephalon) by a stalk (the INFUNDIBULUM) and is covered by DURA MATER (the part of the meninges covering the brain). It is supported by the SELLA TURCICA (concave space) of the SPHENOID BONE. The pituitary gland is exclusively an ENDOCRINE GLAND that secretes a number of hormones that regulate many bodily processes of growth, reproduction and various metabolic activities such as maintenance of water balance, sugar and fat metabolism and regulation of body temperature. The secretions of the pituitary gland are controlled by the HYPOTHALAMUS (a brain organ housing endocrine tissue) and negative feedback inhibition from the TARGET GLANDS.

The pituitary gland is structurally and functionally divided into two lobes, the ANTERIOR LOBE (ADENOHYPOPHYSIS) and the POSTERIOR LOBE (NEUROHYPOPHYSIS). These lobes have different embryonic origins with distinct types of tissues and secrete different hormones regulated by different control systems. The lobes receive hormones from the hypothalamus via the INFUNDIBULAR PORTION of the HYPOPHYSEAL STALK. Each has its own particular circulatory communicating pathway with specific regions of the hypothalamus.

The anterior lobe (adenohypophysis) secretes its own hormones (non-neural secretory cells) as a response from hypothalamic regulating hormones (regulating factors) released from NEUROSECRETORY CELLS in the hypothalamus. Its circulatory pattern involves a hypothalamic-hypophyseal portal (venous) system. The posterior lobe (neurohypophysis) does not secrete its own hormones and receives hormones from specific HYPOTHALAMIC NUCLEI (PARAVENTRICULAR and SUPRAOPTIC), which function as endocrine glands. These hormones pass through axon fibers of the HYPOTHALAMIC-HYPOPHYSEAL TRACT and are stored in the posterior lobe. The posterior lobe of the pituitary is only a storage bin; it is not an endocrine gland like the anterior lobe. It develops embryologically as a downward extension of the diencephalon on the interior aspect of the brain.

PITUITARY GLAND AND HYPOTHALAMUS:

P — Pituitary Gland (Hypophysis)

1 — Adenohypophysis (Anterior Lobe)

Nonneural portion has secretory cells

a — Anterior Lobe (Pars Distalis)

b — Intermediate Lobe (Pars Intermedia)

c — Pars Tuberalis

2 — Neurohypophysis (Posterior Lobe)

Neural portion has no secretory cells

d — Posterior Lobe (Pars Nervosa)

e — Infundibulum

A downward extension of the floor of the third ventricle

3 — Hypophyseal Stalk

The pathway between hypothalamus and hypophysis

c — Pars Tuberalis of Anterior Lobe

e — Infundibulum of Posterior Lobe

Contains nerve fibers and neuroglia-like cells called pituicytes

PITUITARY GLAND LOCATION IN BONE CONCAVITY

4 — Sella Turcica of Sphenoid Bone

X — Hypothalamus
(Hypothalamic Secretions to Pituitary)

X_1 — Median Eminence
(Basal portion of Hypothalamus)

5 — Paraventricular Nucleus

6 — Supraoptic Nucleus

10 — Secretory Neurons

COLOR GUIDELINES:
Arteries = red;
Capillary Plexuses = purple;
Veins = blue.

CIRCULATORY PATHWAYS OF THE PITUITARY

To Neurohypophysis. Storage bins from hypothalamic nuclei hormones via the axons of the hypothalamic-hypophseal track

7 — Capillary Plexus of the Infundibular Process

8 — Hypophyseal Hypophyseal Artery

9 — Posterior Hypophyseal Vein

To Adenohypophysis. Endocrine gland from hypthalamic secretory neuron releasing regulating factors to median eminence of hypothalamus.

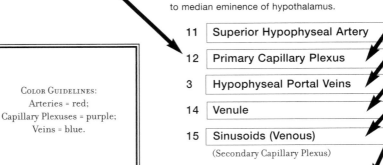

11 — Superior Hypophyseal Artery

12 — Primary Capillary Plexus

3 — Hypophyseal Portal Veins

14 — Venule

15 — Sinusoids (Venous)
(Secondary Capillary Plexus)

16 — Anterior Hypophyseal Vein

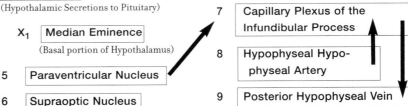

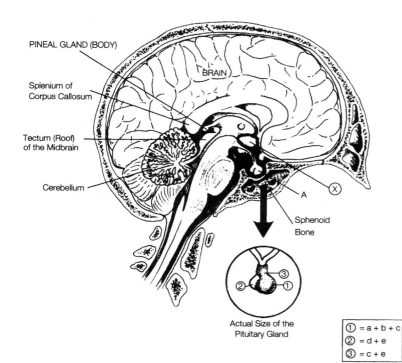

PINEAL GLAND (BODY)

Splenium of
Corpus Callosum

Tectum (Roof)
of the Midbrain

Cerebellum

BRAIN

A

Ⓧ

Sphenoid
Bone

② ③ ①

Actual Size of the
Pituitary Gland

①	= a + b + c
②	= d + e
③	= c + e

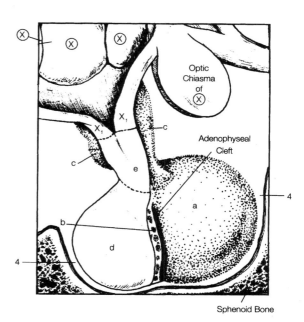

Ⓧ Ⓧ Ⓧ

Optic
Chiasma
of
Ⓧ

X₁

X₁

c

c

e

Adenophyseal
Cleft

a

4

b

4

d

Sphenoid Bone

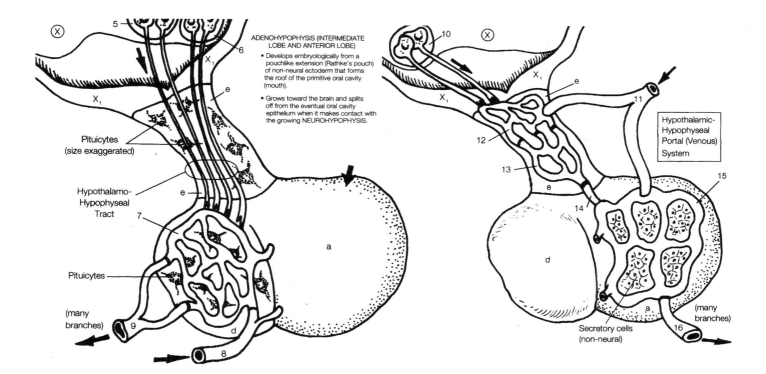

5

6

X₁

e

Pituicytes
(size exaggerated)

Hypothalamo-
Hypophyseal
Tract

7

Pituicytes

(many
branches)

9

d

8

Ⓧ

ADENOHYPOPHYSIS (INTERMEDIATE
LOBE AND ANTERIOR LOBE)

• Develops embryologically from a
pouchlike extension (Rathke's pouch)
of non-neural ectoderm that forms
the roof of the primitive oral cavity
(mouth).

• Grows toward the brain and splits
off from the eventual oral cavity
epithelium when it makes contact with
the growing NEUROHYPOPHYSIS.

a

NEUROHYPOPHYSIS
(Posterior lobe of the pituitary gland)

10

Ⓧ

X₁

X₁

12

13

e

14

11

Hypothalamic-
Hypophyseal
Portal (Venous)
System

15

d

a

16

(many
branches)

Secretory cells
(non-neural)

ADENOHYPOPHYSIS
(Anterior lobe of the pituitary gland)

The HYPOPHYSIS (pituitary gland) and HYPOTHALAMUS are strongly coordinated in the regulation of hormones of specific ENDOCRINE GLANDS and a number of metabolic functions. There are nine major hormones secreted by the PITUITARY GLAND; the ADENOHYPOPHYSIS secretes seven of them and the NEUROHYPOPHYSIS secretes two.

The secretory cells are the ACIDOPHILS, BASOPHILS and CHROMOPHOBES cells. Most of the hormones secreted are called TROPHIC HORMONES; they make their target organs HYPERTROPHY (increase in size). The secretory cells are stimulated (or inhibited) by regulating factors formed from SECRETORY NEURONS in the hypothalamus. Their axon endings do not enter the anterior lobe, but end in the median eminence of the hypothalamus. From here, the hormones are drained by VENULES in the INFUNDIBULUM, then to a SECONDARY CAPILLARY PLEXUS (SINUSOIDS) in the lobe, receiving venous blood. This vascular link with the MEDIAN EMINENCE forms a HYPOTHALAMIC-HYPOPHYSEAL PORTAL SYSTEM (similar to the HEPATIC PORTAL SYSTEM in the digestive system).

The regulating factors (and thus the trophic hormones) are controlled by hormones produced by the target organs. Thus, the anterior pituitary and hypothalamus are not "master glands." A chain of specificity exists for each trophic hormone which goes to a target organ(s) to a target organ hormone, and then to a regulating factors (hypothalamic hormones) and finally back to the specific trophic hormone.

The neurohypophysis secretes two hormones, though it contains no secretory cells, and receives its hormones from the hypothalamus along axon fibers into the lobe. Secretion of these hormones is controlled by NEUROENDOCRINE REFLEXES.

The BLOOD SUPPLY of the pituitary gland is extremely rich and is furnished by the CIRCLE OF WILLIS.

Hormones of the hypothalamus are self-regulated by negative feedback inhibition from target hormone secretions, and lead to the hormones of the hypophysis. ADH and oxytocin are produced in the hypothalamus by the PARAVENTRICULAR and SUPRAOPTIC NUCLEI. They are transported along the axons of the hypothalamo-

HORMONES OF THE HYPOTHALAMUS

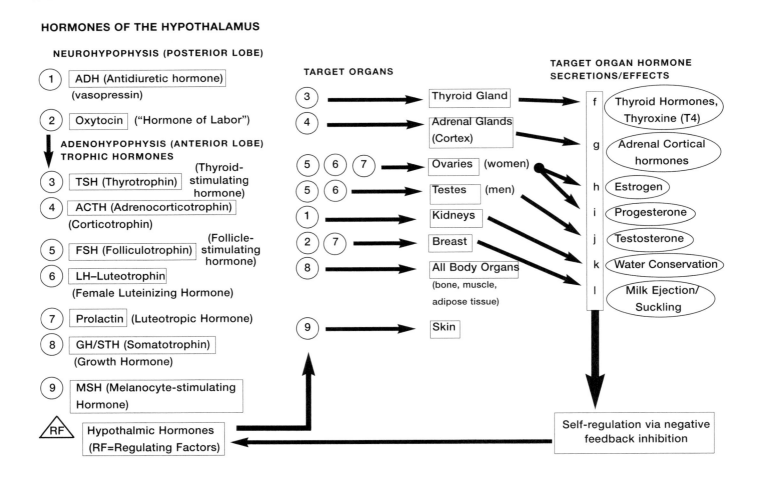

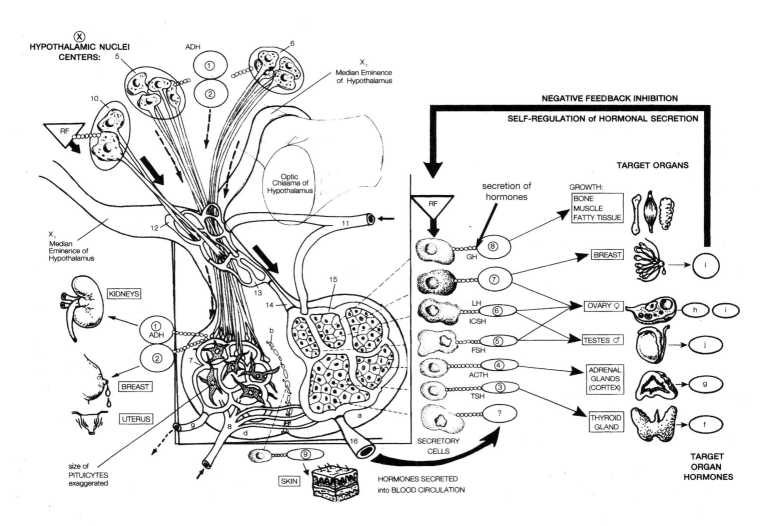

HYPOTHALAMIC NUCLEI
CENTERS:

ADH

X₁
Median Eminence
of Hypothalamus

RF

Optic
Chiasma of
Hypothalamus

X₁
Median
Eminence of
Hypothalamus

KIDNEYS

ADH

BREAST

UTERUS

size of
PITUICYTES
exaggerated

SKIN

HORMONES SECRETED
into BLOOD CIRCULATION

NEGATIVE FEEDBACK INHIBITION

SELF-REGULATION of HORMONAL SECRETION

TARGET ORGANS

secretion of
hormones

GROWTH:
BONE
MUSCLE
FATTY TISSUE

RF

GH

BREAST

LH
ICSH

OVARY ♀

FSH

TESTES ♂

ACTH

ADRENAL
GLANDS
(CORTEX)

TSH

THYROID
GLAND

?

SECRETORY
CELLS

TARGET
ORGAN
HORMONES

hypophyseal nerve tract and then stored in the posterior lobe. The posterior lobe has no secretory cells and is not an endocrine gland, but a storage warehouse for those two hormones. Hypothalamic hormones (regulating factors) from the hypothalamus stimulate production of the seven trophic hormones in the adenohypophysis. All nine hormones from the pituitary travel to their specific target organs, eliciting specific hormone secretions and effects from them, which in turn act as the self-regulating negative feedback inhibition for further hypothalamic hormone secretion.

There are three types of secretory cells in the adenohypophysis: acidophils, basophils and chromophobes (see illustration).

COLORING NOTES:
Pick nine of your favorite colors and first do the
nine hormones: 1–9.
Then color in the target organ hormone
secretions f–i (7 of them).
Refer to previous spread for other coloring guidelines.

The THYROID GLAND is the largest of the endocrine glands, and weighs about 1 ounce (25 grams). It is located just below the LARYNX in front of the TRACHEA (at the second through fourth TRACHEAL RINGS), and is covered by a thin connective tissue capsule. Microscopically, the thyroid gland consists of many spherical hollow sacs, or THYROID FOLLICLES. The interior of each follicle is filled with a protein-rich fluid, COLLOID. In this fluid is THYROGLOBULIN, a stored form of the principal thyroid hormones T_4 and T_3. The principal follicular cells form a simple epithelial wall around the thyroid follicles, and reach the lumen of their hollow sacs. They synthesize T_4 and T_3, which are extremely important in controlling overall body metabolic rate and are essential for healthy growth and development. The wall cells between the FOLLICULAR CELLS that do not reach the inner lumen are the C-CELLS, or PARAFOLLICULAR CELLS. The thyroid gland has a rich supply of blood. It receives 80–120 milliters of blood per minute. The nerve supply consists of postganglionic fibers from the superior and middle CERVICAL SYMPATHETIC GANGLIA.

The hormonal secretions of the principal follicular cells are THYROXIN (T_4) and TRIIODOTHYRONINE (T_3). The secretion effects of both hormones are increases in the rate of protein synthesis, and the rate of energy release from carbohydrates, and increases in oxygen consumption in all tissues. They are also instrumental in the regulation of the rate of total growth and development, and they stimulate maturity of the nervous system. Hormonal regulation sources for T_3 and T_4 are the hypothalamic influences upon the release of TSH from the anterior lobe of the pituitary. The parafollicular cells secrete thyrocalcitonin, which regulates calcium homeostasis, inhibits release of calcium from bone tissue (thus lowering blood calcium levels) and antagonizes the action of the parathyroid hormone and vitamin D_3. The hormonal regulating source for TCT is calcium in the blood.

C | The Thyroid Gland

STRUCTURE

a | Connective Tissue Capsule

b | Right and Left Lateral Lobes

c | Isthmus

d | Pyramidal Lobe of the Isthmus

SPHERICAL HOLLOW SACS

e | Thyroid Follicles

1 | Colloid/Thyroglobulin

SIMPLE CUBOIDAL EPITHELIAL WALL AROUND FOLLICLES

f | Principal Follicular Cells

2 | Thyroxine (T_4)
Teyraiodothyronine

3 | Triiodothyronine (T_3)

g | Parafollicular Cell (C-cells)

4 | Thyrocalcitonin (TCT) (Calcitonin)

BLOOD CIRCULATION

1 | Superior Thyroid Artery
Branch of external carotid artery

2 | Inferior Thyroid Artery
Branch of subclavian artery

3 | Superior Thyroid Vein

4 | Middle Thyroid Vein
3 and 4 pass into the internal jugular veins

5 | Inferior Thyroid Vein
Joins the brachiocephalic vein

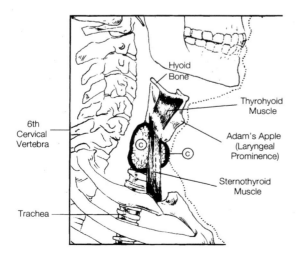

6th Cervical Vertebra

Trachea

Hyoid Bone

Thyrohyoid Muscle

Adam's Apple (Laryngeal Prominence)

Sternothyroid Muscle

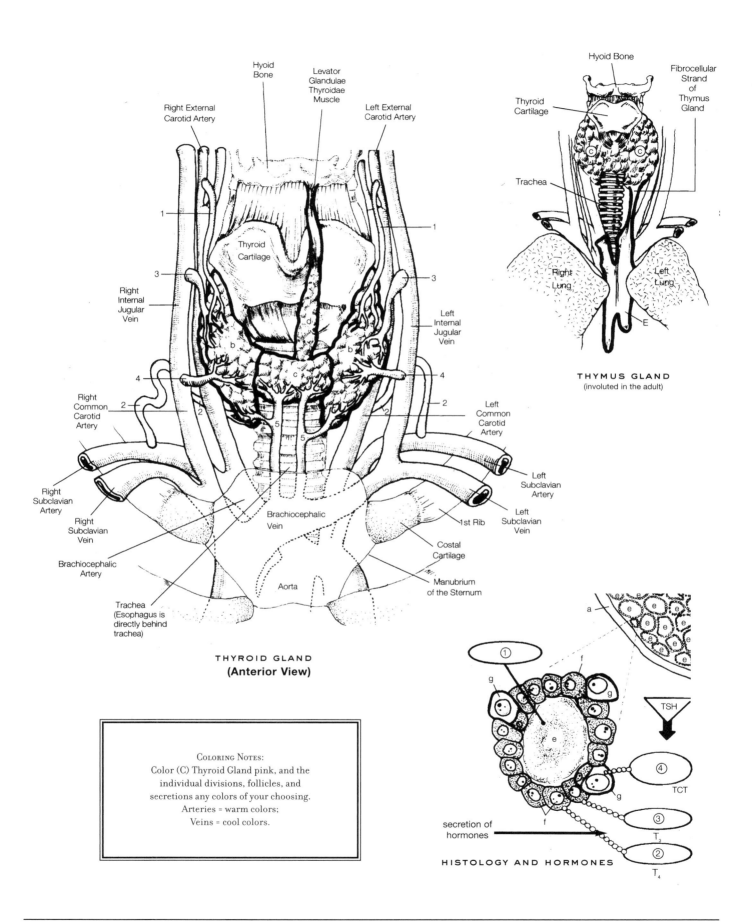

Hyoid Bone

Levator Glandulae Thyroidae Muscle

Right External Carotid Artery

Left External Carotid Artery

1

1

Thyroid Cartilage

3

3

Right Internal Jugular Vein

Left Internal Jugular Vein

4

4

Right Common Carotid Artery

2

2

2

2

Left Common Carotid Artery

5

5

Right Subclavian Artery

Left Subclavian Artery

Right Subclavian Vein

1st Rib

Left Subclavian Vein

Brachiocephalic Artery

Brachiocephalic Vein

Costal Cartilage

Manubrium of the Sternum

Trachea (Esophagus is directly behind trachea)

Aorta

THYROID GLAND
(Anterior View)

Hyoid Bone

Fibrocellular Strand of Thymus Gland

Thyroid Cartilage

Trachea

Right Lung

Left Lung

E

THYMUS GLAND
(involuted in the adult)

COLORING NOTES:
Color (C) Thyroid Gland pink, and the individual divisions, follicles, and secretions any colors of your choosing.
Arteries = warm colors;
Veins = cool colors.

a

e

①

f

g

g

e

TSH

④

TCT

g

③

T₃

secretion of hormones

f

②

T₄

HISTOLOGY AND HORMONES

The four PARATHYROID GLANDS are small, rounded masses of tissue embedded in the posterior surface of each lateral lobe of the THYROID GLAND. They are 0.1–0.3 inches long, 0.07–0.2 inches wide and 0.02–0.07 inches thick. Histologically, two kinds of epithelial cells are immersed in CONNECTIVE TISSUE SEPTA. The PRINCIPAL (CHIEF) CELLS synthesize the PARATHYROID HORMONE (PTH), which acts on the bones, kidneys, and intestines, promoting a rise in blood calcium levels. OXYPHIL CELLS support the chief cells, and may synthesize a reserve capacity of PTH. Blood supply and drainage are similar to that of the thyroid gland.

The secretion effects of PTH are that it regulates homeostasis of BLOOD CALCIUM levels (increases) and BLOOD PHOSPHATE levels (decreases). The bones release calcium (stimulation of disintegration and resorption of bone tissue). Kidneys conserve calcium (through decreased output in the urine). Intestinal walls absorb calcium. Increased calcium levels stimulate negative feedback inhibition of PTH. In addition, PTH also antagonizes the action of TCT from the thyroid gland.

THE PARATHYROID GLANDS

D	The 4 Parathyroid Glands (2 Pairs)
a	Left and Right Superior Parathyroids
b	Left and Right Inferior Parathyroids

RELATED STRUCTURES

c	Thyroid Gland
d	Pharynx
e	Esophagus
f	Trachea

HISTOLOGY

| g | Connective Tissue Septa |

TWO TYPES OF EPITHELIAL CELLS

h	Principal (Chief) Cells
I	Parathyroid Hormone (PTH) (Parathormone)
i	Oxyphil Cells

Reserve quantities of PTH

BLOOD CIRCULATION

| 1 | Branch of Superior Thyroid Artery |

Branch of external carotid artery

| 2 | Inferior Thyroid Artery |

Branch of subclavian artery

| 3 | Superior Thyroid Vein and |
| 4 | Middle Thyroid Vein |

Pass into the internal jugular veins

| 5 | Inferior Thyroid Vein |

Joins the brachiocephalic vein

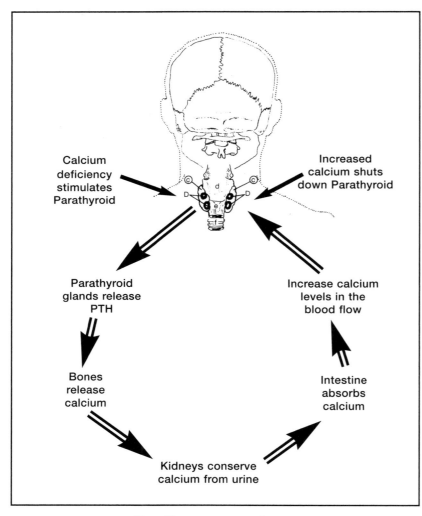

Calcium deficiency stimulates Parathyroid

Increased calcium shuts down Parathyroid

Parathyroid glands release PTH

Increase calcium levels in the blood flow

Bones release calcium

Intestine absorbs calcium

Kidneys conserve calcium from urine

THE PARATHYROID CYCLE

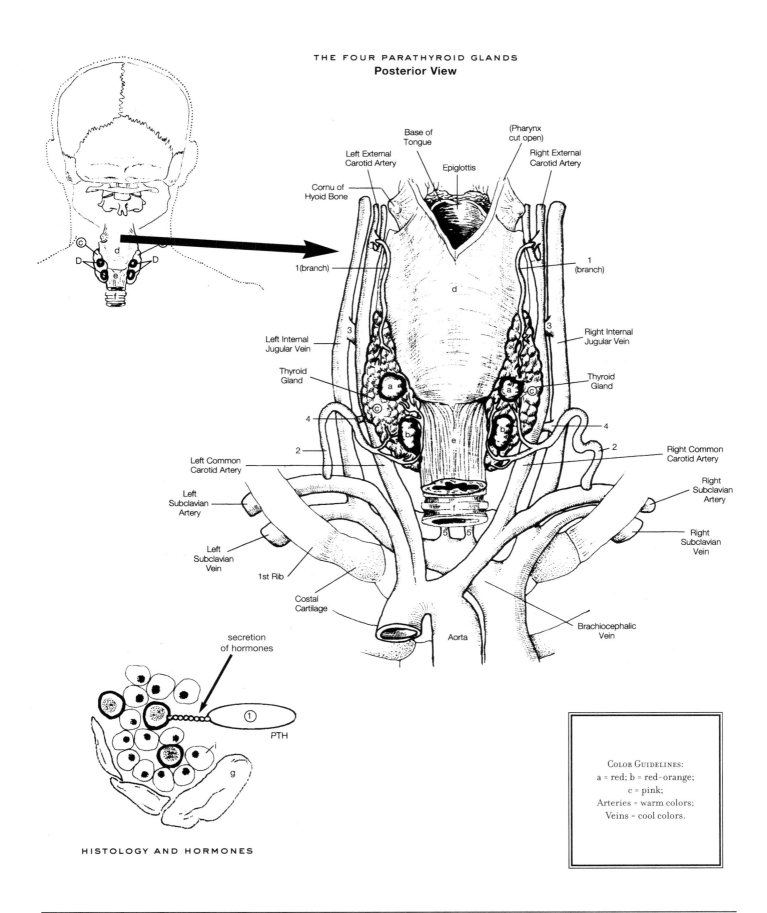

THE FOUR PARATHYROID GLANDS
Posterior View

Base of Tongue

(Pharynx cut open)

Left External Carotid Artery

Epiglottis

Right External Carotid Artery

Cornu of Hyoid Bone

1(branch)

1 (branch)

d

Left Internal Jugular Vein

3

3

Right Internal Jugular Vein

Thyroid Gland

a

a

c

Thyroid Gland

c

4

c

b

b

4

2

e

2

Left Common Carotid Artery

Right Common Carotid Artery

Left Subclavian Artery

Right Subclavian Artery

f

Left Subclavian Vein

5   5

Right Subclavian Vein

1st Rib

Costal Cartilage

Aorta

Brachiocephalic Vein

secretion of hormones

① PTH

i

g

HISTOLOGY AND HORMONES

COLOR GUIDELINES:
a = red; b = red-orange;
c = pink;
Arteries = warm colors;
Veins = cool colors.

The pair of ADRENAL (SUPRARENAL) GLANDS lie atop the superior and medial borders of the kidneys (within the RENAL FASCIA). Like the KIDNEYS, they are RETROPERITONEAL ORGANS. The adrenal glands are actually two different glands (CORTEX and MEDULLA) in a fibrous capsule. They have different and separate hormonal secretions, functions in the body, control mechanisms and embryological germ layers. The ADRENAL CORTEX is divided into three zones that secrete different types of corticosteroid hormones, which help regulate mineral balance (ZONA GLOMERULOSA), energy balance and stress (ZONA FASCICULATA); and reproductive function (ZONA RETICULARIS). The adrenal medulla secretes hormones called catecholamines (mono-amines), derived from tyrosine (amino acid); these hormones come from specialized postganglionic nerve cells called CHROMAFFIN CELLS. Each cluster of chromaffin cells receives pregaanglionic autonomic innervation, arranged near blood vessels. It affects the effects of the SYMPATHETIC NERVOUS SYSTEM in that it stimulates the metabolic rate, increasing energy in response to acute stress.

The nerve supply (innervation) of the adrenal medulla come from the PREGANGLIONIC FIBERS of the greater SPLANCHNIC NERVES (and the fibers of the CELIAC PLEXUS and associated sympathetic plexuses).

With regard to the blood circulation, the adrenal glands are highly vascular. Each gland is supplied by three separate suprarenal arteries (superior, middle, and inferior), and drained by two separate veins a renal and a suprarenal.

ALDOSTERONE is the mineralocorticoid hormone secreted by the zona glomerulosa of the adrenal cortex. The secretion effect of aldosterone is the regulation of the concentration of extracellular electrolytes (sodium resorption and potassium excretion). The hormonal regulation source of aldosterone are the blood enzyme of the renin-angiotensin system (angiotensin II) and blood potassium.

The secretion effects of the zona fasciculata are to inhibit immune response, regulate normal metabolism and regulate stress resistance (metabolism of carbohydrates, proteins and fats, vasoconstriction and decreasing of edema) in response to stress. The regulation source for the adrenal control hormones (as well as the sex steroids) is ACTH from the anterior lobe of the pitutitary gland, called the PITUITARY-ADRENAL axis.

The secretion effects of glucocorticoid hormones (HYDROCORTISONE or CORTISOL) of the zona reticularis is the supply of low-level (weak) supplements of the sex hormones from the gonads to hormonally regulate ACTH. The secretions of the chromaffin cells increase and mobilize blood pressure, respiratory rate, muscular contraction, blood glucose level, blood fatty acid level and the release of ACTH and TSH. Epinephrine (adrenalin) and norepinephrine create "fight or flight" reactions in response to life-threatening situations.

The embryological origins of the adrenal gland are the three-zoned ADRENAL CORTEX, which develops from the same non-neural mesodermal ridge from which the gonads develop.

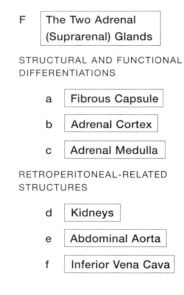

F | The Two Adrenal (Suprarenal) Glands

STRUCTURAL AND FUNCTIONAL DIFFERENTIATIONS

a | Fibrous Capsule

b | Adrenal Cortex

c | Adrenal Medulla

RETROPERITONEAL-RELATED STRUCTURES

d | Kidneys

e | Abdominal Aorta

f | Inferior Vena Cava

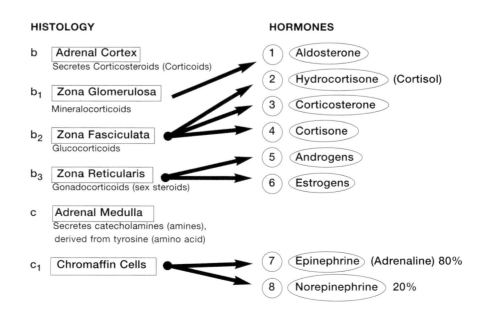

HISTOLOGY

b | Adrenal Cortex
Secretes Corticosteroids (Corticoids)

b_1 | Zona Glomerulosa
Mineralocorticoids

b_2 | Zona Fasciculata
Glucocorticoids

b_3 | Zona Reticularis
Gonadocorticoids (sex steroids)

c | Adrenal Medulla
Secretes catecholamines (amines),
derived from tyrosine (amino acid)

c_1 | Chromaffin Cells

HORMONES

1 | Aldosterone
2 | Hydrocortisone (Cortisol)
3 | Corticosterone
4 | Cortisone
5 | Androgens
6 | Estrogens

7 | Epinephrine (Adrenaline) 80%
8 | Norepinephrine 20%

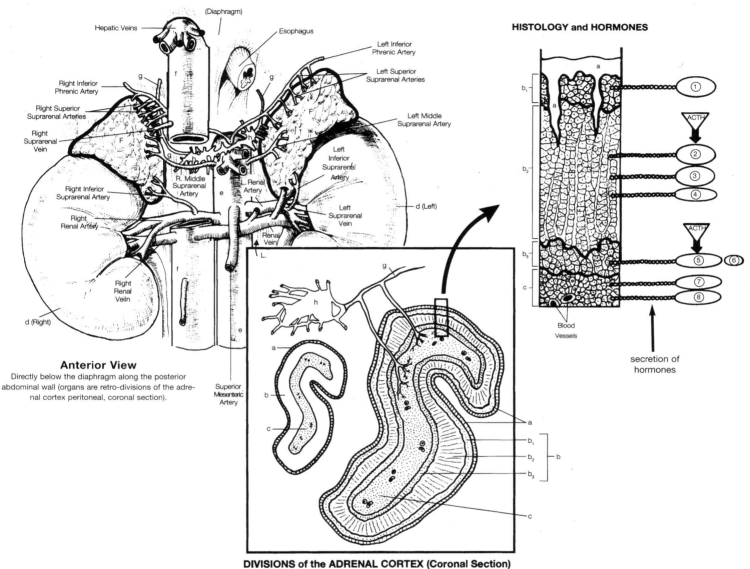

HISTOLOGY and HORMONES

ACTH

ACTH

secretion of
hormones

Blood
Vessels

Anterior View
Directly below the diaphragm along the posterior
abdominal wall (organs are retro-divisions of the adrenal cortex peritoneal, coronal section).

Hepatic Veins

(Diaphragm)

Esophagus

Left Inferior
Phrenic Artery

Left Superior
Suprarenal Arteries

Left Middle
Suprarenal Artery

Left
Inferior
Suprarenal
Artery

Left
Suprarenal
Vein

d (Left)

Right Inferior
Phrenic Artery

Right Superior
Suprarenal Arteries

Right
Suprarenal
Vein

Right Inferior
Suprarenal Artery

Right
Renal Artery

Right
Renal
Vein

d (Right)

R. Middle
Suprarenal
Artery

L. Renal Artery

Renal
Vein

L.

Superior
Mesenteric
Artery

DIVISIONS of the ADRENAL CORTEX (Coronal Section)

It gradually completely encapsulates the developing adrenal medulla. The
INNER ADRENAL MEDULLA develops embryologically from the neuroectodermal
cells, which derive from the NEURAL CREST of the NEURAL TUBE. (Most of the
PNS—the cranial and spinal nerves—derives from the neural crest.)

Like the pituitary, each adrenal gland is a single organ with a nonneural
part and a neural part. The pituitary and adrenal glands have two distinct
types of tissues and two different embryonic origins. They secrete different
hormones and are regulated by different control systems.

COLOR GUIDELINES:
F = light brown
Arteries = warm colors
Veins = cool colors

In the abdominal region of the body, gastrointestinal organs secrete hormones from their mucosa, and the pancreas secretes both endocrine hormonal and exocrine enzymatic substances from different areas of the organ. The STOMACH and SMALL INTESTINE secrete many GASTROINTESTINAL HORMONES that act on the stomach itself, the DUODENUM of the small intestine itself, the GALLBLADDER and the PANCREAS. These hormones are small proteins (polypeptides) that are secreted by the GASTROINTESTINAL MUCOSA. These hormones work together with other hormones and the AUTONOMIC NERVOUS SYSTEM (ANS) to regulate and coordinate the secretory activities of the MUCOSAL WALLS of the DIGESTIVE TRACT (stomach and duodenum), BILE (gallbladder) and PANCREATIC JUICE (pancreas).

The pancreas is both an EXOCRINE GLAND and an ENDOCRINE GLAND, secreting both hormones and enzymes respectively. Within its exocrine portion, ACINI CELLS secrete pancreatic juice (water, bicarbonate, and PANCREATIC ENZYMES). In the endocrine portion, ISLETS (ISLANDS) OF LANGERHANS secrete hormones from the ALPHA, BETA and DELTA CELLS (the alpha cells secrete glucagons, beta cells secrete insulin and delta cells secrete [human growing hormone inhibiting factor] HGHIF).

The arterial blood supply of the pancreas comes from the pancreaticoduodenal arteries (superior and inferior), the pancreatic artery, the splenic artery and the superior mesenteric artery.

Gastrin stimulates secretion of HCI (HYDROCHLORIC ACID): PARIETAL CELLS, stimulates secretion of PEPSINOGEN (PEPSIN): chief cells, stimulates the release of insulin from the pancreas (moderate) and maintains structure of gastric mucosa. Three important hormones secreted by the duodenal mucosa of the small intestine are CHOLECYSTOKININ-PANCREOZYMIN (CCK-P2), SECRETIN and GASTRIC INHIBITORY PEPTIDE (GIP). cholecystokinin-pancreozymin stimulates contractions of the gallbladder and bile flow, stimulates secretion of pancreatic juice enzymes, stimulates secretion of insulin from the pancreas (moderate), potentiates action of secretion on the pancreas and maintains the structure of the pancreas (EXOCRINE ACINI CELLS). Secretin stimulates secretion of water and bicarbonate ions in pancreatic juice, stimulates secretion of insulin from the pancreas (moderate) and potentiates the action of CCK-PZ on the pancreas. Gastric inhibitory peptide inhibits gastric emptying and inhibits gastric acid secretion, and stimulates secretion of insulin from the pancreas (moderate).

The glandular cini cells of the pancreas produce four major active enzymes: AMYLASE (digests starch), TRYPSIN (digests protein), LIPASE (digests triglycerides [fatty acids and glycerol]) and CHYMOTRYPSIN HYDROLYZES (digests proteins to peptones, or further).

The ISLETS OF LANGERHANS are stimulated by stress factors. The alpha cells stimulate the liver to break down sugar storage

---

GASTROINTESTINAL ORGANS

PYLORUS OF THE STOMACH (GASTRIC MUCOSA)

(1) ➡ Gastrin

SMALL INTESTINE (DUODENAL MUCOSA)

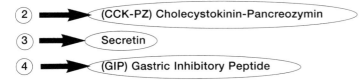

(2) ➡ (CCK-PZ) Cholecystokinin-Pancreozymin

(3) ➡ Secretin

(4) ➡ (GIP) Gastric Inhibitory Peptide

PANCREATIC SECRETIONS

EXOCRINE SECRETIONS = ENZYMES AND SODIUM COMPOUNDS

a   Glandular Acini Cells

(5) ➡ Pancreatic enzymes

ENDOCRINE SECRETIONS = HORMONES

ISLETS OF LANGERHANS

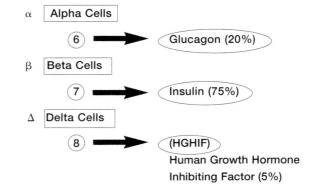

α   Alpha Cells

(6) ➡ Glucagon (20%)

β   Beta Cells

(7) ➡ Insulin (75%)

Δ   Delta Cells

(8) ➡ (HGHIF)
Human Growth Hormone
Inhibiting Factor (5%)

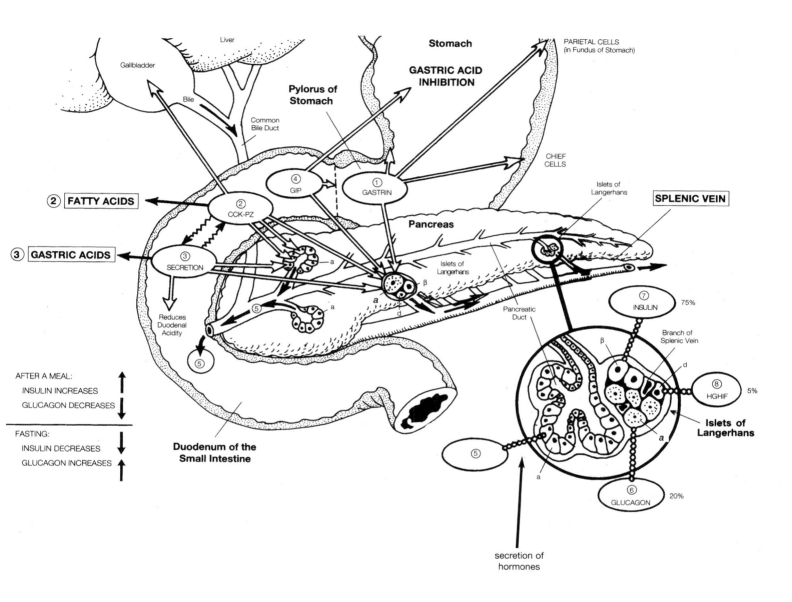

AFTER A MEAL:

INSULIN INCREASES ↑

GLUCAGON DECREASES ↓

FASTING:

INSULIN DECREASES ↓

GLUCAGON INCREASES ↑

(glycogen) into glucose, which increases the blood sugar level; they also promote lipid breakdown as well as amino acid conversion to glucose. The beta cells stimulate the LIVER to convert glucose into stored sugar (glycogen) and fat, decrease the blood sugar level, and promote the entry of glucose and amino acids into tissue cells. They also promote production of cellular protein and fats. The HGHIF hormone (from delta cells) inhibits the secretion of HGH (human growth hormone). All three pancreatic hormones—glucagon, insulin, and HGHIF—are regulated and secreted by the blood glucose level through negative feedback inhibition in the pancreas. (Insulin is also regulated by the VAGUS NERVE.)

COLOR GUIDELINES:
Stomach and Duodenum = light orange;
Pancreas = flesh;
Gallbladder and
Common Bile Duct = light green;
Liver = reddish brown.

CHAPTER 10: CARDIOVASCULAR SYSTEM

CONTENTS

SYSTEM COMPONENTS

The blood, heart and blood vessels

SYSTEM FUNCTIONS

*Distribute oxygen and nutrients to the cells, carry carbon
dioxide and wastes from the cells, maintain body acid-base
balance, protect against disease, prevent hemorrhage
(through blood clot formation)
and regulate body temperature.*

COLORING NOTES:
In general, arteries = red and related warm colors;
Veins = blue and related cool colors
(except in the pulmonary circulation, where the
arteries are blue and the veins are red
due to the difference in oxygen content).
Capillary networks = purple.

The CARDIOVASCULAR SYSTEM is composed of three basic elements: the HEART, BLOOD VESSELS, and BLOOD. It is the chief transport system of the body. The heart is the center of the system and pumps blood (total volume is 5,000 ml) through the blood vessels. The blood vessels form a tubular network throughout the body (approximately 60,000 miles!), permitting blood to flow from the heart to all living body cells (via forty billion microscopic CAPILLARIES), then back to the heart. Under normal conditions, it takes about one minute for the blood to be circulated from the heart to the most distal extremity and back to the heart again. The BLOOD VASCULAR SYSTEM provides four basic functions essential to the life of the human organism: TRANSPORTATION of NUTRIENT SUBSTANCES, BODY PROTECTION, REGULATION of BODY FUNCTIONS and TEMPERATURE REGULATION.

RED BLOOD CELLS (ERYTHROCYTES) transport oxygen (O_2) to the tissue cells. Oxygen received from inhaled air in the lungs attaches to the hemoglobin molecules in red blood cells. CARBON DIOXIDE (CO_2) is produced by cellular respiration, then released into the blood and carried to the lungs for eventual release via exhaled air.

Nutrient-rich molecules of food are mechanically and chemically broken down in the digestive tract and absorbed through the intestinal wall into the bloodstream (mostly carbohydrates and proteins, with some lipids). These nutritive molecules do not travel straight to the heart for distribution to the body system, but first travel through the HEPATIC PORTAL SYSTEM for filtration through the sinusoids (specialized capillaries) of the LIVER, then enter the heart through the VENOUS SYSTEM. There is also lipid absorption through the lymph vascular system before entrance into the heart (see page 232).

EXCRETORY SUBSTANCES from the cells in the blood consist of metabolic wastes, excess water, and ions released from cellular metabolism, plus other molecules from the fluid portion of the blood (PLASMA), which are filtered through the capillaries of the kidneys (of the urinary system) and excreted through the urine. Hormones and regulatory molecules released from the endocrine system are transported to specific, distant target organs and body tissues, regulating body functions. Body protection is offered through the clotting mechanism provided by blood platelets, which protects against blood loss when blood vessels are damaged. WHITE BLOOD CELLS (LEUKOCYTES) protect against foreign microbes or antigens entering the body. The LYMPH VASCULAR SYSTEM is yet another protective mechanism (against disease-causing viruses and toxins).

TEMPERATURE REGULATION is effected through body heat that comes from cellular metabolism (mostly muscle cells), and through the cardiovascular system working in conjunction with the integumentary system to eliminate excess heat through radiation from dilated blood vessels. Thus, to maintain homeostasis, the blood vascular system works closely and cooperatively with the lymph vascular system, respiratory system, urinary system, digestive system, endocrine system and integumentary system.

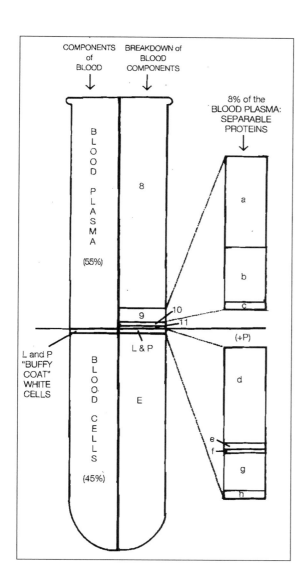

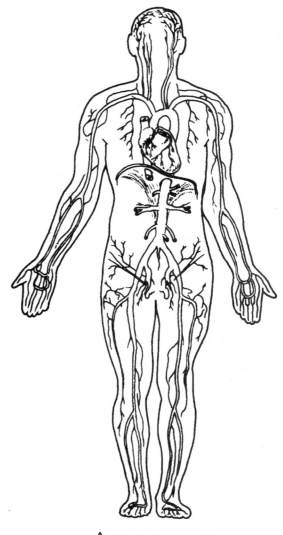

ANTERIOR VIEW
(ARTERIAL SYSTEM ONLY SHOWN)

COMPOSITION OF THE BLOOD

I. FLUID PORTION (55% OF TOTAL)

8   Blood Plasma   (99.0% of I)

9   Proteins   (8.0% of I)

    a   Albumins   (60% of 9)

    b   Globulins   (36% of 9)

       [Alpha-, Beta- and Gamaglobulins]

    c   Fibrinogen   (4% of 9)

10   Organic Acids   (1.0% of I)

11   Salts/Glucose   (0.9/0.0% of I)

       [Salts are sodium ions and compounds]

$$\text{Hematocrit} = \frac{\text{Volume of red blood cells}}{\text{Total blood volume}}$$

II. CELLULAR PORTION (45% OF TOTAL)

TWO TYPES OF BLOOD CELLS:

E   ERYTHROCYTES   (Red Blood Cells)   (44% of II)

L   LEUKOCYTES   (White Blood Cells)   (0.98 of II)

P   PLATELETS [THROMBOCYTES]   (0.02% OF II)

       [Non-nucleated cell fragments]

LEUKOCYTES (Percentage of L)

GRANULAR LEUKOCYTES (69%)   AGRANULAR LEUKOCYTES (31%)

d   Neutrophils (65%)   g   Lymphocytes (25%)

e   Eosinophils (3%)   h   Monocytes (6%)

f   Basophils   (1%)

THE CIRCULATORY SYSTEM: GENERAL SCHEME
BLOOD CIRCULATION THROUGH THE HUMAN BODY

CIRCULATION OF BLOOD THROUGH THE GASTROINTESTINAL TRACT

GENERAL CIRCULATORY ROUTES
Systemic circulation: left ventricle to right atrium

1 Aorta and branches

2 Superior vena cava
   and branches

3 Inferior vena cava and branches

PULMONARY CIRCULATION
Right ventricle to left atrium

4 Pulmonary trunk

5 Pulmonary arteries
   (left and right)

6 Pulmonary veins
   (2 left and 2 right)

HEPATIC PORTAL CIRCULATION

7 Hepatic portal vein

8 Hepatic veins

CAPILLARY AND SINUSOID STATIONS
   (Color the capillary areas on either side of
   these boxes in the illustration light purple)

PULMONARY CIRCULATION

A Lungs

SYSTEMIC CIRCULATION

B Head and neck

C Upper extremities (limbs)

D Thoracic/abdominal wall

E Liver

F Stomach

G Spleen

H Gastrointestinal tract

I Kidneys

J Pelvis and perineum

K Lower extremities (limbs)

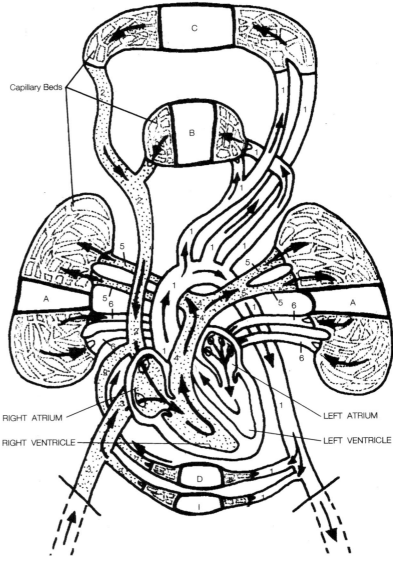

Capillary Beds

RIGHT ATRIUM   LEFT ATRIUM

RIGHT VENTRICLE   LEFT VENTRICLE

The CIRCULATORY SYSTEM of the human body can be generally divided into two systems, the CARDIOVASCULAR SYSTEM (BLOOD VASCULAR SYSTEM) and the LYMPHATIC SYSTEM (LYMPH VASCULAR SYSTEM). The cardiovascular system consists of the HEART, BLOOD and BLOOD VESSELS. The heart is a double muscular pump and consists of FOUR CHAMBERS, the RIGHT and LEFT ATRIUMS and RIGHT AND LEFT VENTRICLES. It can be considered as two separate pumps that pump blood in rhythm simultaneously.

The blood vessels form the two main circulatory routes throughout the entire body. The first is ROUTE 1: PULMONARY CIRCULATION, blood vessels that transport blood to the LUNGS (for gas exchange) and from the lungs back to the HEART. It consists of the right ventricle, PULMONARY TRUNK, PULMONARY ARTERIES (LEFT and RIGHT), PULMONARY CAPILLARIES (in lungs), PULMONARY VEINS (two LEFT and two RIGHT) and the LEFT ATRIUM. ROUTE 2: SYSTEMIC CIRCULATION is composed of all of the blood vessels in the body that are not part of the pulmonary circulation. In route 2, blood leaves the heart and travels through the large AORTA and its arterial branches to all the body regions, through the CAPILLARIES, and

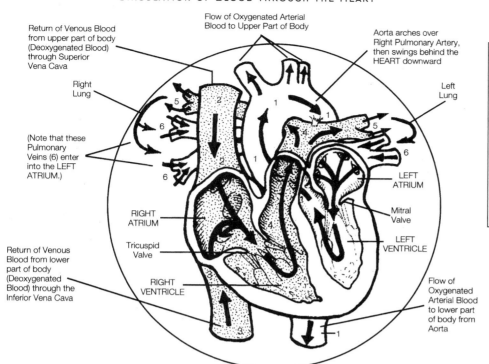

Return of Venous Blood from upper part of body (Deoxygenated Blood) through Superior Vena Cava

Flow of Oxygenated Arterial Blood to Upper Part of Body

Aorta arches over Right Pulmonary Artery, then swings behind the HEART downward

Right Lung

Left Lung

(Note that these Pulmonary Veins (6) enter into the LEFT ATRIUM.)

LEFT ATRIUM

RIGHT ATRIUM

Mitral Valve

Tricuspid Valve

LEFT VENTRICLE

Return of Venous Blood from lower part of body (Deoxygenated Blood) through the Inferior Vena Cava

RIGHT VENTRICLE

Flow of Oxygenated Arterial Blood to lower part of body from Aorta

COLOR GUIDELINES:
All capillary beds = purple;
1 = bright red; 2,3 = blues;
4 = green; 5 = light blue;
6 = pink; 7,8 = blue-greens.

SCHEMATIC DIAGRAM OF THE GENERAL CIRCULATORY ROUTES OF THE VASCULAR SYSTEM

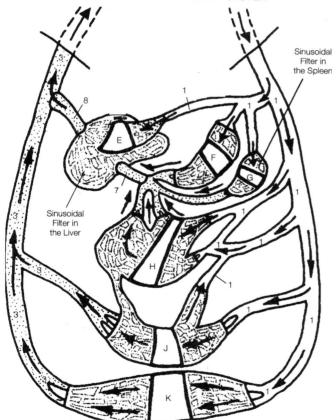

Sinusoidal Filter in the Spleen

Sinusoidal Filter in the Liver

Oxygen Content in the Blood Vessels

■ = DEOXYGENATED BLOOD   □ = OXYGENATED BLOOD

back to the heart via the VENOUS VESSELS (VEINS), and culminating back to the heart through the large veins (venae cavae).

Systemic Circulation can be said to begin in the heart at the left ventricle, then through the AORTA and AORTIC BRANCHES. Then ending with the large venae cavae to the right atrium in the heart, where the blood then drops to the left ventricle, then making its way to the PULMONARY CIRCULATION route.

Blood from the stomach, gastrointestinal tract and spleen travels through the HEPATIC PORTAL SYSTEM, a secondary CAPILLARY NETWORK (SINUSOIDAL FILTER) inside the liver, before returning to the heart via the INFERIOR VENA CAVA (see pages 218–19).

Within the heart proper, the right heart receives deoxygenated venous blood from the systematic circulation through the vena cavae. Blood enters the right atrium and is pumped by the right ventricle through the PULMONARY TRUNK and PULMONARY ARTERIES to both lungs, where it is oxygenated. The left heart receives oxygenated blood from the lungs through the pulmonary veins. Blood enters the left atrium and is pumped by the left ventricle through the aorta to the entire body.

The microscopic CAPILLARY and SINUSOID STATIONS mediate all exchanges of materials between blood and tissue cells occurring across the walls of the single-layered, endothelial-tubed capillaries, the thinnest and most numerous of all blood vessels. SINUSOID SPACES (lined with phagocytic kupffer cells) act as capillary substitutes in the liver and spleen. They are a discontinuous type of capillary in contrast to the continuous type found throughout the body, and are larger than capillaries. They are highly permeable and allow direct contact of liver and spleen cells with the blood.

THE CIRCULATORY SYSTEM: GENERAL SCHEME, CONTINUED
STRUCTURE OF BLOOD VESSELS

TYPES OF BLOOD VESSELS
(SIZE TRANSITION RATIOS)

1 Artery (3-10)

2 Arteriole (100)

3 Capillaries (500)

4 Venules (100)

5 Veins (10-3)

3a Sinusoids

(Discontinuous capillary spaces lined with phagocytic Kupffer cells [act as a capillary substitution in the liver and spleen].)

MICROCIRCULATION

6 Metarterioles

Arteriovenous anastomoses

7 Precapillary sphincter muscles

GENERAL WALL STRUCTURE OF BLOOD VESSELS

THREE MAIN LAYERS
Outer (longitudinal arrangement)

8 Tunica externa (adventitia)

A Fibrous connective tissue

B Vasa vasorum

Tiny blood vessels distributed to large arteries and veins

Middle (circular arrangement)

9 Tunica media

C Areolar and external elastic tissue

D Smooth muscle cells

Inner (longitudinal arrangement)

10 Tunica interna intima

E Areolar and internal elastic tissue

F Endothelium

Faces the inner lumen of the vessel

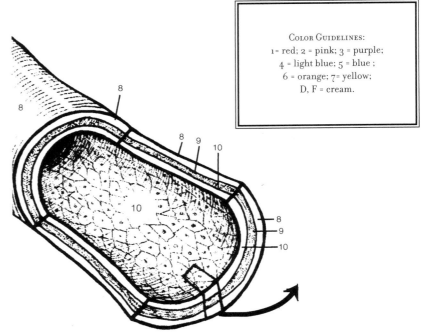

GENERAL 3-WALLED STRUCTURE OF A BLOOD VESSEL

COLOR GUIDELINES:
1= red; 2 = pink; 3 = purple; 4 = light blue; 5 = blue; 6 = orange; 7= yellow; D, F = cream.

In general, ARTERIES (in order: large, medium, and small) transport blood away from the heart to the microscopic CAPILLARIES, where the export oxygen and import waste products in conjunction with the surrounding tissue cells (in the systemic circulation, arterial blood is oxygenated); VEINS transport blood toward the heart from the capillaries (in the systemic circulation, venous blood is deoxygenated). The thin-walled structure of the capillaries enables exchange of plasma fluid, dissolved molecules and gases between blood and the surrounding tissues (tissue cells). Capillaries are only one-layer thick, consisting of flat squamous epithelium, called endothelium. Blood transported away from the heart travels in vessels of progressively decreasing diameters—from ELASTIC ARTERIES to MUSCULAR ARTERIES to ARTERIOLES to capillaries. Capillaries are microscopic blood vessels surrounding tissue cells that connect the arterial flow into them to the venous flow out of them. Blood returning to the heart travels in venous vessels of progressively increasing diameters: capillaries to VENULES to MUSCULAR VEINS to GREAT VEINS.

Arteries have more muscle and smaller diameters than veins when vessels of similar size are compared. Arteries are rounder, and veins are more collapsed in nature (they are not usually filled and function as reservoirs since they can stretch to receive more blood). Most veins have semilunar, one-way valves (arteries do not have valves).

Concerning levels of microcirculation, in some tissues, small arterioles (20–30 millimeters in diameter) bypass capillary beds and transport blood directly to venules via shunts: METARTERIOLES (or ARTERIOVENOUS ANASTOMOSES). PRECAPILLARY SPHINCTER MUSCLES regulate the flow of blood from the arterioles through the capillaries. Arterioles also have more smooth muscle than elastin, providing a narrow consistent lumina, and the greatest resistance to blood flow through the arterial system.

The critical phase of circulation occurs in the capillaries (one cell wall thick), because only from capillaries can blood give up food and oxygen to body tissues, and receive waste products and carbon dioxide from body tissues.

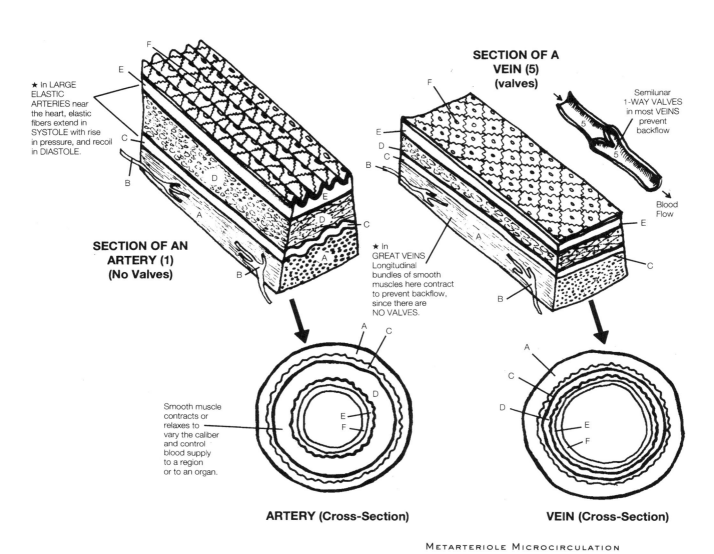

★ In LARGE ELASTIC ARTERIES near the heart, elastic fibers extend in SYSTOLE with rise in pressure, and recoil in DIASTOLE.

SECTION OF AN ARTERY (1)
(No Valves)

SECTION OF A VEIN (5)
(valves)

Semilunar 1-WAY VALVES in most VEINS prevent backflow

Blood Flow

★ In GREAT VEINS Longitudinal bundles of smooth muscles here contract to prevent backflow, since there are NO VALVES.

Smooth muscle contracts or relaxes to vary the caliber and control blood supply to a region or to an organ.

ARTERY (Cross-Section)

VEIN (Cross-Section)

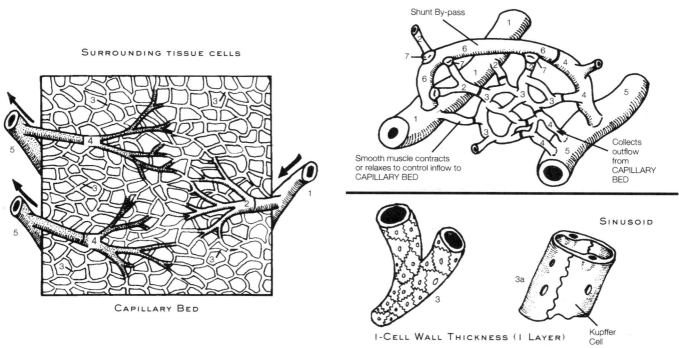

SURROUNDING TISSUE CELLS

METARTERIOLE MICROCIRCULATION

Shunt By-pass

Smooth muscle contracts or relaxes to control inflow to CAPILLARY BED

Collects outflow from CAPILLARY BED

CAPILLARY BED

SINUSOID

1-CELL WALL THICKNESS (1 LAYER)

Kupffer Cell

LOCATION OF THE HEART

ORIENTATION

1 Base

2 Apex

3 Axis

4 Midline body axis

BORDERS

5 Left

6 Superior

7 Right

8 Inferior

SURFACE

9 Sternocostal (anterior)

10 Diaphragmatic (inferior)

A Pericardium: pericardial sac

A₁ FIbrous pericardium layer (outer)

A₂ Serous pericardium layer (inner)

B Pericardial cavity/ Pericardial fluid

Potential space between the parietal pericardium and the visceral pericardium with lubricating watery fluid

THREE LAYERS OF THE HEART WALL

C Epicardium (Visceral Pericardium) (external)

D Myocardium (Middle)

E Endocardium (Inner)

MEDIASTINUM

SM Superior

AM Anterior

MM Middle

PM Posterior

FOUR CHAMBERS OF THE HEART

RIGHT HEART

RA Right atrium

RV Right ventricle

LEFT HEART

LA Left atrium

LV Left ventricle

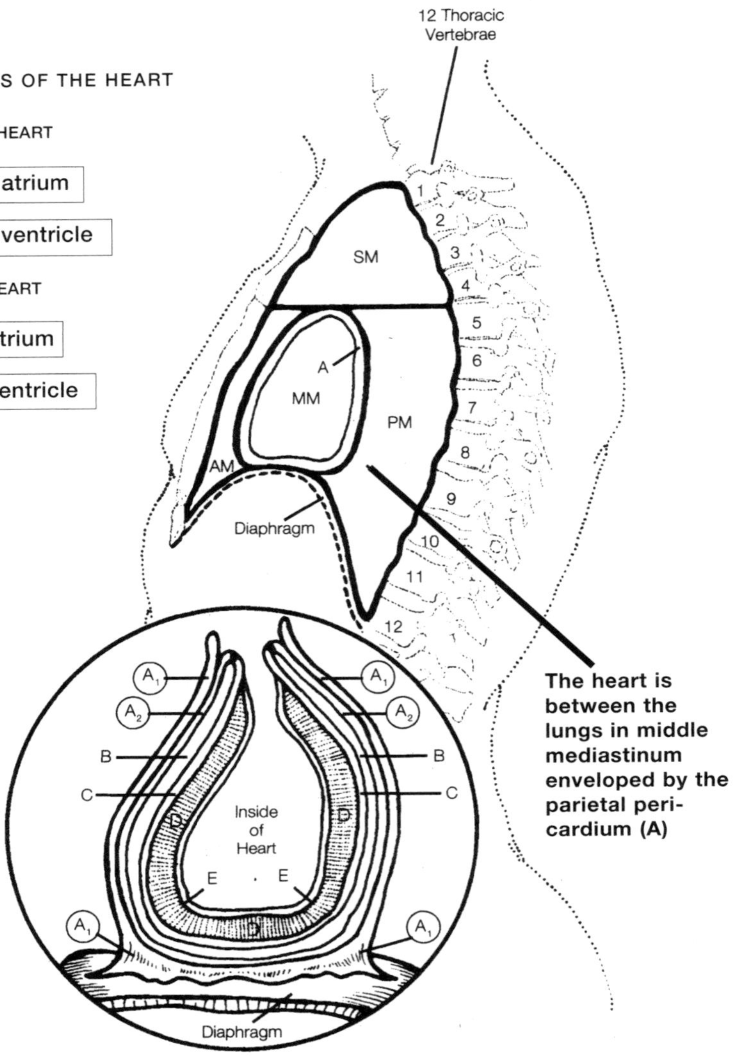

The heart is between the lungs in middle mediastinum enveloped by the parietal pericardium (A)

MIDSAGITTAL SECTION OF PERICARDIUM AND HEART

The HEART is a hollow four-chambered muscular organ. It functions as a double pump—two pumps in one, each with a different function, but working together contracting simultaneously. The RIGHT PUMP collects blood low in oxygen from the body (via the SUPERIOR and INFERIOR VENA CAVAE) and pumps it to the LUNGS (via the PULMONARY TRUNK). The LEFT PUMP collects blood high in oxygen from the lungs (via the PULMONARY VEINS) and pumps it to the body via the aorta. The heart is located in the THORACIC cavity, in the mediastinum between the lungs. Two-thirds of the organ is situated left of the midline body axis. The apex points downward and to the left and comes in contact with the diaphragm. The base is the superior aspect of the heart where the large vessels are attached.

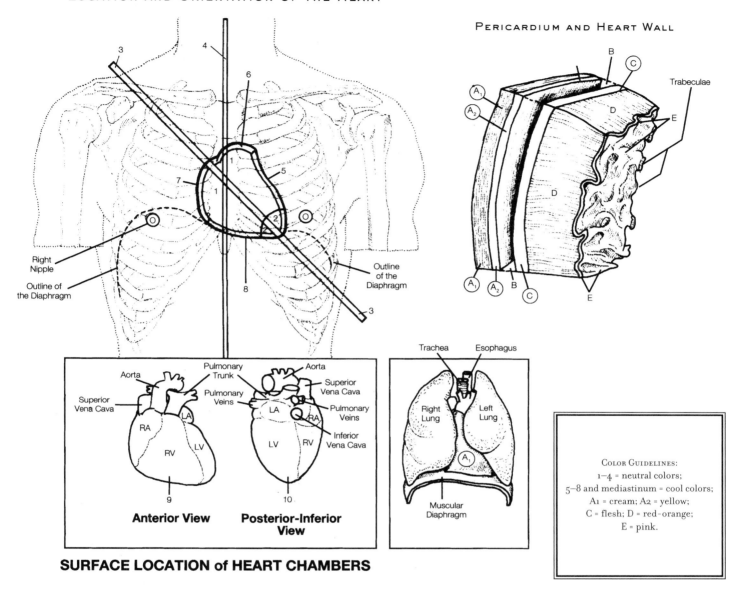

Trabeculae

Right Nipple

Outline of the Diaphragm

Outline of the Diaphragm

Aorta

Pulmonary Trunk

Aorta

Pulmonary Veins

Superior Vena Cava

Superior Vena Cava

Pulmonary Veins

Inferior Vena Cava

RA

LA

LA

RA

RV

LV

LV

RV

9

10

Anterior View

Posterior-Inferior View

SURFACE LOCATION of HEART CHAMBERS

Trachea

Esophagus

Right Lung

Left Lung

Muscular Diaphragm

COLOR GUIDELINES:
1–4 = neutral colors;
5–8 and mediastinum = cool colors;
A1 = cream; A2 = yellow;
C = flesh; D = red-orange;
E = pink.

The PERICARDIUM (PARIETAL PERICARDIUM) is also called the PERICARDIAL SAC. It is a double membranous, fibroserous sac enclosing the heart and the origin of the great blood vessels and protecting them from the rest of the thoracic organs. It consists of two layers, the FIBROUS PERICARDIUM (outer), a tough, fibrous connective tissue that prevents overdistension of the heart, provides protection, and anchors the heart in the MEDIASTINUM; and the SEROUS PERICARDIUM (INNER), which is a thin, delicate and continuous with the EPICARDIUM (VISCERAL PERICARDIUM), the outer layer of the wall of the heart, located at the base of the heart and around the great blood vessels. Between the serous pericardium and the epicardium is the PERICARDIAL CAVITY, containing some watery fluid (secreted by the serous pericardium). This PERICARDIAL FLUID prevents friction between the membranes during heart movement.

There are three layers of the heart wall: the EPICARDIUM/VISCERAL PERICARDIUM (external), the MYOCARDIUM (middle) and the ENDOCARDIUM (inner). The epicardium is a thin, transparent, outer serous layer continuous with the thin, inner serous layer of the pericardium. The myocardium is the cardiac muscle tissue responsible for heart contraction. It is the bulk of the heart. The cardiac muscle fibers are INVOLUNTARY, STRIATED, AND BRANCHED; and arranged in interlacing bundles. The endocardium is a thin layer of serous ENDOTHELIUM that lines the inner surface and cavities of the heart, lining the inside of the myocardium. It covers the valves of the heart and the tendons that hold them. It is continuous with the endothelium or TUNICA INTERNA of the blood vessels attached to the heart.

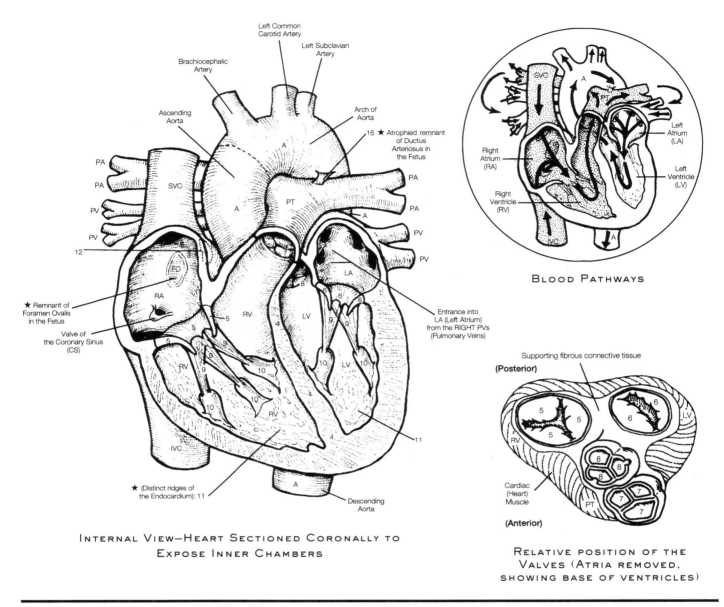

INTERNAL VIEW—HEART SECTIONED CORONALLY TO EXPOSE INNER CHAMBERS

BLOOD PATHWAYS

RELATIVE POSITION OF THE VALVES (ATRIA REMOVED, SHOWING BASE OF VENTRICLES)

The HEART is a four-chambered double pump. Simultaneously, both UPPER ATRIA from each pump contract simultaneously and empty blood into the lower ventricles, and the lower ventricles contract and push blood out of the heart at the upper base. The RIGHT VENTRICLE pushes blood (low oxygen) through the PULMONARY TRUNK and the LEFT VENTRICLE pushes blood (high oxygen) through the AORTA. The partition between the the atria, the INTERATRIAL SEPTUM, is thin and muscular, while the partition between ventricles, the INTERVENTRICULAR SEPTUM, is thick and muscular. The valves between the atria and ventricles are ATROVENTRICULAR (CUSPID) VALVES. The MITRAL (BICUSPID) VALVE is between LEFT ATRIA and left ventricle, and the TRICUSPID VALVE is between RIGHT ATRIA and right ventricle. The valves at the base of the PULMONARY TRUNK and aorta are SEMILUNAR VALVES.

The SULCI (or grooves) on the surface of the heart are indicative of the partitions between the chambers. They contain the coronary arteries. The ATRIOVENTRICULAR (CORONARY) SULCUS encircles the heart, and indicates the partition between atria and ventricles. The ANTERIOR and POSTERIOR INTERVENTRICULAR SULCI indicate partitioning between both ventricles.

The cusps (leaflets) of the atrioventricular valves are each held in position by strong, tendinous cords, or CHORDAE TENDINEAE. These cords in turn are secured to the ventricular walls by cone-shaped PAPILLARY MUSCLES. These cords and muscles prevent the atrioventricular valves from inverting into the atria when the ventricles contract. The LIGAMENTUM ARTERIOSUM is a remnant of the ductus arteriosus shunt in the fetus.

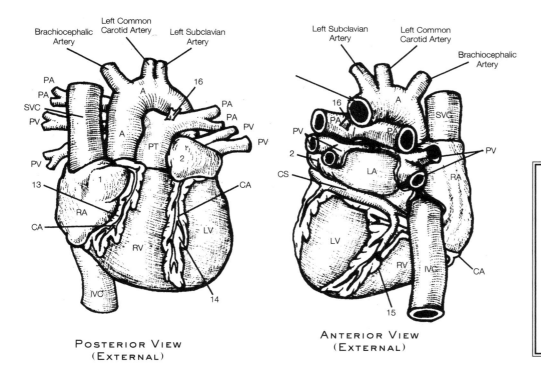

POSTERIOR VIEW
(EXTERNAL)

ANTERIOR VIEW
(EXTERNAL)

COLOR GUIDELINES:
Heart chambers = warm colors (light);
Aorta = reds;
Vena Cavae = blues;
Sulci = yellows, creams;
PT = green-blue;
PA = light blue;
PV = pink.

GROSS ANATOMY OF THE HEART
CHAMBERS

RA | Right atrium

LA | Left atrium

RV | Right ventricle

LV | Left ventricle

AURICLES
Conical pouches of the atria that project from the upper anterior portions

1 | Right auricle

2 | Left auricle

SEPTA (PARTITIONS)

3 | Interatrial septum

FO | Fossa ovalis
Only in right atrium

4 | Interventricular septum

VALVES
ATRIOVENTRICULAR VALVES

5 | Tricuspid valve

6 | Bicuspid (mitral) valve

SEMILUNAR VALVES

7 | Pulmonary semilunar valve

8 | Aortic semilunar valve

ACCESSORY MUSCLES AND CORDS

9 | Chordae tendineae

10 | Papillary muscles

11 | Trabeculae carneae

12 | Musculi pectinati
Anterior auricular combed bundles

SULCI (GROOVES)

13 | Atrioventricular (coronary) sulcus

14 | Anterior interventricular sulcus

15 | Posterior interventricular sulcus

LIGAMENTS

16 | Ligamentum arteriosum

GREAT BLOOD VESSELS: VENOUS

SVC | Superior vena cava

IVC | Inferior vena cava

CS | Coronary sinus

PV | Pulmonary veins
Two from each lung

GREAT BLOOD VESSELS: ARTERIAL

PT | Pulmonary trunk

PA | Pulmonary arteries (right and left)

CA | Coronary arteries

A | Aorta

The AORTA is the major systemic vessel arising from the heart, and begins as the ascending aorta, ascending from the LEFT VENTRICLE. Just beyond the segment of the AORTIC SEMILUNAR VALVE are the LEFT and RIGHT AORTIC SINUSES. These aortic sinuses, also known as the SINUS OF VALSALVA, are small dilatations of the aorta that precede the only branches of the ascending aorta—the LEFT and RIGHT CORONARY ARTERIES.

The CORONARY ARTERIES supply the heart itself, serving the MYOCARDIUM. They lie within a groove bedding that encircles the heart, called the ATRIOVENTRICULAR SULCUS (indicative of the inner partition between the atria and ventricles). The branches of the left coronary artery are the ANTERIOR INTERVENTRICULAR, which serves both ventricles, and the CIRCUMFLEX ARTERY, which serves the left chambers (the left atrium and the left ventricle). It is the branch of the circumflex artery that supplies the left ventricle. Branches of the right coronary artery are the POSTERIOR INTERVENTRICULAR ARTERY, which serves both ventricles, and the MARGINAL ARTERY, which serves the right chambers (mainly the right ventricle).

The CORONARY VEINS parallel the coronary arteries. Blood enters them from the capillaries in the myocardium. They are more superficial than arteries, with thinner walls. Principal veins are the GREAT CARDIAC VEIN that drains the anterior aspect of the heart; the MIDDLE CARDIAC VEIN that drains posterior aspect of the heart; and the SMALL CARDIAC VEIN that drains the anterior-inferior aspect of the heart. The CORONARY SINUS is a large venous channel on the posterior aspect of the heart that receives venous blood from all the cardiac (coronary) veins and then empties into the right atrium.

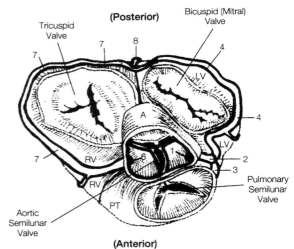

RELATIVE POSITION OF THE VALVES
Atria removed, showing base of the Ventricles

(Posterior)

Tricuspid Valve

Bicuspid (Mitral) Valve

Aortic Semilunar Valve

Pulmonary Semilunar Valve

(Anterior)
Superior View
(Atria removed, showing the base of the ventricles)

COLORING NOTES:
Arteries = reds and warm colors
Veins = blues and cool colors.

CORONARY CIRCULATION OF THE HEART
VESSELS SUPPLYING THE HEART
The aortic sinuses and the coronary arteries

1	Left aortic sinus
2	Left coronary artery
3	Anterior interventricular branch
4	Circumflex artery (branch)
5	Posterior artery of the left ventricle
6	Right aortic sinus (Sinus of Valsalva)
7	Right coronary artery
8	Posterior interventricular branch
9	Marginal branch

VESSELS DRAINING THE HEART
THE CARDIAC (CORONARY) VEINS
(13 drains into 12; 10, 12, 14–17 drain into 11)

10	Anterior cardiac veins
11	Coronary sinus
12	Great cardiac vein
13	Left marginal vein
14	Small cardiac vein
15	Middle cardiac vein
16	Posterior vein of the left ventricle
17	Oblique vein of the left atrium

The
CORONARY
ARTERIES

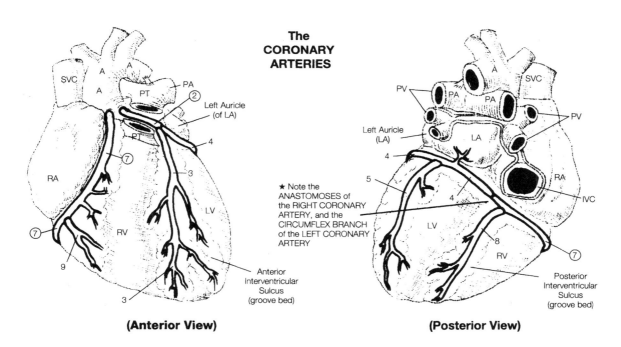

★ Note the ANASTOMOSES of the RIGHT CORONARY ARTERY, and the CIRCUMFLEX BRANCH of the LEFT CORONARY ARTERY

(Anterior View)

(Posterior View)

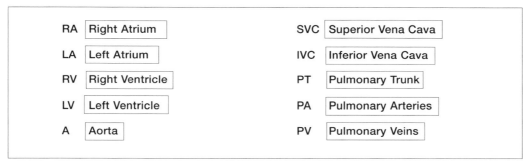

RA	Right Atrium	SVC	Superior Vena Cava
LA	Left Atrium	IVC	Inferior Vena Cava
RV	Right Ventricle	PT	Pulmonary Trunk
LV	Left Ventricle	PA	Pulmonary Arteries
A	Aorta	PV	Pulmonary Veins

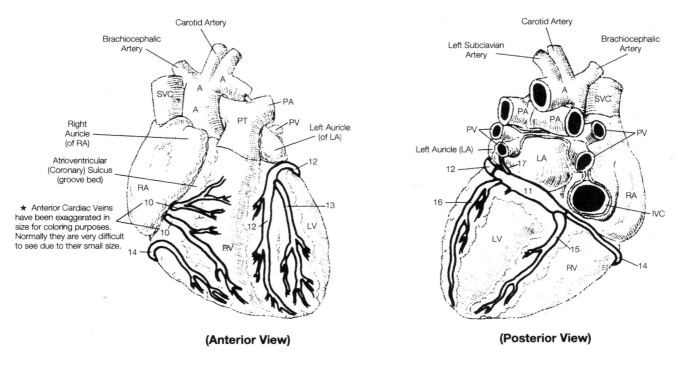

★ Anterior Cardiac Veins have been exaggerated in size for coloring purposes. Normally they are very difficult to see due to their small size.

(Anterior View)

(Posterior View)

The two COMMON CAROTID ARTERIES pass upward along either side of the trachea in the NECK (the RIGHT COMMON CAROTID ARTERY is shown here). Small branches of the common carotid artery supply various neck structures: the larynx, thyroid gland, anterior neck muscles and lymph glands. The common carotid artery bifurcates the INTERNAL CAROTID ARTERY (entering the base of the skull and supplying the brain) and the EXTERNAL CAROTID ARTERY (mainly supplying structures in the neck and head area external to the skull). The main branches of the EXTERNAL CAROTID ARTERY include the SUPERIOR THYROID ARTERY (hyoid muscles, larynx and vocal cords and thyroid gland), the ASCENDING PHARYNGEAL ARTERY (pharyngeal area and lymph nodes), LINGUAL ARTERY (tongue [extensive] and sublingual salivary gland). The FACIAL ARTERY passes through a notch on the inferior side of the mandible and reaches the pharyngeal area, palate, chin, lips and nose); the final main branches are the OCCIPITAL ARTERY (posterior scalp, meninges covering the brain, mastoid process of temporal bone and posterior neck muscles) and the POSTERIOR AURICULAR ARTERY (the ear and scalp over the ear). The external carotid artery divides into two large arteries near the mandibular condyle: the SUPERFICIAL TEMPORAL ARTERY (parotid salivary gland and superficial structures on the sides of the head) and the MAXILLARY ARTERY (teeth and gums, muscles of mastication, nasal cavity, eyelids and meninges).

Note that three major vessels arise from the aortic arch: the BRACHIOCEPHALIC ARTERY (TRUNK), LEFT COMMON CAROTID ARTERY and LEFT SUBCLAVIAN ARTERY. The BRACHIOCEPHALIC ARTERY bifurcates into two vessels, the RIGHT COMMON CAROTID ARTERY (supplying the neck and head on the right side) and the RIGHT SUBCLAVIAN ARTERY (supplying the right arm).

FIRST VESSEL TO BRANCH FROM THE AORTIC ARCH (VEERING RIGHT):

| B | Brachiocephalic (innominate) artery |

TWO BIFURCATIONS OF THE BRACHIOCEPHALIC

| S | Right subclavian artery |
| CC | Right common carotid artery |

BRANCHES OF THE RIGHT SUBCLAVIAN A:

IT	Internal thoracic artery (internal mammary artery.)
V	Vertebral artery
TT	Thyrocervical trunk
1	Inferior thyroid artery
2	Suprascapular artery
3	Transverse cervical artery
CT	Costocervical trunk
4	Deep cervical artery
5	Highest intercostal artery

BRANCHES OF THE RIGHT COMMON CAROTID ARTERY:

| IC | Internal carotid artery |
| EC | External carotid artery |

BRANCHES OF THE EXTERNAL CAROTID ARTERY

St	Superior thyroid artery
AP	Ascending pharyngeal artery
L	Lingual artery
F	Facial artery
6	Superior and inferior labial branch
7	Lateral nasal branch
8	Angular branch
9	Submental branch
O	Occipital artery
PA	Posterior auricular artery

EXTERNAL CAROTID ARTERY TERMINATES BY DIVIDING INTO TWO SEPARATE ARTERIES:

ST	Superficial temporal artery
10	Transverse facial branch
11	Zygomatic-orbital branch
M	Maxillary artery
12	Superior and inferior alveolar branch
13	Infraorbital branch
14	Middle meningeal branch
15	Buccal branch

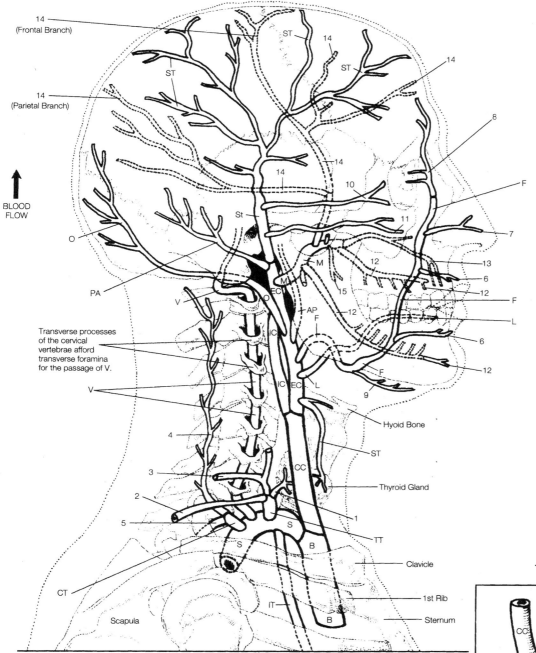

14
(Frontal Branch)

14
(Parietal Branch)

ST

ST

ST

14

14

14

8

10

14

11

F

7

BLOOD
FLOW

O

St

M

M

EC

V

AP
F

13

6

12

F

L

6

12

PA

15

12

12

Transverse processes
of the cervical
vertebrae afford
transverse foramina
for the passage of V.

IC

IC EC

L

F

9

V

Hyoid Bone

4

ST

3

CC

Thyroid Gland

2

1

5

S

TT

S

B

CT

Clavicle

IT

1st Rib

B

Scapula

Sternum

ARTERIES SUPPLYING the HEAD and NECK (Right Side)

RIGHT LATERAL VIEW

COLORING NOTES:
Use strong, bright warm colors
for major arteries and major
branches and lighter, warm
colors and neutral colors (or no
colors) for minor branches.

DOTTED LINES
INDICATE
ARTERIES
INTERNAL TO
THE SKULL

THE AORTIC ARCH

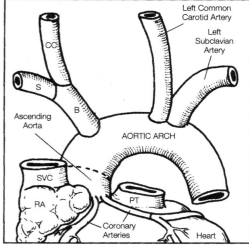

Left Common
Carotid Artery

Left
Subclavian
Artery

CC

S

B

Ascending
Aorta

AORTIC ARCH

SVC

RA

PT

Coronary
Arteries

Heart

RIGHT CEREBRAL HEMISPHERE (RIGHT LATERAL VIEW)

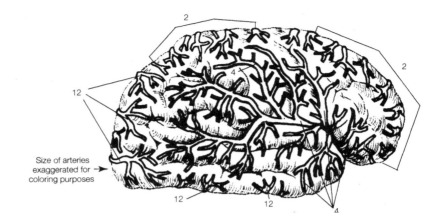

Size of arteries exaggerated for coloring purposes →

MAIN ARTERIAL SUPPLY TO THE BRAIN:

- (IC) Internal carotid arteries (paired)
- (V) Vertebral arteries (paired)

BRANCHES OF THE TWO INTERNAL CAROTID ARTERIES:

- 1 Anterior choroid artery (2)
- 2 Anterior cerebral artery (2)
- 3 Anterior communicating artery (1)
- 4 Middle cerebral artery (2)

BRANCHES OF THE TWO VERTEBRAL ARTERIES:

- 5 Anterior spinal artery (1)
- 6 Posterior inferior cerebellar artery (2)
- 7 Basilar artery (1)

Within the brain case, the two vertebral arteries unite to form one large basilar artery.

BRANCHES OF THE BASILAR ARTERY (7):

- 8 Anterior inferior artery (2)
- 9 Labyrinthine artery (2) (internal auditory artery)
- 10 Pontine branches (many)
- 11 Superior cerebellar artery (2)
- 12 Posterior cerebral artery (2)
- 13 Posterior communicating arteries (2)

RIGHT CEREBRAL HEMISPHERE (RIGHT MEDIAL VIEW)

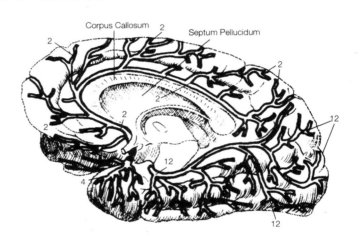

Corpus Callosum   Septum Pellucidum

The blood to the BRAIN is supplied by two INTERNAL CAROTID ARTERIES and two VERTEBRAL ARTERIES (four vessels). They all eventually unite in a unique circular arrangement of vessels on the inferior aspect of the brain surrounding the pituitary gland (hypophysis) called the CIRCLE OF WILLIS. The paired vertebral arteries enter the skull through the foramen magnum, and once inside the brain case, unite to form the BASILAR ARTERY at the level of the pons. The main branches of the vertebral arteries are the two POSTERIOR CEREBRAL ARTERIES, which supply the posterior portion of the cerebrum, and the two POSTERIOR COMMUNICATING ARTERIES, which branch from the posterior cerebral arteries and form part of the arterial circle of Willis by joining with the INTERNAL CAROTID ARTERIES. The internal carotid arteries enter the base of the skull through the carotid canal of the temporal bone.

The circle of Willis, the arterial circle surrounding the pituitary gland (hypophysis), is an anastomoses center: if one of the four main vessels becomes occluded, the other vessels may offer limited alternate routes for passage of blood to the brain (although, practically and functionally speaking, they are not very effective). The circle of Willis may, however, relieve pressure from larger vessels to the smaller branches of the circle (mainly those to the pituitary gland).

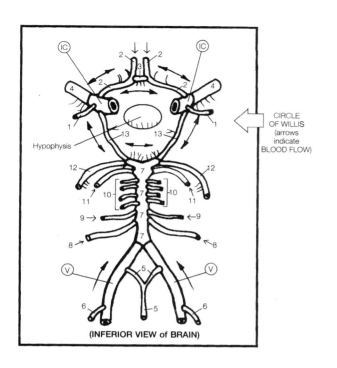

IC   IC

CIRCLE
OF WILLIS
(arrows
indicate
BLOOD FLOW)

Hypophysis

V   V

(INFERIOR VIEW of BRAIN)

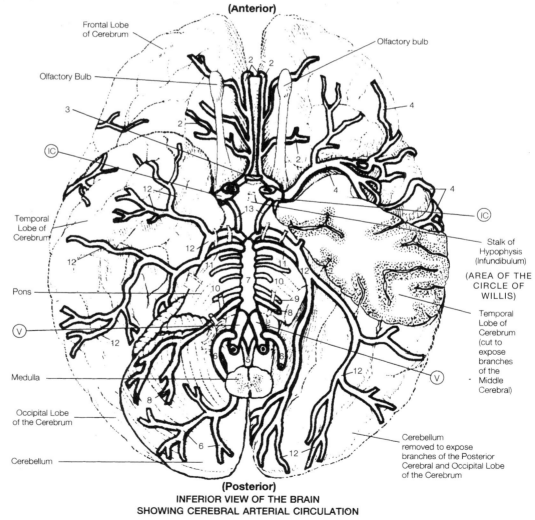

(Anterior)

Frontal Lobe
of Cerebrum

Olfactory bulb

Olfactory Bulb

Temporal
Lobe of
Cerebrum

IC

IC

Stalk of
Hypophysis
(Infundibulum)

(AREA OF THE
CIRCLE OF
WILLIS)

Pons

Temporal
Lobe of
Cerebrum
(cut to
expose
branches
of the
Middle
Cerebral)

V

V

Medulla

Occipital Lobe
of the Cerebrum

Cerebellum

Cerebellum
removed to expose
branches of the Posterior
Cerebral and Occipital Lobe
of the Cerebrum

(Posterior)
INFERIOR VIEW OF THE BRAIN
SHOWING CEREBRAL ARTERIAL CIRCULATION

On the right side of the body, the RIGHT SUBCLAVIAN ARTERY branches from the BRACHIOCEPHALIC ARTERY, whereas on the left side of the body, the LEFT SUBCLAVIAN ARTERY branches directly from the aortic arch.

The subclavian arteries pass laterally deep to the clavicle, and as they enter the axillary region, they become the AXILLARY ARTERIES (from the outer border of the first rib to the lower border of the teres major muscle). Small branches supply the tissues of the upper thorax and the tissues of the shoulder region.

As the AXILLARY ARTERIES enter the brachial region, they become the BRACHIAL ARTERIES as they continue along the medial side of the humerus. The major brachial artery is the DEEP BRACHIAL ARTERY, which curves posteriorly (near the radial nerve) and supplies the triceps muscle.

Just below the cubital fossa on the anterior side of the arm, the brachial artery bifurcates into a LATERAL RADIAL ARTERY and a MEDIAL ULNAR ARTERY, which supply the forearm and all of the hands and digits. The first and largest branch of the radial artery is the RADIAL RECURRENT ARTERY, which supplies the elbow region. The first and largest branches of the ulnar artery are the ANTERIOR and POSTERIOR ULNAR RECURRENT ARTERIES.

At the wrist, the RADIAL and ULNAR ARTERIES anastomose to form the PALMAR ARCHES, which branch into the PALMAR DIGITAL ARTERIES to the fingers.

BLOOD PRESSURE POINTS are at the brachial artery, on the medial side of the humerus. PULSATION POINTS are at the subclavian artery (just above the medial portion of the clavicle), radial artery (lateral side of wrist at base of thumb) and ulnar artery (medial side of wrist).

The THORACOACROMLAL ARTERY (2) has four main branches important in the collateral circulation of the shoulder region: the pectoral branch, clavicular branch, acromial branch and deltoid branch.

FIRST BRANCH OF THE AORTIC ARCH

B | Brachiocephalic (innominate) artery

TWO BRANCHES OF THE BRACHIOCEPHALIC ARTERY

CC | Right common carotid artery

S | Right subclavian artery

AS THE SUBCLAVIAN ARTERY PASSES THROUGH THE AXILLARY REGION IT BECOMES THE

AX | Axillary artery

AS THE AXILLARY ARTERY CONTINUES THROUGH THE BRACHIAL REGION IT BECOMES THE

BR | Brachial artery

THE BRACHIAL ARTERY BIFURCATES AT THE ELBOW INTO TWO MAIN ARTERIES

R | Radial artery (lateral)

U | Ulnar artery (medial)

THE RADIAL AND ULNAR ARTERIES ANASTOMOSE IN THE PALM TO FORM TWO PALMAR ARCHES

SP | Superficial palmar arch

DP | Deep palmar arch

ANASTOMOSES

PC | Palmar carpal arch

DC | Dorsal carpal arch

19 | Dorsal metacarpal arteries

20 | Dorsal digital arteries

BRANCHES OF THE AXILLARY ARTERY

1 | Superior thoracic artery

2 | Thoracoacromial artery

3 | Lateral thoracic artery

4 | Subscapular artery

5 | Anterior and posterior humeral circumflex arteries

BRANCHES OF THE BRACHIAL ARTERY

6 | Profunda brachii (deep brachial)

7 | Superior ulnar collateral artery

8 | Inferior ulnar collateral artery

BRANCHES OF THE ULNAR ARTERY

9 | Anterior ulnar recurrent artery

10 | Posterior ulnar recurrent artery

11 | Common interosseous artery

a | Anterior interosseous artery

b | Posterior interosseous artery

c | Interosseous recurrent artery

12 | Dorsal carpal branch

BRANCHES OF THE RADIAL ARTERY

13 | Radial recurrent artery

14 | Dorsal carpal branch

d | Dorsalis pollicis

e | Dorsalis indicis

BRANCHES OF THE DEEP PALMAR ARCH

15 | Princeps pollicus (principal artery of the thumb)

16 | Radialis indicis

17 | Palmar metacarpal arteries (4)

BRANCHES OF THE SUPERFICIAL PALMAR ARCH

18 | Palmar digital arteries

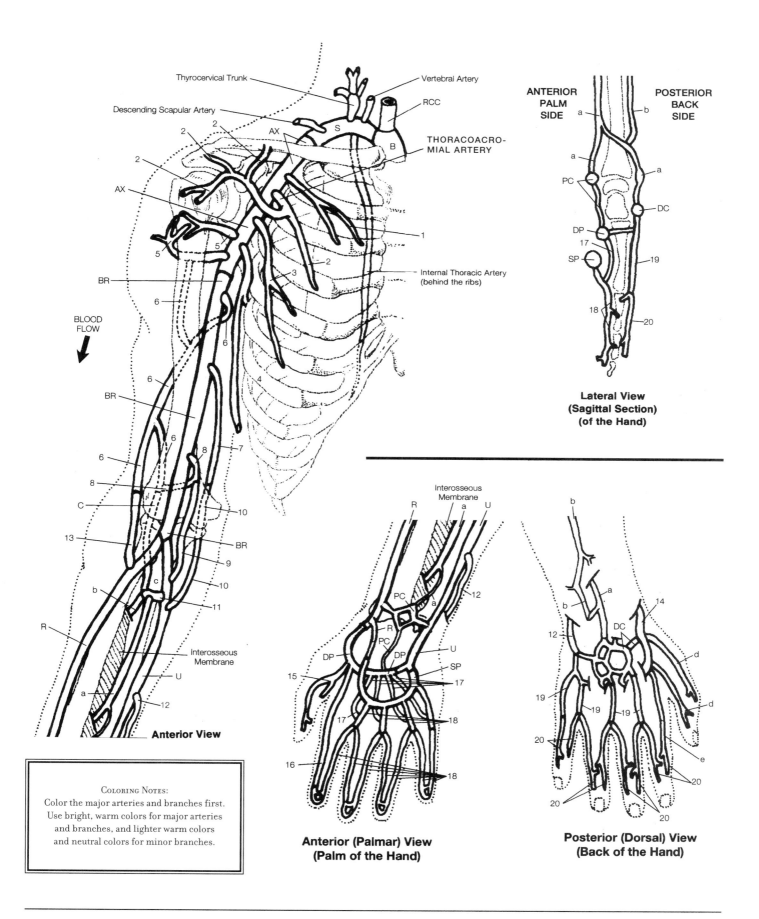

Thyrocervical Trunk

Descending Scapular Artery

Vertebral Artery

RCC

THORACOACRO-
MIAL ARTERY

AX

S

B

2

2

2

AX

5

5

BR

6

BLOOD
FLOW

6

6

BR

6

6

8

8

C

13

9

10

b

c

11

R

Interosseous
Membrane

a

U

12

Anterior View

1

Internal Thoracic Artery
(behind the ribs)

2

3

4

7

10

BR

ANTERIOR
PALM
SIDE

POSTERIOR
BACK
SIDE

a

b

a

a

PC

DC

DP

17

SP

19

18

20

Lateral View
(Sagittal Section)
(of the Hand)

COLORING NOTES:
Color the major arteries and branches first.
Use bright, warm colors for major arteries
and branches, and lighter warm colors
and neutral colors for minor branches.

Interosseous
Membrane

R

a

U

PC

a

R

PC

DP

DP

U

SP

15

17

17

18

16

18

12

Anterior (Palmar) View
(Palm of the Hand)

b

a

b

12

DC

14

19

19

19

20

20

20

20

20

d

d

e

Posterior (Dorsal) View
(Back of the Hand)

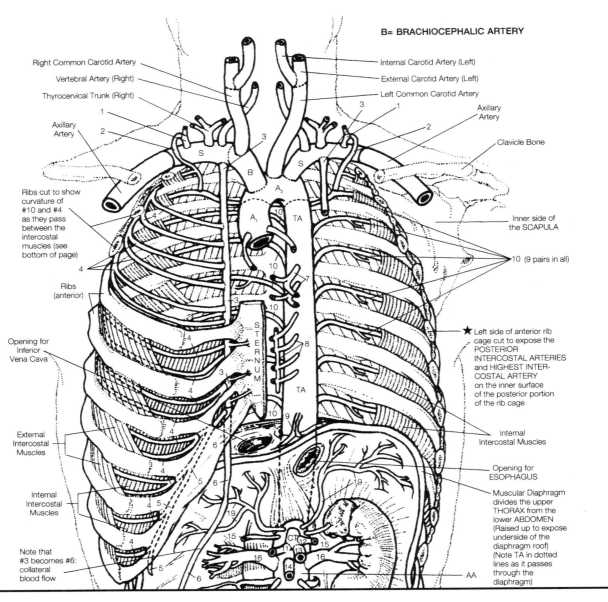

B= BRACHIOCEPHALIC ARTERY

Right Common Carotid Artery

Vertebral Artery (Right)

Thyrocervical Trunk (Right)

Axillary Artery

Ribs cut to show curvature of #10 and #4 as they pass between the intercostal muscles (see bottom of page)

Ribs (anterior)

Opening for Inferior Vena Cava

External Intercostal Muscles

Internal Intercostal Muscles

Note that #3 becomes #6: collateral blood flow

Internal Carotid Artery (Left)

External Carotid Artery (Left)

Left Common Carotid Artery

Axillary Artery

Clavicle Bone

Inner side of the SCAPULA

10 (9 pairs in all)

★ Left side of anterior rib cage cut to expose the POSTERIOR INTERCOSTAL ARTERIES and HIGHEST INTERCOSTAL ARTERY on the inner surface of the posterior portion of the rib cage

Internal Intercostal Muscles

Opening for ESOPHAGUS

Muscular Diaphragm divides the upper THORAX from the lower ABDOMEN (Raised up to expose underside of the diaphragm roof) (Note TA in dotted lines as it passes through the diaphragm)

AA

As the DESCENDING AORTA enters the abdomen, it passes through the muscular diaphragm and becomes the ABDOMINAL AORTA. The abdominal aorta descends along the posterior wall of the abdomen and ends at the level of the fourth lumbar vertebra.

The arteries that serve the thoracic wall are SEGMENTAL POSTERIOR and ANTERIOR INTERCOSTAL ARTERIES, which serve the external and intercostal muscles and structures of the thoracic wall. Note the blood flow from the THORACIC AORTA to the POSTERIOR INTERCOSTAL ARTERIES, around the rib cage to the anterior intercostals arteries, and then into the INTERNAL THORACIC ARTERIES, which become the SUPERIOR EPIGASTRIC ARTERIES.

The branches of the COMMON ILIAC ARTERIES are the EXTERNAL ILIAC ARTERIES, consisting of twenty-two INFERIOR EPIGASTRIC ARTERIES); and the INTERNAL ILIAC ARTERIES (HYPOGASTRIC ARTERIES), consisting of twenty-three LATERAL SACRAL arteries, twenty-four SUPERIOR GLUTEAL arteries and twenty-five ILIOLUMBAR arteries.

The vertebral column and spinal cord are supplied by branches of the POSTERIOR INTERCOSTAL ARTERIES, VERTEBRAL ARTERY, LUMBAR ARTERIES and LATERAL SACRAL ARTERIES.

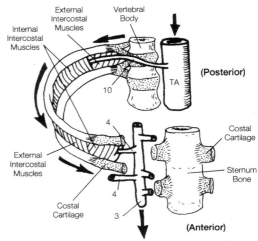

External Intercostal Muscles

Vertebral Body

Internal Intercostal Muscles

(Posterior)

TA

External Intercostal Muscles

Costal Cartilage

Costal Cartilage

Sternum Bone

(Anterior)

ARTERIES THAT SERVE THE THORACIC WALL

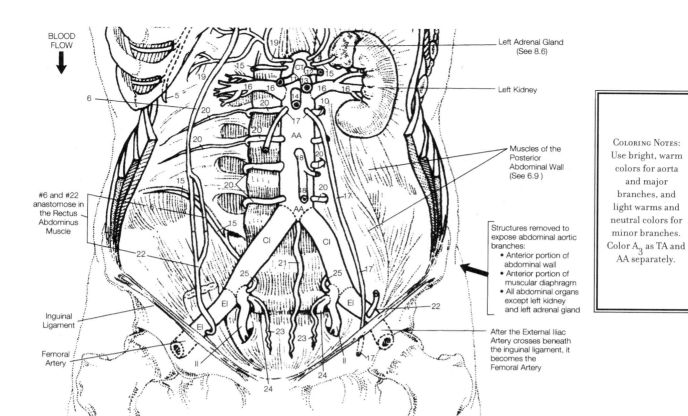

BLOOD FLOW

Left Adrenal Gland (See 8.6)

Left Kidney

Muscles of the Posterior Abdominal Wall (See 6.9)

#6 and #22 anastomose in the Rectus Abdominus Muscle

Inguinal Ligament

Femoral Artery

Structures removed to expose abdominal aortic branches:
• Anterior portion of abdominal wall
• Anterior portion of muscular diaphragm
• All abdominal organs except left kidney and left adrenal gland

After the External Iliac Artery crosses beneath the inguinal ligament, it becomes the Femoral Artery

COLORING NOTES:
Use bright, warm colors for aorta and major branches, and light warms and neutral colors for minor branches. Color A_3 as TA and AA separately.

THE AORTA IS GENERALLY DIVIDED INTO THREE MAIN SECTIONS:

A_1 Ascending aorta

A_2 Aortic arch

A_3 Descending aorta
   (TA and AA)

THE DESCENDING AORTA IS DIVIDED INTO TWO MAIN SECTIONS:

TA   Thoracic aorta

AA   Abdominal aorta

BRANCHES OF THE AORTIC ARCH SUPPLYING THE THORAX:

B   Brachiocephalic Artery

5   Subclavian arteries
   (left and right)

1   Costocervical trunks

2   Highest intercostal artery

3   Internal thoracic arteries

4   Anterior intercostal arteries

5   Musculophrenic artery

6   Superior epigastric artery

BRANCHES OF THE DESCENDING AORTA SUPPLYING THE THORAX: THORACIC AORTA

(The thoracic aorta passes downward from the 4th–12th thoracic vertebrae and ends at the diaphragm.)

VISCERAL BRANCHES (supply viscera)

7   Bronchial arteries
   (2 left, 1 right)

8   Esophageal arteries (4–5)

   (not shown in drawing: pericardial, mediastinal and muscular visceral branches)

PARIETAL BRANCHES
Supply body wall structures

9   Superior phrenic arteries

10   Posterior intercostal arteries
   (9 pairs)

   (not shown in drawing: subcostal arteries, similar to the posterior intercostals)

At the level of the 4th lumbar vertebra, the abdominal aorta bifurcates into 2 large common iliac arteries:

CI   Common iliac arteries
   (left and right)

BRANCHES OF THE DESCENDING AORTA SUPPLYING THE ABDOMEN: ABDOMINAL AORTA

VISCERAL BRANCHES

CT   Celiac trunk (coeliac trunk)

11   Common hepatic artery

12   Left gastric artery

13   Splenic artery

14   Superior mesenteric artery

15   Suprarenal (adrenal)
   arteries (left and right)

16   Renal arteries (left and right)

GONADAL ARTERIES (LEFT AND RIGHT)

17   Testicular (male) arteries;
   Ovarian (female) arteries

18   Inferior mesenteric artery

PARIETAL BRANCHES

19   Inferior phrenic arteries

20   Lumbar arteries

21   Middle sacral
   (midsacral) artery

There are three principal abdominal arteries supplying the digestive organs as they branch off from the ABDOMINAL AORTA (see pages 208–09): the CELIAC TRUNK, SUPERIOR MESENTERIC ARTERY, and INFERIOR MESENTERIC ARTERY. The celiac trunk is a thick, short artery that immediately divides into three arteries: the COMMON HEPATIC ARTERY, the SPLENIC ARTERY and the LEFT GASTRIC ARTERY. The common hepatic artery has three main branches: the left and right hepatic arteries (liver and gallbladder), the GASTRODUODENAL ARTERY (stomach, body of pancreas and duodenum), and the RIGHT GASTRIC ARTERY (stomach). The splenic artery has three main branches: the LEFT GASTROEPIPLOIC ARTERY (stomach), PANCREATIC ARTERY (tail of the pancreas) and POLARS, superior and inferior (spleen). The splenic artery also ends in numerous direct branches to the spleen and some offshoots to the stomach. The left gastric artery is in the lesser curvature of the stomach and the esophagus.

The SUPERIOR MESENTERIC ARTERY is an unpaired vessel arising anteriorly from the abdominal aorta, just below the celiac trunk, with numerous branches supplying abdominal organs: the pancreas; duodenum; small intestine; cecum and appendix; and the ascending and transverse colons of the large intestine. The INFERIOR MESENTERIC ARTERY is an unpaired vessel arising anteriorly from the abdominal aorta, just above the bifurcation of the aorta, supplying lower abdominal organs: the descending and sigmoid colons of the large intestine, and the rectum.

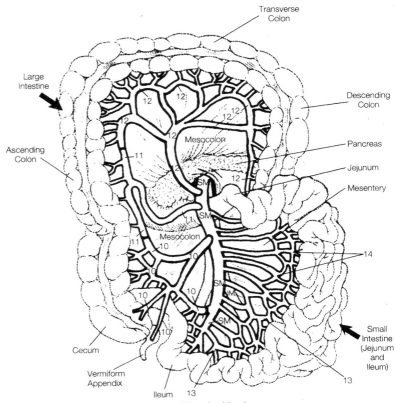

(Anterior View)
GASTROINTESTINAL TRACT SPREAD to SHOW ARTERIAL SUPPLY
(Greater Omentum Apron Removed)

THREE UNPAIRED VESSELS SUPPLYING THE ORGANS OF DIGESTION IN THE ABDOMEN:
CELIAC TRUNK, SUPERIOR MESENTERIC ARTERY AND INFERIOR MESENTERIC ARTERY

CT | CELIAC TRUNK

- **CH** | Common hepatic artery
  - **H** | Hepatic arteries (left and right)
    - **1** | Cystic artery from right hepatic artery
  - **G** | Gastroduodenal artery
    - **2** | Anterior superior pancreaticoduodenal artery
    - **3** | Right gastroepiploic artery
      - **3a** | Omental branches
    - **4** | Posterior superior pancreaticoduodenal artery (retroduodenal)
  - **R-G** | Right gastric artery
- **S** | Splenic artery
  - **5** | Left gastroepiploic artery
  - **6** | Pancreatic artery
  - **7** | Superior and inferior polar arteries
- **L-G** | Left gastric artery

SM | SUPERIOR MESENTERIC ARTERY

Branches to Pancreas and duodenum:
- **9** | Inferior pancreaticoduodenal artery

Branches to large intestine:
- **10** | Ileocolic artery
  - **a** | Anterior cecal branch
  - **b** | Posterior cecal branch
  - **c** | Ileal branch
  - **d** | Appendicular branch
- **11** | Right colic artery
- **12** | Middle colic artery

Branches to small intestine:
- **13** | Ileal intestinal artery
- **14** | Jejunal intestinal artery

IM | INFERIOR MESENTERIC ARTERY

Branches to rectum:
- **15** | Superior rectal artery

Branches to large intestine:
- **16** | Left colic artery
- **17** | Sigmoid artery

> COLORING NOTES:
> Use bright, warm colors for major arteries.
> Use lighter warm and neutral colors for minor branches.

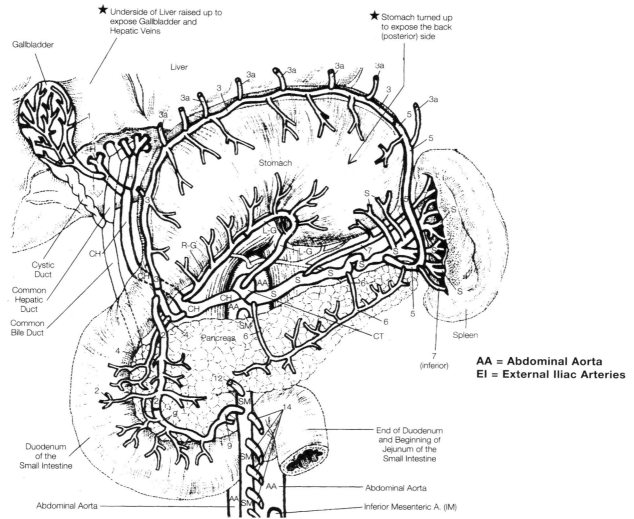

★ Underside of Liver raised up to expose Gallbladder and Hepatic Veins

★ Stomach turned up to expose the back (posterior) side

Gallbladder

Liver

Stomach

Cystic Duct

Common Hepatic Duct

Common Bile Duct

Pancreas

Spleen

Duodenum of the Small Intestine

AA = Abdominal Aorta
EI = External Iliac Arteries

End of Duodenum and Beginning of Jejunum of the Small Intestine

Abdominal Aorta

Abdominal Aorta

Inferior Mesenteric A. (IM)

ARTERIAL SUPPLY OF MAJOR ABDOMINAL ORGANS

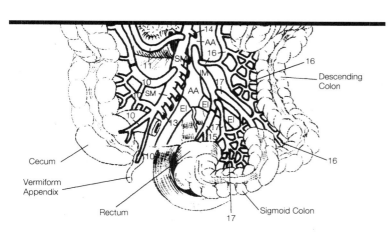

Descending Colon

Cecum

Vermiform Appendix

Rectum

Sigmoid Colon

SMALL INTESTINE REMOVED TO EXPOSE ARTERIAL SUPPLY TO THE SIGMOID COLON and RECTUM. (Note cut branches from the JEJUNAL (14) and ILEAL (13) INTESTINAL ARTERIES of the SUPERIOR MESENTERIC ARTERY.)

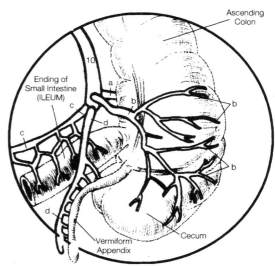

Ascending Colon

Ending of Small Intestine (ILEUM)

Cecum

Vermiform Appendix

Posterior View
VERMIFORM APPENDIX and the
CECUM of the LARGE INTESTINE

The ABDOMINAL AORTA bifurcates into the left and right COMMON ILIAC ARTERIES at the level of the 4th lumbar vertebra. The common iliac arteries divide two to three inches inferiorly into two main branches, the EXTERNAL ILIAC ARTERIES (left and right) and the INTERNAL ILIAC ARTERIES (left and right). The internal iliac arteries are the principal supply of the PELVIS and PERINEUM. Its branches are the ILIOLUMBAR and LATERAL SACRAL ARTERIES (pelvic wall and muscles [PSOAS MAJOR, QUADRATUS LUMBORUM]), the MIDDLE RECTAL ARTERY (the internal pelvic organs), the SUPERIOR, MIDDLE and INFERIOR VESICULAR ARTERIES (the urinary bladder), the SUPERIOR and INFERIOR GLUTEAL ARTERIES (the buttocks) and the OBTURATOR ARTERY (upper medial thigh muscles) and the INTERNAL PUDENDAL ARTERY (external genitalia in the PERINEAL area). The internal pudendal artery is an important vessel in sexual activity: penile erection

and female genital engorgement are vascular expressions (controlled by the AUTONOMIC NERVOUS SYSTEM).

The two external iliac arteries become the FEMORAL ARTERIES as they exit the pelvic cavity and cross the INGUINAL LIGAMENT. Two branches arise from the external iliac arteries just before this change: the INFERIOR EPIGASTRIC ARTERY (skin and abdominal wall muscles) and the DEEP ILIAC CIRCUMFLEX ARTERY (muscles of the iliac fossa). The femoral arteries send branches back into the pelvic region to supply the genitals and abdominal wall.

The femoral arteries and the inferior mesenteric artery assist the internal iliac arteries in supplying the pelvis and perineum.

In regard to the female pelvic region, the same nomenclature is identical to juxtaposed veins: all these vessels and their names can generally be applied to the venous system as well, where the COMMON ILIAC VEIN drains into the INFERIOR VENA CAVA.

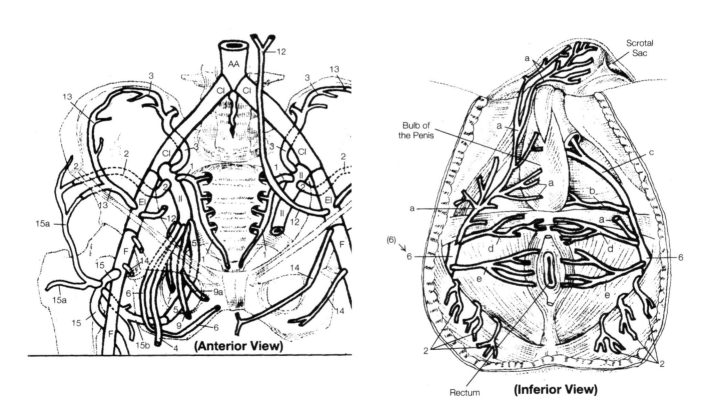

(Anterior View)

(Inferior View)

MALE PERINEUM

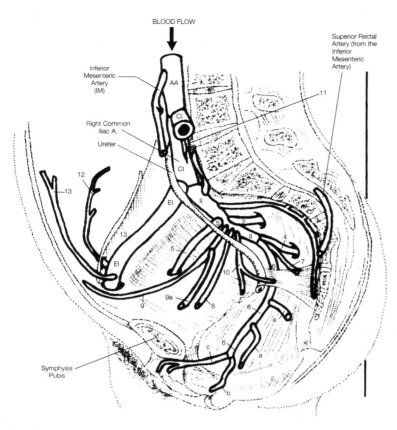

BLOOD FLOW

Inferior Mesenteric Artery (IM)

Superior Rectal Artery (from the Inferior Mesenteric Artery)

Right Common Iliac A.

Ureter

Symphysis Pubis

(Sagittal View)
Viewing branches of the Right Internal Iliac Artery
FEMALE PELVIC REGION

COLORING NOTES:
Repeat colors used for aorta, common iliac and femoral arteries from previous pages. Use light warm and neutral colors for minor branches.

PRIMARY BLOOD SUPPLY OF THE PELVIS AND PERINEUM: INTERNAL ILIACS

II   Internal iliac arteries (left and right) (hypogastric arteries)

POSTERIOR DIVISION OF INTERNAL ILIACS

1   Lateral sacral artery

2   Superior gluteal artery

3   Iliolumbar artery

ANTERIOR DIVISION OF INTERNAL ILIACS

PARIETAL BRANCHES

4   Inferior gluteal artery

5   Obturator artery

6   Internal pudendal artery

a   Posterior scrotal (labial) artery

b   Artery of the bulb of the penis (vestibule)

c   Arteries of penis (clitoris)

d   Perineal artery

e   Inferior rectal artery

VISCERAL BRANCHES

7   Middle rectal artery

8   Inferior vesicle artery

9   Umbilical artery

9a   Superior vesicle artery

10   Vaginal/uterine artery

SECONDARY BLOOD SUPPLY OF THE PELVIS AND PERINEUM

AA   Abdominal aorta

11   Midsacral artery

C₁   Left and right common iliac arteries

EI   External iliac arteries (left and right)

12   Inferior epigastric artery

13   Deep iliac circumflex artery

F   Femoral arteries (left and right)

14   External pudendal arteries

15   Deep femoral arteries (profunda femoris)

15a   Lateral femoral circumflex artery

15b   Medial femoral circumflex artery

Reviewing, the EXTERNAL ILIAC ARTERIES diverge through the pelvis and enter the thigh region, becoming the femoral arteries.

Regarding the FEMORAL TRIANGLE: superiorly, the FEMORAL ARTERIES are close to the surface. A triangular area around this point serves as an important pressure point in diagnosis and palpation. This area is formed by the inguinal ligament, sartorius muscle and adductor longus muscle.

The branches of the femoral artery include the DEEP FEMORAL ARTERY (PROFUNDIS FEMORIS), the largest branch that supplies the hamstring muscles (passing posteriorly). MEDIAL and LATERAL FEMORAL CIRCUMFLEX ARTERIES supply muscles in the proximal thigh via branches encircling the FEMUR BONE. The FEMORAL ARTERY passes down the medial and posterior side of the thigh at the back of the knee joint, where it becomes the POPLITEAL ARTERY. It serves small branches to the knee joint (geniculars), then divides below the knee into two branches. The ANTERIOR TIBIAL ARTERY (traverses over the interosseous membrane and serves the anterior aspect of the leg (and leg muscles); and at the ankle, it becomes the DORSALIS PEDIS, serving the ankle and dorsum of the foot. The POSTERIOR TIBIAL ARTERY continues down the back of the leg between the knee. It sends off a large PERONEAL ARTERY, serving the peroneal leg muscles. At the ankle, it bifurcates into the LATERAL and MEDIAL PLANTAR ARTERIES, which serve the sole (or bottom of the foot).

The LATERAL PLANTAR ARTERY anastomoses with the DORSAL PEDIS ARTERY (not shown) to form the PLANTAR ARCH.

AA	Abdominal Aorta
CI	Right Common Iliac artery
II	Internal Iliac artery
1	Superior Gluteal artery
2	Inferior Gluteal artery
3	Obturator artery
EI	External Iliac artery

As it enters the thigh, the external iliac becomes the femoral artery:

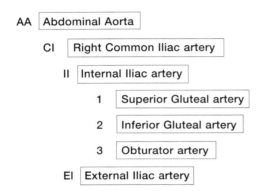

F	Femoral Artery
4	Deep Femoral artery (Profunda Femoris)
a	Perforating Branches
b	Medial Femoral Circumflex artery
c	Lateral Femoral Circumflex artery
c_1	Descending Branch
5	Descending Genicular artery

As it crosses the posterior aspect of the knee, the femoral artery becomes the popliteal artery:

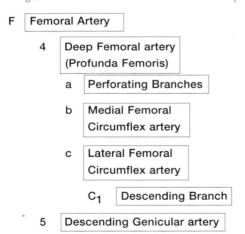

P	Popliteal Artery
6	Arteries (Superior, Middle, and Inferior)

Below the knee, the popliteal artery divides into the anterior tibial and posterior tibial:

AT	Anterior Tibial Artery

At the ankle, the anterior tibial becomes the dorsalis pedis artery:

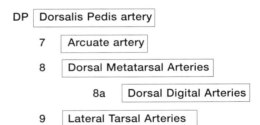

DP	Dorsalis Pedis artery
7	Arcuate artery
8	Dorsal Metatarsal Arteries
8a	Dorsal Digital Arteries
9	Lateral Tarsal Arteries

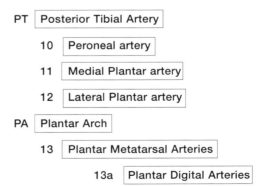

PT	Posterior Tibial Artery
10	Peroneal artery
11	Medial Plantar artery
12	Lateral Plantar artery
PA	Plantar Arch
13	Plantar Metatarsal Arteries
13a	Plantar Digital Arteries

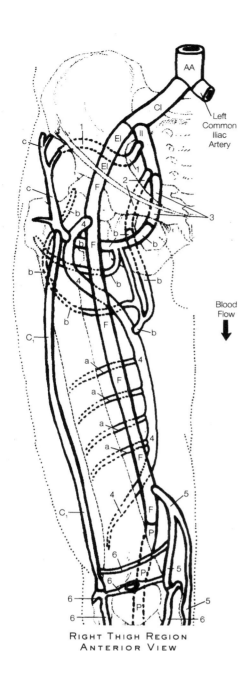

RIGHT THIGH REGION
ANTERIOR VIEW

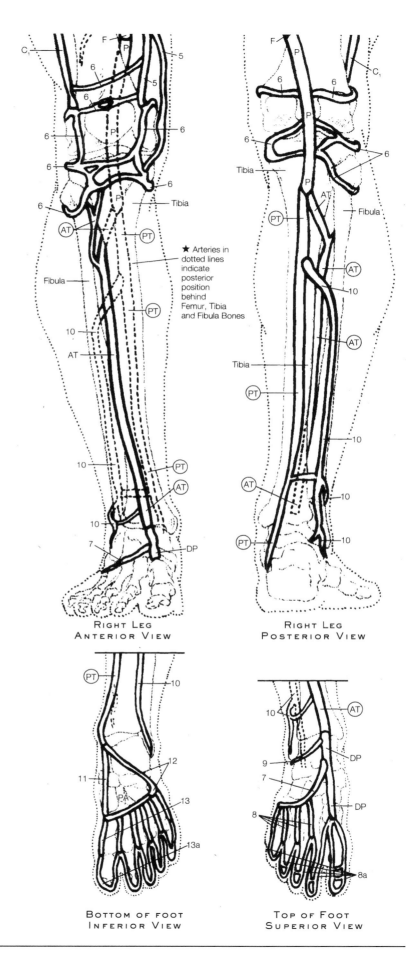

Left
Common
Iliac
Artery

Blood
Flow

★ Arteries in
dotted lines
indicate
posterior
position
behind
Femur, Tibia
and Fibula Bones

Tibia

Fibula

RIGHT LEG
ANTERIOR VIEW

Tibia

Fibula

RIGHT LEG
POSTERIOR VIEW

BOTTOM OF FOOT
INFERIOR VIEW

TOP OF FOOT
SUPERIOR VIEW

CORLORING NOTES:
Check previous section for same
vessels with different reference
numbers and use same colors
for consistency.
Use light warm and neutral colors
for minor branches.

PRINCIPAL VEINS OF THE BODY
VEINS OF THE HEAD AND NECK

COLORING NOTES:
For this and the next page, use bright, cool colors for major tributaries. Use light, cool colors and neutral colors for minor tributaries.

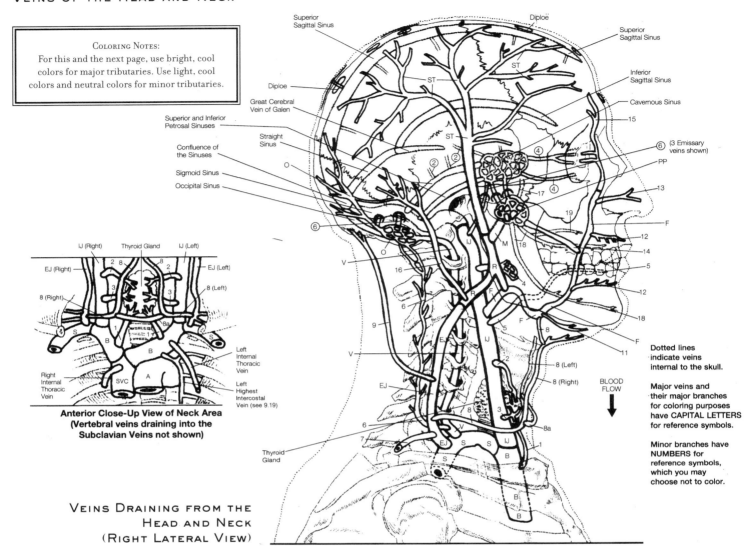

Anterior Close-Up View of Neck Area
(Vertebral veins draining into the Subclavian Veins not shown)

VEINS DRAINING FROM THE
HEAD AND NECK
(RIGHT LATERAL VIEW)

Dotted lines indicate veins internal to the skull.

Major veins and their major branches for coloring purposes have CAPITAL LETTERS for reference symbols.

Minor branches have NUMBERS for reference symbols, which you may choose not to color.

MAJOR VEINS OF THE NECK REGION (AND THEIR TRIBUTARIES)

B Brachiocephalic (innominate) vein
- 1 Inferior thyroid vein
- IJ Internal jugular vein
  - 2 Superior thyroid vein
  - 3 Middle thyroid vein
  - 4 Pharyngeal plexus
  - 5 Lingual vein
S Sunclavian vein
- V Vertebral vein
  - 6 Deep cervical vein

EJ External jugular vein
- 7 Tranverse cervical vein
- 8 Anterior jugular vein
  - 8a Jugular venou arch
- 9 Posterior external jugular vein

MAJOR VEINS OF THE FACE

F Facial vein
- 11 Submental vein
- 12 Superior and inferior veins Labial veins
- 13 External nasal vein
- 14 Deep facial vein

15 Angular vein
ST Superficial temporal vein
R Retromandibular vein
- 16 Posterior angular vein
O Occipital vein

DEEP VEINS

M Maxillary vein
PP Pterygoid plexus
- 17 Middle meningeal vein
- 18 Superior and inferior alveolar vein
- 19 Infraorbital vein

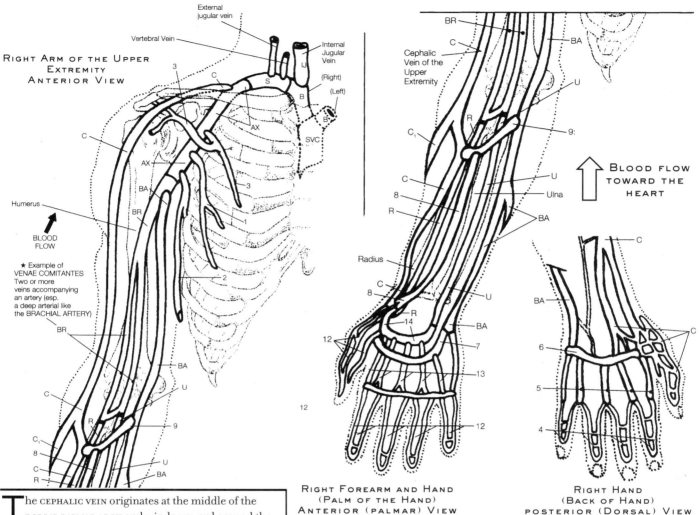

RIGHT ARM OF THE UPPER EXTREMITY ANTERIOR VIEW

Humerus

BLOOD FLOW

★ Example of VENAE COMITANTES Two or more veins accompanying an artery (esp. a deep arterial like the BRACHIAL ARTERY)

External jugular vein

Vertebral Vein

Internal Jugular Vein

(Right)
(Left)

SVC

BLOOD FLOW TOWARD THE HEART

Cephalic Vein of the Upper Extremity

Ulna

Radius

RIGHT FOREARM AND HAND (PALM OF THE HAND) ANTERIOR (PALMAR) VIEW

RIGHT HAND (BACK OF HAND) POSTERIOR (DORSAL) VIEW

The CEPHALIC VEIN originates at the middle of the DORSAL PALMAR ARCH and winds upward around the radial borde of the forearm. It joins the BASILIC VEIN via the MEDIA CUBITAL VEIN, in the cubital fossa in front of the elbow, then unites with the ACCESSORY CEPHALIC VEIN to form the cephalic vein of the upper extremity, and empties into the AXILLARY VEIN just above the clavicle, where it pierces the fascia. The BASILIC VEIN's origin is at the ulnar end of the dorsal arch, and it passes upward along the medial side of the arm (the ulnar side of forearm). It merges with the BRACHIAL VEIN just below the head of the HUMERUS to form the AXILLARY VEIN.

DEEP VEINS of the upper extremity are: RADIAL VEINS (lateral side of the forearm) and ULNAR VEINS (medial side of the forearm), which drain the DEEP and SUPERFICIAL PALMAR (VOLAR) ARCHES of the hand. They join in the cubital fossa to form the BRACHIAL VEIN. The brachial vein continues up the medial side of the BRACHIUM, merging with the basilic vein near the head of the humerus to form the AXILLARY VEIN, becoming the SUBCLAVIAN VEIN.

SUBCLAVIAN VEINS (left and right) join INTERNAL JUGULAR VEINS in the neck to form the BRACHIOCEPHALIC VEINS.

SVC	Superior vena cava		**8**	Median Vein of the Forearm (Median Anterbrachial)
B	Brachiocephalic Vein		**9**	Median Cubital Vein
S	Subclavian Vein		**C**	Cephalic Vein
AX	Axillary Vein		**C₁**	Accessory Cephalic Vein
1	Lateral Thoracic Vein		**BA**	Basilic Vein
2	Subscapular Vein			
3	Thoracoacromial Vein			

SUPERFICIAL VEINS OF THE UPPER EXTREMITY
DORSAL NETWORK

4 Dorsal Digital Veins

5 Dorsal Metacarpal Veins

6 Dorsal Palmar Arch

7 Superficial Palmar Arch (Volar Arch)

DEEP VEINS OF THE UPPER EXTREMITY
DEEP PALMAR NETWORK

12 Palmar Digital Veins

13 Palmar Metacarpal Veins

14 Deep Palmar Arch (Volar Arch)

U Ulnar Vein

R Radial Vein

BR Brachial Vein

The THORACIC REGION is where the blood flow in the veins is from smaller vessels (tributaries) that drain organs and structures in the thoracic region, ultimately emptying into the large BRACHIOCEPHALIC VEINS (left and right). The left and right brachiocephalic veins merge slightly to the right at the top of the STERNUM (MANUBRIUM) to form the SUPERIOR VENA CAVA. The superior vena cava and the brachiocephalic veins do not have valves though they are large vessels.

The superior vena cava, in addition to collecting blood from the brachiocephalic veins and their tributaries, also collects blood from vessels along the POSTERIOR THORACIC WALL: the AZYGOS VENOUS SYSTEM. The azygos vein is a single vein arising in the abdomen as a branch of the RIGHT ASCENDING LUMBAR VEIN. It passes upward through the AORTIC HIATUS of the DIAPHRAGM (through which the aorta passes into the abdomen) into the THORAX, then along the right side of the vertebral column to the level of the fourth thoracic vertebra, where it turns and enters the superior vena cava in the MEDIASTINUM. If the inferior vena cava becomes obstructed, the azygos vein is the principal vein by which blood can return to the heart from the abdomen and lower extremities.

In the thorax, the azygos vein receives the HEMIAZYGOS, ACCESSORY AZYGOS and BRONCHIAL VEINS (plus the right posterior and highest INTERCOSTAL VEINS, and the RIGHT SUBCOSTAL VEINS).

The abdominal region consists of the left and right COMMON ILIAC VEINS which merge at the level of the fifth lumbar vertebra and form the INTERIOR VENA CAVA. The INFERIOR VENA CAVA is the vessel with the largest diameter in the body. It ascends through the abdominal cavity, parallel and to the right of the ABDOMINAL AORTA. It receives an abundance of tributaries (similar in name and position to corresponding arteries). The VERTEBRAL COLUMN and SPINAL CORD are drained by tributaries of the POSTERIOR INTERCOSTAL VEINS, VERTEBRAL VEIN, LUMBAR VEINS and LATERAL SACRAL VEINS.

The inferior vena cava does not receive blood directly from the GASTROINTESTINAL TRACT, SPLEEN or the PANCREAS. The venous blood from these organs passes through a filtering system in the liver (the HEPATIC PORTAL SYSTEM) before entering the IVC.

The veins draining the PELVIS and PERINEUM are parallel to and carry the same names as arteries supplying those regions. The main difference lies in the fact that the COMMON ILIAC ARTERIES are branches of the ABDOMINAL AORTA, but the common iliac veins are tributaries that join to form the inferior vena cava.

Venous Tributaries of the Superior and Inferior Vena Cavae

SVC | Superior Vena Cava

BRACHIOCEPHALIC TRIBUTARIES OF THE SUPERIOR VENA CAVA:

B | Brachiocephalic Vein

IT | Internal Thoracic

1 | Anterior Intercostal Vein

2 | Musculophrenic Vein

3 | Superior Epigastric Vein

4 | Superior Phrenic Vein

S | Subclavian Vein

AX | Axillary Vein

5 | Thoracoepigastric Vein

* Collateral Venous Flow via 3 and 5
Not shown: Pericardial, Mediastinal and Muscular Tributaries

THE AZYGOS VENOUS SYSTEM PRINCIPAL AZYGOS VEINS

AZ | Azygos Vein (Azygous Vein)

AL | Ascending Lumbar Veins

HAZ | Hemiazygos Vein (Drains left side of Vertebral Column)

AAZ | Accessory Azygos Vein

TRIBUTARIES OF THE PRINCIPAL AZYGOS VEINS

L | Lumbar Veins (Drain Posterior Abdominal, Vertebral Column and Spinal Cord)

6 | Highest Intercostal Veins

7 | Posterior Intercostal Veins

8 | Bronchial Veins

9 | Esophageal Veins

(left and right subcostals not shown, similar to [7])

IVC | Inferior Vena Cava

ABDOMINAL TRIBUTARIES OF THE INFERIOR VENA CAVA

10 | Hepatic Vein (Left and Right)

(originate in Capillary Sinusoids of the Liver)

11 | Renal Vein (Left and Right)

12 | Suprarenal (Adrenal) Veins

13 | Inferior Phrenic Veins

14 | Testicular-Ovarian (gonadal) Veins

TWO LARGE ABDOMINAL TRIBUTARIES JOIN TO FORM THE INFERIOR VENA CAVA

CI | Common Iliac Veins (Left and Right)

ABDOMINAL TRIBUTARIES OF THE COMMON ILIAC VEINS:

6 | Midsacral Vein

EI | External Iliac Veins

7 | Inferior Epigastric Veins

II | Internal Iliac Veins (Hypogastric Veins) for Branches (Described in Arterial form)

F | Femoral Veins

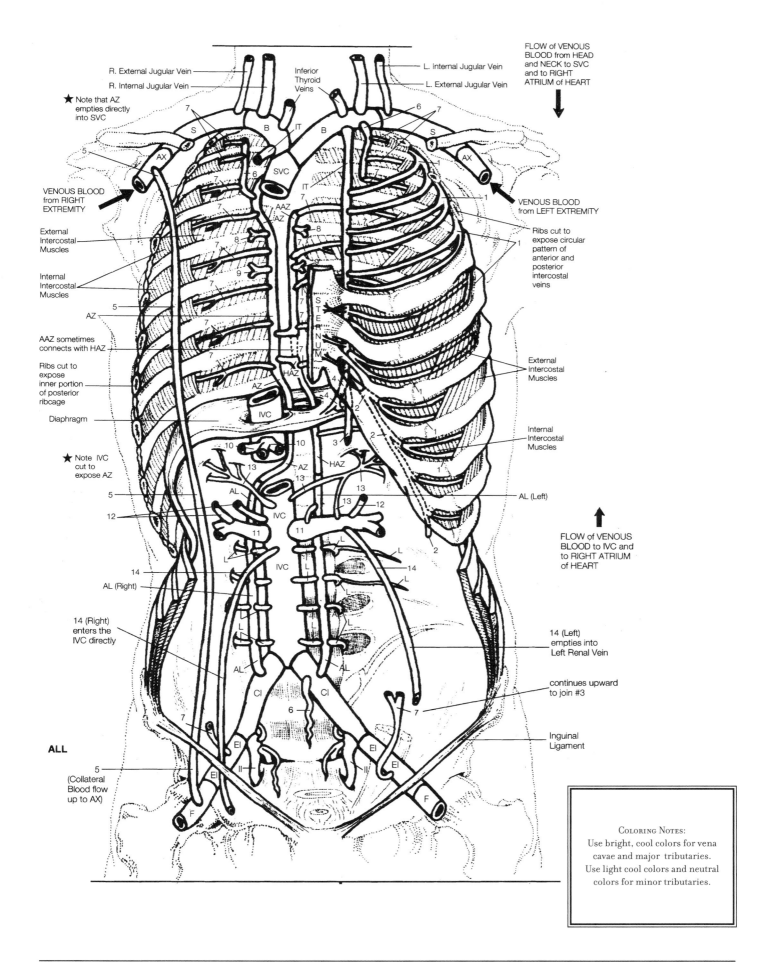

R. External Jugular Vein

R. Internal Jugular Vein

Inferior Thyroid Veins

L. Internal Jugular Vein

L. External Jugular Vein

FLOW of VENOUS BLOOD from HEAD and NECK to SVC and to RIGHT ATRIUM of HEART

★ Note that AZ empties directly into SVC

VENOUS BLOOD from RIGHT EXTREMITY

External Intercostal Muscles

Internal Intercostal Muscles

AAZ sometimes connects with HAZ

Ribs cut to expose inner portion of posterior ribcage

Diaphragm

★ Note IVC cut to expose AZ

VENOUS BLOOD from LEFT EXTREMITY

Ribs cut to expose circular pattern of anterior and posterior intercostal veins

External Intercostal Muscles

Internal Intercostal Muscles

AL (Left)

FLOW of VENOUS BLOOD to IVC and to RIGHT ATRIUM of HEART

14 (Right) enters the IVC directly

AL (Right)

14 (Left) empties into Left Renal Vein

continues upward to join #3

Inguinal Ligament

ALL

5 (Collateral Blood flow up to AX)

COLORING NOTES:
Use bright, cool colors for vena cavae and major tributaries.
Use light cool colors and neutral colors for minor tributaries.

A portal system is specialized venous flow consisting of two capillary beds joined by a vein. Under normal conditions, veins draining a capillary network drain blood directly into systemic veins. In a VENOUS PORTAL SYSTEM, veins draining one capillary network deliver blood to another capillary network. Coming from the second capillary network, blood then continues to the usual systemic veins and enters the venae cavae (and ultimately the right atrium of the heart).

In the HEPATIC PORTAL SYSTEM, the hepatic portal vein collects blood from the capillaries of the abdominal viscera (containing the absorbed nutrients of digestion) and conveys it to the sinusoids in the liver (discontinuous spaces lined with specialized capillary cells called KUPFFER CELLS) for filtration through the hepatocyte liver cells. From the liver, the blood then passes through the hepatic veins to the inferior vena cava, then to the heart.

The HEPATIC PORTAL VEIN receives blood from the digestive organs, and is formed by the union of two main vessels, the SUPERIOR MESENTERIC VEIN and the SPLENIC VEIN. The superior mesenteric vein drains blood from the small intestine. The splenic (lienal) vein drains blood from the spleen and from four tributaries: the INFERIOR MESENTERIC VEIN (from the large intestine), PANCREATIC VEIN (from the pancreas); and the LEFT and RIGHT GASTROEPIPLOIC VEINS (from the stomach). The hepatic portal vein also drains blood from the right and left GASTRIC VEINS (from stomach) and the cystic vein (from the gallbladder through its right branch).

The liver sinusoids receive blood from two sources. It is a mixture of oxygenated and deoxygenated "blue" blood—oxygen-rich "red" blood from the hepatic artery—and nutrient-rich "blue" blood from the hepatic portal vein. The sinusoids have a unique ability to modify the chemical nature of the venous blood from the digestive tract.

DRAINAGE OF THE ORGANS OF DIGESTION: THE HEPATIC PORTAL SYSTEM

PORTAL SYSTEM OF VEINS

HV	Hepatic Veins Right and Left
S	Sinusoids
HP	Hepatic Portal Vein
HP$_1$	Right Branch of the Hepatic Portal Vein
HP$_2$	Left Branch of the Hepatic Portal Vein

COLLATERAL ANASTOMOTIC BACK-UP: SYSTEMIC CAVAL SYSTEM CONNECTIONS

IVC	Inferior Vena Cava
E	Esophageal Vein
CI	Common Iliac Vein
1	Ascending Lumbar Vein
2	Lumbar vein
EI	External Iliac Vein
II	Internal Iliac Vein
3	Middle Rectal Vein
4	Inferior Rectal Vein
P-U	Para-Umbilicals

TRIBUTARIES OF THE HEPATIC PORTAL VEIN

HP$_1$	Right Branch of the Hepatic Portal Vein
5	Cystic Vein
HP$_2$	Left Branch of the Hepatic Portal Vein
LT	Ligamentum Teres (Round Ligament)
LV	Ligamentum Venosum
SM	Superior Mesenteric Vein
6	Middle Colic Vein
7	Right Colic Vein
8	Illeocolic Vein
9	Small Intestinal Vein
	Jejunal and Ileal
10	Right Gastroepiploic Vein
11	Pancreaticoduodenal Vein

SP	Splenic (Lienal) Vein
12	Short Gastric Vein
13	Left Gastroepiploic Vein
14	Pancreatic Vein
IM	Inferior Mesenteric Vein
15	Superior Rectal Vein
16	Sigmoid Vein
17	Left Colic Vein
G$_2$	Left Gastric (Coronary) Vein
G$_1$	Right Gastric (Pyloric) Vein

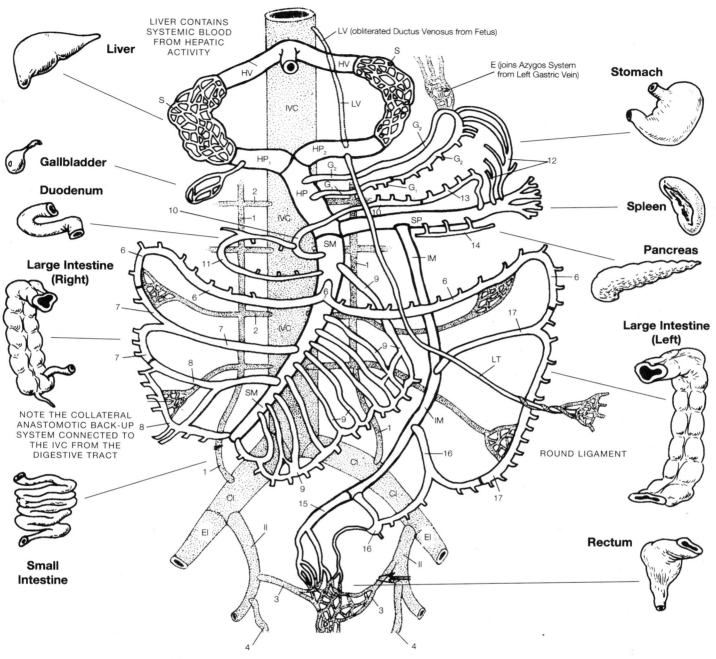

Liver

LIVER CONTAINS SYSTEMIC BLOOD FROM HEPATIC ACTIVITY

LV (obliterated Ductus Venosus from Fetus)

E (joins Azygos System from Left Gastric Vein)

Stomach

Gallbladder

Duodenum

Large Intestine (Right)

Spleen

Pancreas

Large Intestine (Left)

NOTE THE COLLATERAL ANASTOMOTIC BACK-UP SYSTEM CONNECTED TO THE IVC FROM THE DIGESTIVE TRACT

ROUND LIGAMENT

Small Intestine

Rectum

VENOUS TRIBUTARIES OF THE INFERIOR VENA CAVA
THE HEPATIC PORTAL SYSTEM

COLORING NOTES:
Repeat color choices for collateral anastomotic back-up vessels from previous pages, but now use lighter tones for them—or refrain from coloring them altogether!
Use strong, bright, cool colors for the hepatic portal system and tributaries.

The veins of the lower extremities are generally divided into two groups, the SUPERFICIAL GROUP and the DEEP GROUP. DEEP VEINS have more valves than the SUPERFICIAL VEINS. In deep veins, the ANTERIOR and POSTERIOR TIBIAL VEINS drain blood from the deep veins of the foot and join below the knee to form the POPLITEAL VEIN. The popliteal vein becomes the FEMORAL VEIN just above the knee. The femoral vein drains blood from the deep femoral vein and LATERAL-MEDIAL CIRCUMFLEX VEINS in the upper thigh. As the femoral vein approaches the inguinal ligament, it drains blood from the GREAT SAPHENOUS VEIN, and then becomes the EXTERNAL ILIAC VEIN after entering the abdominal cavity. The external iliac vein merges with the INTERNAL ILIAC VEIN at the level of the sacroiliac joint in the pelvic region to form the COMMON ILIAC VEIN. The left and right common iliac veins merge at the level of the fifth vertebra to form the large INFERIOR VENA CAVA.

In superficial veins, the DORSAL DIGITAL and METATARSAL VEINS enter the DORSAL VENOUS ARCH, which gives rise to two large superficial veins, the GREAT and SMALL SAPHENOUS VEINS. The great saphenous vein is the longest vessel in the body. It ascends along the MEDIAL aspect of the leg and enters the femoral vein. The small saphenous vein ascends along the lateral aspect of the foot, posteriorly up the calf of the leg, and passes deep to enter the popliteal vein in the POSTERIOR FOSSA behind the knee. The GENICULARS (the superior, medial, and inferior genicular veins) drain the knee region and enter the popliteal vein in the posterior knee region.

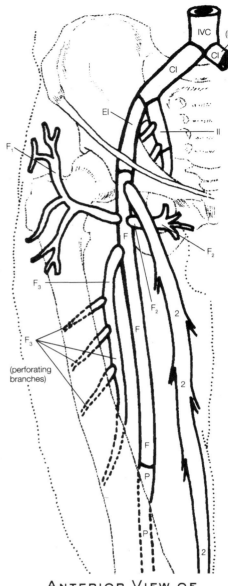

ANTERIOR VIEW OF RIGHT THIGH

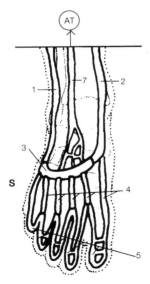

DORSAL-TOP SURFACE OF THE FOOT
SUPERIOR VIEW

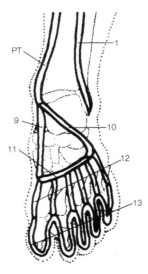

VENTRAL-BOTTOM SURFACE OF THE FOOT
INFERIOR VIEW

COLORING NOTES:
Use bright, cool colors for
major tributaries.
(Repeat colors for IVC, Cl, IT, EI.)
Use light, cool colors and neutral colors
for minor tributaries.

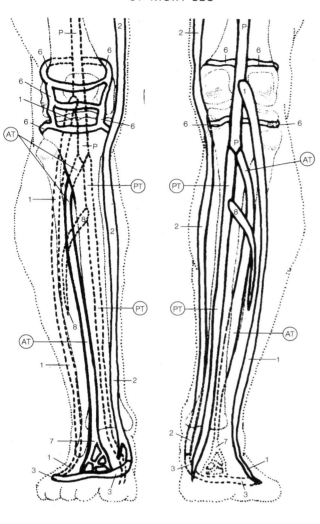

ANTERIOR AND POSTERIOR VIEW OF RIGHT LEG

POSTERIOR AND ANTERIOR VIEW OF SUPERFICIAL VEINS

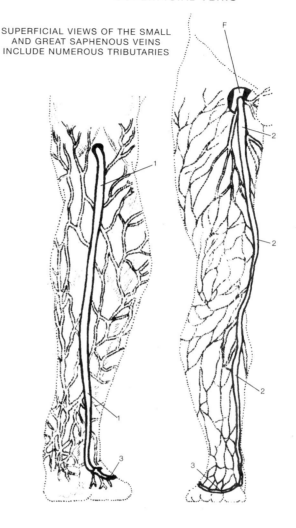

SUPERFICIAL VIEWS OF THE SMALL AND GREAT SAPHENOUS VEINS INCLUDE NUMEROUS TRIBUTARIES

SUPERFICIAL VEINS OF THE LOWER EXTREMITY

1	Small saphenous vein
	(Enters popliteal vein)
2	Great saphenous vein
	(Enters femoral vein)
3	Dorsal venous arch
4	Dorsal metatarsal veins
5	Dorsal digital veins
6	Genicular veins

IVC	Inferior vena cava
CI	Common Iliac veins (left and right)
II	Internal iliac vein
EI	External iliac vein

As it approaches the thigh, the femoral vein becomes the external iliac vein.

F	Femoral vein
F₁	Lateral femoral circumflex vein
F₂	Medial femoral circumflex vein
F₃	Deep femoral vein

Just above the knee, the POPLITEAL VEIN becomes the FEMORAL VEIN. Just below the knee, the POPLITEAL VEIN is formed by the junction of the anterior and posterior tibial veins.

P	Popliteal vein
AT	Anterior tibial vein
7	Dorsalis pedis vein
PT	Posterior tibial vein
8	Peroneal vein
9	Medial plantar vein
10	Lateral plantar vein
11	Plantar venous arch
12	Plantar metatarsal veins
13	Plantar digital veins

CHAPTER 11: LYMPH VASCULAR SYSTEM

CONTENTS

GENERAL ORGANIZATION
General Scheme of Lymph Fluid Drainage and Circulation

BODY AREAS AND PRINCIPAL CHANNELS OF LYMPH DRAINAGE

FOUR MAJOR LYMPHOID ORGANS (GLANDS)
Lymph Nodes, Spleen, Tonsils and Thymus Gland

SYSTEM COMPONENTS

*Lymph, lymph nodes, vessels, ducts and glands
(i.e.: spleen, tonsils and thymus gland).*

SYSTEM FUNCTION

Return of protein and fluid to the blood vascular system.

Transportation of fats from the digestive system to the blood vascular system

*Filtration of blood
Production of white blood cells
Protection against disease*

The LYMPH VASCULAR SYSTEM is made up of a series of small masses of lymphoid tissue called LYMPH NODES, connected by LYMPHATIC VESSELS through which flows LYMPH FLUID, and a variety of organs (glands) of which there are three major ones: SPLEEN, TONSILS, and THYMUS. The lymphatic vessels originate as blind-end tubes called LYMPH CAPILLARIES that begin in spaces between cells. There are only four areas where lymph capillaries do not originate throughout the entire body: the AVASCULAR TISSUE (such as cartilage), CENTRAL NERVOUS SYSTEM, SPLENIC PULP and the BONE MARROW. Lymph capillaries are slightly larger and more permeable than blood capillaries. In the following chapter, the Lymphoid Immune System, you will see that it is the BONE MARROW and the THYMUS GLAND that are the two primary lymphoid organs which produce the precursor disease-fighting lymphocyte cells that eventually migrate throughout the entire body to defend it against microorganisms and damaged/abnormal/infected cells the body does not recognize as "self."

The lymph vascular system prevents the accumulation of tissue fluid around the body tissue cells (a condition called EDEMA). It also aids in the necessary return eventually through the left and right subclavian veins to the two large brachiocephalic veins in the thorax) of escaped extravascular fluids to the CARDIOVASCULAR SYSTEM, increasing the amount of fluid returned to the heart for requisite constant fluid circulation. Micro-scopic, unconnected lymph capillaries (ending blindly in swollen or rounded ends) begin draining protein-containing fluid from the intercellular tissue (interstitial spaces). Some of this interstitial fluid has previously escaped through the highly permeable blood capillaries of the cardiovascular system.

The lymph vascular system collects interstitial fluid as LYMPH (with much the same makeup of blood and plasma), and ultimately returns this fluid to the venous system of the cardiovascular system through a series of progressively larger lymphatic vessels (purifying and filtering the lymph of foreign particulate matter, especially bacteria, at periodic filtering stations called lymph nodes). It arises from VEINS in the developing embryo, is associated with veins throughout the body and drains lymph into the two large BRACHIOCEPHALIC VEINS at the rate of two liters per day. It does not have a separate heart to circulate lymph, and relies on the alternating contraction and relaxation of nearby skeletal muscles to accomplish lymph flow.

The lymph vascular system also transports into the cardiovascular system large aggregations of FAT MONOMERS (FATTY ACIDS and GLYCEROL) and other aggregations that are unable to be directly absorbed through the digestive process.

GENERAL FLOW OF LYMPH

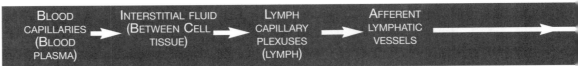

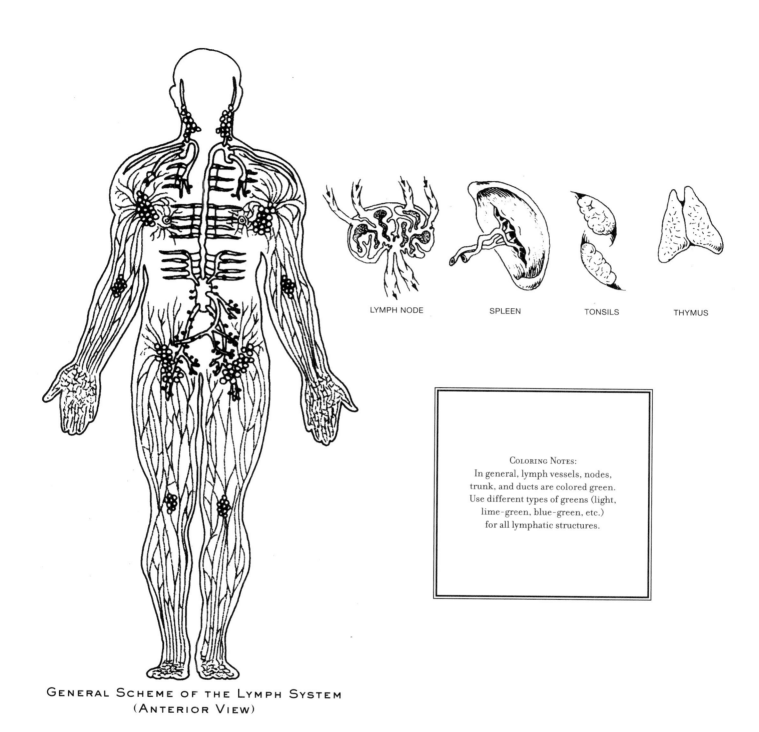

LYMPH NODE   SPLEEN   TONSILS   THYMUS

Coloring Notes:
In general, lymph vessels, nodes,
trunk, and ducts are colored green.
Use different types of greens (light,
lime-green, blue-green, etc.)
for all lymphatic structures.

GENERAL SCHEME OF THE LYMPH SYSTEM
(ANTERIOR VIEW)

→ LYMPH NODES → EFFERENT LYMPHATIC VESSELS → LYMPH TRUNKS → THORACIC DUCT AND RIGHT LYMPHATIC DUCTS → VENOUS SYSTEM (SUBCLAVIAN VEINS)

The LYMPH VASCULAR SYSTEM is not a closed-loop system (as is the BLOOD VASCULAR SYSTEM). It originates in a vast network (plexus) of LYMPH CAPILLARIES, which are BLIND or CLOSED-ENDED microscopic tubes that begin in the INTERCELLULAR SPACES between tissue cells. It ends in the subclavian veins of the venous system of the blood vascular system.

Lymph is usually a clear, transparent colorless alkaline fluid, however, in vessels draining the intestines it may appear milky (and is called CHYLE) due to the presence of absorbed fats. Within the villi of the small intestine, lymph capillaries, called LACTEALS, which transport the products of fat absorption away from the digestive tract.

The lymph capillaries are composed of highly permeable simple squamous epithelium. LYMPH FLUID, which can easily enter them, is formed as a filtrate of plasma through blood capillaries. Think of lymph as blood minus the red corpuscles with a lower protein content. INTERSTITIAL FLUID and lymph are basically the same as blood plasma, but they have a lower concentration of protein. Large proteins are held back by filtration.

The "pumping" of lymph is due to skeletal muscle contractions, respiratory movements and lymph valves.

The general flow of lymph (the general direction of flow in the body is upward, toward the neck and thoracic region) starts in the blood capillaries (as blood plasma) and escapes into the interstitial fluid (between tissue cells). From there, it is absorbed into the blind-pouched lymph capillary plexuses (as lymph), then carried to the afferent lymphatic vessels, where it is filtered through the lymph node stations, and then released into the efferent lymphatic vessels. It then travels to the larger lymph trunks, then to the thoracic duct and right lymphatic ducts, eventually entering the venous system through the subclavian veins en route to the brachiocephalic veins, then the superior vena cavae towards the heart "pump."

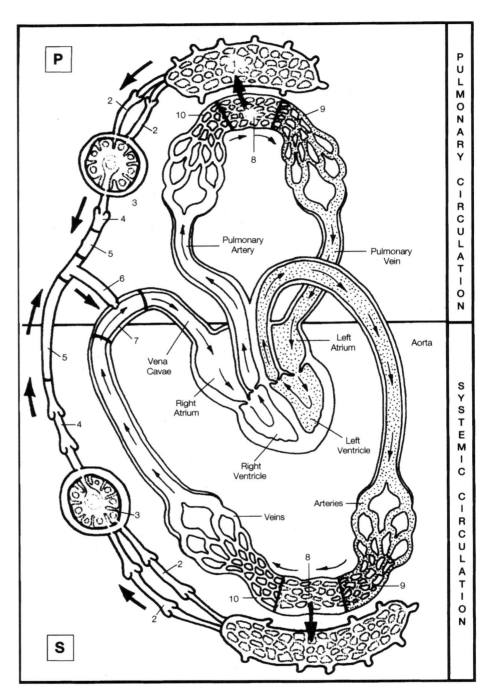

GENERAL SCHEME of DRAINAGE and CIRCULATION of LYMPH (Small arrows = Blood flow; Large arrows = Lymph flow)

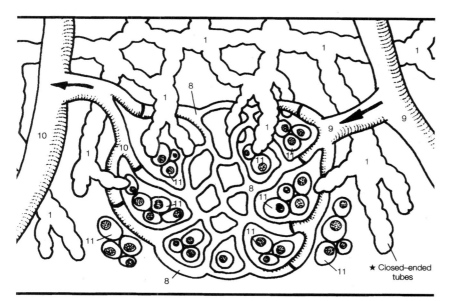

ORIGINATION of LYMPH VASCULAR SYSTEM
(Arrows indicate Blood Flow)

★ Closed–ended tubes

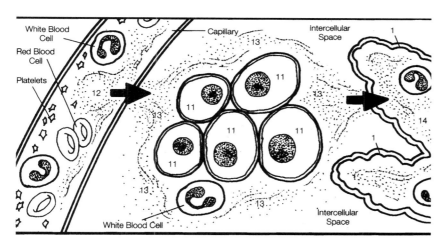

White Blood Cell

Red Blood Cell

Platelets

Capillary

Intercellular Space

White Blood Cell

Intercellular Space

FILTRATION of BLOOD FLUID
(Arrows indicate Blood Plasma Flow)

CoLor GuidELines:
All Lymphatic Structures should be colored in Greens (yellow-greens, blue-greens, light and deep greens, etc.)
Oxygenated blood = red ; Capillary beds = purple;
Deoxygenated blood = blue; Arterioles = red;
Venules = blue.

GENERAL SCHEME OF LYMPH DRAINAGE

1	Lymph capillaries/ plexuses
2	Afferent lymphatic vessels
3	Lymph nodes
4	Efferent lymphatic vessels
5	Lymph trunks
6	Lymph ducts
7	Subclavian veins

BLOOD CIRCULATION

S	Systemic circulation
P	Pulmonary circulation

SCHEME OF BLOOD OXYGENATION

	Oxygenated blood
	Deoxygenated blood

ORIGINATION OF LYMPH VASCULAR SYSTEM

1	Lymph capillaries/ plexuses
8	Capillary beds
9	Arterioles
10	Venules
11	Tissue cells

FILTRATION OF BLOOD FLUID
From blood capillaries into lymph capillaries

In blood capillary

12	Blood plasma

In intercellular spaces

13	Interstitial fluid

In lymph capillary

14	Lymph

BODY AREAS AND PRINCIPAL CHANNELS OF LYMPH DRAINAGE

TWO MAIN BODY AREAS DRAINED BY TWO PRINCIPAL LYMPH DUCTS

THE PRINCIPAL DEEP CHANNELS (TRUNKS AND DUCTS) OF LYMPH DRAINAGE

R	Right lymphatic duct
1	Right jugular trunk
2	Right subclavian trunk
3	Right bronchomediastinal trunk
L	Large thoracic duct (Left lymphatic duct)
4	Left jugular trunk
5	Left subclavian trunk
6	Left bronchomediastinal trunk
7	Lower intercostal trunks
7a	Upper intercostal lymph vessels
8	Cisterna chyli
9	Right and left lumbar trunks
10	Intestinal trunks

THE PRINCIPAL SUPERFICIAL GROUPS OF NODE CENTER CLUSTERS

NECK

11	Cervical Nodes (submaxillary)

CHEST

12	Thoracic nodes

UPPER EXTREMITY

13	Axillary nodes
14	Cubital nodes

SMALL INTESTINE

15	Peyer's patches

LOWER EXTREMITY

16	Inguinal nodes
17	Popliteal nodes

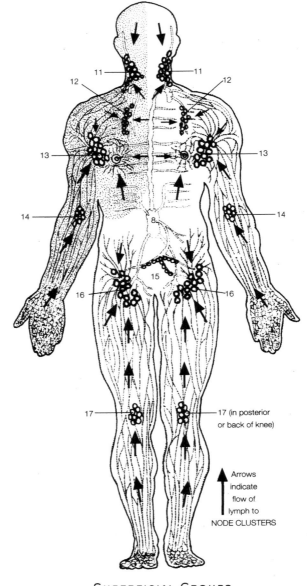

17 (in posterior or back of knee)

Arrows indicate flow of lymph to NODE CLUSTERS

SUPERFICIAL GROUPS
NODE CENTER CLUSTERS

The basic LYMPHATIC SYSTEM consists of LYMPHATIC VESSELS, LYMPH NODES and LYMPH. The three basic functions of the lymphatic system are: to return lost blood filtrate—called INTERSTITIAL (TISSUE) FLUID—back to the blood stream (venous system); transport absorbed fats from the intestine to the blood; and to help distribute LYMPHOCYTE CELLS, throughout the body from diseases (see chapter 12: Lymphoid Immune System).

LYMPH FLUID differs from BLOOD in that red corpuscles are absent and the protein content is lower. (Thus, lymph is similar to BLOOD PLASMA except for the low protein content.) Lymph contains PROTEINS (serum albumin, serum globulin, serum fibrinogen), ORGANIC SUBSTANCES (urea, glucose, neutral fats, creatinine), water, and LYMPHOCYTE CELLS (produced in lymphoid organs). INTESTINAL LYMPH (CHYLE) contains fats absorbed from the intestine.

Lymph is formed in tissue spaces all over the body (from blood filtrate that escapes through the thin-walled blood capillaries). It is gathered into small vessels that carry it centrally. All lymph eventually enters either the RIGHT LYMPHATIC DUCT or the LARGE (LEFT) THORACIC DUCT.

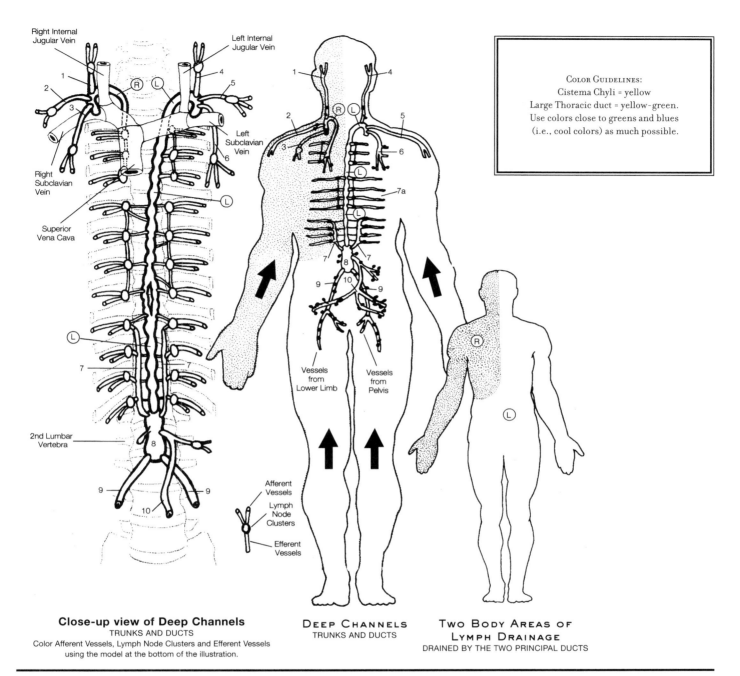

COLOR GUIDELINES:
Cisterna Chyli = yellow
Large Thoracic duct = yellow-green.
Use colors close to greens and blues
(i.e., cool colors) as much possible.

Right Internal
Jugular Vein

Left Internal
Jugular Vein

1

2

3

4

5

Left
Subclavian
Vein

6

Right
Subclavian
Vein

Superior
Vena Cava

7a

L

L

7

7

7

7

8

9

9

10

2nd Lumbar
Vertebra

8

9

9

10

1

4

2

3

R L

5

6

L

Vessels
from
Lower Limb

Vessels
from
Pelvis

R

L

Afferent
Vessels

Lymph
Node
Clusters

Efferent
Vessels

Close-up view of Deep Channels
TRUNKS AND DUCTS
Color Afferent Vessels, Lymph Node Clusters and Efferent Vessels
using the model at the bottom of the illustration.

DEEP CHANNELS
TRUNKS AND DUCTS

**TWO BODY AREAS OF
LYMPH DRAINAGE**
DRAINED BY THE TWO PRINCIPAL DUCTS

The thoracic duct begins in the abdomen as a dilated sac, the CISTERNA CHYLI, which receives lymph vessels from the PELVIS, LOWER LIMBS, INTESTINE, and the DIGESTIVE ORGANS.

The lymph is usually a clear, transparent, colorless fluid. In vessels draining the intestines, however, it appears milky, owing to the presence of absorbed fats, and is then called CHYLE. The absorption of fats takes place through the EPITHELIAL CELLS of the INTESTINE and those of the VILLI. These cells carry it to the LACTEALS (the lymph vessels of the small intestine). These lacteals take up the chyle and pass it to the lymph circulation and, via the thoracic duct, to the bloodstream.

The right lymphatic duct returns lymph to the venous system by emptying into the RIGHT SUBCLAVIAN VEIN (at the junction of the right INTERNAL JUGULAR VEIN). The large thoracic duct returns

lymph to the venous by emptying into the LEFT SUBCLAVIAN VEIN (at the junction of the left internal jugular vein).

LYMPH NODES usually appear in NODE CLUSTERS in specific regions of the body. PEYER'S PATCHES in the submucosa of the small intestine contain many scattered LYMPHOCYTES, large aggregations of lymphatic tissue and lymphatic nodules.

Lymph from LYMPH CAPILLARIES is passed to lymphatic vessels. Afferent lymphatic vessels pass toward and penetrate lymph nodes. Lymph is filtered through the lymph nodes. Efferent lymphatic vessels pass from the lymph nodes: they either run with afferent vessels to another node of the same group (cluster) or pass on to another group of nodes. Efferent vessels from the most proximal group of each chain unite to form LYMPHATIC TRUNKS.

Four Major Lymphoid Organs (Glands)
Lymph Nodes, Spleen, Tonsils and Thymus Gland

STROMA (FRAMEWORK) FOR LYMPH NODES, SPLEEN AND THYMUS

A Capsule

Dense (spleen) and fibrous (thymus gland)

B Trabeculae of capsule

C Hilus (lymph nodes and spleen)/Lobules (thymus gland)

PARENCHYMA OF LYMPH NODES
OUTER CORTEX

1 Cortical sinuses

2 Cortical nodules

Densely packed lymphocytes

3 Germinal centers

INNER MEDULLA

4 Medullary sinuses

5 Medullary cords

Lymphocytes arranged in strands

PARENCHYMA OF A SPLEEN: SPLENIC PULP
(lymphoid tissue arranged around arteries)

6 Malpighian corpuscles

Splenic nodules and densely packed lymphocytes

7 Splenic artery and branches

RED PULP:
Pulp infiltrated with red blood cells associated with veins

8 Billroth's cords (splenic cords)

9 Splenic vein

And venous sinuses filled with blood

PARENCHYMA OF A THYMUS LOBULE
OUTER CORTEX

10 Thymocytes (T-cells)

11 Epithelioreticular fibrous tissue

INNER MEDULLA

10 Thymocytes (T-Cells)

12 Epithelioreticular cells

13 Thymic (Hassall's) corpuscles

THREE PAIRS OF TONSILS—THE "RING OF WALDEYER"

14 Pharyngeals (adenoids)

15 Palatines

16 Linguals

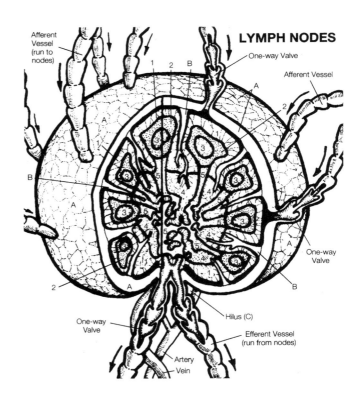

LYMPH NODES

The four LYMPHOID ORGANS are the LYMPH NODES, SPLEEN, TONSILS and THYMUS GLAND. In passing from any region of the body to the two main lymph ducts, lymph must pass through lymph vessels that lead to the lymph nodes. The lymph travels from the numerous afferent vessels to the CORTICAL SINUSES just under the capsule, then to the MEDULLARY SINUSES, then out through one or two efferent vessels. The lymph nodes filter lymph, freeing it of foreign particulate matter, especially bacteria. GERMINAL CENTERS produce LYMPHOCYTES— agranular LEUKOCYTES or WHITE BLOOD CELLS. The nodes lie between afferent vessels and efferent vessels and are 1–25 millimeters (0.04–1 inch) in length.

The spleen is an elongated, dark red ovoid body lying posterior and inferior to the stomach. It is the largest collection of reticuloendothelial (power to ingest or phagocytose bacteria or colloidal particles) cells in the body. It is not a vital body organ. It functions in various stages of development in BLOOD FORMATION, BLOOD STORAGE, and BLOOD FILTRATION. White pulp produces lymphocytes (white blood cells). It contains efferent vessels only via the white pulp. Since there are no afferent vessels, it does not filter lymph, only blood.

The thymus gland is an unpaired bilobed organ located in the mediastinal cavity anterior to and above the heart. Each lobe is divided by the trabeculae into many lobules. It is very large in fetuses and children, regresses during puberty, and becomes degenerate and involuted in the adult. It is essential for maturation of the thymic lymphoid cells, or the T-CELLS (THYMOCYTES), important in the body's cellular immune response. It is important in the development of immune response in children.

The tonsils are masses of lymphatic tissue located in depressions of the mucous membrane of the FAUCES and the PHARYNX. Their filtering action protects the body from the invasion of bacteria and aids in the formation of LEUKOCYTES.

THYMUS GLAND

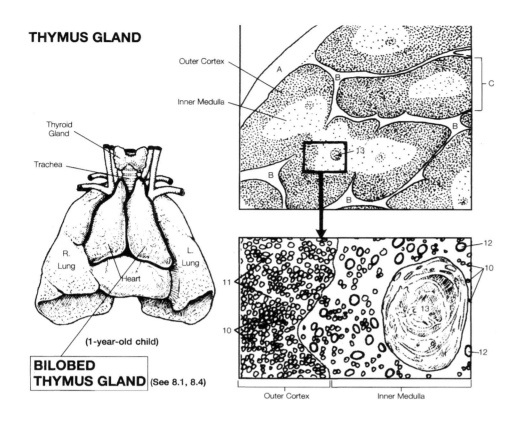

Outer Cortex

Inner Medulla

A

B

B

B

B

C

13

Thyroid
Gland

Trachea

R.
Lung

L.
Lung

Heart

(1-year-old child)

**BILOBED
THYMUS GLAND** (See 8.1, 8.4)

12

10

11

10

13

12

Outer Cortex

Inner Medulla

TONSILS

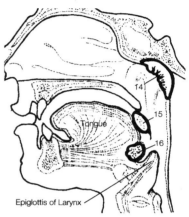

14

15

16

Tongue

Epiglottis of Larynx

SPLEEN

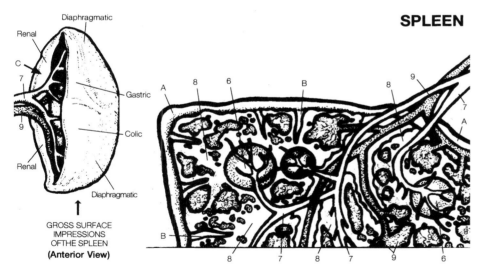

Diaphragmatic

Renal

C

7

9

Renal

Gastric

Colic

Diaphragmatic

GROSS SURFACE
IMPRESSIONS
OF THE SPLEEN
(Anterior View)

A

8

6

B

8

9

7

A

B

8

7

8

7

9

6

"RING of WALDEYER"

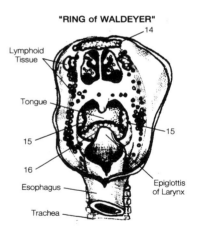

14

Lymphoid
Tissue

Tongue

15

15

16

Esophagus

Trachea

Epiglottis
of Larynx

(Posterior View, pharynx cut open)

COLOR GUIDELINES:
Spleen = color body dark red;
Lymph vessels, lymph capsule, and trabeculae = greens;
Parenchyma = yellows and blues; Tonsils = warm colors;
Thymus = light brown.

CHAPTER 12: LYMPHOID IMMUNE SYSTEM

In the previous chapter on the lymph vascular system, we briefly discussed some of the major lymphoid organs (all glands) that are part of the body's lymphoid immune system providing defensive security immune responses to attacks from material (antigen) the body deems as foreign, and thus threatening. The goal is to destroy invading microorganisms that have entered the body, or to obliterate and remove damaged, infected or abnormal cells, or cell parts that the body no longer considers part of "itself."

The aforementioned lymphoid organs/glands (lymph nodes, spleen, tonsils/adenoids, and thymus) in conjunction with other isolated, aggregated pockets and areas of lymphoid tissue located throughout the body (bone marrow, appendix, and mucosal associated lymphoid tissue or M.A.L.T.), in total comprise the gross anatomical components of the lymphoid immune system, notwithstanding the population of lymphocyte cells that reside in the blood.

All the lymphoid organs/glands and tissues are structurally composed of B-lymphocyte cells, T-lymphocyte cells or their cellular derivatives, and are most often supported by intricate mesh-like frameworks of reticular fibers and cells. [Note that the lymphocyte cells/derivatives also compose a very small percentage (<0.1%) of the non-plasma "formed elements" of blood tissue.]

PRIMARY LYMPHOID ORGANS

The origin of all types of committed, active Lymphocytes comes from two primary organs, the RED BONE MARROW and the THYMUS GLAND.

All LYMPHOCYTE PRECURSOR CELLS are contained within the soft, unencapsulated red bone marrow of cancellous, spongy bone and from there are spread throughout the body via venous sinusoids and vessels, to eventually reside in the blood (20-45% of white blood cells) and lymphoid organs/tissues. These precursors (which mingle within the red marrow landscape of reticular fibers and cells, fat cells, phagocytes, and a wide variety of maturing blood cells) follow three main paths: some mature and differentiate inside the marrow to become B-lymphocytes (for "bone marrow-derived"), and are involved in the so-called humoral antibody-mediated immunity; some larger ones enter the circulation from the marrow and become natural killer cells (<5% of blood lymphocyte population), which are neither b or t cells; and some leave the marrow partly differentiated as "pre-T lymphocytes" and migrate directly to the thymus gland via the blood.

Once inside the encapsulated thymus, the uncommitted early "pre-T lymphocytes" differentiate into T-lymphocytes (the "T" stands for "thymus-derived). From these, three main T-cell subpopulations are further differentiated, and are involved in the so-called cellular-mediated immunity which are helper T-cells, suppressor T-cells or cytotoxic killer T-cells. (The greater part of this T-cell proliferation occurs during embryonic and fetal life, and into the first ten years after birth.) From the thymus, these T-cell groups re-enter the circulation (both systemic and lymph) and migrate to establish residence in the secondary lymphoid organs and tissues.

SECONDARY LYMPHOID ORGANS AND TISSUES

The "action" takes place within all the secondary lymphoid organs and tissues. The body's immune system responds (lymphocytic activation) to antigenic challenges, the so-called acquired immunity (the humoral and cellular immunity responses). Recall that the organs/tissues are packed and populated by lymphocytes that have migrated from the primary lymphoid organs (red bone marrow and thymus).

The lymph nodes and spleen (both complex, encapsulated structures) act as fluid-filtering/screening/processing "watchdog" towers, keeping a vigilant eye out for potential invading troublemakers and aged/damaged cells. The spleen processes blood (but not lymph); in much the same way, the lymph nodes process lymph. The lymph node is the site where both humoral antibody-mediated immunity (B-cell) and cell-mediated immunity (T-cell) occur.

The partially encapsulated tonsils and adenoids (engorged tonsils) protect the region of the pharynx, the important crossroads that provides "customs security" for the intake from both the nasal (nose) and oral (mouth) cavities.

The mucosa of the unencapsulated vermiform appendix (the worm-like projection from the blind end of the cecum of the large intestine) is packed with multiple lymphoid follicles (nodular masses). The mucosal layers of all open visceral cavities—especially of the digestive, respiratory, and urinary tracts—house unencapsulated, variably-sized groups of lymphoid follicles and also diffuse lymphocytic distributions, collectively called mucosal-associated lymphoid tissue or M.A.L.T. (in the gut region, they are called G.A.L.T.).

ACQUIRED AND NATURAL IMMUNITY

Acquired immunity is a state of security against disease that involves the presence of an antigen, resulting in specific forms of lymphocyte reactions (the immune response). Whether a B-cell or T-cell is activated, memory cells from each are formed which increase rapid responses to subsequent antigenic exposures.

There are two main types of acquired immunity: humoral antibody-mediated and cell-mediated. The humoral-antibody-mediated immunity is tissue fluid-based, and involves the antigen-antibody reaction, and the activation of B-cells. The antigen physically activates the B-cell, resulting in proliferation of (1) B-memory cells, (2) secreting antibody specific to the antigen, and (3) differentiated plasma cells, which secrete more specific antibody.

With cell-mediated immunity, the antigen is presented to the T-cell for activation within the grasp of a phagocyte or any infected cell. T-cell activation results in proliferation of the three T-cell subpopulations mentioned above. The helper T-cells form T-memory cells, enhance humoral immunity by activating B-cells, boost the natural inflammatory response, and activate phagocytes through lymphokine factors. (The suppressor T-cells suppresses humoral immunity). The cytotoxic killer T-cells form T-memory cells and binds to and destroys infected, damaged or abnormal cells with lysing cytotoxins.

PRIMARY ORGANS

SECONDARY ORGANS AND TISSUES

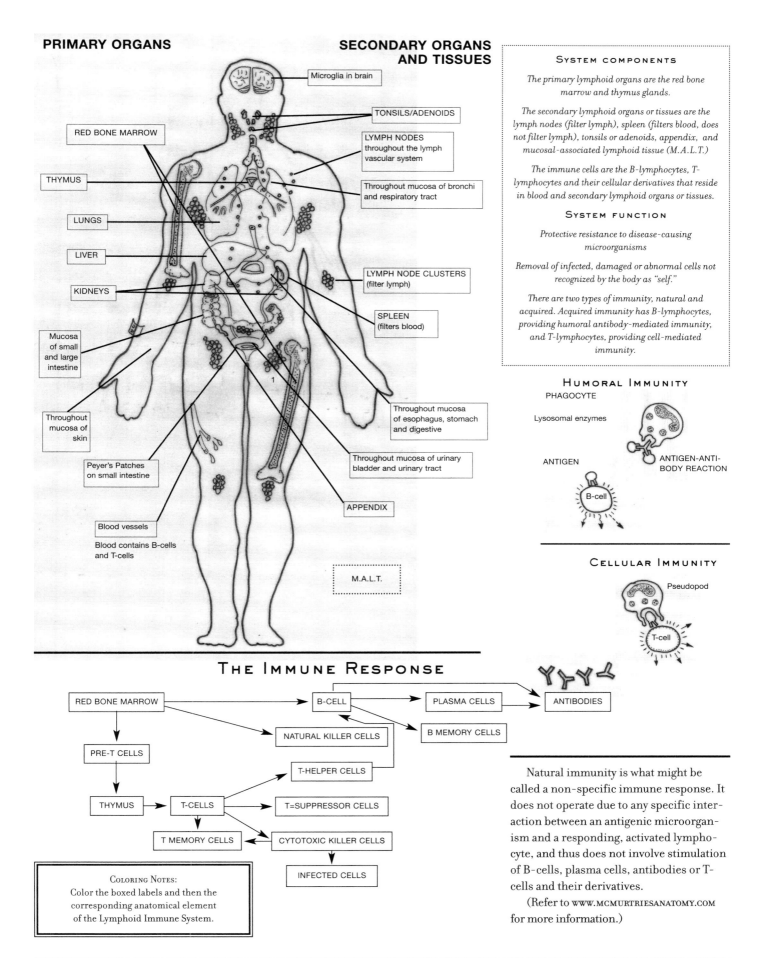

Microglia in brain

RED BONE MARROW

TONSILS/ADENOIDS

THYMUS

LYMPH NODES
throughout the lymph
vascular system

LUNGS

Throughout mucosa of bronchi
and respiratory tract

LIVER

KIDNEYS

LYMPH NODE CLUSTERS
(filter lymph)

Mucosa
of small
and large
intestine

SPLEEN
(filters blood)

Throughout
mucosa of
skin

Throughout mucosa
of esophagus, stomach
and digestive

Peyer's Patches
on small intestine

Throughout mucosa of urinary
bladder and urinary tract

Blood vessels

Blood contains B-cells
and T-cells

APPENDIX

M.A.L.T.

SYSTEM COMPONENTS

The primary lymphoid organs are the red bone marrow and thymus glands.

The secondary lymphoid organs or tissues are the lymph nodes (filter lymph), spleen (filters blood, does not filter lymph), tonsils or adenoids, appendix, and mucosal-associated lymphoid tissue (M.A.L.T.)

The immune cells are the B-lymphocytes, T-lymphocytes and their cellular derivatives that reside in blood and secondary lymphoid organs or tissues.

SYSTEM FUNCTION

Protective resistance to disease-causing microorganisms

Removal of infected, damaged or abnormal cells not recognized by the body as "self."

There are two types of immunity, natural and acquired. Acquired immunity has B-lymphocytes, providing humoral antibody-mediated immunity, and T-lymphocytes, providing cell-mediated immunity.

HUMORAL IMMUNITY

PHAGOCYTE

Lysosomal enzymes

ANTIGEN

ANTIGEN-ANTI-
BODY REACTION

B-cell

CELLULAR IMMUNITY

Pseudopod

T-cell

THE IMMUNE RESPONSE

RED BONE MARROW → B-CELL → PLASMA CELLS → ANTIBODIES

RED BONE MARROW → NATURAL KILLER CELLS

B-CELL → B MEMORY CELLS

RED BONE MARROW → PRE-T CELLS → THYMUS → T-CELLS

T-CELLS → T-HELPER CELLS

T-CELLS → T=SUPPRESSOR CELLS

T-CELLS → T MEMORY CELLS

T-CELLS → CYTOTOXIC KILLER CELLS → INFECTED CELLS

CYTOTOXIC KILLER CELLS → T MEMORY CELLS

COLORING NOTES:
Color the boxed labels and then the corresponding anatomical element of the Lymphoid Immune System.

Natural immunity is what might be called a non-specific immune response. It does not operate due to any specific interaction between an antigenic microorganism and a responding, activated lymphocyte, and thus does not involve stimulation of B-cells, plasma cells, antibodies or T-cells and their derivatives.

(Refer to WWW.MCMURTRIESANATOMY.COM for more information.)

Chapter 13: Respiratory System

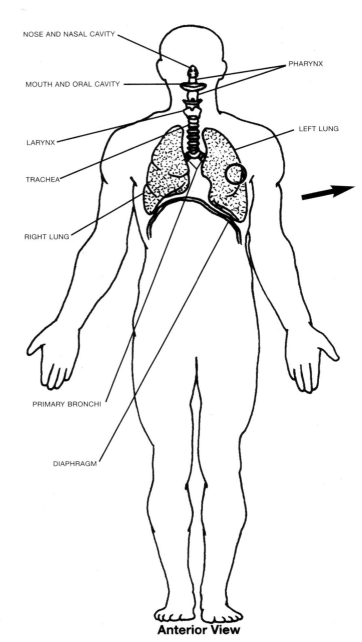

NOSE AND NASAL CAVITY
PHARYNX
MOUTH AND ORAL CAVITY
LARYNX
LEFT LUNG
TRACHEA
RIGHT LUNG
PRIMARY BRONCHI
DIAPHRAGM

Anterior View

Cells need a constant supply of OXYGEN (O$_2$) for the release of energy in carrying out vital cell activities. These cell activities release CARBON DIOXIDE (CO$_2$), which becomes toxic to the cell (the carbon dioxide in the cell environment forms carbonic acid, which in turn ionizes, yielding hydrogen ions and leading to an acid condition). Thus, the presence of carbon dioxide in the cell environment must be immediately and efficiently eliminated. All living cells require elimination of carbon dioxide to the fluid around them; they also obtain oxygen from it. The RESPIRATORY SYSTEM and the CARDIOVASCULAR SYSTEM are the two systems in the body that eliminate carbon dioxide and supply oxygen. They both share in the function of RESPIRATION.

The respiratory system consists of a series of passageways and two pliable LUNGS that exchange GASES between the external atmosphere and the BLOOD. The cardiovascular system transports the gases in the blood between the lungs and the BODY TISSUE CELLS. Respiration is the overall exchange of gases among the atmosphere, the blood, and the CELLS, and involves three basic processes. These are VENTILATION, GAS EXCHANGE and OXYGEN UTILIZATION. Ventilation (breathing) is the movement of air between the atmosphere and the lungs.

Gas exchange consists of EXTERNAL RESPIRATION, or exchange of oxygen and carbon dioxide between the lungs and the blood, and INTERNAL RESPIRATION, or exchange of oxygen and carbon dioxide between the blood and the TISSUES (CELLS). It is not possible to have a normal respiratory exchange of gases (oxygen and carbon dioxide) in the lungs unless the pulmonary tissue is adequately perfused with blood.

Oxygen utilization occurs when cell respiration releases energy needed for tissue metabolism by reacting with food substances in the cytoplasm and mitochondria of the cell. These reactions are called GLYCOLYSIS, the KREBS CITRIC-ACID CYCLE, and the ELECTRON TRANSPORT SYSTEM.

SYSTEM COMPONENTS

The lungs and a series of passageways (the NOSE, MOUTH, PHARYNX, LARYNX, TRACHEA, *and* BRONCHI) *in and out of the lungs.*

SYSTEM FUNCTION

Supply oxygen, eliminate carbon dioxide and regulate the body ACID-BASE BALANCE.

CONTENTS of a LOBULE IN LUNG TISSUE

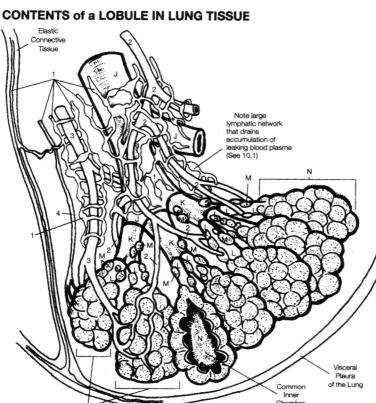

Elastic Connective Tissue

Note large lymphatic network that drains accumulation of leaking blood plasma (See 10.1)

Visceral Pleura of the Lung

Common Inner Chamber

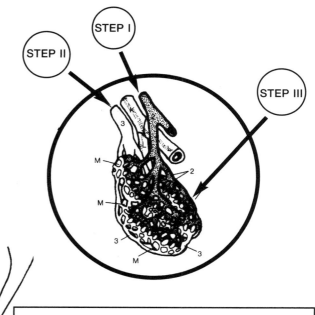

STEP I

STEP II

STEP III

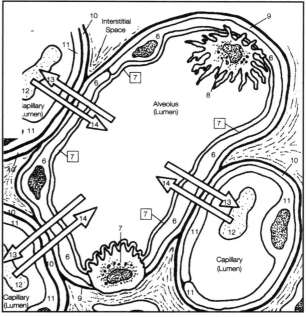

GAS EXCHANGE IN THE ALVEOLUS

STEP I: The Pulmonary artery brings venous (deoxygenated) blood from the right ventricle of the heart—the pulmonary arteriole branches into capillaries that surround each alveolus.

STEP II : Oxygen and carbon-dioxide are exchanged across the alveolar-capillary membrane in the capillaries.

STEP III: Capillaries link to form the pulmonary venule tributaries of the Pulmonary veins, which transport (oxygenated) arterial blood to the left auricle of the heart, which then pumps it to the left ventricle, and from there to the rest of the body.

Interstitial Space

Alveolus (Lumen)

Capillary (Lumen)

Capillary (Lumen)

Capillary (Lumen)

ALVEOLAR-CAPILLARY MEMBRANE "AIR-BLOOD" BARRIER (2 MICRONS)

END OF THE CONDUCTION ZONE

J Terminal bronchiole

Each bronchopulmonary segment of a lung is broken up into numerous small compartments called LOBULE, each of which is wrapped in an elastic connective tissue covering containing the following:

K Respiratory bronchiole

L Alveolar ducts   atria

M Alveoli   (singular: alveolus)

N Alveolar sacs

CELLS OF ALVEOLAR (EPITHELIAL) WALL:

6 S.P.E. Cells [Type I]

7 Septal Cells [Type II]

7 Surfactant layer

8 Alveolar macrophage

THE ALVEOLAR-CAPILLARY MEMBRANE:

9 Alveolar basement membrane

10 Capillary basement membrane

11 Capillary endothelium

12 Red blood cell

13 Diffusion of O_2

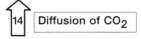

14 Diffusion of CO_2

GENERAL SCHEME

RESPIRATORY SYSTEMS AND ORGANS

The respiratory system consists of organs that exchange gases between the atmosphere and the blood. These organs can be conveniently divided into two RESPIRATORY TRACTS: the UPPER RESPIRATORY TRACT (URT), consisting of the NOSE, NASAL CAVITY and PHARYNX, and the LOWER RESPIRATORY TRACT (LRT), consisting of the LARYNX, TRACHEA, BRONCHIAL TREE (the BRONCHI and their increasingly smaller subdivisions) and LUNGS.

The CARDIOVASCULAR SYSTEM transports the gases in the blood between the lungs and the cells. Both the respiratory system and the cardiovascular system supply oxygen and eliminate carbon dioxide. Both systems are equal partners in respiration (the overall exchange of gases between the atmosphere, blood, and cells). If either system fails, there is rapid death of cells from oxygen starvation and disruption of homeostasis.

There are two alternating phases to the RESPIRATORY CYCLE—the first process of respiration is INSPIRATION (the inhalation or intake of air to lungs) and the second, EXPIRATION (the exhalation or expulsion of air from the lungs). Rhythmic bellows-like movements that aid the cycle are achieved through the ALTERNATING VOLUME CHANGES within the THORAX and lungs. These volume changes are the result of ALTERNATING MUSCLE CHANGES, principally from the DIAPHRAGM and INTERCOSTAL MUSCLES, and the expansion of the RIB CAGE. The movement of air in and out of the lungs is due to differences in pressure between the air outside of the body and the air inside the lungs and thorax. For example: the diaphragm lowers, the rib cage elevates, the thorax expands, thoracic volume increases, pulmonary pressure decreases, and air enters the lungs (inspiration).

All structures of the respiratory system aside from the PRIMARY BRONCHI (including the BRONCHIAL TREE and ALVEOLI) are contained within the lungs. Most of the mass of the lungs is comprised of the ALVEOLAR DUCTS and their terminal alveoli and ALVEOLAR SACS. The MOUTH is an organ of secondary importance for the entrance and exit of air.

The respiratory system can also be viewed as being divided into two zones: the CONDUCTION ZONE and the RESPIRATION ZONE.

The CONDUCTION ZONE consists of permanently opened passageways for the conduction of air to lung tissue. It is responsible for the transportation of air to the RESPIRATORY ZONE in the lungs. Various epithelia along these passageways warm, cleanse (filter) and humidify the air. There is no exchange of gases in the conduction zone, the structure of which include (from outside to inside) the nose, nasal cavity, pharynx, larynx, trachea, PRIMARY BRONCHI, SECONDARY BRONCHI, TERTIARY BRONCHI, BRONCHIOLES AND TERMINAL BRONCHIOLES. In simpler terms, the conduction zone consists of the larynx, trachea, and the bronchial tree.

The RESPIRATION ZONE is responsible for the oxygen-carbon dioxide exchange with the blood, the structures of which includes the RESPIRATORY BRONCHIOLES, ALVEOLAR DUCTS (ATRIA), ALVEOLI and ALVEOLAR SACS (the latter two being the functional units of the lungs where actual gas exchange occurs). (See www.mcmurtriesanatomy.com for more information.)

In VOICE PRODUCTION, the larynx creates sounds (pitch varies with tension of vocal cords). The mouth, nose, nasal cavity, SINUSES, pharynx and thorax act as resonators to affect quality and volume of sounds; the walls of the pharynx produce vowel sounds; and the LIPS, TONGUE and TEETH convert sounds into speech.

THE RESPIRATORY SYSTEM CAN BE CONCEPTUALLY DIVIDED INTO TRACTS—UPPER AND LOWER; OR ZONES—CONDUCTION AND RESPIRATION.

UPPER RESPIRATORY TRACT (URT)

A | Nose
(External Nose [Covered with Skin and Supported by Paired Nasal Bones and Cartilage])

B | Nasal Cavity (Internal Nose)

C | Pharynx (throat or gullet)

LOWER RESPIRATORY TRACT (LRT)

D | Larynx (Voice Box)

E | Trachea (Windpipe)

THE BRONCHIAL TREE

F | Primary Bronchi (2)

G | Secondary (Lobar) Bronchi

H | Tertiary (Segmental) Bronchi

I | Bronchioles

J | Terminal Bronchioles

RESPIRATION ZONE

K | Respiratory Bronchioles

L | Alveolar Ducts (atria)

M | Alveoli

N | Alveolar Sacs

O | Lungs

COLORING NOTES:
In general, color bronchial tree in blues and cool colors. Lungs = light brown or purple.
Remember, pulmonary arterioles should be colored blue (not red), because they carry deoxygenated blood to the alveoli,
and pulmonary venules are colored red (not blue) because they carry oxygenated blood to the heart!
URT = warm colors (flush, light oranges); LRT = cool colors (blues, purples, etc.);
Lungs = light brown or light purple; A–O will be used throughout the entire chapter.

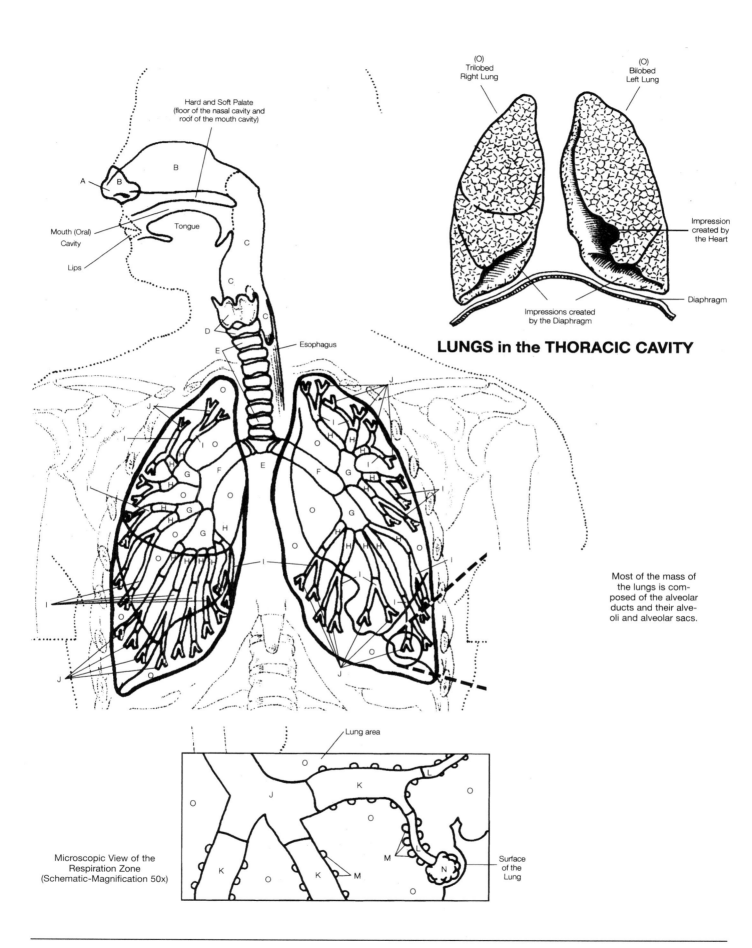

Hard and Soft Palate
(floor of the nasal cavity and
roof of the mouth cavity)

B

A
B

Mouth (Oral)
Cavity

Tongue

Lips

C

C

C

D

E

Esophagus

(O)
Trilobed
Right Lung

(O)
Bilobed
Left Lung

Impression
created by
the Heart

Diaphragm

Impressions created
by the Diaphragm

LUNGS in the THORACIC CAVITY

O
J
I
H
G
F
E
O
H

J
I
H
G
F
G
O
H

Most of the mass of
the lungs is com-
posed of the alveolar
ducts and their alve-
oli and alveolar sacs.

I

J
O

Lung area

O

L

Microscopic View of the
Respiration Zone
(Schematic-Magnification 50x)

O

J

K

O

K

O

M

K

M

N

Surface
of the
Lung

O

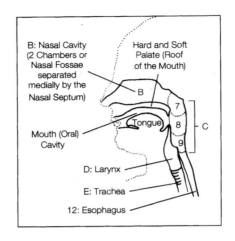

B: Nasal Cavity (2 Chambers or Nasal Fossae separated medially by the Nasal Septum)

Hard and Soft Palate (Roof of the Mouth)

B

7

8

9

C

Tongue

Mouth (Oral) Cavity

D: Larynx

E: Trachea

12: Esophagus

LATERAL WALL OF THE NASAL CAVITY:
ANTERIOR PORTION (OF NOSTRILS)

1 | Vestibule |

POSTERIOR PORTION

2 (2) | Superior Concha | (and Meatus)

3 (3) | Middle Concha | (and Meatus)

4 (4) | Inferior Concha | (and Meatus)

OPENINGS OF THE NASAL CAVITY: THE NASAL APERTURES
FRONT: EXTERNAL (ANTERIOR) NARES

5 | Nostrils |

BACK: INTERNAL (POSTERIOR) NARES

6 | Choanae |

THE PHARYNX

THREE DIVISIONS OF THE PHARYNX

7 | Nasopharynx |

8 | Oropharynx |

9 | Laryngopharynx |

OPENINGS INTO THE PHARYNX
NASOPHARYNX

10 | Eustachian Tube Opening |

Opens from lateral wall

6 | Choanae |

Two openings from the nasal cavity separated by the nasal septum

OROPHARYNX

11 | Fauces |

Opening from oral (mouth) cavity

LARYNGOPHARYNX

D1 | Epiglottis Opening of Larynx |

12a | Beginning of Esophagus |

The nasal cavity is the cavity between the floor of the cranium and the roof of the mouth, separated into two nasal fossae chambers by a medial nasal septum. As inspired air travels from the nose into the cavity, the NASAL EPITHELIUM warms, moistens, and filters the air. (Inhalation can also take place through the mouth, with lesser effect). The cavity acts as a resonating chamber for voice phonetics, while the olfactory epithelium in the ethmoid bone (in the roof of the cavity) provides a sense of smell. Three scroll-like turbinate bones called CONCHAE project medially from each lateral wall of the nasal cavity. Each of these three conchae overlies a meatus (tubelike passageway). The SUPERIOR and MIDDLE CONCHAE are projections of the lateral mass of the ethmoid bone. The INFERIOR CONCHA is its own (facial) bone. The NOSTRILS, OR NARES, are the two EXTERNAL APERTURES into the nasal cavity. The choanae are the two INTERNAL APERTURES, which provide a funnel-shaped opening for communication between the NASAL FOSSAE and the pharynx.

The pharynx is a funnel-shaped passageway for air into the LARYNX, and food and fluid into the ESOPHAGUS. It connects both the nasal and ORAL CAVITIES with the larynx (at the base of the skull). The lumen is lined with a mucous membrane and the supporting pharyngeal walls are composed of skeletal muscles. The pharynx acts as a resonating chamber for certain speech sounds constriction and relaxation of these wall muscles (constrictors) produce the VOWEL SOUNDS of SPEECH. If air from the pharynx enters the esophagus instead of the larynx (or if gas enters the esophagus from the stomach), a belch (burp) may occur. If food or fluid enters the larynx, coughing often results.

The three divisions of the pharynx are the NASOPHARYNX, OROPHARYNX and LARYNGOPHARYNX. The nasopharynx is the section above the PALATE. It contains the PHARYNGEAL TONSILS and is lined with pseudostratified ciliated epithelium with GOBLET CELLS (continuation of the nasal cavity epithelium). The oropharynx is the section between the palate and the HYOID BONE. It contains the PALATINE and the LINGUAL TONSILS and is the transition to STRATIFIED COLUMNAR EPITHELIUM. The LARYNGOPHARYNX is the section below the hyoid bone and above the beginning of the esophagus. It is lined with STRATIFIED SQUAMOUS EPITHELIUM.

The pharynx communicates (has openings) with five different structures: the EUSTHACIAN TUBES, CHOANAE, FAUCES, EPIGLOTTIS, opening of the LARYNX and the beginning of the ESOPHAGUS.

RIGHT LATERAL WALL OF THE NASAL CAVITY, PHARYNX AND ASSOCIATED STRUCTURES

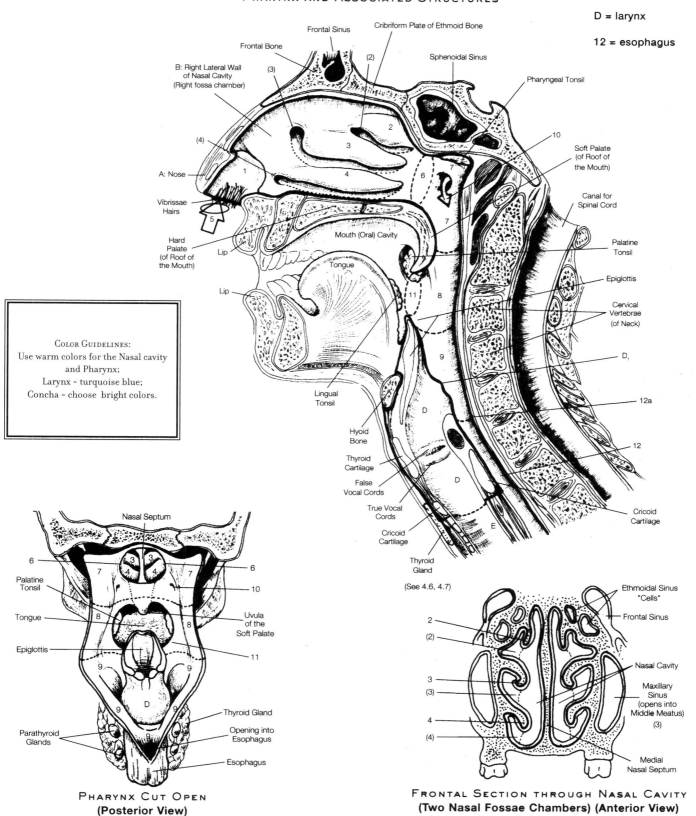

D = larynx

12 = esophagus

Frontal Sinus

Frontal Bone

Cribriform Plate of Ethmoid Bone

(2)

B: Right Lateral Wall
of Nasal Cavity
(Right fossa chamber)

(3)

Sphenoidal Sinus

Pharyngeal Tonsil

(4)

2

10

A: Nose

3

1

Soft Palate
(of Roof of
the Mouth)

Vibrissae
Hairs

4

6

7

5

Canal for
Spinal Cord

Hard
Palate
(of Roof of
the Mouth)

Mouth (Oral) Cavity

7

Palatine
Tonsil

Lip

Epiglottis

Tongue

11

8

Cervical
Vertebrae
(of Neck)

Lip

D₁

COLOR GUIDELINES:
Use warm colors for the Nasal cavity
and Pharynx;
Larynx = turquoise blue;
Concha = choose bright colors.

9

12a

Lingual
Tonsil

D

12

Hyoid
Bone

Thyroid
Cartilage

D

False
Vocal Cords

D

True Vocal
Cords

Cricoid
Cartilage

Cricoid
Cartilage

E

Thyroid
Gland

(See 4.6, 4.7)

PHARYNX CUT OPEN
(Posterior View)

Nasal Septum

6

3 3

7

4 4

7

6

Palatine
Tonsil

8

8

10

Tongue

Uvula
of the
Soft Palate

Epiglottis

11

9

9

Thyroid Gland

Parathyroid
Glands

D

Opening into
Esophagus

Esophagus

FRONTAL SECTION THROUGH NASAL CAVITY
(Two Nasal Fossae Chambers) (Anterior View)

Ethmoidal Sinus
"Cells"

Frontal Sinus

2

(2)

Nasal Cavity

3

(3)

Maxillary
Sinus
(opens into
Middle Meatus)

(3)

4

(4)

Medial
Nasal Septum

PARANASAL AIR SINUSES are PAIRED AIR SPACES located within four bones of the skull—3 CRANIAL BONES (the FRONTAL, SPHENOID and ETHMOID) and 1 FACIAL bone (the MAXILLA). They produce mucus, give structural strength to the skull, and decrease its weight. Each paranasal air sinus communicates within the nasal cavity via drainage ducts on its own side (the openings are through the lateral walls of the nasal cavity). They are lined with mucous membranes that are continuous with the nasal cavity. Tears secreted from the lacrimal gland drain first into the nasolacrimal duct and then into the nasal cavity through the inferior meatus. Therefore, it is understandable why it is natural to "blow the nose" after crying.

There are three functions of the paranasal sinuses associated with the nasal cavity. They warm, moisten and filter air; they assist the enhancement of the SENSE OF SMELL from the olfactory epithelium; and they provide voice phonetics as a SOUND RESONATING CHAMBER.

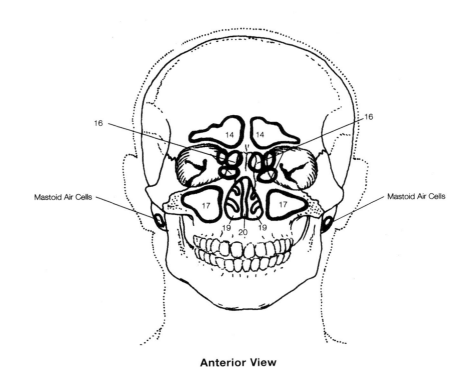

Anterior View

THREE FUNCTIONS FOR THE PARANASAL SINUSES ASSOCIATED WITH THE NASAL CAVITY:

1. Warms, moistens and cleans air;

2. Sense of smell from olfactory epithelium;

3. Sound resonating chamber (voice phonetics).

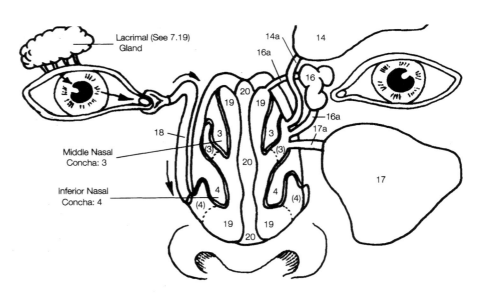

Anterior View (Schematic)

Duct openings of the Ethmoidal Sinus Cells open into both the Superior and Middle Meatuses.

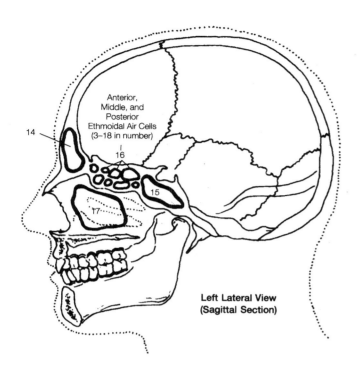

Anterior,
Middle, and
Posterior
Ethmoidal Air Cells
(3–18 in number)

14

16

15

17

**Left Lateral View
(Sagittal Section)**

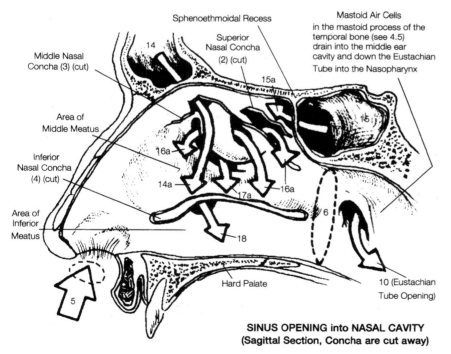

Middle Nasal
Concha (3) (cut)

Sphenoethmoidal Recess

Superior
Nasal Concha
(2) (cut)

Mastoid Air Cells

in the mastoid process of the
temporal bone (see 4.5)
drain into the middle ear
cavity and down the Eustachian
Tube into the Nasopharynx

Area of
Middle Meatus

Inferior
Nasal Concha
(4) (cut)

Area of
Inferior
Meatus

14

15a

15

16a

14a

17a

16a

6

18

5

Hard Palate

10 (Eustachian
Tube Opening)

**SINUS OPENING into NASAL CAVITY
(Sagittal Section, Concha are cut away)**

CoLORING NOTES:
Choose a bright color scheme for the four
sinuses: yellow, blue, green, and red.
Conduction Zone: Paranasal air sinuses
and openings into the nasal cavity.
Use the same colors for the sinuses
and their duct openings.
Nasal septum = light flesh color.

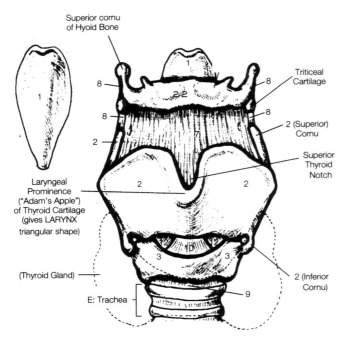

ANTERIOR VIEW OF LARYNX

NINE CARTILAGES OF THE LARYNX
THREE LARGE (SINGLES)

1 | Epiglottis

2 | Thyroid cartilage

"Adam's Apple"

3 | Cricoid cartilage

SIX SMALL (3 PAIRED)

4 | Arytenoid cartilage

5 | Corniculate cartilage

6 | Cuneiform cartilage

LIGAMENTS AND MEMBRANES
EXTRINSIC

7 | Thyrohyoid membrane

8 | Thyrohyoid ligament

9 | Cricotracheal ligament

INTRINSIC

10 | Cricothyroid ligament

11 | Thyroepiglottic ligament

INNER MEMBRANE OF TRUE VOCAL CORDS

12 | Conus elasticus

INNER LIGAMENT OF TRUE VOCAL CORDS

13 | Vocal ligament

LARYNGEAL FOLDS

14 | Interarytenoid fold

15 | Aryepiglottic fold

16 | True vocal cords | Vocal cords

17 | False vocal cords | Ventricular folds

LARYNGEAL OPENING

18 | Glottis (rima glottidis)

LARYNGEAL CAVITIES

19 | Vestibule

20 | Ventricle

21 | Infraglottic larynx

Inferior entrance to the glottis

22 | Hyoid bone

The LARYNX or VOICE BOX is the enlarged upper end of the TRACHEA below the root of the tongue, shaped like a triangular box. It forms the entryway into the LOWER RESPIRATORY TRACT (LRT) and connects the PHARYNX (lower portion, LARYNGOPHARYNX) with the trachea. Its musculocartilaginous structure is lined with a mucous membrane that forms the organ of voice. It is located in the anterior midline of the neck between the fourth to sixth cervical vertebrae.

Its structure consists of nine CARTILAGES (three large single and three small paired) bound together by an elastic membrane and moved by muscles. These are extrinsic muscles, which include the INFRAHYOIDS and others; and the intrinsic muscles. The cavity of the larynx is divided into three regions: an upper VESTIBULE, a lower INFRAGLOTTIC LARYNX, and a middle VENTRICLE, which is the space between the TRUE and FALSE VOCAL CORDS (FOLDS). The mucous membrane of the larynx is arranged into two pairs of folds. The upper pair are VENTRICULAR FOLDS (FALSE VOCAL CORDS which are not used in sound production) but help support the lower pair of vocal folds (TRUE VOCAL CORDS) which are used in sound production. These two pairs of strong connective tissue bands are stretched across the upper opening of the larynx from anterior to posterior.

Under the mucous membrane of the true vocal cords lie the CONUS ELASTICUS bands, which stretch between the large THYROID CARTILAGE and the small ARYTENOID CARTILAGES.

The space (slit) between the true vocal cords is the GLOTTIS (RIMA GLOTTIDIS), or the LARYNGEAL OPENING. The spoon-shaped EPIGLOTTIS lies on top of the larynx behind the root of the tongue and aids in closing the glottis during swallowing. The "stem" portion is hinged to the thyroid cartilage by the THYROEPIGLOTTIC LIGAMENT. The "leaf" portion is free and can move up and down like a trap door. During swallowing, the larynx elevates (due to the contraction of the EXTRINSIC MUSCLES), causing the free end of

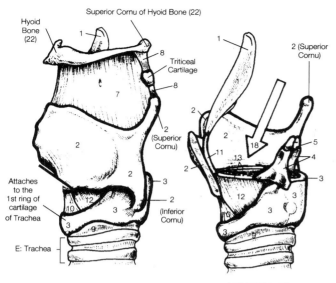

Hyoid Bone (22)
Superior Cornu of Hyoid Bone (22)
Triticeal Cartilage
2 (Superior Cornu)
(Inferior Cornu)
Attaches to the 1st ring of cartilage of Trachea
E: Trachea

Left Lamina of
Thyroid Cartilage (2)
Removed to Expose Inner Contents

Left Lateral Views of Larynx

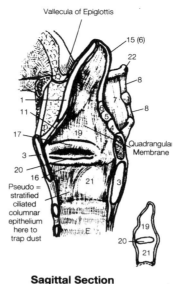

Vallecula of Epiglottis
Quadrangular Membrane
Pseudo = stratified ciliated columnar epithelium here to trap dust

Sagittal Section

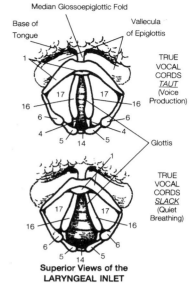

Median Glossoepiglottic Fold
Base of Tongue
Vallecula of Epiglottis
TRUE VOCAL CORDS *TAUT* (Voice Production)
Glottis
TRUE VOCAL CORDS *SLACK* (Quiet Breathing)

**Superior Views of the
LARYNGEAL INLET**

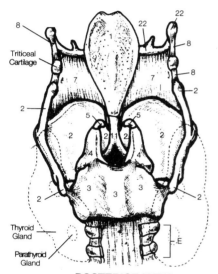

Triticeal Cartilage
Thyroid Gland
Parathyroid Gland

POSTERIOR VIEW
(Note that 1 and 11 are positioned
in front of 5 and 4 (See LATERAL VIEW)

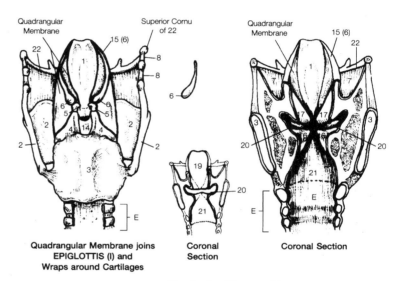

Quadrangular Membrane
Superior Cornu of 22
Quadrangular Membrane

**Quadrangular Membrane joins
EPIGLOTTIS (I) and
Wraps around Cartilages**

Coronal Section

Coronal Section

Posterior Views of Larynx

the epiglottis to form a lid over the glottis, which closes the glottis. (Thus it is impossible to breathe and swallow at the same time.)

There are two principal functions of the larynx: to permit passage of air to pass and sound production via the vocal chords controlled by INTRINSIC MUSCLES. The larynx is also involved with sound production via the vocal cords, which are controlled by INTRINSIC MUSCLES.

The intrinsic skeletal muscles of the larynx, when contracted, change the length, position, and tension of the vocal cords. These muscles are attached internally to the thyroid cartilage, the pyramid-shaped arytenoid cartilages, and to the vocal cords themselves. When the muscles contract, they pull the conus elasticus "strings" tight and stretch the cords out into the air passageways, narrowing the glottis. If air is directed from the lungs against the folds, they vibrate and create sound waves in the column of air in the pharynx, nasal cavity, and oral cavity. (These three latter cavities and the PARANASAL SINUSES act as RESONATING CHAMBERS, giving the voice its individual, human quality.) During breathing, the true vocal cords are ABDUCTED; during coughing, they are closed, then rapidly released; during PHONATION (speech, sounds) they are ADDUCTED with a thin glottis opening.

THE BRONCHIAL TREE

E | Trachea

("trunk" of the upside-down bronchial tree)

TWO PRIMARY BRONCHI
TRILOBED RIGHT LUNG AND BILOBED LEFT LUNG

F_1 | Right primary bronchus

F_2 | Left primary bronchus

3 SECONDARY (LOBAR) RIGHT BRONCHI AND 10 TERTIARY (SEGMENTAL) RIGHT BRONCHI

G_1 | Right superior lobar bronchus

1 | Apical

2 | Posterior

3 | Anterior

G_2 | Right middle lobar bronchus

4 | Medial

5 | Lateral

G_3 | Right inferior lobar bronchus

6 | Superior

7 | Medial basal

8 | Anterior basal

9 | Lateral basal

10 | Posterior basal

TWO SECONDARY (LOBAR) LEFT BRONCHI AND EIGHT TERTIARY (SEGMENTAL) LEFT BRONCHI

G_4 | Left superior lobar bronchus

1 and 2 Apicoposterior

3 | Anterior

4 | Superior lingular

5 | Inferior lingular

G_5 | Left inferior lobar bronchus

6 | Superior

7 and 8 | Anterior medial basal

9 | Lateral basal

10 | Posterior basal

I | Bronchioles

J | Terminal bronchioles

The conduction zone ends with the terminal bronchioles.

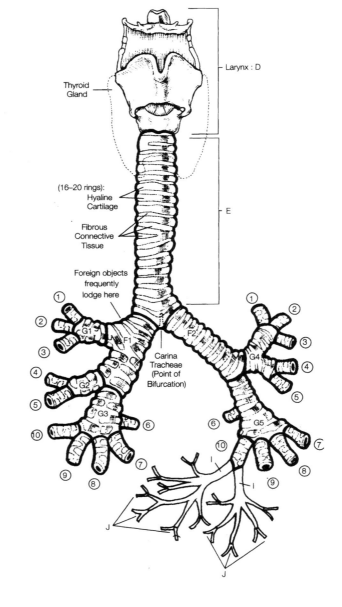

Larynx : D

Thyroid Gland

(16–20 rings): Hyaline Cartilage

Fibrous Connective Tissue

Foreign objects frequently lodge here

E

Carina Tracheae (Point of Bifurcation)

COLORING NOTES:
Trachea, Primary and Secondary Bronchi = color in Blues;
(Fibrous connective tissue on Trachea = light brown);
Tertiary Bronchi = open up and have fun!
Remember to color the bronchopulmonary segments of the lungs the same as their corresponding bronchial segments.

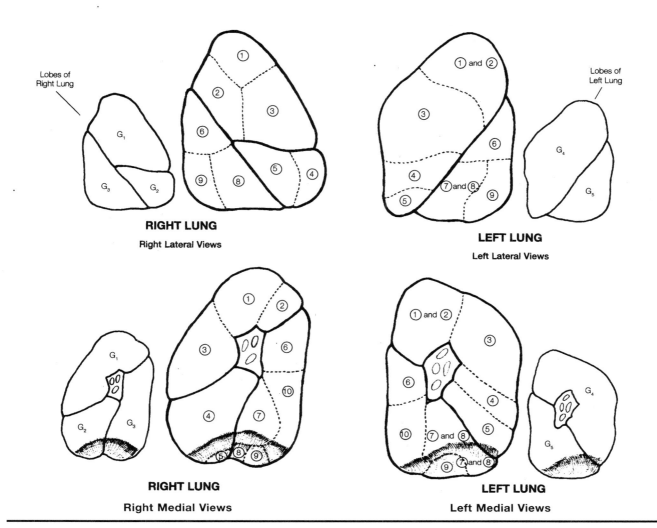

RIGHT LUNG

Right Lateral Views

LEFT LUNG

Left Lateral Views

Lobes of
Right Lung

Lobes of
Left Lung

RIGHT LUNG

Right Medial Views

LEFT LUNG

Left Medial Views

The CRICOID CARTILAGE of the larynx forms a ring and attaches to the first ring of cartilage of the TRACHEA. The trachea (WIND-PIPE) is a cylindrical, cartilaginous, rigid tube left permanently open for the passage of air that connects the larynx to the BRONCHIAL TUBES. It is 11.3 cm (4 1/2 in.) in length and 2.5 cm (1 in.) in diameter. It is supported by a stack of c-shaped rings of HYALINE CARTILAGE (interspersed with fibrous connective tissue) that are positioned anteriorly, with the open end of the c's in posterior position. It extends from the larynx to the STERNAL ANGLE in the THORAX (CHEST), (from the sixth cervical vertebra to the fifth thoracic vertebra), where it divides at a point called the CARINA into the RIGHT and LEFT PRIMARY BRONCHI, one leading to each lung.

The BRONCHIAL TREE contains the right primary bronchus, which is shorter and more vertical than the left primary bronchus. After entering the lung, each primary bronchus divides further into increasingly smaller tubes, terminating in microscopic BRONCHIOLES. The continuous branching of the trachea into primary bronchi, SECONDARY BRONCHI, TERTIARY BRONCHI, bronchioles, and finally into terminal bronchioles resembles an upside-down tree trunk with its branches; thus it is commonly referred to as the BRONCHIAL TREE.

The lungs are conveniently divided into BRONCHOPULMONARY SEGMENTS according to the bronchial branches that penetrate them. As they enter the lungs, the primary bronchi divide to form smaller bronchi, the secondary (LOBAR) bronchi, one for each lobe of the lung. (The right lung has three lobes, the left lung has two lobes.) The secondary bronchi branch into smaller tertiary (SEGMENTAL) bronchi, which branch into still smaller bronchioles, which in turn branch into even smaller TERMINAL BRONCHIOLES.

The structural changes through the bronchial tree are RINGS OF CARTILAGE (in E and F), replaced by PLATES OF CARTILAGE (in G and H), which disappear in I and J; as cartilage decreases, smooth muscle content increases; pseudostratified ciliated columnar epithelium (in E, F, G, and H) changes to simple cuboidal epithelium in I and J.

Bronchioles are analogous in function to the ARTERIOLES in the VASCULAR SYSTEM; they provide the greatest resistance to air flow in the conducting passages, as they contain thick smooth muscle for contraction and dilation.

Similar to the lower cavity of the larynx, the inner cavity (lumen) of the trachea and all bronchi is lined with a mucosa of pseudostratified ciliated columnar epithelium. It contains many goblet cells, which secrete mucus. Dust particles cling to the mucus, are swept up by the cilia, and are expelled from the body by a cough reflex. The posterior portion of the trachea consists of connective tissue (imbedded with smooth trachealis muscle) in contact with the esophagus. This soft area of tissue allows the esophagus to expand as swallowed food passes toward the stomach.

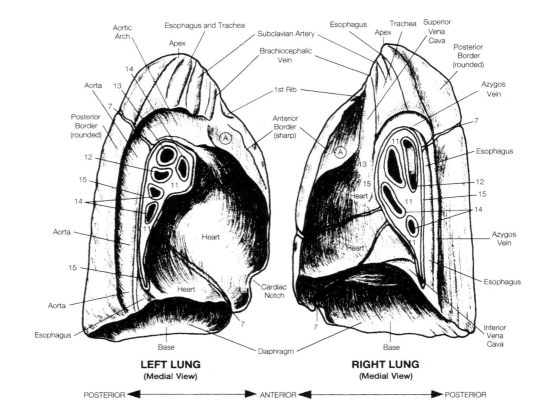

Aortic Arch · Esophagus and Trachea · Apex · Subclavian Artery · Brachiocephalic Vein · 1st Rib · Anterior Border (sharp) · Esophagus · Trachea · Superior Vena Cava · Apex · Posterior Border (rounded) · Azygos Vein · Aorta · Posterior Border (rounded) · Esophagus · Azygos Vein · Aorta · Heart · Cardiac Notch · Heart · Esophagus · Aorta · Heart · Inferior Vena Cava · Esophagus · Base · Diaphragm · Base

LEFT LUNG (Medial View)   **RIGHT LUNG** (Medial View)

POSTERIOR ← → ANTERIOR ← → POSTERIOR

LOBES OF THE LUNGS
TRILOBED RIGHT LUNG

1	Right superior lobe
2	Right middle lobe
3	Right inferior lobe

BILOBED LEFT LUNG

| 4 | Left superior lobe |
| 5 | Left inferior lobe |

FISSURES OF THE LUNGS

| 6 | Horizontal fissure |
| 7 | Oblique fissure |

PLEURAE OF THE LUNGS

Serous membranes that enclose and protect the lungs

| 8 | Visceral pleura |

Pulmonary pleura

| 9 | Parietal pleura |
| 10 | Pleural cavity |

LEFT AND RIGHT MEDIASTINAL SURFACES OF THE LUNGS

11	Root (hilum)
12	Primary bronchus
13	Pulmonary artery
14	Pulmonary vein
15	Pulmonary ligament

The bending of the visceral pleurae to become the parietal pleura

The RIGHT LUNG and LEFT LUNG are separately contained in PLEURAL MEMBRANES. There are three functions of the PLEURAE: lubricant secreted by the pleurae in pleural cavities allows the lungs to slide along the chest wall, pressure in the pleural cavities is maintained as less than the pressure in the lungs, which is required for ventilation, and the pleurae provide effective, tight separation of major thoracic organs.

The VISCERAL PLEURA covers the lungs and cannot be separated from them. (It also invaginates the interlobular fissures.) The PARIETAL PLEURA lines (continuous with the VISCERAL PLEURAE) and adheres to the rib cage, the diaphragm, and the pericardium of the heart. It is an extension of the visceral pleura (at the root of the lung) as it bends around and out. The PLEURAL CAVITY is the potential space between the two pleural layers. It is filled with a lubricating fluid secreted by the cells of the pleurae.

Each lung has four borders or surfaces that match the contour of the thoracic cavity. The APEX (CUPULA) is the anterior border that extends above the level of the clavicle bone.

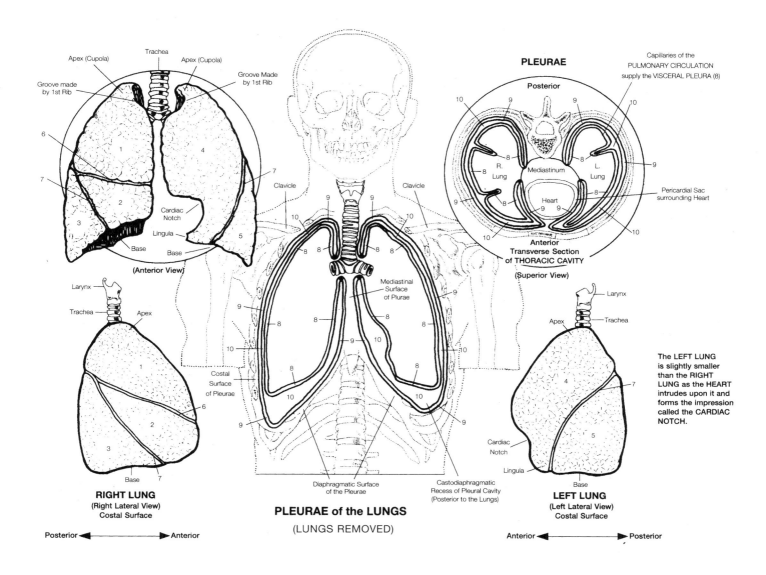

PLEURAE

Posterior

Capillaries of the
PULMONARY CIRCULATION
supply the VISCERAL PLEURA (8)

Mediastinum

R. Lung

L. Lung

Heart

Pericardial Sac
surrounding Heart

Anterior

**Transverse Section
of THORACIC CAVITY**

(Superior View)

Apex (Cupola)   Trachea   Apex (Cupola)

Groove made
by 1st Rib

Groove Made
by 1st Rib

Cardiac
Notch

Lingula

Base   Base

(Anterior View)

Clavicle

Clavicle

Mediastinal
Surface
of Plurae

RIGHT LUNG
(Right Lateral View)
Costal Surface

Larynx

Trachea   Apex

Costal
Surface
of Pleurae

Diaphragmatic Surface
of the Pleurae

PLEURAE of the LUNGS

(LUNGS REMOVED)

Castodiaphragmatic
Recess of Pleural Cavity
(Posterior to the Lungs)

Posterior ◄──────► Anterior

LEFT LUNG
(Left Lateral View)
Costal Surface

Larynx

Apex   Trachea

Cardiac
Notch

Lingula

Base

Anterior ◄──────► Posterior

The LEFT LUNG
is slightly smaller
than the RIGHT
LUNG as the HEART
intrudes upon it and
forms the impression
called the CARDIAC
NOTCH.

The BASE is the inferior border whose concave form results from the convex dome of the diaphragm. The COSTAL SURFACE is the broad, rounded surface of the lungs in contact with the membranes covering the ribs, and the MEDIASTINAL SURFACE is the medial surface of the lungs as they hug the lateral aspects of the MEDIASTINUM. The mediastinal surface (or MEDIAL SURFACE) of each lung contains a vertical slit called a ROOT (HILUS), through which bronchi, PULMONARY VESSELS, and nerves enter and exit.

The lungs are very spongy and pliable. Permanent impressions from surrounding organs in the thoracic cavity help mold their form, especially along the mediastinal surface. Studying these impressions helps in understanding the relationships of the organs in the thoracic cavity. The left lung is slightly smaller than the right lung as the heart intrudes upon it and forms the impression called the CARDIAC NOTCH.

COLORING NOTES:
Pulmonary artery = blue;
Pulmonary vein = red;
Heart, Aorta, and Arteries = warms
(reds, oranges);
Veins, Vena Cavae = Cools (blues, greens);
Color the Lung lobes the same colors used
on the Secondary (Lobar Bronchi
and Lung lobes).

CHAPTER 14: DIGESTIVE SYSTEM

SYSTEM COMPONENTS

A tubular passageway and associated organs (i.e.: salivary glands, liver, gallbladder and pancreas)

SYSTEM FUNCTIONS

Physical and chemical breakdown of food for cell usage

Elimination of solid wastes

INGESTION of organic food substances from the outside environment is a vital requirement for maintaining life for the human organism and other animals (unlike plants, which can form them from inorganic compounds within their cells). The three basic organic food substances are CARBOHYDRATES (MONOSACCHARIDES), FATS (LIPIDS) and PROTEINS (AMINO ACIDS), which come in a wide variety of edible plant and animal forms. They provide the source for VITAL CELL ACTIVITIES in the cells of the body tissues for LIFE ENERGY (calories) and the GROWTH and REPAIR OF BODY TISSUES (especially protein, which contains nitrogen).

Initial ingestion of these organic food substances is in the form of aggregations of large molecules called POLYMERS—forms not suitable for ready use by the tissue cells. The processes of digestion (both MECHANICAL AND CHEMICAL) break down these long chains of POLYMERS into their subunit MONOMERS—forms readily absorbable across the intestinal wall, entry into the vascular system, and ultimately accessible for use by the tissue cells. These free monomers are the end products of digestion.

Simple foods that can be absorbed unchanged are SALT, the SIMPLEST SUGARS (such as glucose), CRYSTALLOIDS (in general) and WATER.

Starches, fats, more complex sugars and proteins are not absorbable until split into these smaller monomer molecules. The digestion or breaking down of these food molecules occurs as a series of catalytic HYDROLYSIS REACTIONS (which occur with water and a wide variety of digestive enzymes). Each enzyme acts in an ACID or ALKALINE or NEUTRAL juice according to its peculiar properties.

The location of the fully active digestive enzymes is limited primarily to the cavity (lumen) of the GASTROINTESTINAL TRACT (GI TRACT), the STOMACH and INTESTINES. The process of ABSORPTION transports the formed monomers across the wall of the INTESTINES to the CARDIOVASCULAR and LYMPH VASCULAR SYSTEMS for distribution to the tissue cells.

The monomers of carbohydrates and the monomers of proteins are absorbed into the cardiovascular system, as are small aggregations of fat monomers (FATTY ACIDS and GLYCEROL). Larger aggregations of fat monomers are absorbed into the LYMPH VASCULAR SYSTEM, with wider lymphatic vessels designed to accept them.

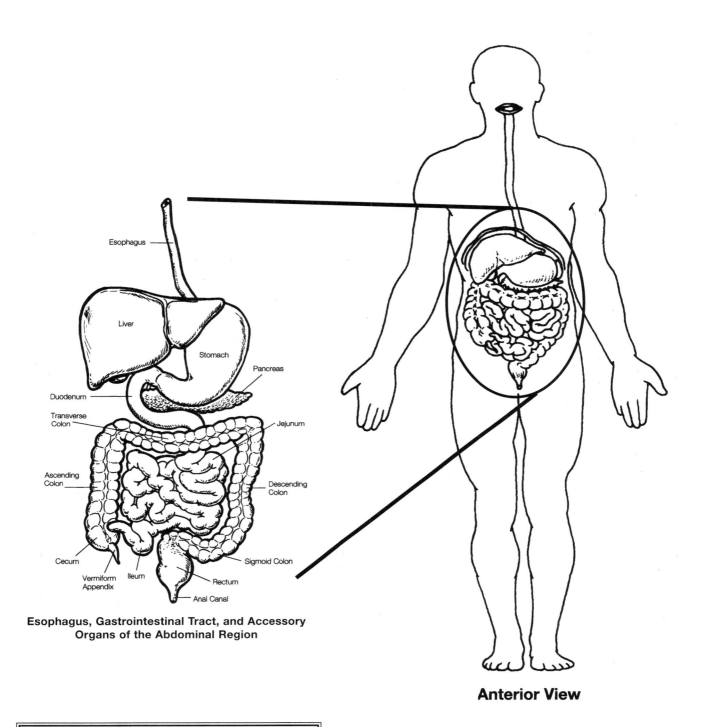

**Esophagus, Gastrointestinal Tract, and Accessory
Organs of the Abdominal Region**

Esophagus

Liver

Stomach

Pancreas

Duodenum

Transverse
Colon

Jejunum

Ascending
Colon

Descending
Colon

Cecum

Sigmoid Colon

Vermiform
Appendix

Ileum

Rectum

Anal Canal

Anterior View

COLORING GUIDELINES:
Liver = reddish brown;
Esophagus and Gastrointestinal tract = light oranges, flesh;
Gallbladder = yellow-green;
Pancreas = yellow-orange.

The DIGESTIVE SYSTEM can be anatomically and functionally divided into two main divisions, the ALIMENTARY CANAL and ACCESSORY ORGANS. The alimentary canal or TRACT is a continuous digestive tube transversing the VENTRAL BODY CAVITY and extending from the MOUTH to the ANUS, including the ORAL (BUCCAL) CAVITY, PHARYNX, ESOPHAGUS, STOMACH, SMALL and LARGE INTESTINES and the RECTUM (considered part of the large intestine). The GASTROINTESTINAL TRACT (or GI TRACT) is a subdivision of the alimentary canal and consists of the STOMACH and intestines (small and large). It is in the GI tract where most of the fully active digestive enzymes are located, produced by the "brush-border" enzymes of the intestinal mucosal lining facing the inner lumen, or entering into the tract within juice secreted by ACCESSORY ORGANS (the LIVER, GALL BLADDER, and PANCREAS).

The alimentary canal is a one-way transport system open at both ends. It is therefore continuous with the outside world, and is not in truth part of the internal environment. Thus, indigestible material that passes from one end (mouth) to the other (anus) without being absorbed across the lining of the tract never enters the body, and is considered to occur apart from the body.

The length of the alimentary canal is 30 feet (9 meters) in a cadaver, but slightly shorter in a living person due to muscle tone in tract walls. The alimentary canal contains the food from the time it is ingested until it is digested and prepared for DEFECATION. As the food progresses along the canal, it is subject to CHEMICAL and MECHANICAL changes that break it down into suitable units for ABSORPTION and ultimate ASSIMILATION (utilization of these simple absorbed units by all the cells of the body). The different regions of the alimentary canal are specialized for different functions of the digestive process. The basic functions are INGESTION, MASTICATION, DEGLUTITION, PERISTALSIS, DIGESTION, ABSORPTION and DEFECATION.

CHEMICAL DIGESTION involves enzymatic secretions produced by cells along the tract which, combined with enzymatic secretions from cells in the accessory organs break down the large food molecules (PROTEINS, CARBOHYDRATES AND LIPIDS) into usable units (AMINO ACIDS, MONOSACCHARIDES, GLYCEROL and FATTY ACIDS, respectively). These units are small enough to pass through the walls of the digestive organs (mainly the small intestine) into the blood capillaries (and in the case of larger fat molecules into lymph capillaries), and then through venous blood entering and exiting the heart, which distributes them to all the body's tissues and then through the plasma membranes of the body's cells, into the cell interior for energy transformation and utilization.

Mechanical digestion involves various movements that aid in chemical digestion. The teeth physically break down the food (mastication) before it can be swallowed. Muscular contractions of the smooth muscles in the walls of the stomach and small intestine physically break down the food by churning it; the food is thus thoroughly mixed with the enzymes involved in the catabolic reactions of chemical digestion.

FOUR LAYERS (TUNICS) OF THE GI TRACT

a — Mucosa
Tunica mucosa

b — Submucosa
Tunica submucosa

c — Muscularis
Tunica muscularis

d — Serosa/Visceral Peritoneum
Tunica serosa

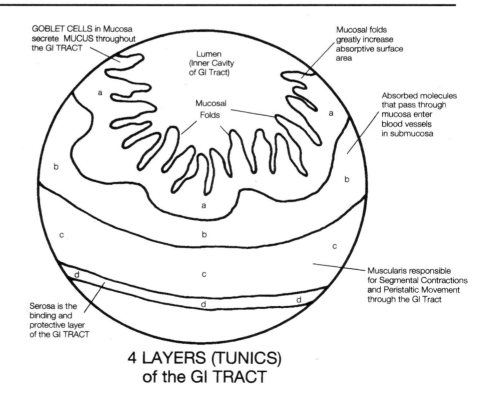

GOBLET CELLS in Mucosa secrete MUCUS throughout the GI TRACT

Lumen (Inner Cavity of GI Tract)

Mucosal Folds

Mucosal folds greatly increase absorptive surface area

Absorbed molecules that pass through mucosa enter blood vessels in submucosa

Muscularis responsible for Segmental Contractions and Peristaltic Movement through the GI Tract

Serosa is the binding and protective layer of the GI TRACT

4 LAYERS (TUNICS)
of the GI TRACT

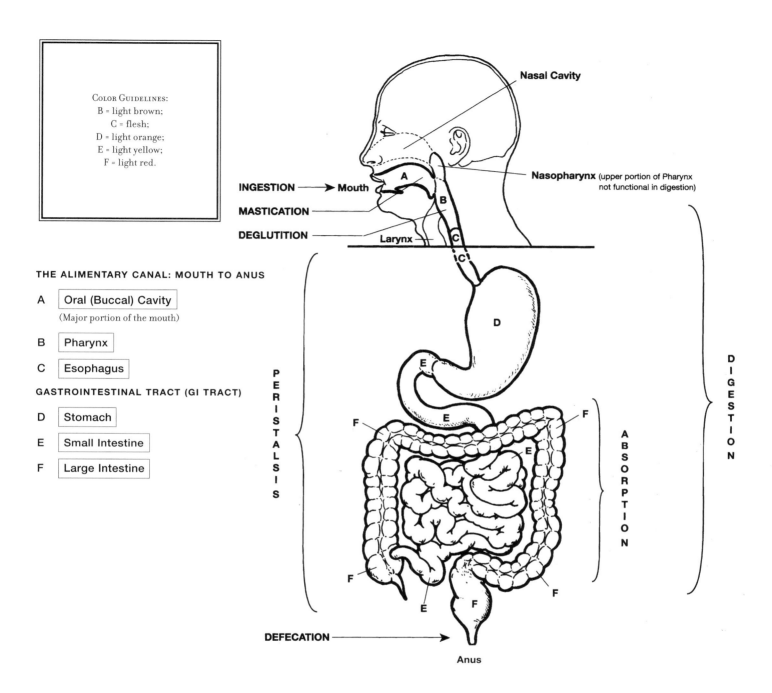

Color Guidelines:
B = light brown;
C = flesh;
D = light orange;
E = light yellow;
F = light red.

Nasal Cavity

INGESTION → **Mouth**

MASTICATION

DEGLUTITION

Nasopharynx (upper portion of Pharynx not functional in digestion)

Larynx

THE ALIMENTARY CANAL: MOUTH TO ANUS

A ☐ Oral (Buccal) Cavity

(Major portion of the mouth)

B ☐ Pharynx

C ☐ Esophagus

GASTROINTESTINAL TRACT (GI TRACT)

D ☐ Stomach

E ☐ Small Intestine

F ☐ Large Intestine

PERISTALSIS

ABSORPTION

DIGESTION

DEFECATION →

Anus

Regarding the histology of the alimentary canal walls, the oral cavity and pharynx are three-layered, and do not contain the four tunic layers specified for digestion. They are lined with nonkeratinized, stratified squamous epithelium (mucosa) and do not contain a serosa layer. Therefore, the four tunic layers of digestion are only continuous from the beginning of the esophagus down to the end of the anal canal. The inner layer of the four tunics—the MUCOSA—is a layer of columnar epithelial cells in close contact with the contents of the inner lumen cavity. Its specialized functions include secretion, absorption of nutrients and hormones into the bloodstream, and mobility—folds of the mucosa form projections into the lumen that continually change shape and surface area. The SUBMUCOSA is a thick, vascular layer of connective tissue serving the mucosa, gorged with a rich blood supply and immersed in a Meissner nerve plexus involved in the coordination of motor and secretory activities. The muscularis substantiates movement, controlling tube diameter, mixing and moving contents in peristaltic action along the tube, and serves nervous coordination through the Auerbach's nerve plexus. The outer serosa is continuous with the abdominal visceral peritoneum membrane and covers many digestive and accessory organs in the abdominal cavity, which reduces friction between contacting organs. The serosa also carries many blood and lymph vessels and nerves to and from the MESENTERY (see pages 252–53).

The secondary group of the digestive organs comprising the other anatomical and functional division of the digestive system are the ACCESSORY ORGANS (TEETH, TONGUE, SALIVARY GLANDS, LIVER, GALLBLADDER, PANCREAS and GASTROINTESTINAL GLANDS).

Accessory Organs inside the Alimentary Canal

The teeth protrude into the alimentary canal (oral cavity) and aid in the physical breakdown of food. The tongue comprises the floor of the oral cavity. It is a freely movable muscular organ whose digestive function is manipulation of the FOOD BOLUS in MASTICATION and DEGLUTITION (also involved in speech production and taste). The numerous BUCCAL GLANDS are located in the mucous membranes lining the oral cavity (in the palatal region). The GASTRIC and INTESTINAL GLANDS are located mainly in submucosal layers of the GI tract with openings into the gastrointestinal lumen.

The accessory glands outside the alimentary canal are the salivary glands, the liver, the gallbladder and the pancreas. These accessory organs produce or store secretions that aid in the chemical breakdown of food. These secretions are released into the alimentary tract through DUCTS. Ducts of the abdominal accessory organs carry their secretions into the DUODENUM via the common AMPULLA of VATER, where the common BILE DUCT and MAIN PANCREATIC DUCT meet.

The large complex food molecules of CARBOHYDRATES, PROTEINS, and LIPIDS are broken down into usable units by combinations of enzymatic secretions of the GI tract and the accessory organs. In the STOMACH, the ingested food bolus is converted into a pasty CHYME. When chyme reaches the SMALL INTESTINE, carbohydrates and proteins have only been partly digested, while LIPID digestion has not actually begun to any extent. In the small intestine, INTESTINAL JUICE (SUCCUS ENTERICUS) contains enzymes which complete the digestion of carbohydrates and proteins; lipids are digested in the small intestine by the pancreatic lipase enzyme from PANCREATIC JUICE.

Carbohydrates

Carbohydrate digestion begins in the MOUTH with the action of SALIVARY AMYLASE (from salivary glands), converting cooked STARCH (a POLYSACCHARIDE) first into DEXTRIN and then into MALTOSE (a DISACCHARIDE). The action of salivary amylase is stopped in the stomach by the low pH of GASTRIC JUICE. Ingested SUCROSE (table sugar) and LACTOSE (milk sugar)—two DISACCHARIDES—are not acted upon until they reach the small intestine. Most carbohydrate digestion occurs in the duodenum of the small intestine by the action of pancreatic amylase, from PANCREATIC JUICE, which cleaves the STRAIGHT, LARGE POLYSACCHARIDE CHAINS into simpler glucose chains of DISACCHARIDES (MALTOSE), TRISACCHARIDES (MALTRIOSE), and short-chained OLIGOSACCHARIDES. These three simple glucose chains, together with sucrose and lactose, are then hydrolyzed to monosaccharides in the small intestine by "brush-border" enzymes located in the microvilli of the lining epithelium. The monosaccharides are then absorbed across the microvilli and secreted unaltered from the epithelial cells into blood capillaries within the villi, and then into the hepatic portal vein en route to the liver.

Protein digestion begins in the stomach with the action of acidic pepsin secreted in the gastric JUICE from the STOMACH WALL. PEPSIN breaks down ENTWINED LARGE POLYPEPTIDE CHAINS into shorter-chained PEPTONES or PROTEOSES. (The action of pepsin is stopped in the duodenum by alkaline pancreatic juice). Most protein digestion occurs in the duodenum and jejunum of the small intestine by pancreatic juice enzymes, TRYPSIN, CHYMOTRYPSIN and ELASTASE, resulting in simpler DIPEPTIDES, TRIPEPTIDES (and some FREE AMINO ACIDS). These all are absorbed into the DUODENAL EPITHELIAL CELLS, and within these cells, the peptides eventually hydrolyze into more free amino acids. The free amino acids are then absorbed and secreted unaltered from the epithelial cells into the blood capillaries that carry them to the hepatic portal vein en route to the liver.

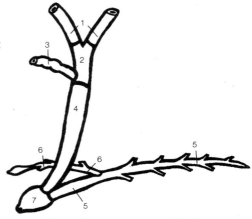

DUCT SYSTEM OF THE ABDOMINAL ACCESSORY ORGANS

ACCESSORY STRUCTURES OF THE ORAL CAVITY

G | Teeth

H | Tongue

MAJOR SALIVARY GLANDS (PAIRED GLANDS WITH DUCTS)

I | Parotid Glands (2)

J | Submandibular Glands (2) (Submaxillary)

K | Sublingual Glands (2)

ACCESSORY ORGANS OF THE ABDOMINAL REGION

L | Liver

M | Gallbladder

N | Pancreas

Gastric and Intestinal Glands (not shown)

DUCT SYSTEM OF THE ABDOMINAL ACCESSORY ORGANS

1 | Right and Left Hepatic Ducts

2 | Common Hepatic Duct

3 | Cystic Duct

4 | Common Bile Duct

5 | Main Pancreatic Duct (Duct of Wirsung)

6 | Accessory Pancreatic Duct (Duct of Santorini)

7 | Ampulla of Vater

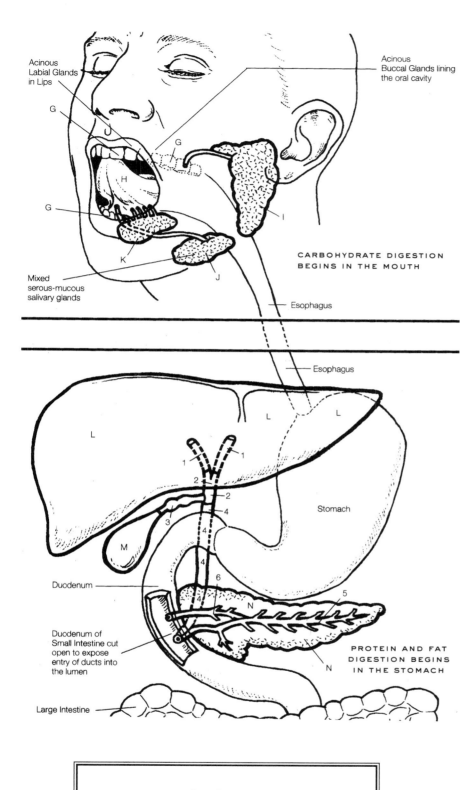

Acinous
Labial Glands
in Lips

G

G

H

G

K

J

Mixed
serous-mucous
salivary glands

Acinous
Buccal Glands lining
the oral cavity

**CARBOHYDRATE DIGESTION
BEGINS IN THE MOUTH**

Esophagus

Esophagus

L

L

L

L

1

1

2

2

3

4

M

4

4

6

Stomach

Duodenum

Duodenum of
Small Intestine cut
open to expose
entry of ducts into
the lumen

5

N

N

**PROTEIN AND FAT
DIGESTION BEGINS
IN THE STOMACH**

Large Intestine

Color Guidelines:
Salivary glands = neutral colors; L = reddish brown;
M = yellow-green; N = yellow-orange;
Duct system = cool colors.

FATS (LIPIDS)

Fat digestion begins in the stomach with the action of LINGUAL LIPASE (from lingual salivary gland), at a very limited level. Most of the ingested FAT enters the duodenum undigested as fat globules. These globules are not digested (the bonds joining the subunits are not hydrolyzed), but emulsified into finer droplets by bile salts from the liver released into the duodenum—which increases the surface area for subsequent digestion. Almost all fat digestion occurs in the small intestine by the action of pancreatic lipase—the only active fat-digesting enzyme in the adult body. Complex TRIGLYCERIDES are hydrolyzed into FREE FATTY ACIDS and MONOGLYCERIDES. (PHOSPHOLIPASE breaks up PHOSPHOLIPIDS into free fatty acids and LYSOLECITHIN). These simpler components then dissolve into MICELLE PARTICLES (AGGREGATIONS) of BILE SALTS, CHOLESTEROL and LECITHIN secreted by the LIVER, producing "MIXED MICELLES" which move to the "brush-border" of the small intestinal epithelium.

Unlike carbohydrates and protein monomers which pass through intestinal epithelial cells unaltered, these lipid products, after having been absorbed through the membrane of the microvilli, resynthesize into triglycerides and phospholipids within the epithelial cells. They (and cholesterol) are then combined with protein inside the epithelial cells to form tiny CHYLOMICRONS.

These chylomicrons are too large to pass into the hepatic portal blood, and bypass the liver by secreting into the lymphatic capillaries of the intestinal villi. They later enter venous blood by way of the LEFT THORACIC DUCT of the LYMPHATIC SYSTEM, to the heart, then to the liver via the hepatic artery. Once in the blood, triglycerides are removed from the chylomicrons and hydrolyzed by LIPOPROTEIN LIPASE (found in blood plasma), producing free fatty acids and GLYCEROL for use by tissue cells.

THE PERITONEUM: THE VISCERAL AND PARIETAL PERITONEUM

1 | **Parietal Peritoneum** |

2 | **Visceral Peritoneum** | (serosa)

PERITONEAL CAVITIES

3 | **Greater Peritoneal Cavity** |

4 | **Omental Bursa** |

(Lesser peritoneal cavity within the greater omentum)

5 | **Epiploic Foramen of Winslow** |

(The opening between the greater and lesser peritoneal cavities)

EXTENSIONS AND FOLDS OF THE PERITONEUM:

6 | **Falciform Ligament** |

(Attaches liver to anterior abdominal wall and diaphragm)

DOUBLE FOLDS BETWEEN ORGANS

7 | **Lesser Omentum** | (Two layers between organs)

8 | **Greater Omentum ("Fatty Apron")** |

(Four layers between organs)

MESENTERY

9 | **Dorsal Mesentery** |

(Double fold between posterior abdominal wall and small intestine)

10 | **Mesocolon** |

(Double fold between posterior abdominal wall and large intestine)

11 | **Mesoappendix** |

(Double fold between small intestine and appendix)

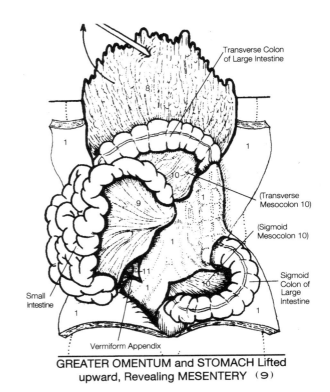

GREATER OMENTUM and STOMACH Lifted upward, Revealing MESENTERY (9)

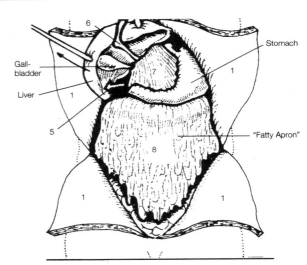

Torso cut open, Liver lifted up, Revealing the LESSER OMENTUM and FALCIFORM LIGAMENT

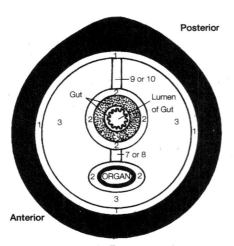

Schematic Representation of the SEROUS MEMBRANES (Cross-section through abdominal cavity)

The organs of the GI tract (the STOMACH and INTESTINES) and the ABDOMINAL ACCESSORY DIGESTIVE ORGANS, located in the abdominal cavity, are covered and supported by the PERITONEUM—the largest serous membrane in the body. It is composed of SIMPLE SQUAMOUS EPITHELIUM with portions reinforced with connective tissue. (This epithelium is also referred to as MESOTHELIUM, as it derives from the MESODERM lining the primitive body cavity in the embryo). Lesser serous membranes are

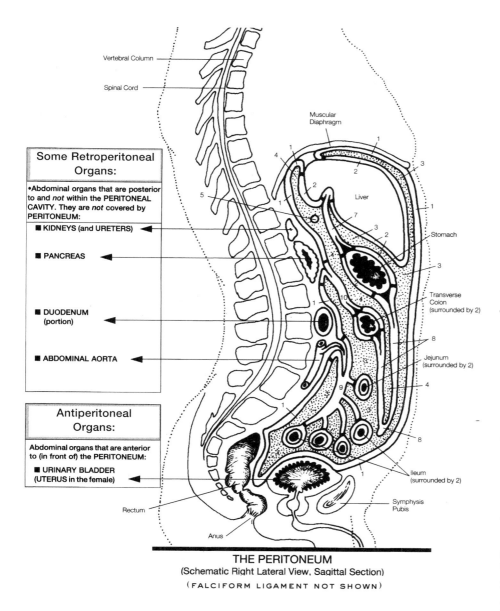

Some Retroperitoneal Organs:

•Abdominal organs that are posterior to and *not* within the PERITONEAL CAVITY. They are *not* covered by PERITONEUM:

■ **KIDNEYS (and URETERS)** ◄

■ **PANCREAS** ◄

■ **DUODENUM** (portion) ◄

■ **ABDOMINAL AORTA** ◄

Antiperitoneal Organs:

Abdominal organs that are anterior to (in front of) the PERITONEUM:

■ **URINARY BLADDER** (UTERUS in the female) ◄

Vertebral Column

Spinal Cord

Muscular Diaphragm

Liver

Stomach

Transverse Colon (surrounded by 2)

Jejunum (surrounded by 2)

Ileum (surrounded by 2)

Symphysis Pubis

Rectum

Anus

THE PERITONEUM
(Schematic Right Lateral View, Sagittal Section)
(FALCIFORM LIGAMENT NOT SHOWN)

COLOR GUIDELINES:
1 = light yellow;
2 = light orange;
3 = light blue.

found in the thoracic cavity: the PERI-CARDIUM surrounds the heart, and the PLEURAE surround the lungs.

Unlike the pericardium and the pleurae, the peritoneum contains large folds that weave in and out of the abdominal organs, binding the organs together and to the wall of the abdominal cavity. The folds contain blood vessels, lymph vessels and nerves that supply the abdominal organs.

All serous membranes are made up of two continuous portions— a PARIETAL POR-TION that lines the body wall, and a VIS-CERAL PORTION that covers the internal organs. The PARIETAL PERITONEUM lines the walls of the abdominal cavity. Along the dorsal aspect of the cavity (back portion)

each side of the parietal peritoneum joins together to form a double-layered fold, called the MESENTERY, which projects into the cavity and supports the GI tract. The DORSAL MESENTERY acts as a pendulum for the free-moving small intestine during peristaltic movements. A specific portion of the mesentery—the MESOCOLON—supports the large intestine. The peritoneal covering continues around most of the abdominal organs as the VISCERAL PERI-TONEUM. (SEROSA TUNIC). The space between the parietal peritoneum and the visceral peritoneum IS called the PERITONEAL CAVITY, which contains a small amount of lubricating fluid secreted by the peritoneum. This fluid minimizes the friction created

as the viscera glide on each other, or against the wall of the abdominal cavity.

Extensions of the parietal peritoneum include the falciform ligament, and the lesser and greater omenta. The lesser omentum suspends the stomach and DUO-DENUM from the LIVER. The greater omentum is called the "FATTY APRON," because it hangs like an apron over the front of the intestines arising from the serosa of the stomach. It stores large quantities of fat and protects against the spread of infections by numerous lymph nodes.

It is extremely important for the surgeon to understand the relationships within the ABDOMINAL CAVITY of the RETROPERITONEAL, INTRAPERITONEAL, and ANTIPERITONEAL organs and structures. Retroperitoneal (behind the peritoneum) include the SUPRARENAL GLANDS (PAIRED), KIDNEYS AND URETERS (PAIRED), PANCREAS, DUODENUM of SMALL INTESTINE except the ASCENDING and DESCENDING COLONS of the LARGE INTESTINE and the ABDOMINAL AORTA and INFERIOR VENA CAVA (and their branches). INTRAPERITONEAL (within the peritoneum) include the LIVER, GALLBLADDER, STOMACH, SPLEEN, SMALL INTESTINE, TRANSVERSE and SIGMOID COLONS of the large intestine and MESENTERIES packed with NERVES and GANGLIA, BLOOD VESSELS, LYMPHATIC VESSELS and NODES. ANTIPERITONEAL structures (in front of the peritoneum) include the URI-NARY BLADDER (in the male) and (in the female) the UTERUS.

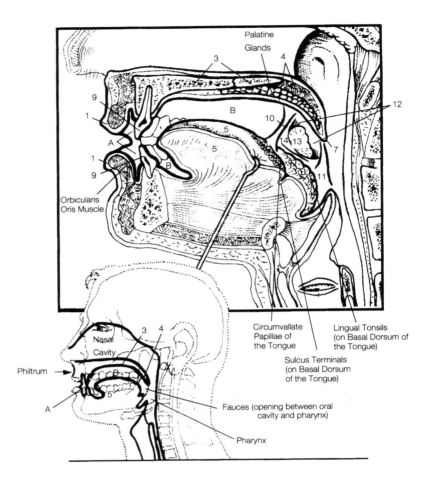

Palatine Glands

Orbicularis Oris Muscle

Circumvallate Papillae of the Tongue

Lingual Tonsils (on Basal Dorsum of the Tongue)

Sulcus Terminals (on Basal Dorsum of the Tongue)

Nasal Cavity

Philtrum

Fauces (opening between oral cavity and pharynx)

Pharynx

THE ORAL ORIFICE = OPENING OF THE MOUTH

THE MOUTH

A | Vestibule |

B | Oral Cavity Proper | (extends from the vestibule to the fauces)

BOUNDARIES OF THE MOUTH

1 | Lips | (Anterior)

2 | Cheeks | (Lateral walls)

3 | Hard Palate | (Superior [Roof])

4 | Soft Palate | (Posterior [Roof])

5 | Tongue | (Inferior [Floor])

ASSOCIATED STRUCTURES OF THE MOUTH

6 | Labial Frenulum |

7 | Uvula of the Soft Palate |

8 | Gingiva | (Gums)

9 | Teeth |

CHEEK

2a | Buccinator Muscle | (Cheek)

2b | Buccal Fat | (Cheek)

THE FAUCES = OPENING BETWEEN THE ORAL CAVITY AND THE PHARYNX

ANTERIOR "PILLAR" OF FAUCES

10 | Palatoglossal Arch | (Glossopalatine Arch)

POSTERIOR "PILLAR" OF FAUCES

11 | Palatopharyngeal Arch | (Pharyngopalatine Arch)

12 | Tonsillar Fossa |

13 | Palatine Tonsil | (Between the Two Arches)

14 | Triangular Fold |

The process of digestion begins with the ingestion of food into the ORAL CAVITY of the MOUTH. Chewing movements of the LIPS, CHEEKS, TONGUE, TEETH, and LOWER JAW break down the food in a destructive process. Then the food is mixed with saliva secreted by the SALIVARY GLANDS and is rolled into a bolus, or soft moist mass that becomes suitable for swallowing through the PHARYNX.

The mouth consists of the VESTIBULE and the ORAL CAVITY. The vestibule is an anterior slitlike cavity (depression) bounded externally by the lips and cheeks and internally by the GUMS and teeth. The oral cavity is that part of the mouth enclosed by the teeth. It is thus a slightly smaller area than the mouth itself, excluding the vestibule. The mouth is lined with NONKERATINIZED STRATIFIED SQUAMOUS EPITHELIUM (mucosa). Imbedded in this mucosal membrane are thousands of mucous glands that, along with the salivary glands, keep the mouth constantly moist. The mouth also assists secondarily in the passage of air into the lungs (see chapter 13, Respiratory System).

The lips are the anterior terminal portion of the cheeks. On the outside they are covered by skin, on the inside by a mucous membrane. The transition zone between the two is called the vermilion border. Functions of the lips include SPEECH, SUCKLING, MANIPULATION of FOOD and KEEPING FOOD BETWEEN UPPER and LOWER TEETH.

The boundaries of the FAUCES (ISTHMUS)—the opening between the oral cavity and the pharynx—are the inferior sulcus terminalis (on the DORSUM of the TONGUE), the superior soft palate and the LATERAL PALATOGLOSSAL ARCH and PALATOPHARYNGEAL ARCH.

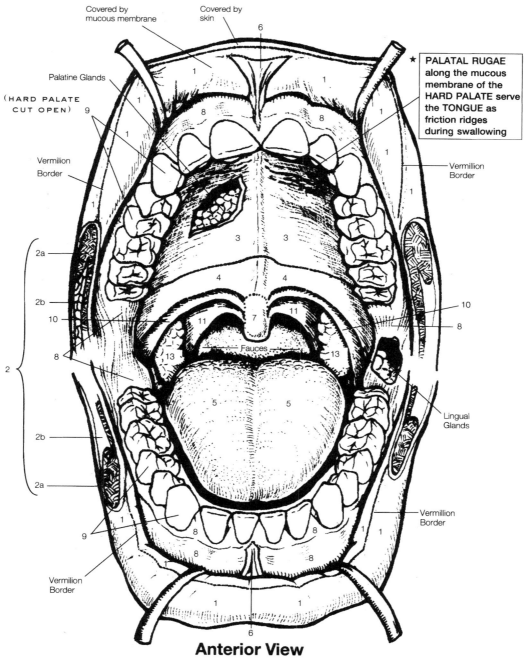

Covered by
mucous membrane

Covered by
skin

6

Palatine Glands

(HARD PALATE
CUT OPEN) 9

Vermilion
Border

★ PALATAL RUGAE
along the mucous
membrane of the
HARD PALATE serve
the TONGUE as
friction ridges
during swallowing

Vermillion
Border

1

1'

1

8

8

3   3

4   4

2a

2b

10

11   7   11

10

8

8

Fauces

2

13   13

Lingual
Glands

5   5

2b

2a

9

1

8   8

1   1

Vermillion
Border

Vermilion
Border

8

8

8   8

1   1

6

Anterior View

**Oral Cavity as Seen Through Oral Orifice
(The Opening of the Mouth)**

COLOR GUIDELINES:
1= red; 4 = pink ; 5 = orange;
8 = light red; 2b = yellow.

THE TEETH (DENTES)

MAJOR PORTIONS OF A TOOTH

1 Crown (enamel)

2 Neck/Cervix

3 Root (anchor)

STRUCTURE OF TOOTH

4 Enamel (Covers dentin)

5 Dentin
   (Bulk of Tooth (Harder than Bone))

6 Odontoblast Layer

7 Cementum (Covers root)

PULP CAVITY: CENTRAL REGION

8 Pulp Chamber (Hollow core inside dentin)

9 Pulp
   (Connective tissue, blood vessels, lymph vessels, and nerves)

10 Root Canal

SUPPORTING STRUCTURES OF A TOOTH

11 Gingivae (Gums)

12 Periodontal Membrane
   (Lines alveolus socket [fibers insert into cementum])

13 Alveolus Socket (Pocket for root)

DENTITION:
The TYPE, NUMBER, ARRANGEMENT OF TEETH in the DENTAL ARCH.
(Humans are DIPHYODONT, meaning they develop two sets of teeth during a lifetime.)

FIRST DENTITION: Milk or baby teeth

14 Deciduous 20

SECOND DENTITION: Replaces baby teeth

15 Permanent 32

TYPES OF TEETH

I   Central incisor and Lateral incisor

C   Canine Cuspid (eyetooth)

P   1st Premolar and 2nd Premolar
    Bicuspids

WISDOM TEETH

M   Molars (3rd)

M2 2nd molar

M3 3rd molar (wisdom teeth)

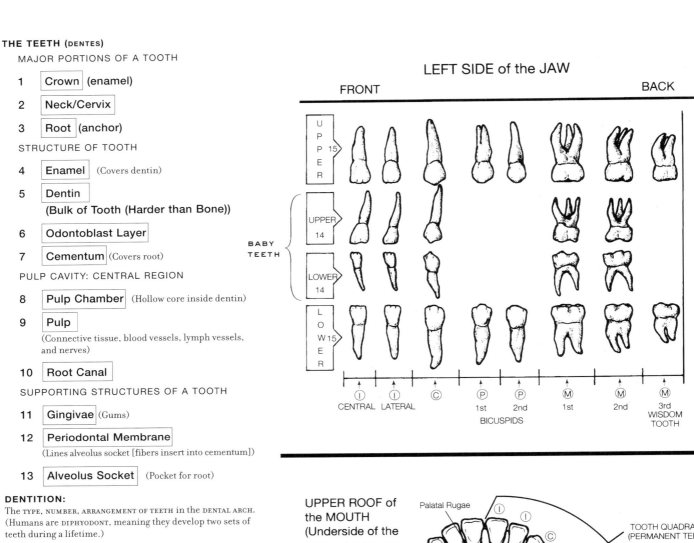

LEFT SIDE of the JAW

FRONT   BACK

U P P E R 15

UPPER 14

LOWER 14

BABY TEETH

L O W E R 15

I CENTRAL I LATERAL C P 1st P 2nd M 1st M 2nd M 3rd WISDOM TOOTH

BICUSPIDS

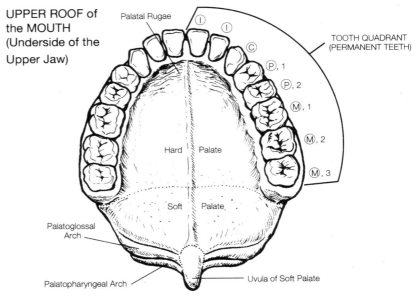

UPPER ROOF of the MOUTH (Underside of the Upper Jaw)

Palatal Rugae

TOOTH QUADRANT (PERMANENT TEETH)

I I C P, 1 P, 2 M, 1 M, 2 M, 3

Hard Palate

Soft Palate

Palatoglossal Arch

Palatopharyngeal Arch

Uvula of Soft Palate

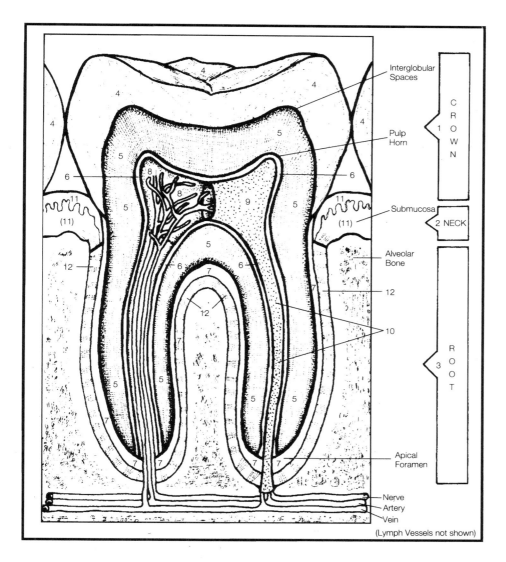

Interglobular
Spaces

Pulp
Horn

Submucosa

Alveolar
Bone

Apical
Foramen

Nerve
Artery
Vein
(Lymph Vessels not shown)

CROWN 1

2 NECK

ROOT 3

Color Guidelines:
1 = white ;
5 = light pink ;
7 = purple ;
9 = light yellow;
11 = light red;
Artery = red;
Vein = blue;
Nerve = grey.

CROSS-SECTION of a LOWER MOLAR

The mechanical action of teeth begins the destruction and breakdown of ingested food, known as the process of mastication, under voluntary control.

A typical tooth consists of three major portions, the CROWN (above the level of the gums), ROOT (consisting of one to three projections embedded in the socket [pockets in the alveolar processes of the mandible and maxillae bones]) and the NECK or CERVIX (a general constricted area between the crown and root [in the area of the GUMS]). The primary component of a tooth is a bonelike substance called DENTIN. It encloses a cavity. The enlarged part of the cavity lies in the crown and is called the PULP CAVITY or PULP CHAMBER. It is filled with PULP. Extensions of the pulp cavity into the root are the narrow ROOT CANALS, which end at an opening at the base of the root called the APICAL FORAMEN.

The tooth receives nourishment from blood vessels, and also receives lymph vessels and nerves, which traverse the apical foramen and become pulp. The dentin of the crown is covered by ENAMEL, the hardest substance in the body, being ninety-nine percent mineral (calcium phosphate and calcium carbonate). Enamel protects the tooth from the wear of chewing. It cannot be replaced after a tooth erupts.

BABY TEETH erupt at about six months of age, beginning with the incisors (eruption ends at two and a half years). PERMANENT TEETH begin to replace the baby teeth at about age six, in sequence from the INCISORS to the molars, continuing to about age seventeen. WISDOM TEETH (3rd MOLARS) are the last to erupt (between the ages of seventeen to twenty-five).

THE SALIVARY GLANDS

A | Parotid Glands | (2)

B | Submandibular Glands | (2)
   (Submaxillary)

C | Sublingual Glands | (2)

DUCTS OF THE SALIVARY GLANDS
PAROTID GLANDS

A1 | Parotid (Stensen's) Ducts

SUBMANDIBULAR GLANDS

B1 | Wharton's Ducts

SUBLINGUAL GLANDS

C1 | Rivinus's Ducts

C2 | Bartholin's Ducts

C3 | Sublingual Caruncles

SECRETORY CELLS OF SALIVARY GLANDS

1 | Mucous Cells
   Secrete thick, slimy, stringy mucus

2 | Serous Cells
   Secrete watery fluid containing digestive enzymes

3 | Serous Demilunes
   Crescent-shaped serous cells capping mucous acini cells.
   Demilunes of Giannuzzi

D | Tongue (Lingua)

4 | Lingual Frenulum

5 | Median Sulcus

6 | Foramen Caecum

7 | Sulcus Terminalis

PAPILLAE CONTAINING THE NERVE ENDINGS OR
TASTE BUDS

8 | Filiform Papillae
   Do not contain many taste buds

9 | Fungiform Papillae

10 | Circumvallate Papillae

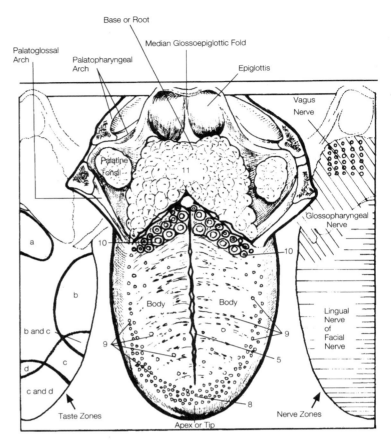

FILIFORM and FUNGIFORM PAPILLAE

TASTE ZONES OF THE TASTE BUDS

a | Bitter

b | Sour

c | Salt

d | Sweet

11 | Lingual Tonsils

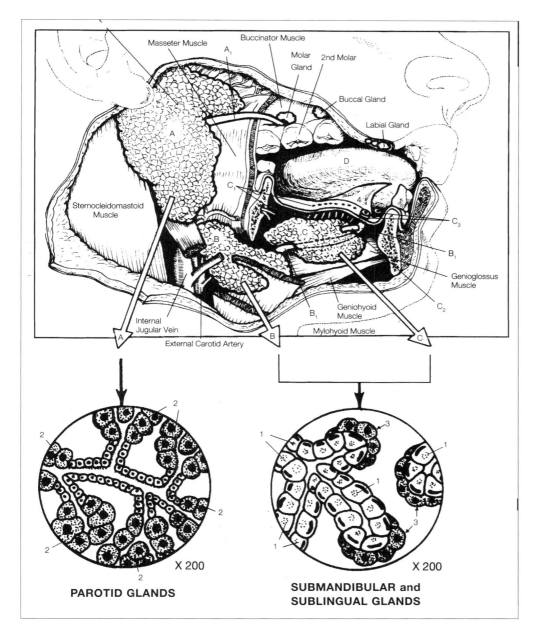

PAROTID GLANDS

SUBMANDIBULAR and SUBLINGUAL GLANDS

X 200 (each)

Color Guidelines:
Glands = light browns, greys;
Ducts = cool colors;
8-10 = warm colors;
11 = yellow.

Labels in figure: Masseter Muscle, Buccinator Muscle, Molar Gland, 2nd Molar, Buccal Gland, Labial Gland, Sternocleidomastoid Muscle, Internal Jugular Vein, External Carotid Artery, Genioglossus Muscle, Geniohyoid Muscle, Mylohyoid Muscle

The chemical activity of SALIVA (through the enzyme SALIVARY AMYLASE or PTYALIN) initiates the breakdown of INGESTED CARBOHYDRATES (or POLYSACCHARIDES) ultimately into MONOSACCHARIDES. This is the only chemical digestion that occurs in the mouth. (actually, only 3–5% of carbohydrates are reduced to DISACCHARIDES in the mouth since food is swallowed so quickly.) Most of the SALIVA is produced by 3 pairs of SALIVARY GLANDS outside the ORAL CAVITY and transported to it by SALIVARY DUCTS. The PAROTID GLANDS secrete the digestive enzyme PTYALIN (or SALIVARY AMYLASE), which starts to turn cooked STARCH to dextrins to maltose, a DISACCHARIDE that is later digested lower in the GI tract.

The other two salivary glands are the SUBMANDIBULAR and SUBLINGUAL GLANDS. SALIVA is a slightly acid solution of SALTS and ORGANIC SUBSTANCES, whose secretion is REFLEX and involuntary. Saliva also functions as a solvent in cleansing the teeth and in removing food particles from the oral cavity.

Solid substances must be dissolved in watery SALIVA to stimulate the TASTE BUDS on the TONGUE. The NERVE ENDINGS (TASTE BUDS) in the PAPILLAE of the TONGUE react to the food placed on them and they activate the secretion of the SALIVARY GLANDS.

The TONGUE is a highly specialized, striated muscular organ covered with a mucous membrane. It manipulates food during mastication, is involved with speaking, cleansing of teeth, and swallowing and it contains taste buds that sense food tastes. Only the anterior two-thirds of the TONGUE lie in the ORAL CAVITY. The remaining one-third lies in the pharynx and is attached to the HYOID BONE. The tongue is divided into symmetrical lateral halves by a MEDIAN SEPTUM (dorsally seen as the MEDIAN SULCUS). Each lateral half consists of an identical complement of EXTRINSIC and INTRINSIC muscles. Rounded masses of LINGUAL TONSILS can be found on the dorsal surface of the base of the TONGUE. Roughened PAPILLAE aid in the handling of food.

The PHARYNX is a fibromuscular pouch, posterior to the MOUTH, that connects the ORAL CAVITY and NASAL CAVITY with the ESOPHAGUS and LARYNX. It is therefore a common passageway for both the DIGESTIVE SYSTEM and the RESPIRATORY SYSTEM. It extends from the base of the skull (attached by PHARYNGOBASILAR FASCIA) to the level of the 6th cervical vertebra, where it is continuous with the ESOPHAGUS. Regarding the role of the pharynx in the digestive process, it does *not* produce digestive enzymes and does *not* carry on absorption. It secretes mucus continually from the lining of the mucous membrane, which along with saliva keeps the PHARYNX constantly moist for efficient passage of the food bolus. The pharynx aids in the transport of the bolus to the ESOPHAGUS.

The PHARYNX is three-layered (with an inner MUCOSA, SUBMUCOSA, and skeletal, voluntary MUSCULARIS) but does not contain outer SEROSA, as does the rest of the ALIMENTARY CANAL. Around the pharynx and around the thoracic portion of the ESOPHAGUS, in place of a true SEROSA, is the ADVENTITIA. This is a layer of connective tissue blending without a noticeable demarcation into the underlying surrounding tissue of the muscularis.

The ESOPHAGUS is a collapsible muscular tube approximately 25 cm long (10 in.) that conveys ingested food and fluid from the MOUTH (and PHARYNX) to the STOMACH. The esophagus originates at the LARYNX posterior to the TRACHEA, as a continuation of the LARYNGOPHARYNX. At its lower end, it opens through the DIAPHRAGM at the ESOPHAGEAL HIATUS and terminates at the CARDIA of the stomach. The anterior portion of the ESOPHAGUS is connected to the posterior portion of the TRACHEA by connective tissue imbedded with smooth muscularis muscle. This tissue attaches the open ends of the C-rings of the tracheal hyaline cartilage. This soft area of tissue allows the ESOPHAGUS to expand as swallowed food passes toward the STOMACH.

The esophagus does *not* secrete digestive enzymes and does *not* carry on absorption. It is involved in the DEGLUTITION REFLUX (SWALLOWING) and also downward PERISTALTIC movement of the food bolus. The ESOPHAGUS is the first digestive organ to contain all four TUNIC LAYERS. The MUCOSA (inner lining) is nonkeratinized STRATIFIED SQUAMOUS COLUMNAR EPITHELIUM; a thick,

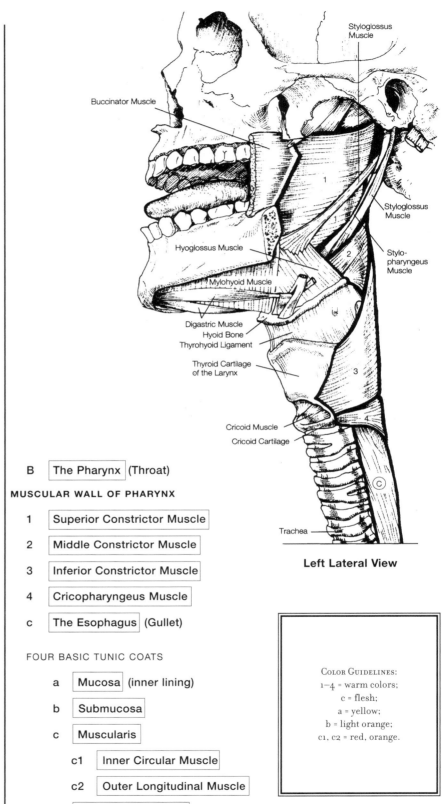

Left Lateral View

| B | The Pharynx | (Throat) |

MUSCULAR WALL OF PHARYNX

1	Superior Constrictor Muscle
2	Middle Constrictor Muscle
3	Inferior Constrictor Muscle
4	Cricopharyngeus Muscle
c	The Esophagus	(Gullet)

FOUR BASIC TUNIC COATS

a	Mucosa	(inner lining)
b	Submucosa
c	Muscularis
c1	Inner Circular Muscle
c2	Outer Longitudinal Muscle
d	Adventitia/"Serosa"

COLOR GUIDELINES:
1–4 = warm colors;
c = flesh;
a = yellow;
b = light orange;
c1, c2 = red, orange.

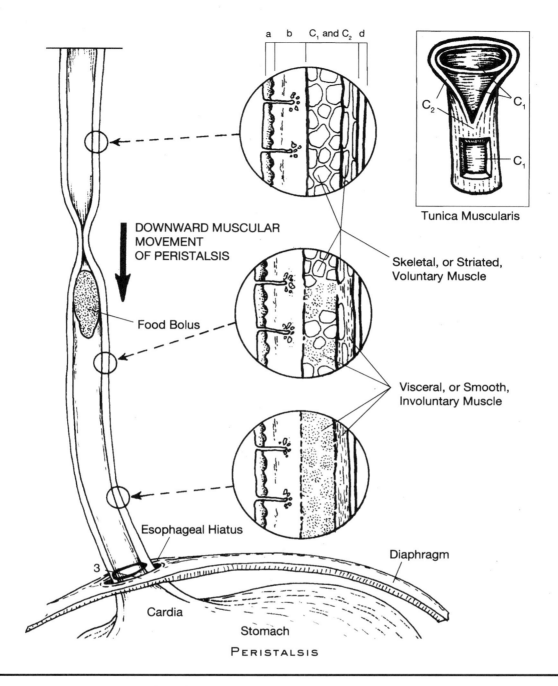

DOWNWARD MUSCULAR MOVEMENT OF PERISTALSIS

a b C₁ and C₂ d

Tunica Muscularis

Skeletal, or Striated, Voluntary Muscle

Food Bolus

Visceral, or Smooth, Involuntary Muscle

Esophageal Hiatus

Diaphragm

Cardia

Stomach

PERISTALSIS

protective surface layer open to the LUMEN (CANAL). The SUBMUCOSA contains mucous glands that secrete MUCUS along DUCTS to the inner mucosa for lubrication. The MUSCULARIS is arranged in two layers with an INNER CIRCULAR MUSCLE and an OUTER LONGITUDINAL MUSCLE. In the upper third of the esophagus. The muscle is SKELETAL (STRIATED, or VOLUNTARY). In the middle third the muscle is a mixture of SKELETAL and VISCERAL. In the lower third, the muscle is VISCERAL (SMOOTH, or INVOLUNTARY). The SEROSA is an outer serous coat combined with connective tissue that blends with the TRACHEA above and the stomach below. Except during the passage of food, the ESOPHAGUS is closed and flattened with its mucosa thrown into many longitudinal folds.

Just above the level of the DIAPHRAGM, the ESOPHAGUS is slightly narrowed due to the CARDIAC (GASTROESOPHAGEAL) SPHINCTER. This is known as the LOWER ESOPHAGEAL CARDIAC SPHINCTER. During swallowing, RELAXATION of the CARDIAC SPHINCTER permits the food bolus to enter the STOMACH. After the food bolus passes into the STOMACH, this sphincter CONSTRICTS to prevent reflux of stomach contents back up into the ESOPHAGUS. (This reflux would normally occur due to pressure differentials in the THORACIC and ABDOMINAL CAVITIES during RESPIRATION.) A common problem, "heartburn", is caused by esophageal refluxing. The movement of the DIAPHRAGM against the STOMACH during RESPIRATION also helps prevent refluxing of gastric contents.

PERISTALSIS is progressive, wavelike movement of smooth muscle (longitudinal and circular layers), especially in hollow tubes of alimentary canal.

FOUR REGIONS OF THE STOMACH

A | Cardia

B | Fundus

C | Body

D | Pylorus ("gatekeeper")

PORTIONS OF THE PYLORUS

D1 | Pyloric Sphincter (pyloric valve [circular muscle])

D2 | Pyloric Vestibule

D3 | Pyloric Antrum

RELATED STRUCTURES

E | Esophagus

F | Cardiac Sphincter

G | Duodenum of Small Intestine

H | Diaphragm

TWO SURFACES (BROADLY ROUNDED)

1 | Anterior Surface

2 | Posterior Surface

TWO CURVATURES (BORDERS)

3 | Greater Curvature (lateral convex)

4 | Lesser Curvature (medial concave)

INDENTATIONS

5 | Cardiac Incisura

6 | Angular Incisura

7 | Pyloric Intermediate Sulcus

FOLDS

8 | Gastric Rugae Folds of the Mucosa

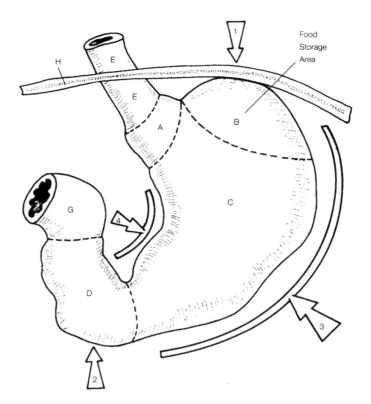

The STOMACH is the most distensible part of the GI TRACT. It has two openings: it is continuous with the ESOPHAGUS superiorly at the UPPER CARDIAC ORIFICE and empties into the DUODENUM of the SMALL INTESTINE through the LOWER PYLORIC ORIFICE inferiorly.

The FUNCTIONS OF THE STOMACH include: acting as a RESERVOIR (STORAGE BASE) for food as the food is mechanically CHURNED and mixed with GASTRIC SECRETIONS (acid and enzymes) made by the STOMACH CELLS. The food bolus is changed into a pasty, liquid CHYME, aided by peristaltic mixing waves and muscular contractions; the stomach also delivers chyme in small quantities (through REGULATORY MOVEMENTS OF THE PYLORUS) to the DUODENUM. (Pyloric movements also inhibit backflow of CHYME back into the STOMACH.). The stomach also INITIATES DIGESTIONS OF PROTEINS (PARTIAL DIGESTION of PROTEINS by the action of the enzyme PEPSIN).

CARBOHYDRATES ARE NOT DIGESTED AT ALL IN THE STOMACH. Fats are very slightly digested. (The complete digestion and absorption of food molecules occurs when the CHYME enters the SMALL INTESTINE.) Saliva swallowed into the STOMACH continues to act on cooked starch changing it into a form of SUGAR. The STOMACH wall has limited absorption. It is impermeable to the passage of most substances into the blood. (The only commonly ingested substances are ALCOHOL, ASPIRIN, GLUCOSE, and some water and salts.) It is important in ACID-BASE EQUILIBRIUM of the body (especially when vomiting removes ELECTROLYTES). It regulates DILUTION/CONCENTRATION of FLUIDS so they are the same concentration as the body's own fluids. The STOMACH ACID kills a large amount of the MICROBES present in food. The stomach secretes INTRINSIC FACTOR, which is needed for vitamin B12 (an EXTRINSIC FACTOR) absorption in the intestine.

The STOMACH is the beginning of the GASTROINTESTINAL TRACT (GI TRACT) (The term GASTRO refers to the stomach). It is a J-shaped, dilated, sac-like bulge in the alimentary canal located directly below the DIAPHRAGM to the right of the SPLEEN, partly under the LIVER. The STOMACH is generally found in the EPIGASTRIC and LEFT HYPOCHONDRIAC REGIONS of the abdomen, although its position and size vary continually due to: INDIVIDUALITY, RESPIRATORY MOVEMENTS and the DEGREE OF FULLNESS (stomach distends/stretches as swallowed food collects in it).

STOMACH Partly Sectioned at Both Orifices to Expose SPHINCTER MUSCLES, Portions of the PYLORUS, and GASTRIC RUGAE FOLDS of the MUCOSA

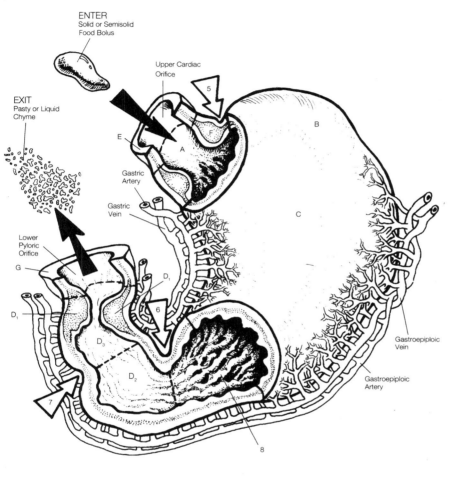

ENTER
Solid or Semisolid
Food Bolus

Upper Cardiac
Orifice

5

EXIT
Pasty or Liquid
Chyme

B

E

F

A

Gastric
Artery

Gastric
Vein

C

Lower
Pyloric
Orifice

G

D₁

D₁

6

D₃

Gastroepiploic
Vein

D₂

Gastroepiploic
Artery

7

8

Anterior View

COLORING NOTES:
A–D = oranges, reds, pinks;
E = flesh;
Arrows = cool colors;
Bolus = ochre.

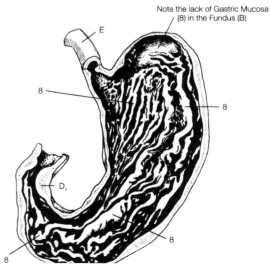

Note the lack of Gastric Mucosa
(8) in the Fundus (B)

E

8

8

8

D₁

8

FOUR BASIC STOMACH LAYERS

a | Stomach Mucosa

b | Stomach Submucosa

c | Stomach Muscularis

d | Stomach Serosa

STOMACH MUCOSA

a1 | Simple Columnar Lining Epithelium

a2 | Lamina Propria

a3 | Muscularis Mucosa

STOMACH MUSCULARIS (SMOOTH MUSCLE)

c1 | Inner Oblique Layer

c2 | Middle Circular Layer

c3 | Outer Longitudinal Layer

DIFFERENTIATIONS OF LINING EPITHELIUM

1 | Surface Epithelial Cells

G | Gastric Pit/Gland

SECRETORY CELLS AND SECRETIONS (GASTRIC JUICE)

2 | Goblet Cells

Mucus (mucin) and intrinsic factor (of the antianemic principle)

3 | Parietal/Oxyntic Cells

hydrochloric acid (HCL)

4 | Chief/Zymogenic Cells

Pepsinogen (inactive pro-enzyme): converts into active pepsin when mixed with hydrochloric acid in the gastric juice.

5 | Argentaffen Cells

Serotonin (vasoconstrictor) and histamine (pharmacologic action): increases gastric secretion, among other things.

6 | Pyloric G Cells

Gastrin (hormone): stimulates secretion of gastric acid. Also affects secretory activity of the gallbladder, pancreas, and small intestine.

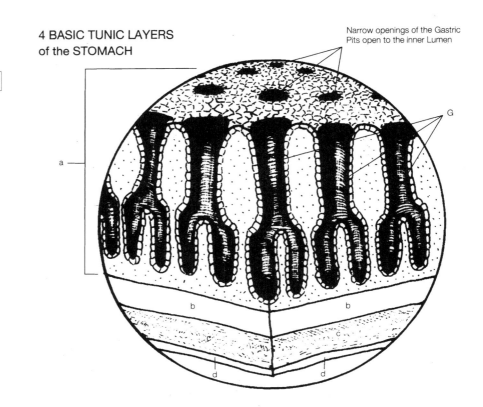

4 BASIC TUNIC LAYERS of the STOMACH

Narrow openings of the Gastric Pits open to the inner Lumen

The STOMACH wall is composed of the four basic layers as the rest of the alimentary canal (GI TRACT), but with specific structural and functional modifications.

The STOMACH MUCOSA is shaped into many large, longitudinal folds, called GASTRIC RUGAE (visible to the naked eye) when the stomach is empty. When the stomach fills, it distends and stretches out as the rugae smooth out and disappear. Microscopically, it consists of two layers of SIMPLE COLUMNAR EPITHELIUM folded downward to form numerous GASTRIC PITS (GASTRIC GLANDS) lined with an assortment of secreting cells. Secretions of the gastric glands together are called GASTRIC JUICE, a strongly acid (Ph 0.9–1.6), thin, colorless fluid containing: pepsin (initiates protein digestion), hydrochloric acid, mucin, gastric lipase, intrinsic factor, and small quantities of inorganic salts.

The STOMACH SUBMUCOSA connects mucosa to muscularis, composed of loose areolar connective tissue.

The STOMACH MUSCULARIS is composed of three smooth muscle layers, compared to the normal two layers in the other areas of the alimentary canal, with the addition of an oblique layer inside the circular layer. This triple arrangement of fibers allows a wide variety of gastric contractions to churn food, break it into smaller particles, mix it with gastric juice, and then pass it into the duodenum.

Differentiated Cells of the Lining Epithelium

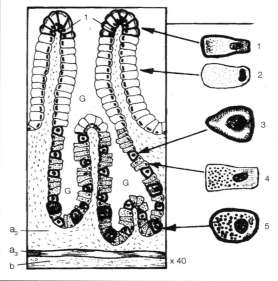

× 40

Pyloric Glands Secrete
Alkaline Mucus

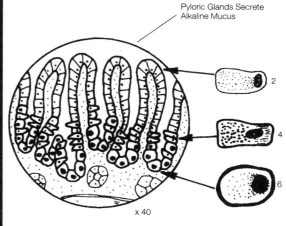

× 40

SURFACE
EPITHELIUM OF
MUCOSA

1 (of a₁)

Narrow openings
of the Gastric Pits

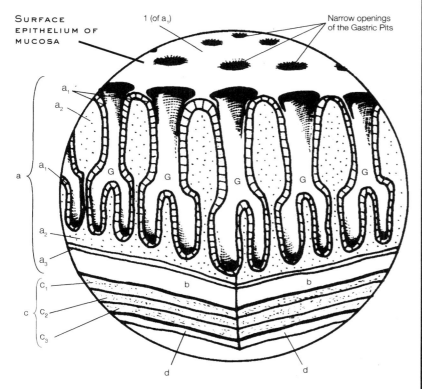

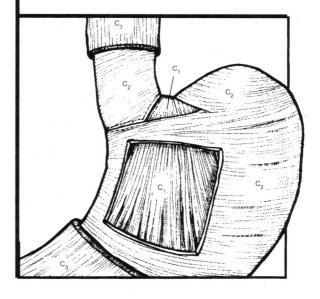

The STOMACH SEROSA is part of the visceral peritoneum. At the lesser curvature, the two layers of the visceral peritoneum come together to form the lesser omentum, which extends upward to the liver. At the greater curvature, the two layers of the visceral peritoneum come together as the greater omentum, which continues downward as an "apron" hanging over the intestines.

Remember, the chief chemical activity of the stomach is to begin the digestion of proteins, achieved primarily through the enzyme PEPSIN. Pepsin cannot digest the proteins in the chief cells that produce it because it is secreted in an inactive form called pepsinogen. It is converted into pepsin when activated by the HYDROCHLORIC ACID secreted by the parietal cells. Once pepsin has been activated by HCl, the goblet cells secrete mucus, which lines the mucosa and forms a protective barrier between the stomach lining and the acidic gastric juices to prevent self-digestion within the stomach epithelium. It has not been determined with any degree of certainly whether intrinsic factor is secreted by goblet, chief, or argentaffen cells.

THREE REGIONS OF THE SMALL INTESTINE

A | Duodenum |

B | Jejunum | (About 2/5)

C | Ileum | (About 3/5)

5 PORTIONS OF THE DUODENUM

A1 | Superior |

(Pars superior duodenal cap/bulb)

A2 | Descending | (Pars descendens)

A3 | Horizontal | (Pars inferior)

A4 | Ascending | (Pars ascendens)

A5 | Duodenojejunal Flexure |

SUPPORTIVE STRUCTURE

1 | Dorsal Mesentery |

OPENINGS INTO THE DESCENDING DUODENUM

2 | Accessory Pancreatic Duct |

3 | Hepatopancreatic Ampulla |

(Ampulla of Vater)

ELEVATION FOR ENTRY OF ACCESSORY PANCREATIC DUCT

4 | Lesser Duodenal Papilla |

ELEVATION FOR ENTRY OF HEPATOPANCREATIC AMPULLA

5 | Greater Duodenal Papilla |

ENTRY INTO MEDIAL SIDE OF THE CECUM (OF THE COLON) FROM THE TERMINAL PORTION OF THE ILEUM

6 | Ileocecal Valve (Colic Valve) |

Color Guidelines:
A = light flesh;
B = light yellow;
C = light orange;
Ducts = cool colors.

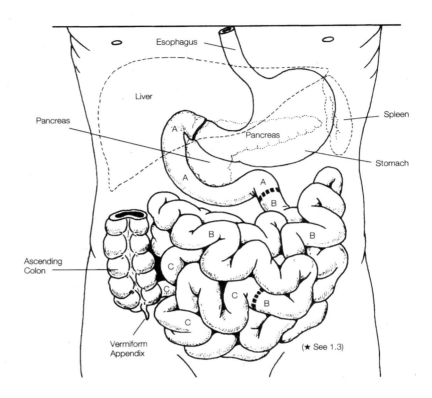

The INTESTINE is the portion of the alimentary canal extending from the PYLORUS of the STOMACH to the ANUS. It is divided into the SMALL INTESTINE and the COLON (LARGE INTESTINE).

The SMALL INTESTINE is a long, muscular tube (during life it is approximately 7 meters, or 23 feet long). It begins at the PYLORIC VALVE of the STOMACH, coils repeatedly through the central and lower parts of the abdominal cavity, and ends at the ILEOCECAL VALVE opening into the LARGE INTESTINE. The small intestine is the MAJOR SITE FOR DIGESTION and ABSORPTION of INGESTED ORGANIC FOOD SUBSTANCES. It is named so because of its small, 1 inch diameter. It is divided into three SEGMENTS, or REGIONS: the DUODENUM, the JEJUNUM, and ILEUM. The DUODENUM is the widest and most fixed region. The short first part 20–28 cm long (8–11 inches), from the PYLORIC SPHINCTER (PYLORIC VALVE) to the DUODENOJEJUNAL FLEXURE. It is retroperitoneal, except for a beginning part of the SUPERIOR PORTION near the STOMACH. It has no mesentery. A crucial section of the ALIMENTARY CANAL, it receives secretions from the LIVER, GALLBLADDER, and PANCREAS through a SYSTEM OF DUCTS opening into the DESCENDING PORTION (see pages 270–71). It mixes ACID CHYME it receives in small quantities from the STOMACH with BILE from the LIVER and GALLBLADDER, PANCREATIC JUICE from the PANCREAS, and intestinal juices secreted by BRUNNER'S GLANDS and CRYPTS OF LIEBERKUHN from the DUODENAL WALL. CHYLE is formed in the duodenum from the pasty liquid chyme formed in the stomach. The JEJUNUM extends from the duodenum to the ileum. It is the second region, 2.4 meters (8 feet) long, with a slightly larger lumen and more internal fixed folds than the ileum. It is the PRIMARY SITE OF ABSORPTION (although some occurs in the DUODENUM and ILEUM).

The ILEUM is the third and last region of the small intestine. It varies in length in the adult (9.6 to 4.7 meters) = (31½ to 15½ feet). It has an abundance of lymphatic tissue in ileum walls and joins the large intestine at the ILEOCECAL VALVE.

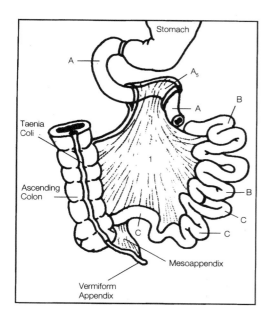

Stomach

A

A₅

A

B

Taenia Coli

Ascending Colon

1

B

C

C

Mesoappendix

Vermiform Appendix

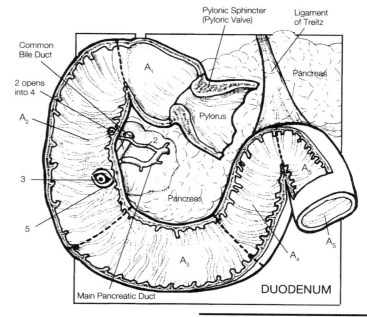

Pylonic Sphincter (Pyloric Valve)

Ligament of Treitz

Common Bile Duct

A₁

Pancreas

2 opens into 4

A₂

Pylorus

Pancreas

3

A₅

5

A₅

A₃

A₄

Main Pancreatic Duct

DUODENUM

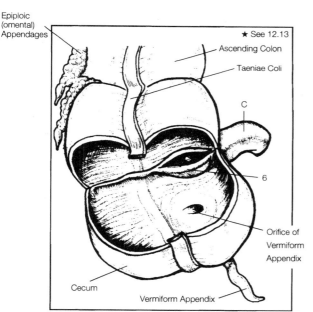

Epiploic (omental) Appendages

★ See 12.13

Ascending Colon

Taeniae Coli

C

6

Orifice of Vermiform Appendix

Cecum

Vermiform Appendix

CECUM Cut Open to Expose
ILEOCECAL VALVE

reduced by sympathetic innervation

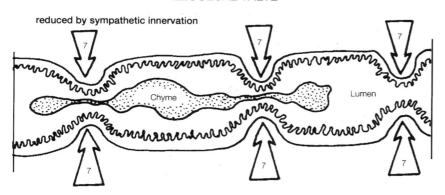

Chyme

Lumen

The SMALL INTESTINE is supported, except for the DUODENAL REGIONS, by the mesentery. Recall that the MESENTERY extends from a TIP from the posterior abdominal wall. Therefore, it functions as a FAN-SHAPED ATTACHMENT allowing MOBILITY and little chance for the coiled organ of the small intestine to become twisted. The mesentery encloses the structures that feed and supply it: BLOOD VESSELS, LYMPHATIC VESSELS, and NERVES.

The ILEOCECAL VALVE: opens and closes during digestion to allow spurts of chyme from the ileum to enter the LARGE INTESTINE. This is initiated when food enters the STOMACH by the GASTROILEAL REFLEX (controlled by the AUTONOMIC NERVOUS SYSTEM). DIGESTION and ABSORPTION of food are usually complete by the time the CHYME RESIDUE reaches the ILEOCECAL VALVE.

There are two MAJOR TYPES of CONTRACTION in the SMALL INTESTINE: PERISTALSIS and SEGMENTATION. PERISTALSIS is much weaker than in the ESOPHAGUS and STOMACH. Movement of chyme through the intestine (INTESTINAL MOTILITY) is very slow and mainly due to greater pressure at the pyloric end of the small intestine than at the distal end. SEGMENTATION is simultaneous contractions of smooth muscle at different intestinal segments, regulated by the ANS. Muscular constrictions of the LUMEN mix CHYME more thoroughly with digestive enzymes and mucus. Contraction is stimulated by parasympathetic (vagus nerve) innervation; contraction is reduced by sympathetic innervation.

FOUR BASIC SMALL INTESTINAL LAYERS

a | Small Int. Mucosa

b | Small Int. Submucosa

c | Small Int. Muscularis

d | Small Int. Serosa

FOLDS OF SMALL INTESTINE
Large, permanent, deep folds in mucosa and submucosa (seen with naked eye)

P | Plicae Circulares (Kerckring's folds)

MICROSCOPIC FINGERLIKE FOLDS OF THE MUCOSA

V | Villi

FOLDINGS OF THE APICAL PLASMA CELL MEMBRANE OF EPITHELIAL CELLS (ONLY SEEN WITH ELECTRON MICROSCOPE)

M | Microvilli

BORDER FORMED BY COLLECTIVE MICROVILLI ON EDGES OF THE COLUMNAR EPITHELIAL CELLS:

B | Brush Border

SMALL INTESTINAL MUCOSA

a1 | Simple Columnar Lining Epithelium

1 | Surface Epithelial Cells
Continuously exfoliated at tip of each villus

2 | Absorptive Cells

3 | Goblet Cells | Secrete mucus

4 | Crypts of Lieberkühn
Secrete alkaline intestinal juice, (succus entericus) containing digestive enzymes (mainly from the secretory cells of Paneth)

a2 | Lamina Propria (Core of each villus)

5 | Lymphatic Vessel/Lacteal

6 | Venule

7 | Arteriole

a3 | Muscularis Mucosa

8 | Mesenteric (Peyer's) Patches

SMALL INTESTINAL SUBMUCOSA

9 | Duodenal Brunner's Glands
Secrete alkaline mucus to protect walls from enzyme action

SMALL INTESTINAL MUSCULARIS

C1 | Inner Circular Layer

C2 | Outer Longitudinal Layer

SMALL INTESTINAL SEROSA (VISCERAL PERITONEUM)
Completely covers the small intestine, except for a major portion of the duodenum

Color Guidelines:
a = yellow;
b = light orange;
c = orange, red;
5 = light yellow;
6 = light orange;
7 = red;
8 = green;
9 = purple.

The products of digestion are absorbed at a very fast rate across the epithelial lining of the intestinal mucosa, particularly in the JEJUNUM OF THE SMALL INTESTINE. Mucosa and SUBMUCOSA are specially adapted for the rapid absorption of nutrients. This high absorptive rate is due to an increased mucosal surface area furnished by a number of folds: the PLICAE CIRCULARES, VILLI and MICROVILLI.

The PLICAE CIRCULARES are unlike the GASTRIC RUGAE FOLDS in the stomach, which change their shape and surface area continually. They are deep, large, permanent folds projecting into the intestinal lumen. They either extend all the way around the intestinal circumference or just part way. They cause the CHYME to SPIRAL as it passes through the long intestinal tube.

Four to five million villi, each 0.5–1 mm high, project into the intestinal lumen, giving the mucosa its velvety, carpet-like appearance. Each VILLUS has a core of LAMINA PROPRIA, the connective tissue layer of the MUCOSA, which is embedded with an ARTERIOLE, a VENULE, a CAPILLARY NETWORK, and a CENTRAL LACTEAL (LYMPHATIC VESSEL). Digested PROTEINS (AMINO ACIDS) and CARBOHYDRATES (MONOSACCHARIDES), and small aggregations (chylomicrons) of FAT (FATTY ACIDS and GLYCEROL) are absorbed into BLOOD CAPILLARIES. Larger fat chylomicrons enter the LACTEAL and mix with the LYMPH, forming a milky, alkaline CHYLE. At the BASE of each VILLUS are vertical pouches or pits lined with glandular epithelium that open through pores to the intestinal lumen. These pits are called CRYPTS OF LIEBERKUHN or the INTESTINAL

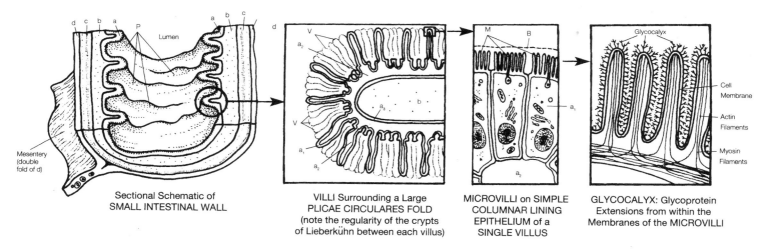

Sectional Schematic of SMALL INTESTINAL WALL

Mesentery (double fold of d)
Lumen

VILLI Surrounding a Large PLICAE CIRCULARES FOLD
(note the regularity of the crypts of Lieberkühn between each villus)

MICROVILLI on SIMPLE COLUMNAR LINING EPITHELIUM of a SINGLE VILLUS

GLYCOCALYX: Glycoprotein Extensions from within the Membranes of the MICROVILLI

Glycocalyx
Cell Membrane
Actin Filaments
Myosin Filaments

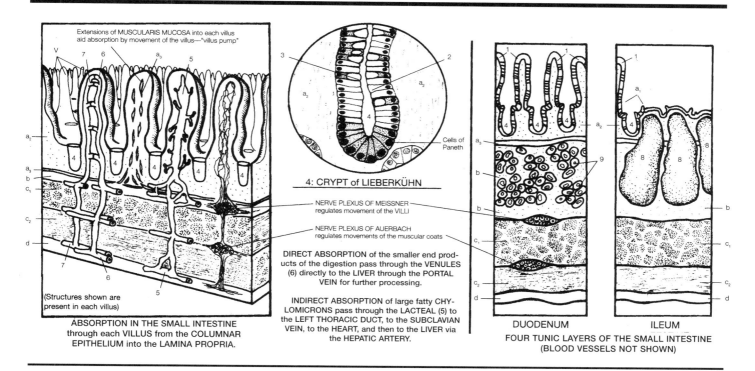

Extensions of MUSCULARIS MUCOSA into each villus aid absorption by movement of the villus—"villus pump"

(Structures shown are present in each villus)

ABSORPTION IN THE SMALL INTESTINE through each VILLUS from the COLUMNAR EPITHELIUM into the LAMINA PROPRIA.

4: CRYPT of LIEBERKÜHN

Cells of Paneth

NERVE PLEXUS OF MEISSNER regulates movement of the VILLI

NERVE PLEXUS OF AUERBACH regulates movements of the muscular coats

DIRECT ABSORPTION of the smaller end products of the digestion pass through the VENULES (6) directly to the LIVER through the PORTAL VEIN for further processing.

INDIRECT ABSORPTION of large fatty CHYLOMICRONS pass through the LACTEAL (5) to the LEFT THORACIC DUCT, to the SUBCLAVIAN VEIN, to the HEART, and then to the LIVER via the HEPATIC ARTERY.

DUODENUM   **ILEUM**

FOUR TUNIC LAYERS OF THE SMALL INTESTINE (BLOOD VESSELS NOT SHOWN)

GLANDS where the intestinal juice (succus entericus) is secreted. New epithelial cells are formed here and are pushed up to the top of each VILLUS as the surface epithelial cells are being continually sloughed off. CELLS OF PANETH deep in the crypts secrete intestinal digestive enzymes. Absorption of most digested food occurs through the BORDER EPITHELIUM covering the VILLI.

The cell membranes of each lining epithelial cell have folded projections of their own called MICROVILLI (collectively, they are called the BRUSH-BORDER, open to the intestinal lumen). They provide an extremely large surface area for absorption. BRUSH-BORDER ENZYMES that stay attached to the cell membrane further hydrolyze CHYME to increase efficiency of rapid absorption. The brush border contain permanently attached enzymes which hydrolyze DISACCHARIDES, POLYPEPTIDES, and other substrates in preparation for absorption into the columnar epithelium.

The SMALL INTESTINAL SEROSA (VISCERAL PERITONEUM) completely covers the small intestine, except for a major portion of the duodenum.

The NERVE PLEXUS OF MEISSNER regulates movement of the villi while the NERVE PLEXUS OF AUERBACH regulates movements of the muscular coats. Direct absorption of the smaller end products of digestion pass through the venules directly to the liver through the portal vein for further processing. Indirect absorption of large, fatty CHYLOMICRONS pass through the lacteal to the left thoracic duct, to the subclavian vein, to the heart, and then to the liver via the hepatic artery.

FOUR REGIONS OF THE LARGE INTESTINE

A | Cecum |
Hangs below ileocecal valve open to the ascending colon

B | Colon | (Major region)

C | Rectum | (Terminal 7½ in. of GI tract)

D | Anal Canal | (Terminal inch of rectum)

THE CECUM BLIND POUCH

1 | Ileocecal Valve |

2 | Vermiform Appendix |

THE COLON: FOUR REGIONS

B1 | Ascending Colon |

B2 | Transverse Colon |

B3 | Descending Colon |

B4 | Sigmoid (Pelvic) Colon |

THE RECTUM

3 | Rectal Transverse Folds |

4 | Pelvic Portion |

5 | Rectal Ampulla |

6 | Anal Columns |

THE ANAL CANAL

7 | Anus |

8 | Internal Anal Sphincter |

9 | External Anal Sphincter |

THE FOUR BASIC TUNIC LAYERS

a | Large Int. Mucosa |

b | Large Int. Submucosa |

c | Large Int. Muscularis |

d | Large Intestinal Serosa |

LARGE INTESTINAL MUCOSA

a1 | Lining epithelium with absorptive cells and goblet cells |

1 **0** | Crypts of Lieberkuhn |

a2 | Lamina Propria |

a3 | Muscularis Mucosa |

LARGE INTESTINAL MUSCULARIS

c1 | Inner Circular Layer |

c2 | Outer Longitudinal Layer |
Taeniae coli

ADAPTATIONS OF THE TUNICAS OF THE LARGE INTESTINE

11 | Haustra | (haustrations)

12 | Taeniae Coli |
3 middle flat sheets from longitudinal muscularis (outer layer)

13 | Epiploic Appendages |
Fat-filled pouches attached to taeniae coli in serous layer

14 | Plicae Semilunaris |
Transverse folds of mucosa lying between haustra

COLOR GUIDELINES:
a = yellow ;
b = light orange;
c = orange, red;
d = gray, neutral;
12 = yellow orange;
A–D = cool colors.

RECTAL AREA (c)

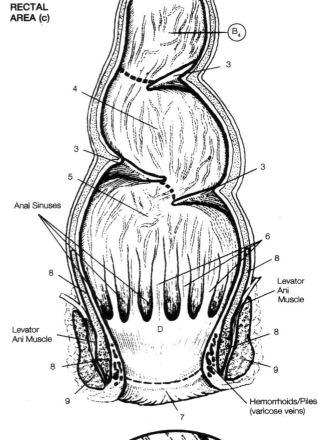

Anal Sinuses

Levator Ani Muscle

Levator Ani Muscle

Hemorrhoids/Piles (varicose veins)

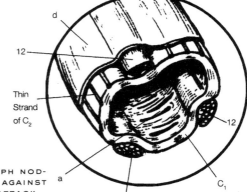

Thin Strand of C_2

SOLITARY LYMPH NODULES DEFEND AGAINST BACTERIAL ATTACK

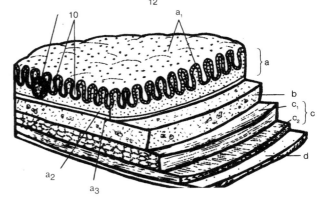

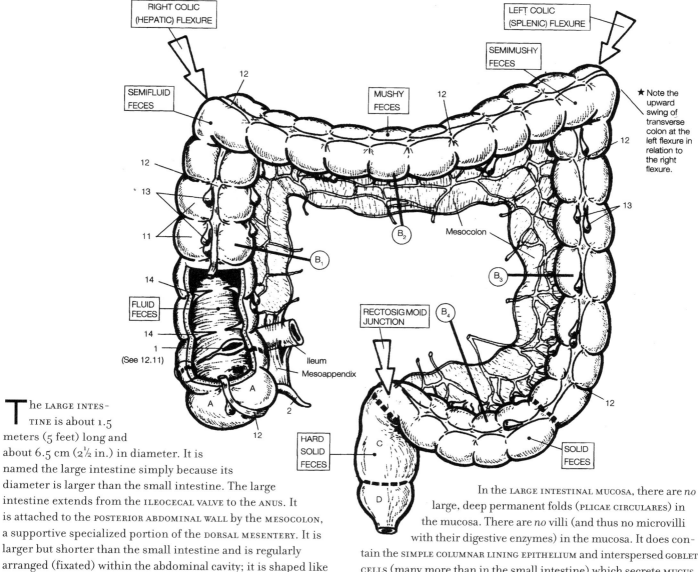

RIGHT COLIC
(HEPATIC) FLEXURE

LEFT COLIC
(SPLENIC) FLEXURE

SEMIMUSHY
FECES

SEMIFLUID
FECES

MUSHY
FECES

★ Note the upward swing of transverse colon at the left flexure in relation to the right flexure.

12

13

11

Mesocolon

B₁

B₂

B₃

14

FLUID
FECES

14

1

(See 12.11)

B₄

RECTOSIGMOID
JUNCTION

Ileum

Mesoappendix

12

A

A

A

12

2

HARD
SOLID
FECES

C

D

SOLID
FECES

The LARGE INTES-
TINE is about 1.5 meters (5 feet) long and about 6.5 cm (2½ in.) in diameter. It is named the large intestine simply because its diameter is larger than the small intestine. The large intestine extends from the ILEOCECAL VALVE to the ANUS. It is attached to the POSTERIOR ABDOMINAL WALL by the MESOCOLON, a supportive specialized portion of the DORSAL MESENTERY. It is larger but shorter than the small intestine and is regularly arranged (fixated) within the abdominal cavity; it is shaped like an upside-down *u*. There is little or no digestive function for the large intestine. NO DIGESTIVE ENZYMES are secreted in the COLON. (Remember, by the time the intestinal contents reach the large intestine, DIGESTION and ABSORPTION are almost totally complete.)

What little chemical digestion does occur is by BACTERIAL ACTION, not ENZYMES, in the LAST STAGES of CHYMAL BREAKDOWN. There are also some VITAMINS synthesized by BACTERIA (B and K). There is a great deal of WATER absorbed in the COLON rather than in the SMALL INTESTINE. Thus, there is an efficient conservation of body fluids during most of digestion and absorption.) Timewise, the RESIDUE OF DIGESTION enters the COLON for the longest part of the digestive journey.

With the absorption of WATER and ELECTROLYTES from the CHYME into the BLOODSTREAM, the contents of the colon are gradually dehydrated and assume the consistency of semisolid or very hard FECES. The large intestine also functions TO FORM AND STORE FECES, and to expel them from the body via the rectum and muscular anal canal, regulated by the anus.

The LARGE INTESTINE contains the same four basic tunicas as the SMALL INTESTINE, although there are many structural differences.

In the LARGE INTESTINAL MUCOSA, there are *no* large, deep permanent folds (PLICAE CIRCULARES) in the mucosa. There are *no* villi (and thus no microvilli with their digestive enzymes) in the mucosa. It does contain the SIMPLE COLUMNAR LINING EPITHELIUM and interspersed GOBLET CELLS (many more than in the small intestine) which secrete MUCUS, which lubricates the contents in the colon as they pass through. It contains ABSORPTIVE CELLS that absorb WATER and ELECTROLYTE SALTS into the BLOOD STREAM (which are then carried by the PORTAL CIRCULATION to the LIVER before they enter into general circulation). Body fluids are conserved and feces dried. CRYPTS OF LIEBERKUHN do not contain secretory PANETH CELLS in the bottom of their pits in the colon, as there is little need for secretory digestive enzymes.

The LARGE INTESTINAL SUBMUCOSA is similar to the rest of the GI tract with loose fibrous tissue with BLOOD VESSELS, LYMPH VESSELS and NERVES. The LARGE INTESTINAL MUSCULARIS is the outer longitudinal incomplete layer; it does not form a continuous sheet around the wall as it does in the rest of the alimentary canal. This smooth muscle layer is broken up into three flat bands or sheets, which run most of the length of the large intestine, placed equidistantly around the circumference of the organ. Since these TAENIAE COLI are not as long as the gut, their contractions gather, the colon into a series of sacculated pouches called HAUSTRA (HAUSTRATIONS). These pouches and a large diameter visually distinguish the LARGE INTESTINE from the SMALL INTESTINE. The LARGE INTESTINAL SEROSA is part of the VISCERAL PERITONEUM.

G   GALLBLADDER

REGIONS OF THE GALLBLADDER

a   FUNDUS

b   BODY

c   Neck

d   Hartmann's Pouch

Necklike pouch; frequent reservoir for lodging of gallstones

SYSTEM OF ACCESSORY ORGAN DUCTS: EXCRETORY DUCT FROM THE GALLBLADDER

1   Cystic Duct

CYSTIC DUCT ENTERS EXCRETORY BILE DUCT FROM THE LIVER

2   Common Hepatic Duct

UNION OF CYSTIC DUCT AND COMMON HEPATIC DUCT

3   Common Bile Duct

Common ampulla where the common bile duct and main pancreatic duct meet

COMMON AMPULLA WHERE THE COMMON BILE DUCT AND THE MAIN PANCREATIC DUCT MEET

4   Ampulla of Vater

ELEVATED AREA IN DESCENDING PORTION OF THE DUODENUM FOR EXIT OF AMPULLA

5   Greater Duodenal Papillae

CONTROLLING SMOOTH MUSCLE AT TERMINAL END OF THE AMPULLA FOR FLUID EXIT:

6   Sphincter of Oddi

P   Pancreas

REGIONS OF THE PANCREAS

e   Head

f   Body

g   Tail

EXOCRINE PORTION OF THE PANCREAS
("external" digestive secretions travel via ducts)

7   Main Pancreatic Duct

Duct of wirsung

8   Accessory Pancreatic Duct

Duct of santorini

9   Alveolar Serous Acini Cells

Secrete pancreatic juice

ENDOCRINE PORTION OF THE PANCREAS
Internal hormonal secretions travel via capillaries to blood stream

10   Islets of Langerhans

α   Alpha Cells

Secrete glucagon hormone

β   Beta Cells

Secrete insulin hormone

Δ   Delta Cells

Secrete HGHIF (human growth hormone inhibiting factor)

The GALLBLADDER, PANCREAS, and LIVER are the three accessory digestive organs in the abdominal cavity that aid in the CHEMICAL BREAKDOWN OF FOOD. The LIVER and a portion of the PANCREAS function as EXOCRINE GLANDS, essential to the digestive process in that their secretions are transported to the GI tract (DUODENUM) via a SYSTEM OF DUCTS.

The GALLBLADDER is a pear-shaped, saclike organ located in a fossa of the visceral (or inferior) surface of the LIVER (RIGHT LOBE). It is 3–4 in. long and contains RUGAE (similar to the stomach) that allow it to expand to its pear shape when filled with bile (up to 2 ounces). IT FUNCTIONS AS A STORAGE SITE FOR BILE. BILE is continuously manufactured by the LIVER. The BILE leaving the LIVER enters the GALLBLADDER via drainage of the right and left. HEPATIC DUCTS, COMMON HEPATIC DUCT, and CYSTIC DUCT. When BILE is needed (as food enters the GI tract) the middle muscularis layer of the GALLBLADDER contracts, ejecting the BILE down the DUCT SYSTEM to the DUODENUM, as ODDI's SPHINCTER opens. When the small intestine is empty of food, the SPHINCTER OF ODDI at the DUODENUM closes, and the BILE is forced back up the duct system into the GALLBLADDER for STORAGE. The gallbladder also CONCENTRATES bile through the loss of fluids absorbed by the GALLBLADDER MUCOSA. The capacity of viscid concentrated bile is 50–75 ml., equivalent to 1½ pints (710 ml.) of LIVER BILE. Among the functions of BILE is included the *emulsification* of undigested FAT GLOBULES into finer fat droplets in the DUODENUM.

The PANCREAS is a soft, lobulated, spongy, glandular organ about 6 in. long and 1 in. thick. It lies posterior to the great curvature of the DUODENAL CURVE, attached to the DUODENUM by the MAIN PANCREATIC DUCT (and sometimes an ACCESSORY PANCREATIC DUCT). It has both EXOCRINE and ENDOCRINE functions. The endocrine function is performed by ISLETS OF LANGERHANS, clusters of cells (alpha, beta, delta cells), that secrete hormones directly into the blood stream (regulates blood sugar level). Within the lobules around the ISLETS are numerous excretory secretory units called ACINI. Each ACINUS consists of a SINGLE LAYER of EPITHELIAL CELLS that empty PANCREATIC JUICE into a LUMEN DUCT, which carries the juice to the PANCREATIC DUCTS, and then to the DUODENUM (COMPLETE DIGESTION of food molecules requires action of BOTH PANCREATIC ENZYMES and BRUSH BORDER ENZYMES). PANCREATIC JUICE contains WATER, PROTEIN, BICARBONATES and mostly inactive DIGESTIVE ENZYMES which are triggered into action when released into the DUODENUM. The THREE IMPORTANT ENZYMES are TRYPSIN (digests PROTEIN PEPTONES into AMINO ACIDS), AMYLASE (digests STARCH into MALTOSE and short chains of glucose molecules) and LIPASE (digests TRIGLYCERIDES cleaving FATTY ACIDS from GLYCEROL).

Secretion of PANCREATIC JUICE into the DUODENUM begins when the acid content of the STOMACH passes through the STOMACH PYLORUS to the DUODENUM, stimulating secretion of two hormones—SECRETIN and CHOLECYSTOKININ from the DUODENAL MUCOSA. Alkaline secretions from the PANCREATIC JUICE neutralizes the acidity of the CHYME entering the DUODENUM from the STOMACH and stops the action of PEPSIN on PROTEINS in the duodenum.

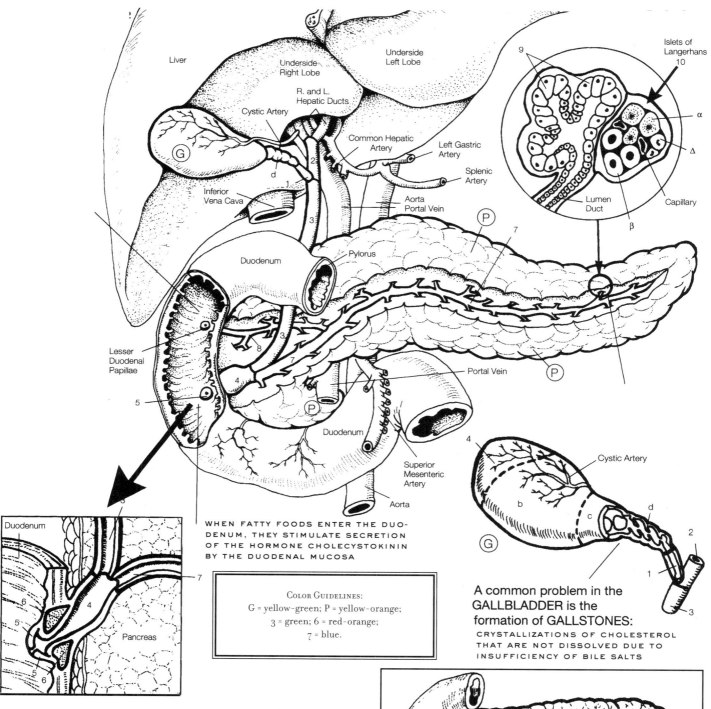

Liver

Underside Right Lobe

Underside Left Lobe

R. and L. Hepatic Ducts

Cystic Artery

Common Hepatic Artery

Left Gastric Artery

Splenic Artery

G

d

Inferior Vena Cava

Aorta Portal Vein

1

2

3

Islets of Langerhans 10

9

α

Δ

Lumen Duct

Capillary

β

Duodenum

Pylorus

P

7

Lesser Duodenal Papillae

8

3

7

Portal Vein

P

4

P

5

Duodenum

Superior Mesenteric Artery

Aorta

Duodenum

7

6

4

5

5

6

Pancreas

WHEN FATTY FOODS ENTER THE DUO-
DENUM, THEY STIMULATE SECRETION
OF THE HORMONE CHOLECYSTOKININ
BY THE DUODENAL MUCOSA

COLOR GUIDELINES:
G = yellow-green; P = yellow-orange;
3 = green; 6 = red-orange;
7 = blue.

Cystic Artery

b

c

d

1

2

G

3

A common problem in the
GALLBLADDER is the
formation of GALLSTONES:

CRYSTALLIZATIONS OF CHOLESTEROL
THAT ARE NOT DISSOLVED DUE TO
INSUFFICIENCY OF BILE SALTS

Most pancreatic enzymes of the PANCREATIC JUICE (excluding PANCRE-
ATIC AMYLASE and PANCREATIC LIPASE) are inactive, called ZYMOGENS, and
are activated in the small intestine by "brush border enzymes."

ENTEROKINASE, a brush border enzyme permanently located in the
cell membrane of small intestine microvilli, activates inactive TRYPSINO-
GEN into TRYPSIN. Trypsin in turn activates the other zymogens of pancreatic juice (CYHMOTRYPSINOGEN converts to CHYMOTRYPSIN, PROELASTASE
converts to ELASTASE and PROPHOSPHOLIPASE converts to PHOSPHOLIPASE).

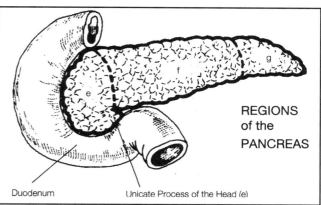

REGIONS
of the
PANCREAS

f

g

e

Duodenum

Unicate Process of the Head (e)

L Liver

AREA NOT INVESTED IN PERITONEUM

BA Bare Area

DIVIDED INTO FOUR LOBES:

TWO MAJOR LOBES

1 Right Lobe

2 Left Lobe

TWO MINOR LOBES ASSOCIATED WITH RIGHT LOBE

3 Quadrate Lobe

Adjacent to gallbladder

4 Caudate Lobe

FIVE CONNECTING LIGAMENTS

5 Falciform Ligament

6 Round Ligament/Ligamentum Teres

Degenerated product of umbilical vein of fetus

7 Coronary Ligament

Becomes parietal peritoneum of diaphragm

8 R. and L. Triangular Ligaments

OTHER RELATED STRUCTURES

9 Lesser Omentum

10 Ligamentum Venosum

Obliterated ductus venosus of the fetus

G Gallbladder and Cystic Duct

IVC Inferior Vena Cava

A Aorta

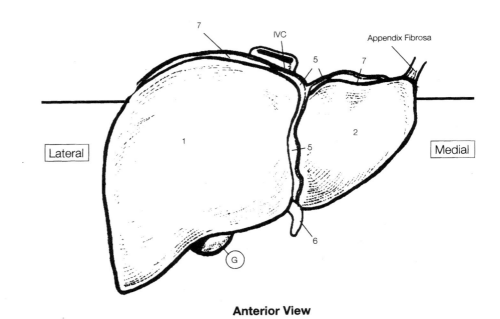

Anterior View

The LIVER is the largest single organ in the body, weighing from 3–4 lbs. in the adult. It is a highly complex organ that we cannot live without. The LIVER is the site of a wide variety of SYNTHETIC, STORAGE, and EXCRETORY FUNCTIONS. It is reddish brown in color, owing to its high vascularity. The LIVER is located directly beneath the DIAPHRAGM on the right side of the body occupying the RIGHT HYPOCHONDRIAC REGION of the ABDOMINAL CAVITY, the upper portion of the EPIGASTRIC REGION, and sometimes extending into the LEFT HYPOCHONDRIAC REGION. The liver is level with the bottom of the STERNUM. Its undersurface is concave. It covers the STOMACH (portions), DUODENUM, RIGHT COLIC (HEPATIC) FLEXURE OF THE COLON, RIGHT KIDNEY, and SUPRARENAL (ADRENAL) CAPSULE. The liver is almost completely covered by the visceral peritoneum (except for a BARE AREA, mainly in the posterior portion of the RIGHT LOBE). It is *completely* covered by a tough, fibrous connective tissue layer, GLISSON'S CAPSULE (not shown), that lies beneath the PERITONEUM. This capsule is thickest at the transverse fissure (hilum) where the blood vessels and hepatic duct enter the organ at the PORTA.

The liver is divided into four lobes, a large RIGHT LOBE, a smaller left lobe, and two very small lobes: QUADRATE and CAUDATE. The LIVER is connected by five ligaments to the undersurface of the DIAPHRAGM, the ANTERIOR ABDOMINAL WALL, and falciform ligament (which divides the RIGHT and LEFT LOBE and attaches to the diaphragm). The ROUND LIGAMENT is contained partially within falciform ligament and extends to the UMBILICUS (navel). The coronary ligament extends over the medial surface of the RIGHT LOBE from the posterior edge of the LIVER and attaches to the DIAPHRAGM). The RIGHT and LEFT TRIANGULAR LIGAMENTS (which lie on the lateral aspects of the RIGHT and LEFT lobes, and pass to the DIAPHRAGM).

In other related liver structure, the LESSER OMENTUM passes from the lesser curvature of the STOMACH to the TRANSVERSE FISSURE of the liver. The LIGAMENTUM VENOSUM lies between the CAUDATE and LEFT LOBES and attaches the left branch of the PORTAL VEIN to the INFERIOR VENA CAVA.

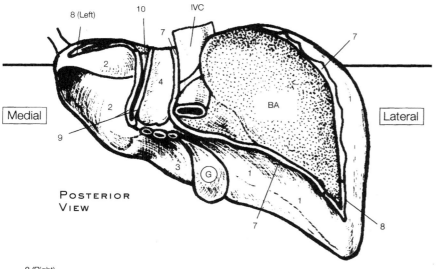

POSTERIOR
VIEW

Medial

Lateral

8 (Left) · 10 · 7 · IVC · 7 · BA · 1 · 2 · 4 · 2 · 9 · 3 · G · 1 · 1 · 7 · 8

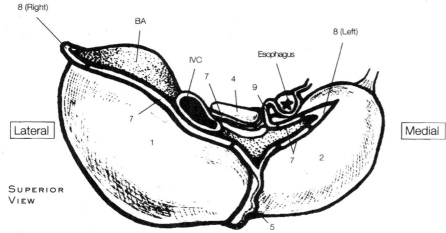

SUPERIOR
VIEW

Lateral

Medial

8 (Right) · BA · IVC · Esophagus · 8 (Left) · 7 · 4 · 9 · 7 · 7 · 1 · 2 · 5

Inferior View Exposes the 2 MINOR LOBES, the HILUS ...here the Blood Vessels and HEPATIC DUCT Enter (and the underlying VISCERAL IMPRESSIONS)

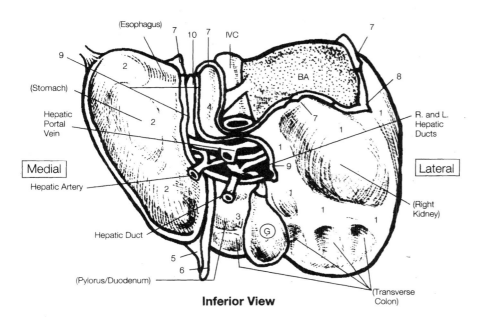

Inferior View

(Esophagus) · 7 · 10 · 7 · IVC · 7
9
(Stomach)
Hepatic Portal Vein
2 · 2 · BA · 8
4 · 7 · 1
R. and L. Hepatic Ducts
Medial · 9 · 1 · Lateral
Hepatic Artery · 2 · 1 · (Right Kidney)
3 · 1
Hepatic Duct · G
5 · 6
(Pylorus/Duodenum)
(Transverse Colon)

HEART

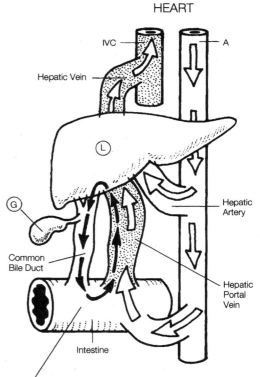

IVC · A
Hepatic Vein
L
Hepatic Artery
G
Common Bile Duct
Hepatic Portal Vein
Intestine

| ENTEROHEPATIC CIRCULATION allows certain compounds excreted in the BILE to return to the LIVER via intestinal absorption and hepatic portal vein. | ENTEROHEPATIC CIRCULATION of UROGLOBIN

Bile pigment (BILIRUBIN) is converted in the intestine into UROBILINOGEN. Some is excreted in the FECES, some is recycled through the BILE, or filtered by the KIDNEYS into the URINE. |
|---|---|

Although the liver is divided into four lobes, the internal structure and overall function of all lobes is the same. The microarchitecture of the liver is very complex, and structure and function are closely related. The structural and functional unit of the liver is the LIVER LOBULE is generally cylindrical, polyhedral units about 2 mm in diameter, with five to seven sides per lobule with strands of connective tissue originating from the surrounding capsule (glisson's capsule) of the entire liver enter the liver parenchyma at the hilus and form the supporting network of the liver. They separate the lobules from one another. These channels of interlobular connective tissue carry a constant trinity of vessels as a branch of the HEPATIC PORTAL VEIN, a branch of the HEPATIC ARTERY and a BILE DUCT (sometimes also lymphatic vessels).

Each lobule has at its center a blood vessel called the central vein. Surrounding the central vein are hepatocyte plates, or cords of hepatic cells, arranged in a more or less radial fashion, which interconnect with one another crosswise. These plates are only 1 to 2 cell-layers thick, allowing direct contact of each hepatocyte cell with the blood. Located at the periphery of each lobule (in general, equally spaced around it) are the six to eight hepatic triads of vessels, which open into sinusoidal spaces between the HEPATOCYTE PLATES.

The arterial blood, and portal venous blood (which contains the absorbed end products of digestion) mix as the blood flows from the periphery of the lobule to the central vein. The central veins of all the lobules converge to form the LARGE HEPATIC VEIN, which drains all blood from the liver. Partly lining the large sinusoidal spaces are fixed or immobile phagocytic kupffer cells which help to remove old red blood cells and bacteria from the blood. There are large intercellular gaps between the KUPFFER CELLS, which permit extreme high permeability of the hepatocyte cells, more so than any other capillaries.

Within each hepatocyte plate are thin channels called bile canaliculi. The hepatocyte cells produce bile and secrete it into these channels, which are drained at the periphery of the lobule by the bile ducts. The bile ducts drain into larger right and left hepatic ducts that carry bile away from the liver (to be stored in the gallbladder, through the cystic duct).

The liver's wide variety of functions are the PRODUCTION OF BILE, CARBOHYDRATE METABOLISM, LIPID METABOLISM, PROTEIN SYNTHESIS, DETOXIFICATION, STORAGE and other activities.

The flow of blood and bile within each liver lobule occurs so that the blood and bile never mix in the liver lobules, the blood and bile travel in opposite directions, the blood flows between hepatocyte plates and the bile flows within hepatocyte plates. Blood flow is from the periphery to the center of the lobule, mixed blood from branches of hepatic artery and portal vein, through the sinusoidal spaces to the central vein, to branches of the hepatic vein, and then to the hepatic vein and the inferior vena cava for general circulation.

The bile flow is from the center of the lobule to the periphery. It is secreted within the hepatocyte plates through thin channels (bile canaliculi) to the bile ducts to the hepatic ducts, in and out of the gallbladder via cystic duct, into the common bile duct to duodenum and the intestines.

In the hepatic triad (trinity) process, "processed" blood—80% "blue" deoxygenated blood from the HEPATIC PORTAL VEIN and 20% "red" oxygenated blood from the HEPATIC ARTERY—drains from each lobule through branches of the hepatic vein to the large hepatic vein.

STRUCTURAL UNITS OF THE LIVER

LL | Liver Lobules

STRUCTURES WITHIN EACH LOBULE

1 | Central Vein of Lobule

2 | Sinusoidal Spaces (sinusoids)

3 | Kupffer Cells (reticuloendothelial lining of sinusoids)

4 | Hepatocyte Plates

5 | Bile Canaliculi

SIX TO EIGHT TRIADS OF VESSELS SURROUNDING EACH LOBULE: EACH HEPATIC TRIAD (HEPATIC TRINITY)

a | Branch of Hepatic Portal Vein

b | Branch of Hepatic Artery

c | Bile Duct

MAIN VESSEL DRAINING ALL LOBULES:

6 | Hepatic Vein (branches)

CONNECTIVE TISSUE SEPARATING LOBULES FROM ONE ANOTHER

7 | Inner Extended Strands of Glisson's Capsule

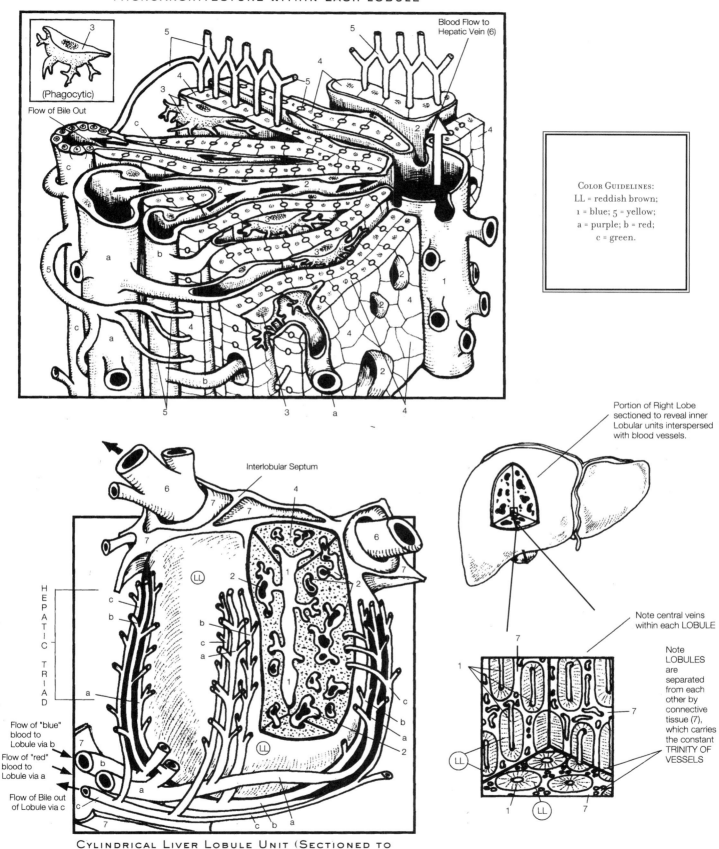

Blood Flow to Hepatic Vein (6)

(Phagocytic)

Flow of Bile Out

COLOR GUIDELINES:
LL = reddish brown;
1 = blue; 5 = yellow;
a = purple; b = red;
c = green.

Portion of Right Lobe sectioned to reveal inner Lobular units interspersed with blood vessels.

Note central veins within each LOBULE

Note LOBULES are separated from each other by connective tissue (7), which carries the constant TRINITY OF VESSELS

Interlobular Septum

HEPATIC TRIAD

Flow of "blue" blood to Lobule via b

Flow of "red" blood to Lobule via a

Flow of Bile out of Lobule via c

CYLINDRICAL LIVER LOBULE UNIT (SECTIONED TO REVEAL CENTRAL VEIN AND SINUSOIDS)

Chapter 15: Urinary System

System Components

The organs of urine production, collection and elimination

System functions

Regulation of blood chemical composition

Elimination of wastes

Regulation of fluid/electrolyte balance and volume

Maintenance of body acid-base balance

Cellular metabolism of nutrients produces a wide variety of by-products, many of which can threaten BODY HOME-OSTASIS. These include CARBON DIOXIDE, excess water and heat, and AMMONIA and UREA, which are toxic nitrogenous wastes of protein catabolism. In addition, many essential ions and other inorganic substances (and water) are ingested through foods and fluids in greater amounts than the body requires. All of the wastes and excess essential materials must be eliminated from the body in order to maintain the body's required equilibrium and health. All of these substances are transported in the blood en route to the filtering station of the KIDNEYS, the main functioning organs of the URINARY SYSTEM.

The primary function of the URINARY SYSTEM is to preserve the homeostasis (dynamic equilibrium) of the body fluids through action of the kidneys by controlling the composition and volume of blood. This is achieved by regulating (removing and restoring) the blood concentrations of water and solutes (solid substances that have been dissolved in a solution) and excreting selected amounts of various solute wastes of metabolism (in the form of urine).

The kidneys are also responsible for maintaining a constant metabolic acid-base balance by regulating the blood levels and bicarbonates.

The KIDNEYS primarily eliminate water, soluble salts from protein catabolism, nitrogenous substances (urea, uric acid, creatine and creatinine) and inorganic minerals and salts; and secondarily eliminate carbon dioxide and heat. The DIGESTIVE ORGANS (GI TRACT) primarily eliminate solid wastes (indigestible residue), secretions and bacteria, and secondarily eliminate carbon dioxide, heat, water and salts. The RESPIRATORY ORGANS (LUNGS) primarily eliminate carbon dioxide and secondarily eliminate water vapor, heat and other gases. The INTEGUMENTARY (SKIN) SYSTEM primarily eliminates heat and secondarily eliminates water, salts and minute quantities of urea (stimulated by kidney inactivity). The urinary system serves more functions than simply forming and discharging urine from the body.

The kidneys are named for their resemblance, in shape, to kidney beans. Due to the size of the liver, the RIGHT KIDNEY is slightly lower than the LEFT. The left and right kidneys are supplied by the LEFT and RIGHT RENAL ARTERIES, respectively (branches of the ABDOMINAL AORTA) and are drained by the LEFT and RIGHT RENAL VEINS (tributaries of the INFERIOR VENA CAVA). The kidneys remove waste from the blood through the filtration and excretion of urine from them, which is then transported passively to the URETERS. By PERISTALSIS and GRAVITATIONAL FORCES, the urine is carried down the ureters to the URINARY BLADDER. Urine is stored in the urinary bladder until it exceeds 200 to 400 ml., and is then expelled (by conscious desire or an unconscious reflex: MICTURITION) through the canal of the URETHRA and discharged from the body.

The kidneys do not only act as organs of excretion of undesired products of metabolism. They also act as organs of conservation of metabolic products; regulators of blood chemical composition, of fluid/electrolyte balance and volume and of body acid-base balance; control blood pressure via the rennin-angiotensin mechanism; and cause release of erythroprotein for red blood cell production. This is all achieved in a state of continuous dynamic equilibrium, called HOMEOSTASIS, where the internal environment of the body is maintained by dynamic processes of feedback and regulation. What is excreted in one instant may be deemed desirable the next, depending on the immediate balance required.

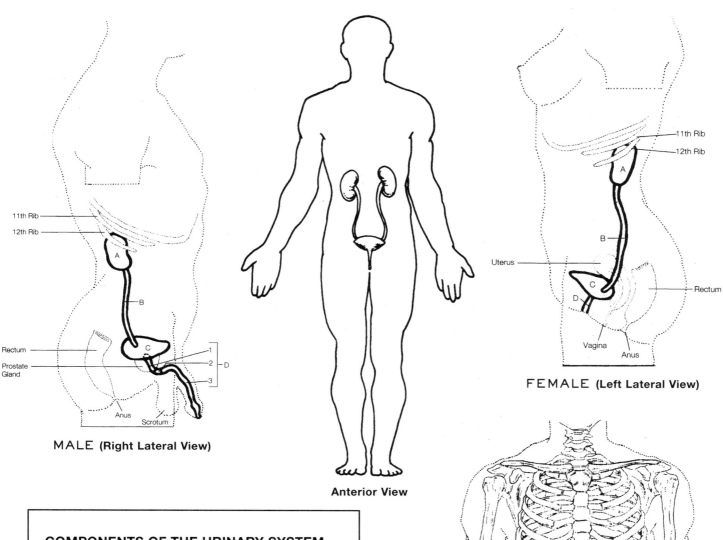

11th Rib
12th Rib

A

B

Rectum

Prostate
Gland

C

1
2 — D
3

Anus

Scrotum

MALE (Right Lateral View)

Anterior View

11th Rib
12th Rib

A

B

Uterus

C

D

Rectum

Vagina   Anus

FEMALE (Left Lateral View)

COMPONENTS OF THE URINARY SYSTEM

A   2 Kidneys

B   2 Ureters

C   1 Urinary bladder

D   1 Urethra

DIVISIONS OF THE URETHRA (MALE)

1   Prostatic urethra

2   Membranous urethra

3   Spongy urethra

COLORING NOTES:
Outer cortex of kidney = deep reddish brown;
Inner medulla of kidney = light reddish brown;
Arterial system = warm colors; Venous system = cool colors.

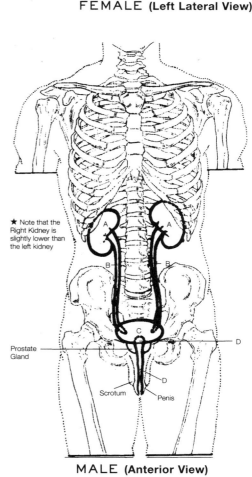

★ Note that the
Right Kidney is
slightly lower than
the left kidney

A   A

B   B

Prostate
Gland

C   D

Scrotum   D

Penis

MALE (Anterior View)

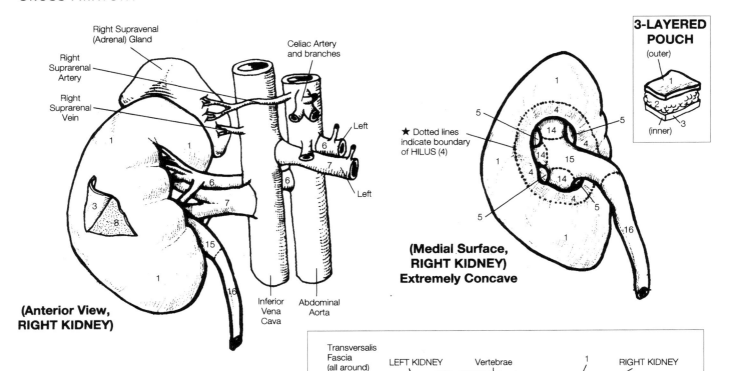

Right Supravenal (Adrenal) Gland

Right Suprarenal Artery

Right Suprarenal Vein

Celiac Artery and branches

Left

Left

1

1

6

6

6

7

3

8

15

1

16

(Anterior View, RIGHT KIDNEY)

Inferior Vena Cava

Abdominal Aorta

3-LAYERED POUCH
(outer)

1

2

3

(inner)

★ Dotted lines indicate boundary of HILUS (4)

1

5

4

14

5

4

1

14

15

4

14

5

4

5

1

16

(Medial Surface, RIGHT KIDNEY) Extremely Concave

Transversalis Fascia (all around)

LEFT KIDNEY

Vertebrae

1

RIGHT KIDNEY

2

3

2

3

1

1

1

Peritoneum

Abdominal Aorta

Each kidney's OUTER PERIPHERAL PART (PARENCHYMA or CORTEX and RENAL PYRAMIDS) contains the PROCESSING CONDUCTORS and the RENAL TUBULES. The renal tubules consist of the NEPHRONS (RENAL CORPUSCLES AND ATTACHED URINARY TUBULES), their VASCULAR COMPONENTS AND COLLECTING TUBULES. The CORTEX (which is reddish brown and granular) extends from the fibrous capsule to the outer bases of the PYRAMIDS (and into the spaces between them). It contains the RENAL CORPUSCLES, the CONVOLUTED PORTIONS (PROXIMAL and DISTAL) of the URINARY TUBULES and vascular components. The MEDULLA (lighter reddish and striated) consists of eight to eighteen pyramids. It contains the COLLECTING TUBULES and the STRAIGHT TUBULES OF NEPHRONS (the ASCENDING and DESCENDING LOOPS OF HENLE), the LOOPS OF HENLE and vascular components. The collecting tubules converge to the APEX of each PYRAMID and open into a minor calyx in the RENAL SINUS. There are thousands of branches of the RENAL ARTERY and RENAL VEIN in both the CORTEX and MEDULLA. This accounts for the reddish brown color throughout the kidneys.

The INNER CENTRAL PART contains the PASSIVE CONDUCTORS, the tubes that conduct formed urine passively to the URETER. Several MINOR CALYCES group together (in different areas) to form MAJOR CALYCES. All the major calyces unite to form one funnel-shaped RENAL PELVIS. The renal pelvis collects urine from the calyces and funnels it into the ureter.

The HILUS is a depression or recess of the kidney where vessels (and nerves) enter and exit, an entrance to the RENAL SINUS.

The ENTERING VESSELS, into each kidney are the RENAL ARTERY and its immediate branches. The EXITING VESSELS include the RENAL VEIN and its tributaries and nerves. The RENAL SINUS is a deep concavity in the center of the medial surface. This inner hollow space goes halfway into the kidney.

The KIDNEYS are exterior and posterior to the PERITONEAL LINING OF THE ABDOMINAL CAVITY. Therefore, they are called RETROPERITONEAL ORGANS. (The ureters and ADRENAL GLANDS are also described as RETROPERITONEAL.) The kidneys are just above the waist between the PARIETAL PERITONEUM and the POSTERIOR ABDOMINAL WALL, upon which they lie. The kidneys, ureters and adrenal glands are the only paired organs in the abdominal cavity. Due to the kidney's retroperitoneal nature, many abdominal organs surround the anterior areas of the kidneys, providing a useful guide when studying locational relationships of abdominal organs. The urinary bladder is an ANTIPERITONEAL organ (located in front of the peritoneal cavity).

Abdominal organs surrounding the anterior kidney are the suprarenal (adrenal) glands, liver, duodenum of small intestine, colon of large intestine, spleen, stomach, pancreas, jejunum of small intestine, diaphragm muscle, psoas muscles, quadratus lumborum muscle and transversus abdominus muscle.

EXTERNAL ANATOMY: EACH KIDNEY IS EMBEDDED IN A THREE-LAYERED FATTY, FIBROUS POUCH

THREE SURROUNDING TISSUE LAYERS

1 Renal fascia (outer)

2 Adipose capsule (middle)

3 Fibrous (renal) capsule (inner)

CAVITIES

4 Hilus (hilum)

5 Renal sinus

BLOOD VESSELS

6 Renal artery

7 Renal vein

INTERNAL ANATOMY: EACH KIDNEY OUTER PERIPHERAL PART

8 Cortex (PRIMARILY VASCULAR)

9 Renal columns

(Cortical substance entering medulla and separating the renal pyramids)

10 Renal Medulla (RENAL PYRAMIDS)

11 Renal papillae

(Apex of renal pyramids)

12 Fat (BELOW RENAL COLUMNS)

INNER CENTRAL PART

13 Minor calyces (CALYX (singular))

14 Major calyces

15 Renal pelvis

16 Ureter

Color Guidelines:
2 = yellow; 8 = reddish ; 9 = pink;
10 = light reddish brown;
12 = yellow orange;
13–14 = cool colors;
16 = cream.

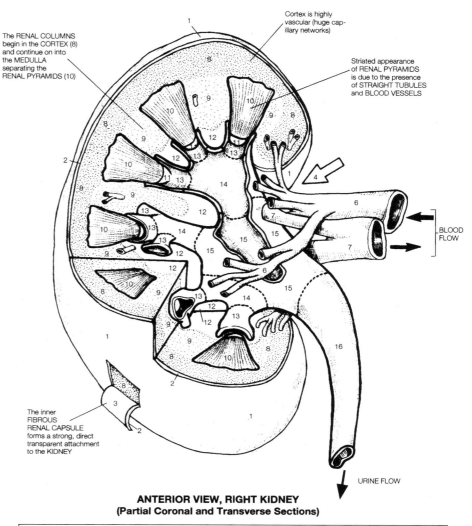

Cortex is highly vascular (huge capillary networks)

The RENAL COLUMNS begin in the CORTEX (8) and continue on into the MEDULLA separating the RENAL PYRAMIDS (10)

Striated appearance of RENAL PYRAMIDS is due to the presence of STRAIGHT TUBULES and BLOOD VESSELS

BLOOD FLOW

The inner FIBROUS RENAL CAPSULE forms a strong, direct transparent attachment to the KIDNEY

URINE FLOW

ANTERIOR VIEW, RIGHT KIDNEY
(Partial Coronal and Transverse Sections)

Hepatic Veins

Esophagus

Celiac Artery and Branches

Diaphragm (9)

Right Renal Artery

Right Renal Vein

Left Renal Artery

Left Renal Vein

LEFT KIDNEY (A)

Superior Mesenteric Artery

Inferior Mesenteric Artery

Left Common Iliac Artery

Left Common Iliac Vein

11th Rib

12th Rib

RIGHT KIDNEY (A)

★ Right Testicular Artery and Vein (Right Testicular Vein empties into Inferior Vena Cava)

Iliacus Muscle

★ Left Testicular Artery and Vein (Left Testicular Vein empties into the Left Renal Vein)

Rectum

Ductus (Vas) Deferens (cross over and in front of the Ureters) (See 14.1)

Symphysis Pubis

A RENAL CORPUSCLE

1 Glomerulus (capillary plexus)

2 Bowman's (glomerular capsule)

A GLOMERULUS

3 Capillary endothelium

4 Basement membrane

A BOWMAN'S CAPSULE (DOUBLE-WALLED CUP

5 Visceral layer/Podocyes (inner layer surrounding capillaries)

6 Bowman's Space (space between the 2 walls or layers)

7 Parietal layer (outer layer)

PODOCYTES (SPECIALIZED EPITHELIUM OF VISCERAL LAYER OF BOWMAN'S CAPSULE)

8 Cell body

9 Primary branch

9a Secondary branch

10 Pedicels (cytoplasmic extensions)

11 Filtration slits (slit pores)

JUXTAGLOMERULAR APPARATUS

12 Juxtaglomerular cells (modified cells of the afferent arteriole)

13 Macula densa (modified cells of the distal convoluted tubule)

14 Afferent arteriole

15 Efferent arteriole

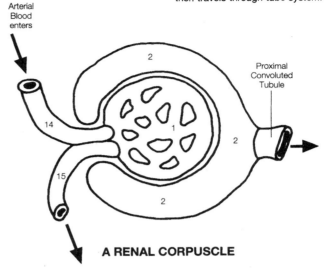

★ Undesired products of metabolism in the Blood Plasma are filtered from 1 into 2 then travels through tube system.

Arterial Blood enters

Proximal Convoluted Tubule

A RENAL CORPUSCLE

Arterial Blood exits (minus filtered plasma)

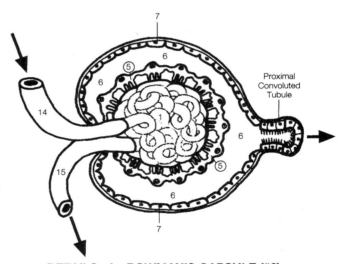

Proximal Convoluted Tubule

DETAILS of a BOWMAN'S CAPSULE (#2)

The undesired products of metabolism in the blood plasma (water and dissolved solutes) are filtered through the glomerulus. They then travel through the tubular system of the kidney. The initial step in urine formation by the nephrons is glomerular filtration. Blood from the afferent arteriole (carrying both wastes and useful nutrients) arrives at the glomerulus at high pressure (60–75 mm Hg). The glomerular capillaries and their surrounding podocytes act as a coarse sieve, permitting pressure-caused passage of materials from the blood vessels into bowman's space according to size or molecular weight (mw). Any substance with a mw of 10,000 or less passes freely through the filtration slits between the podocyte pedicel processes. Beyond 10,000 mw

passage becomes more difficult. Nearly all plasma materials (glucose, amino acids, vitamins, minerals, water and wastes) except formed elements (ie: red and white blood cells and platelets) can filter through the glomerular wall (most plasma proteins, excluding micro-albumin, do not filter). The open pores, or fenestrations, in the capillary endothelium of the glomerulus allow 100 to 400 times the permeability to blood plasma, water and dissolved solutes than the capillaries of skeletal muscles.

The juxtaglomerular apparatus is a structure consisting of myoepitheloid (contractile) cells forming a cuff around the afferent arteriole. It regulates renal blood pressure through the production of renin. (Renin acts on angiotensin to form

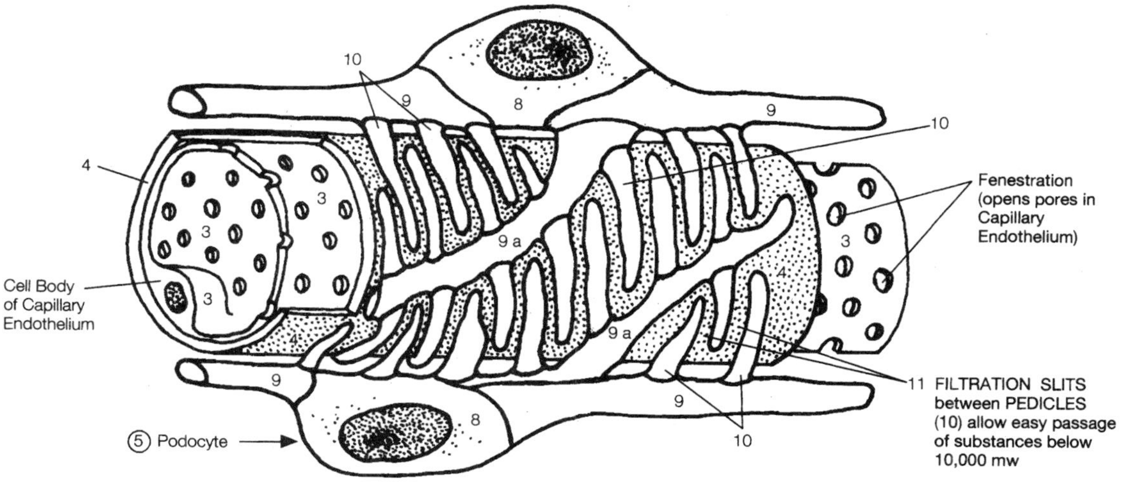
10
9
8
9
10
4
3
3
3
9 a
4
Fenestration
(opens pores in
Capillary
Endothelium)
Cell Body
of Capillary
Endothelium
3
9 a
9
9
8
⑤ Podocyte
10
11 FILTRATION SLITS
between PEDICLES
(10) allow easy passage
of substances below
10,000 mw

ENDOTHELIAL-CAPSULAR MEMBRANE: PODOCYTES ⑤ ENVELOP CAPILLARIES

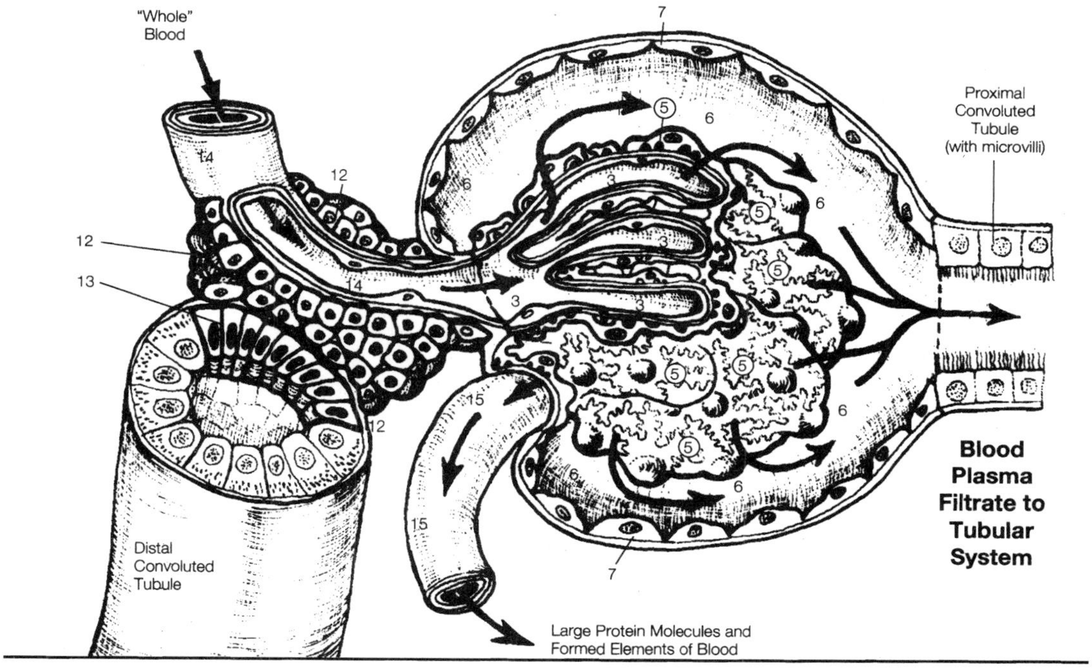
"Whole"
Blood
7
14
12
12
13
Proximal
Convoluted
Tubule
(with microvilli)
6
Blood
Plasma
Filtrate to
Tubular
System
15
Distal
Convoluted
Tubule
7
Large Protein Molecules and
Formed Elements of Blood

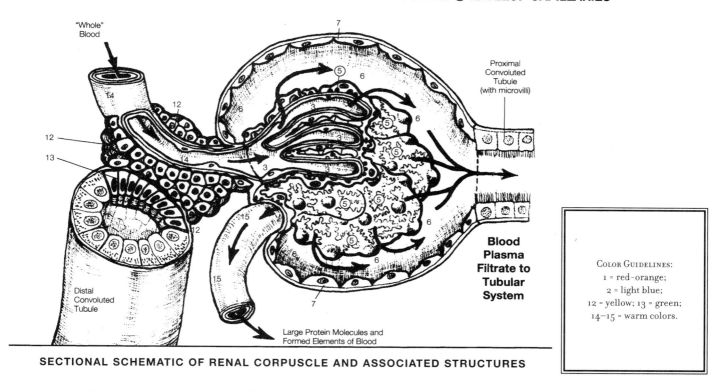

SECTIONAL SCHEMATIC OF RENAL CORPUSCLE AND ASSOCIATED STRUCTURES

Color Guidelines:
1 = red-orange;
2 = light blue;
12 = yellow; 13 = green;
14–15 = warm colors.

angiotensin I, then converted to Angiotensin II, a vasopressor which causes contraction of muscles of capillaries and arteries. This increases resistance to blood flow and elevates blood pressure.) Angiotensin II also stimulates aldosterone secretion which regulates sodium, chloride and potassium metabolism.

The afferent arterioles supply substances to be filtered by the kidney via the endothelial-capsular membrane, and move through the tubular system. The capillary endothelium of the glomerulus filter their substances through open pores called fenestrations. The filtered substances must then pass through the GBM to reach Bowman's space and enter the tubule system. The basement membrane of the glomerulus has no open pores; a dialyzing membrane with myofibrils. There is a negative charge from glycoproteins in the membrane repelling large, negatively charged plasma protein molecules that could pass through the pores. Leading to the epithelium of visceral layer of Bowman's capsule which contains podocytes with pedicel extensions separated by filtration slits. Solutes of blood plasma must pass through these slits to enter the capsule interior called Bowman's space. Large proteins do not cross the visceral layer.

The endothelial-capsular membrane filters water and solutes out of the blood, which pass into Bowman's space and then into the renal tubule. The plasma protein molecules and formed blood elements do not filter through the membrane. This unfiltered blood circulates through the efferent arteriole.

A RENAL CORPUSCLE is a GLOMERULUS and a BOWMAN'S CAPSULE. A NEPHRON is a RENAL CORPUSCLE and an attached URINARY TUBULE (with convoluted and straight portions). A RENAL TUBULE is a NEPHRON (and its VASCULAR SUPPLY) and a COLLECTING TUBULE.

There are about one million nephrons per kidney. The glomerules are located in the CORTEX area of the PARENCHYMA, the outer peripheral part of the kidney. The NEPHRON is the structural and functional unit of the KIDNEY. It consists of a RENAL (MALPIGHIAN) CORPUSCLE (a GLOMERUS cluster of capillary blood vessels enveloped within a thin BOWMAN'S capsule) and its attached URINARY tubule (consisting of a PROXIMAL CONVOLUTED PORTION, a LOOP OF HENLE and a DISTAL CONVOLUTED PORTION). These connect via ARCHED COLLECTING TUBULES with STRAIGHT COLLECTING TUBULES.

URINE is formed by filtration in the RENAL CORPUSCLES (blood enters the GLOMERULAR CAPILLARY PLEXUS in each nephron through the AFFERENT ARTERIOLE). SELECTIVE REABSORPTION and SECRETION is achieved by the cells of the RENAL TUBULE. REABSORPTION involves the active movement of materials OUT OF the filtrate into tubule cells and eventually into blood vessels that return the substances to the circulation. SECRETION involves active movement of materials into the FILTRATE (opposite of REABSORPTION).

Nephrons help regulate the composition and volume of blood (by removing selected amounts of water and solutes). They also remove toxic wastes from the blood, regulate blood pH (acid-base) balance, reabsorb essential nutrients and conserve the water supply.

The DISTAL CONVOLUTED TUBULES of *several nephrons* drain into a single COLLECTING TUBULE or DUCT. In the MEDULLA, the STRAIGHT COLLECTING TUBULES pass through the RENAL PYRAMIDS and open at the RENAL PAPILLAE into the MINOR CALYCES through a number of short, large PAPILLARY DUCTS. (Numerous collecting tubules form a single PAPILLARY DUCT.)

Activities of structures of the renal tubule include GLOMERULAR FILTRATION and ACTIVE TRANSPORT SYSTEMS. In glomerular filtration, the renal corpuscle *filters* blood plasma from the blood (water and dissolved solutes), now called FILTRATE. The ACTIVE TRANSPORT SYSTEMS include the PROXIMAL CONVOLUTED TUBULE, the LOOP OF HENLE, and the DISTAL CONVOLUTED TUBULE (COLLECTING TUBULE). REABSORPTION OF SOLUTES BACK INTO THE BLOOD (by surrounding capillaries preceding from the efferent arteriole) occurs primarily within the proximal convoluted tubule, absorbing almost all glucose, small proteins, amino acids, salts and some vitamins. Eighty-five percent of filtrate is reabsorbed (85% solutes followed by 85% water). *There is little or no secretion into tubule filtrate*

A NEPHRON (= I AND II)

I. RENAL CORPUSCLE (MALPIGHIAN CORPUSCLE)

8	Glomerulus (capillary plexus)
9	Bowman (glomerular) capsule
10	Afferent arteriole
11	Efferent arteriole

II. URINARY TUBULE (1-5)

1	Proximal convoluted tubule (bulk of renal cortex)	
2	Descending limb of the loop of Henle	(straight)
3	Loop of Henle	
4	Ascending limb of the loop of Henle	(straight)
5	Distal convoluted tubule	
6	Collecting tubule (straight) (DETERMINES FINALURINE VOLUME)	
7	Papillary duct (collecting duct)	

(some organic acids, antibiotics, and organic bases), and wastes are not reabsorbed.

The LOOP of HENLE absorbs water into surrounding tissue cells for water conservation and increased concentration of urine solutes (high salts) in the interstitial fluid outside the tubule. The filtrate within the loop is therefore hypotonic with respect to the interstitium.

The reabsorption of 10–15% of nutrients occurs via the DISTAL CONVOLUTED TUBULE. The secretion of hydrogen ions into the tubule filtrate acidifies the filtrate. It is influenced by hormones, such as ADH (antidiuretic hormone), which permits water to pass through collecting tubule membrane to the salt-rich fluid surrounding the tubule, thereby conserving it.

The filtrate to be released becomes two to three times more concentrated than the blood.) Ninety-nine percent of the original filtrate is returned to tissues and 1% is passed on as urine to the calyces.

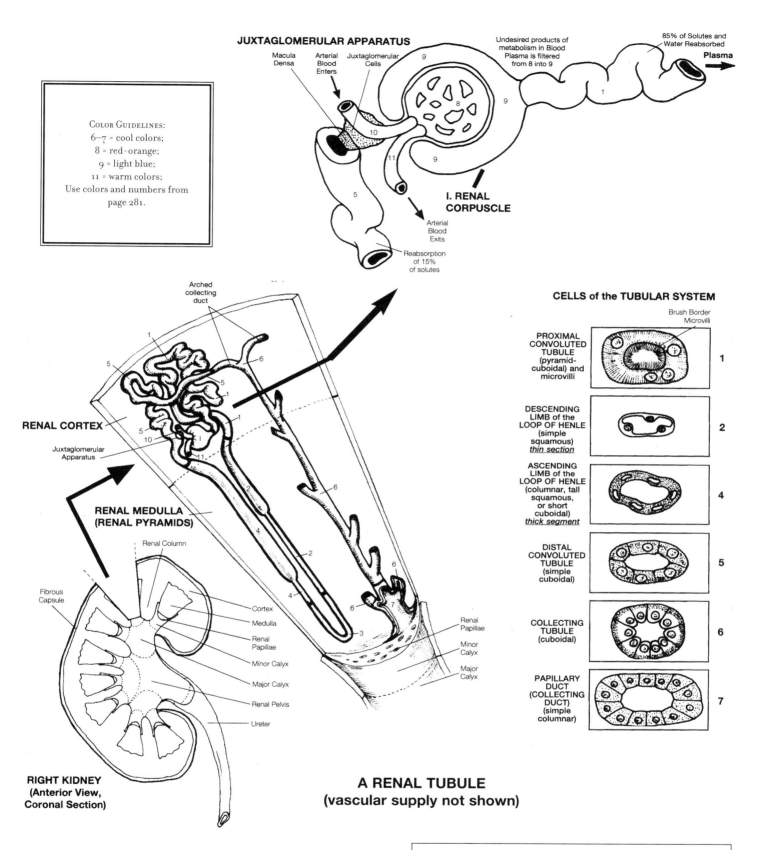

JUXTAGLOMERULAR APPARATUS

Macula
Densa

Arterial
Blood
Enters

Juxtaglomerular
Cells

Undesired products of
metabolism in Blood
Plasma is filtered
from 8 into 9

85% of Solutes and
Water Reabsorbed

Plasma

**I. RENAL
CORPUSCLE**

Arterial
Blood
Exits

Reabsorption
of 15%
of solutes

Arched
collecting
duct

RENAL CORTEX

Juxtaglomerular
Apparatus

**RENAL MEDULLA
(RENAL PYRAMIDS)**

Renal Column

Fibrous
Capsule

Cortex

Medulla

Renal
Papillae

Minor Calyx

Major Calyx

Renal Pelvis

Ureter

Renal
Papillae

Minor
Calyx

Major
Calyx

**RIGHT KIDNEY
(Anterior View,
Coronal Section)**

**A RENAL TUBULE
(vascular supply not shown)**

CELLS of the TUBULAR SYSTEM

Brush Border
Microvilli

PROXIMAL
CONVOLUTED
TUBULE
(pyramid-
cuboidal) and
microvilli

1

DESCENDING
LIMB of the
LOOP OF HENLE
(simple
squamous)
thin section

2

ASCENDING
LIMB of the
LOOP OF HENLE
(columnar, tall
squamous,
or short
cuboidal)
thick segment

4

DISTAL
CONVOLUTED
TUBULE
(simple
cuboidal)

5

COLLECTING
TUBULE
(cuboidal)

6

PAPILLARY
DUCT
(COLLECTING
DUCT)
(simple
columnar)

7

In the selection shown, (1 PYRAMID), there are
approximately 100,000 nephrons
(1 million/kidney); therefore, the renal tubule
shown is greatly exaggerated.

The Ureters, Urinary Bladder and Urethra

Each URETER is a continuation of the RENAL PELVIS of its associated kidney. The URETERS (one ureter comes from each kidney) drain the pelvis. Ureters are only 1/2 inch in diameter at their widest point. The ureters extend from the beginning of each funnel-shaped pelvis, approximately 28 to 34 cm to the URINARY BLADDER. (They are always retroperitoneal in position.) The ureters pass into the urinary bladder (one on either lateral side) for several centimeters. Pressure in the bladder compresses the ureters and prevents backflow during urination. The ureter function is to transport urine from the RENAL PELVIS into the URINARY BLADDER for storage. Urine is transported in the URETERS by peristaltic contraction of ureteric muscular walls, hydrostatic pressure and gravity. Mucus secreted by the inner mucosa of the ureters prevents their cells from coming in contact with urine.

There are THREE COATS OF THE URETERS (AND RENAL CALYCES AND RENAL PELVES of the KIDNEYS). The MUCOSA contains transitional epithelium and lamina propria (thicker in the pelvis and ureter). The MUSCULARIS (smooth) is in the CALYCES, PELVIS and upper two-thirds of the ureters. Its two layers are the INNER CIRCULAR and OUTER LONGITUDINAL, which become progressively thicker toward the ureter. The lower third of the ureter has a third INNER LONGITUDINAL LAYER added. The outermost ADVENTITIA coat blends with the fibrous tunic of the kidney above and the SEROSA of the bladder below. However, this is not a serosa, so no peritoneum is involved.

The FOUR COATS OF THE URINARY BLADDER, from the inside out are: MUCOSA (transitional epithelium and lamina propria), SUBMUCOSA (considered by some as lacking and is in reality a deep loose layer of the underlying lamina propria), DETRUSOR MUSCLE (a very thick circular layer), and the OUTSIDE SEROSA (which is continuous from the peritoneum).

The average capacity of the URINARY BLADDER is 700 to 800 ml. When it exceeds 200 to 400 ml, urine is expelled by conscious desire and unconscious reflex (micturition), both results of nervous impulses. The TRIGONE is the triangle formed by the two URETHRAL ORIFICES and the INTERNAL URETHRAL ORIFICE. The internal sphincter is the middle muscularis of the DETRUSOR MUSCLE at the area opening to the urethra (involuntary). The external sphincter is the (voluntary) SKELETAL MUSCLE.

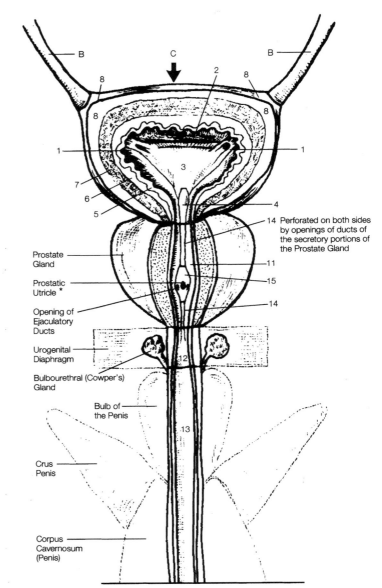

14 Perforated on both sides by openings of ducts of the secretory portions of the Prostate Gland

Prostate Gland

Prostatic Utricle *

Opening of Ejaculatory Ducts

Urogenital Diaphragm

Bulbourethral (Cowper's) Gland

Bulb of the Penis

Crus Penis

Corpus Cavernosum (Penis)

MALE Anterior View, Coronal Section

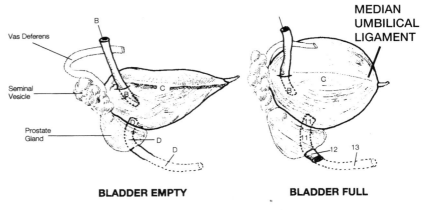

Vas Deferens

Seminal Vesicle

Prostate Gland

MEDIAN UMBILICAL LIGAMENT

BLADDER EMPTY   BLADDER FULL

(Left Lateral Views)

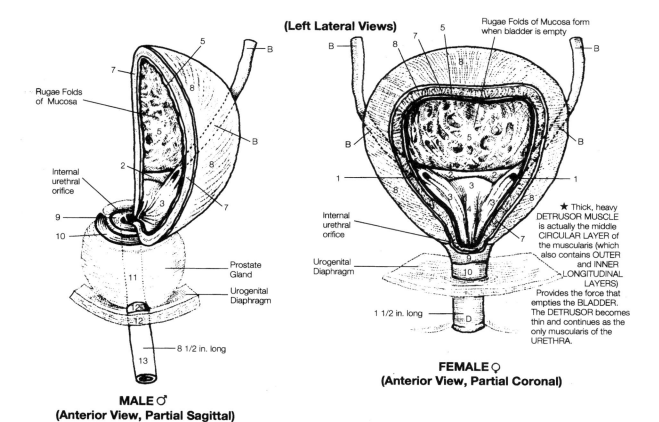

(Left Lateral Views)

Rugae Folds of Mucosa form when bladder is empty

Rugae Folds of Mucosa

Internal urethral orifice

Prostate Gland

Urogenital Diaphragm

8 1/2 in. long

MALE ♂
(Anterior View, Partial Sagittal)

Internal urethral orifice

Urogenital Diaphragm

1 1/2 in. long

★ Thick, heavy DETRUSOR MUSCLE is actually the middle CIRCULAR LAYER of the muscularis (which also contains OUTER and INNER LONGITUDINAL LAYERS) Provides the force that empties the BLADDER. The DETRUSOR becomes thin and continues as the only muscularis of the URETHRA.

FEMALE ♀
(Anterior View, Partial Coronal)

B	2 Ureters
↓ 1	2 Ureteral orifices
C	1 Urinary bladder
2	Interureteric fold
3	Trigone
4	Uvula
D	1 Urethra

FOUR BLADDER COATS:

5 Mucosa (Transitional epithelium and lamina propria)

6 Submucosa

7 Detrusor muscle (thick, heavy circular layer)

8 Serosa (continuous with the peritoneum)

9 Internal sphincter — Sphincter Vesicae

10 External sphincter — Sphincter Urethrae

DIVISIONS OF THE URETHRA (MALE)

11 Prostatic urethra

12 Membranous urethra

13 Spongy (cavernous) urethra (penile urethra)

PORTIONS OF THE PROSTATIC URETHRA (MALE)

13 Spongy (cavernous) urethra (penile urethra)

14 Urethral crest (Longitudinal fold)

15 Seminal colliculus

(ELEVATION OF 14, RECEIVES OPENINGS OF THE EJACULATORY DUCTS AND PROSTATIC UTRICLE)

The URETHRA is the passageway for discharging urine from the body. In the FEMALE, it is 1½ inches long, and is entirely separate from the reproductive organs. In the male, it is about eight inches long, and also acts as a duct for the discharge of semen (reproductive fluid) from the body. It is divided into three parts.

In the male, the urethra consists of three layers: (1) a mucosa— INNER EPITHELIUM IS TRANSITIONAL in (11), PSEUDOSTRATIFIED in (12) and (13), and becomes STRATIFIED SQUAMOUS at the end of (13). (2) A LAMINA PROPRIA, which connects the epithelium to surrounding tissue. In the female, mucosal EPITHELIUM is TRANSITIONAL close to the BLADDER and becomes STRATIFIED SQUAMOUS distally. The LAMINA PROPRIA connects epithelium to SPONGY TISSUE (and contains many veins called *erectile tissue*). (3) MUSCLARIS is continuous with the bladder, forming a circular layer the entire length.

CHAPTER 16: REPRODUCTIVE SYSTEM

SYSTEM COMPONENTS

The organs (testes and ovaries) for production of reproductive cells (sperm and ova) and organs for transportation and storage of those cells.

SYSTEM FUNCTIONS

Reproduction of the organism

Perpetuation of the species

Passage of genetic material from generation to generation

THE REPRODUCTIVE SYSTEM is unique in three aspects from other body systems. Other systems, functional at or close to birth, work to sustain the individual, with minor sexual differences. The reproductive system is latent in development (under hormonal control) and its function is to perpetuate the species. There are major differences between the male and female systems.

The reproductive system of the human organism sustains life by giving rise to offspring through the process of duplication. In the human, this is accomplished by the union of GAMETES (GERM CELLS OR SEX CELLS). Gametes are HAPLOID CELLS—each contains 23 single chromosomes, or a half-complement of the genetic material (each normal somatic cell contains 46 chromosomes in its nucleus). Fertilization of a female gamete (EGG, OR OVUM) by a male gamete (SPERMATOZOA, OR SPERM) produces a normal DIPLOID CELL (23 chromosomes of the ovum being paired with 23 chromosomes of the spermatozoon)—the ZYGOTE. Sex is determined by whether the fertilizing spermatozoon is of x-type (for a female) or of y-type (for a male). It is the random combination of specific genes in the gametes during fusion that results in the unique individual (see cell mitosis, pages 28–29).

The general process of cell division maintains the life of the individual (by single cell duplication, or MITOSIS, of genetic material for growth and repair of the organism) as well as maintains continuance of the species (by reproduction: passing genetic material from generation to generation).

The structures of the male and female reproductive systems can be grouped according to function: The PRIMARY SEX ORGANS or GONADS are mixed glands because they produce sex hormones as well as gametes. The male gonads are called TESTES, which produce SPERMATOZOA through the process of SPERMATOGENESIS. The female gonads are called OVARIES, which produce OVA through the process of OOGENESIS. Both processes involve a special kind of cell division called MEIOSIS.

The SECONDARY SEX ORGANS mature at puberty due to the influence of sex hormones and are essential for successful sexual reproduction in the transportation, aid, and storage of gametes. DUCT SYSTEMS receive, transport, and store gametes while ACCESSORY GLANDS produce substances that support gametes, and the COPULATORY ORGANS provide physical intercourse for gametes.

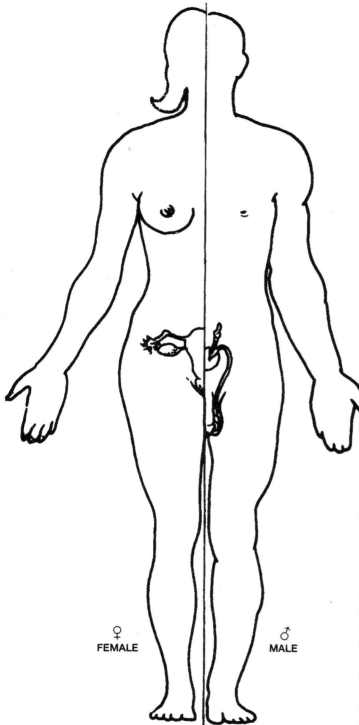

♀
FEMALE

♂
MALE

SPLIT ANTERIOR VIEW

The SECONDARY SEX CHARACTERISTICS are characteristics that are associated with but not essential to reproduction. Their appearance at puberty and maintenance after puberty are due to the appropriately timed activity and proper concentration of sex hormones.

The male grows the testes and the penis; has pubic , facial, underarm and abdominal hair; experiences body growth and growth of the larynx (his voice pitch lowers and his voice deepens); has acne from blocked glands; has a body that is characteristic of the male shape; has growth of bone; and undergoes muscle development of and alteration of skeletal proportions.

The female develops breasts; has patterns of body, pubic, and underarm hair; has a body physique with distribution of subcutaneous fat deposits on her shoulders, hips, and thighs; has a body that is characteristic of the female shape; and has growth of bones, a temporary increase in height and a usually broader pelvis than her male counterpart.

In addition to the development of the SECONDARY SEX ORGANS and SECONDARY SEX CHARACTERISTICS at PUBERTY, psychological changes and sexual instincts associated with adulthood begin to develop.

Male Reproductive System
General Scheme: Puberty; Primary and Secondary Sex Organs

The organs of the MALE REPRODUCTIVE SYSTEM are the PRIMARY SEX ORGANS, THE TESTES (MALE GONADS), and the secondary sex organs, which consist of a SYSTEM OF DUCTS, ACCESSORY GLANDS and SUPPORTIVE STRUCTURES. Secondary sex organs mature at puberty.

Sperm (SPERMATOZOA OR MALE GERM CELLS) is produced in the male testes. The testicular cells secrete ANDROGEN hormones (including the male sex hormone TESTOSTERONE) directly into the bloodstream. At the onset of puberty, androgens cause the appearance of (and after puberty, the maintenance of) spermatogenesis, secondary sex characteristics, secondary sex organ development and function (especially SEMINAL VESICLES, PROSTATE GLAND, and PENIS AND SCROTUM enlargement), and anabolic effects including protein synthesis and red blood cell formation.

The SYSTEM OF DUCTS is for the storage of sperm and transportation of sperm (and secretions contributed by glands) to the exterior. The male duct system technically begins inside each TESTIS (SEMINIFEROUS TUBULES, STRAIGHT TUBULES, RETE TESTIS), then EFFERENT DUCTULES, EPIDIDYMIS, VAS DEFERENS, EJACULATORY DUCT; both EJACULATORY DUCTS empty into the single URETHRA.

The accessory glands (PROSTATE GLANDS, SEMINAL VESICLES, BULBOURETHAL GLANDS) produce secretions that comprise the SEMEN (fluid that carries the sperm).

The supportive structures include the SCROTUM, PENIS and UROGENITAL DIAPHRAGM. The scrotum is the sac of skin that houses and suspends the TESTES, EPIDIDYMIS and a portion of the VAS DEFERENS. The penis contains three columns of erectile tissue. The median column (CORPUS SPONGIOSUM) surrounds the PENILE URETHRA. The penis serves as an INTROMITTENT ORGAN (an organ injecting into a cavity) during sexual intercourse (COITUS OR COPULATION). The UROGENITAL DIAPHRAGM, a musculofascial sheath that lies superficial to the PELVIC DIAPHRAGM, surrounds the MEMBRANOUS URETHRA.

During fetal development, the TESTES arise as paired retroperitoneal organs on the posterior abdominal wall near the paired kidneys. As the body grows, the testes descend into the scrotum, a cutaneously derived pouch, during the 28th to 29th week of development. The scrotum is an outpocketing (herniation) of the abdominal wall, anterior to the PUBIC SYMPHYSIS, which progresses in to the skin posterior to the PENIS. The canal through which the herniation occurs remains as the INGUINAL CANAL that conveys the SPERMATIC CORD from the scrotum to the PELVIC CAVITY.

Puberty in the Male

At the onset of PUBERTY (usually between the ages of 13 to 16), the maturational changes in the HYPOTHALAMUS produce an increase in regulating factor GnRH (GONADOTROPHIN-RELEASING HORMONE) and its secretion into the HYPOTHALAMIC-HYPOPHYSEAL PORTAL VESSELS. GnRH stimulates release of GONADOTROPHIC HORMONES FSH (FOLLICLE-STIMULATING HORMONE, named for its action in the female) and ICSH (INTERSTITIAL CELL-STIMULATING HORMONE) (LUTEINIZING HORMONE) from the ANTERIOR LOBE of the PITUITARY GLAND directly into the blood stream. ICSH stimulate secretion of SEX STEROIDS (ANDROGENS AND ESTROGENS) inside the TESTES. (A decrease in NEGATIVE FEEDBACK INHIBITION by the sex steroids to the sensitivity of the HYPOTHALAMUS and PITUITARY occurs, thus increasing ICSH release.) These sex steroids (especially TESTOSTERONE) cause the appearance and maintenance of SECONDARY SEX CHARACTERISTICS and SECONDARY SEX ORGAN development and function. FSH stimulates the onset and regulation of SPERMATOGENESIS. (FSH receptors are only found in the NURSE, or SERTOLI CELLS. See pages 294–95.)

Primary sex organs: Gonads

A | **PAIRED TESTES**
(singular: testis)

SECONDARY SEX ORGANS

Mature at Puberty
SYSTEM OF DUCTS, ACCESSORY GLANDS, AND SUPPORTING STRUCTURES

SYSTEM OF DUCTS

B | 2 Epididymides
(singular, epididymis)

C | 2 Ductus deferentia
(singular vas or ductus, deferens)

D | 2 common ejaculatory ducts

ONE URETHRA AND THE DIVISIONS OF THE URETHRA, THE CANAL FOR DISCHARGE OF SEMEN (AND URINE).

E | Prostatic urethra

F | Membranous urethra

G | Penile (cavernous) urethra

ACCESSORY GLANDS

H | Prostate gland

I | 2 Seminal vesicles

J | 2 Bulbourethral glands/cowper's glands

SUPPORTIVE STRUCTURES (EXTERNAL GENITAL ORGANS)

K | Scrotum 2 scrotal sacs

L | Penis

M | Urogenital diaphragm

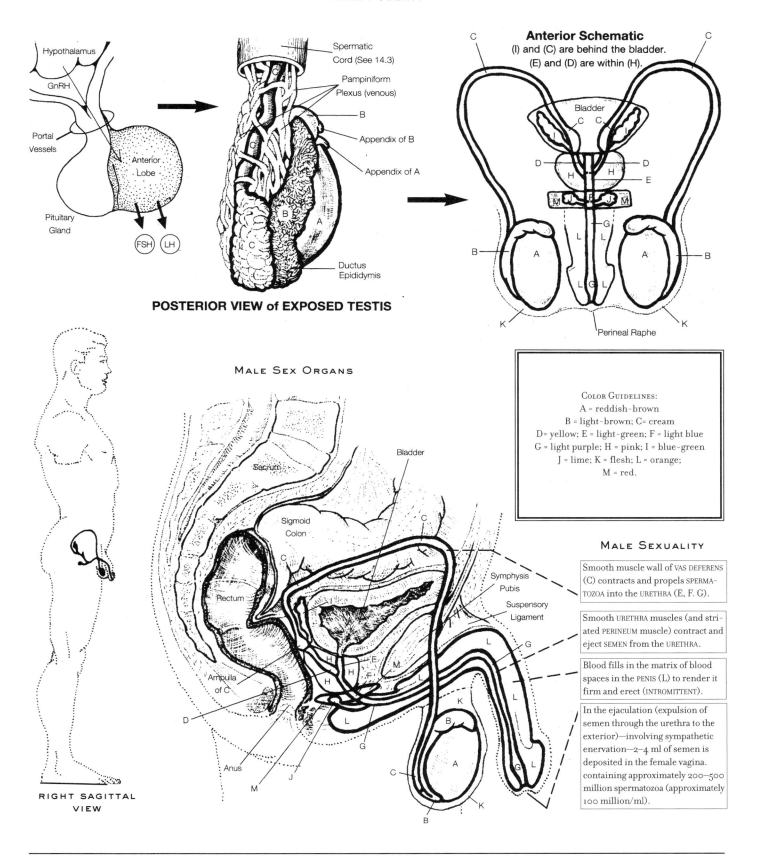

Hypothalamus

GnRH

Portal
Vessels

Anterior
Lobe

Pituitary
Gland

FSH   LH

Spermatic
Cord (See 14.3)

Pampiniform
Plexus (venous)

B

Appendix of B

Appendix of A

Ductus
Epididymis

POSTERIOR VIEW of EXPOSED TESTIS

Anterior Schematic
(I) and (C) are behind the bladder.
(E) and (D) are within (H).

Bladder

Perineal Raphe

MALE SEX ORGANS

Sacrum

Bladder

Sigmoid
Colon

Symphysis
Pubis

Suspensory
Ligament

Rectum

Ampulla
of C

D

Anus

J

M

**RIGHT SAGITTAL
VIEW**

COLOR GUIDELINES:
A = reddish-brown
B = light-brown; C= cream
D= yellow; E = light-green; F = light blue
G = light purple; H = pink; I = blue-green
J = lime; K = flesh; L = orange;
M = red.

MALE SEXUALITY

Smooth muscle wall of VAS DEFERENS
(C) contracts and propels SPERMA-
TOZOA into the URETHRA (E, F, G).

Smooth URETHRA muscles (and striated PERINEUM muscle) contract and
eject SEMEN from the URETHRA.

Blood fills in the matrix of blood
spaces in the PENIS (L) to render it
firm and erect (INTROMITTENT).

In the ejaculation (expulsion of
semen through the urethra to the
exterior)—involving sympathetic
enervation—2–4 ml of semen is
deposited in the female vagina.
containing approximately 200–500
million spermatozoa (approximately
100 million/ml).

SAGITTAL SECTIONS of a TESTIS

Testicular Artery

Pampiniform Plexus

C

B (HEAD)

B (BODY)

B (TAIL)

Appendix of Epididymis

Muscular tissue of C propels sperm to prostatic urethra

Canal of C through which sperm are transported

B (HEAD)

Appendix of Epididymis (B)

Appendix of Testis

Here the spermatozoa increase in fertilizing power. If *not* ejaculated, they will soon degenerate and will be absorbed in these tubules.

Primary sex organs: Gonads
(2 testes)

A SINGLE TESTIS

Two Tunic (tissue) layers

TUNICA VAGINALIS

1 | Tunica albuginea (Tough, Fibrous Membrane)

2 | Septa of the tunica albuginea

3 | Lobules

4 | Seminiferous tubules

5 | Straight portions of seminiferous tubules

Preliminary sperm maturation site

6 | Rete testis

7 | Mediastinum testis

Secondary sex organs (Primary region of sperm maturation)

B | Epididymis

Duct system

8 | Efferent ductules (Coiled, ciliated)

9 | Ductus epididymis

C | Vas (ductus) deferens

The TESTES are paired, whitish, oval organs that measure about 4–5 cm (1½–2 in.) in length and 2.5 cm (1 in.) in diameter. Each weighs between 10 and15 grams. Two tissue layers (tunicas) cover the testes. The TUNICA VAGINALIS is a thin, serous sac derived from the peritoneum during the descent of the testes. The tough, fibrous TUNICA ALBUGINEA directly surrounds each TESTIS. Inward extensions of this tunica divide each testis into a series of 200 to 300 internal compartments called LOBULES. Each lobule contains one to three tightly packed, coiled SEMINIFEROUS TUBULES—the functional units of the testes. (Each tube if uncoiled may exceed 2⅓ feet.) The lobules produce sperm through the process of SPERMATOGENESIS. Endocrine cells between the seminiferous tubules called INTERSTITIAL CELLS OF LEYDIG produce and secrete the male sex hormones (androgens) and to a lesser extent the tubules do also, with a precursor. Thus, the testes are both EXOCRINE GLANDS (SPERM) and ENDOCRINE GLANDS (ENDROGENS).

Once produced, the sperm travel through the lumen of the convoluted seminiferous tubules. Sperm then converge into a partially ciliated duct network (RETE TESTIS) for further maturation within the MEDIASTINUM TESTIS (the thickened portion of the TUNICA ALBUGINEA on the posterior surface of the TESTIS). From the RETE TESTIS, the SPERM are transported to the EPIDIDYMIS via the EFFERENT DUCTULES (uncoiled, the epididymis is 17 feet).

The epididymis (with tall microvilli or stereocilia) houses the final stages of sperm maturation, where the SPERM (SPERMATOZOA) is stored and transported to the VAS DEFERENS. The sperm are morphologically mature in the vas deferens (18 inches or 4.57 cm), which stores and transports sperm for discharge during EJACULATION (using smooth muscular walls).

SPERMATOGENESIS is the process of the formation of mature SPERMATOZOA. It starts just after puberty and normally continues until old age. Production occurs in advancing stages in the GERMINAL EPITHELIUM progressing from outer outer regions of the seminiferous tubule toward the inner lumen of each tubule. Through a process of specialized cell division called meiosis, each diplod spermatonium (46 chromosomes each) is transformed into four haploid spermatozoa (23 chromosomes each).

Seminiferous tubule:

A | Basement membrane |

GERMINAL EPITHELIUM

B | Spermatogonia |

C | Primary spermatocytes |

D | Secondary spermatocytes |

E | Spermatids |

F | Spermatozoa |

G | Nurse (sertoli) cells | (Spermiation)

H | Interstitial cells of Leydig |

T Secrete (Testosterone)

SPERMATOZOÖN (60 μm long)
(Mature sperm cell)

a | Head |

b | Neck |

c | Body |

d | Tail | (flagellum)

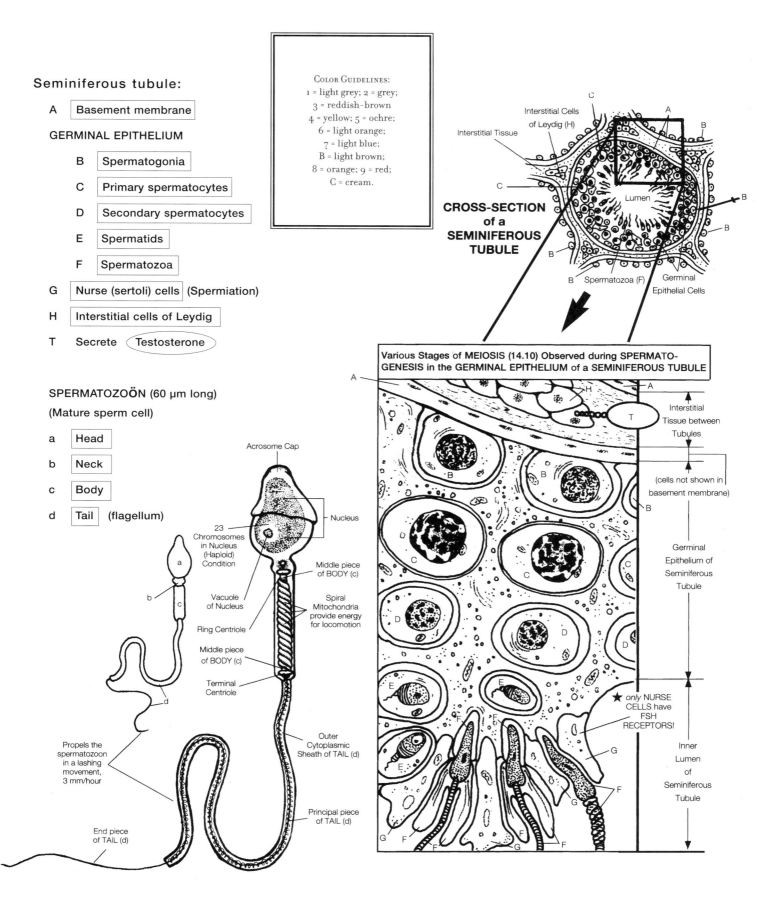

CROSS-SECTION of a SEMINIFEROUS TUBULE

Interstitial Cells of Leydig (H)

Interstitial Tissue

Lumen

Spermatozoa (F)

Germinal Epithelial Cells

Various Stages of MEIOSIS (14.10) Observed during SPERMATO-GENESIS in the GERMINAL EPITHELIUM of a SEMINIFEROUS TUBULE

Interstitial Tissue between Tubules

(cells not shown in basement membrane)

Germinal Epithelium of Seminiferous Tubule

★ *only* NURSE CELLS have FSH RECEPTORS!

Inner Lumen of Seminiferous Tubule

Acrosome Cap

Nucleus

23 Chromosomes in Nucleus (Haploid) Condition

Vacuole of Nucleus

Ring Centriole

Middle piece of BODY (c)

Spiral Mitochondria provide energy for locomotion

Middle piece of BODY (c)

Terminal Centriole

Outer Cytoplasmic Sheath of TAIL (d)

Propels the spermatozoon in a lashing movement, 3 mm/hour

Principal piece of TAIL (d)

End piece of TAIL (d)

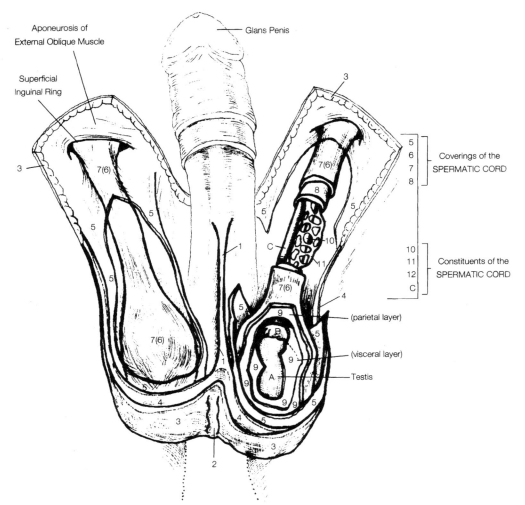

Anterior View (Penis elevated, outer skin layer cut away)

The Scrotum

The scrotum supports and protects the testes and maintains testicle position in relation to the pelvis. The MEDIAN SEPTUM divides the SCROTUM into two sacs; each testis is in its own separate compartment, thus increasing protection. The scrotal sacs are suspended below the PERINEUM, a region anterior to the anal opening and just behind the base of the penis. The coverings of the SPERMATIC CORD continue as the inner layers of the SCROTAL SACS (SCROTUM).

They are outwardly enveloped by the dartos muscle fibers in the subcutaneous tissue and an outer, hairy skin layer.

The DARTOS and CREMASTER MUSCLES maintain a constant temperature of 35°C (95°F), the required temperature for the development and storage of the spermatozoa. In cold weather, the muscles contract and move the testes closer to the warmth of the pelvis. In warm weather, the muscles relax and the testes descend, moving away from the body heat.

In regard to blood circulation, blood supply to the testes is through the TESTICULAR ARTERIES, which arise from the ABDOMINAL AORTA just below the origin of the RENAL ARTERIES. Blood drainage is from the testicular veins—the RIGHT TESTICULAR VEIN enters directly into the INFERIOR VENA CAVA; the left testicular vein enters into the LEFT RENAL VEIN.

THE SPERMATIC CORD

As the VAS (DUCTUS) DEFERENS ascends in the scrotum and passes through the abdominal wall, it travels with the TESTICULAR ARTERY AND VEINS, draining the TESTES (PAMPINIFORM PLEXUS) and various NERVES and LYMPHATIC VESSELS. These form the structures within the spermatic cord.

The spermatic cord coverings are continuations of layers of the abdominal wall. The INTERNAL OBLIQUE MUSCLE and FASCIA of the abdominal wall become the cremaster muscle and fascia.

The INGUINAL CANAL is the slitlike passageway in the anterior abdominal wall through which the SPERMATIC CORD passes into the abdominal cavity. It is a common site for INGUINAL HERNIAS.

The Scrotum

SEPTAL SUBDIVISION

1 | Mediam Septum | (INTERNAL) a continuation of the DARTOS (4)

2 | Perineal raphe | (EXTERNAL RIDGE)

LAYERS OF THE SCROTUM

3 | Outer skin |

4 | Dartos | (smooth muscle fibers)

5 | External spermatic fascia |

6 | Cremasteric fascia* |

7 | Cremaster muscle (skeletal muscle)* |

8 | Internal spermatic fascia |

9 | Tunica vaginalis: |

The outer layer of the TESTES. It is a thin, serous sac (derived from the PERITONEUM during the descent of the TESTES). It folds over itself to form a VISCERAL LAYER and a PARIETAL LAYER, with a CAVITY separating them.

*CREMASTERIC FASCIA and MUSCLE are continuations of the INTERNAL OBLIQUE MUSCLE and FASCIA of the ABDOMINAL WALL.

COVERINGS OF THE SPERMATIC CORD

5 | External spermatic fascia |

6 | Cremasteric fascia |

7 | Cremaster muscle |

8 | Internal spermatic fascia |

CONSTITUENTS OF THE SPERMATIC CORD

C | VAS (DUCTUS) DEFERENS |

10 | TESTICULAR ARTERY |

11 | PAMPINIFORM PLEXUS (venous network) |

12 | Nerves and lymphatics |

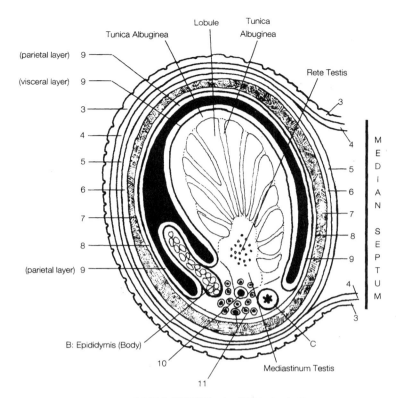

CROSS-SECTION of a SCROTAL SAC and a TESTIS

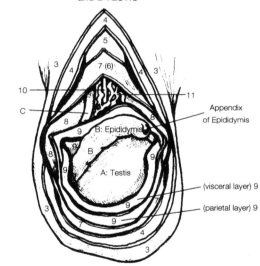

A SCROTAL SAC
(Layers cut away to expose a testicle)
Anterior View

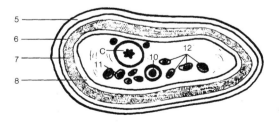

CROSS-SECTION of a SPERMATIC CORD
(One of 2 cords)

The MALE URETHRA is a common tube for the REPRODUCTIVE AND URINARY SYSTEMS. Semen and urine, however, cannot pass through the urethra simultaneously since the SYMPATHETIC EJACULATION REFLEX automatically inhibits the URINARY REFLEX (MICTURITION). The male urethra has three main sections, the PROSTATIC URETHRA passes through the PROSTATE GLAND. The MEMBRANOUS URETHRA passes through the UROGENITAL DIAPHRAGM. The PENILE (CANERNOUS) URETHRA passes through the SHAFT OF THE PENIS in the center of the midvental CORPUS SPONGIOSUM.

The AMPULLA of the VAS DEFERENS and the DUCT of the SEMINAL VESICLE unite to form the EJACULATORY DUCT. The LEFT and RIGHT common EJACULATORY DUCTS pass through the PROSTATE GLAND and open into the PROSTATIC URETHRA. There are many small ducts from the PROSTATE GLAND that open into the PROSTATIC URETHRA. The BULBOURETHRAL (COWPER'S) GLANDS are located in the UROGENITAL DIAPHRAGM. Their ducts open into the PENILE (CAVERNOUS) URETHRA. Mucus is secreted in response to sexual stimulation prior to ejaculation.

The male system of ducts (two VAS DEFERENS with their AMPULLAS and two EJACULATORY DUCTS) stores and transports sperm cells. ACCESSORY GLANDS secrete the liquid portion of the semen. Two SEMINAL VESICLES (60%) secrete a viscous liquid that keeps spermatozoa alive and motile. The PROSTATE GLAND (13–33%) and two BULBOURETHRAL GLANDS (7–27%) add a thin lubricant to semen.

The PENIS is composed of erectile tissue (arranged in three columns), enveloped by skin. The median (midventral) column—the CORPUS SPONGIOSUM—contains the URETHRA. When distended (filled with blood), the penis serves as the COPULATORY ORGAN of the male. When flaccid, it serves as a conduit for urine from the URINARY BLADDER.

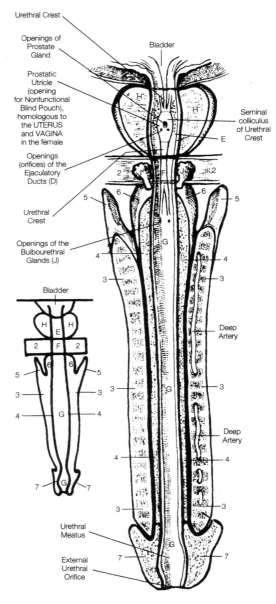

SCHEMATIC CORONAL VIEW

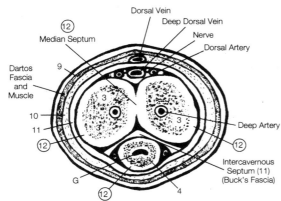

CROSS-SECTION of the SHAFT of the PENIS

COLOR GUIDELINES:
3 = flesh; 4 = red-orange; Fascia = cool colors;
C = cream; 10 = red; 11 = blue.
For other structures, use color guidelines
from previous pages.

The Urethra

3 MAIN REGIONS

E ⬜ Prostatic urethra

F ⬜ Membranous urethra

G ⬜ Penile (cavernous) urethra
(spongy urethra)

SYSTEM OF DUCTS

C ⬜ Vas (ductus) deferens (2)

C1 ⬜ Ampulla of vas deferens (2)

D ⬜ Common ejaculatory duct (2)

ACCESSORY GLANDS

H ⬜ Prostate gland (1)

I ⬜ Seminal vesicles (2)

J ⬜ Bulbourethral glands (2) (Cowper's gland)

1 ⬜ Levator ani (pelvic diaphragm)

2 ⬜ Urogenital diaphragm

The Penis

THREE CYLINDRICAL BODIES OF ERECTILE TISSUE

Two Lateral: 3 Corpus cavernosum
(Corpus cavernosum penis)

One Midventral: 4 Corpus spongiosum
(Corpus canvernosum urethrae)

ROOT PORTION OF CORPUS CAVERNOSUM PENIS

5 ⬜ Crus of the penis

ROOT AND FRONT PORTIONS
OF CORPUS SPONGIOSUM (Corpus Cavernosum Urethrae)

6 ⬜ Bulb of the penis (bulbous urethrae)

7 ⬜ Glans of the penis

SKIN COVERINGS AROUND PENIS

8 ⬜ Prepuce (foreskin) (around glans)

9 ⬜ Outer skin

INNER BINDING FASCIA

10 ⬜ Superficial fascia (loose areolar)

11 ⬜ Deep fascia

12 ⬜ Tunica albuginea

Between the corpora CAVERNOSA PENIS, the deep fascia of the TUNICA ALBUGINEA forms a MEDIAN SEPTUM

RIGHT SAGITTAL VIEW

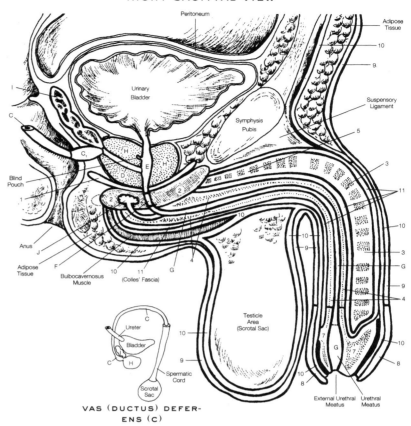

VAS (DUCTUS) DEFER-
ENS (C)

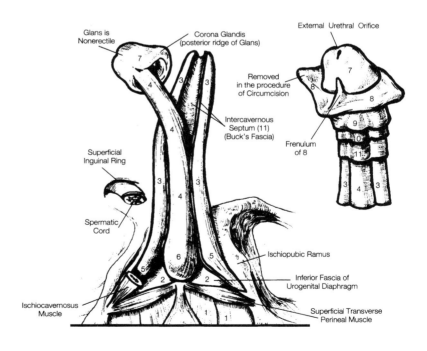

The organs of the FEMALE REPRODUCTIVE SYSTEM consist of the PRIMARY SEX ORGANS—the OVARIES (FEMALE GONADS), and the SECONDARY SEX ORGANS— the INTERNAL ACCESSORY ORGANS (two FALLOPIAN TUBES, the UTERUS and VAGINA), EXTERNAL STRUCTURES (the VULVA) and accessory glands (the PARAURETHRAL and VESTIBULAR GLANDS and the BREASTS or MAMMARY GLANDS).

The FEMALE REPRODUCTIVE STRUCTURES are: PAIRED OVARIES, A SYSTEM OF CHANNELS (INTERNAL ORGANS), EXTERNAL STRUCTURES (VULVA: PUDENDUM) AND ACCESSORY GLANDS. The paired ovaries produce, develop and eventually expel the OVA (EGG CELLS)—FEMALE GERM CELLS, as well as secrete the ESTROGENIC HORMONES (the female sex hormones, ESTRADIOL and ESTRONE). At the onset of puberty, estrogens are responsible for the development of secondary sex characteristics, secondary sex organs and cyclical changes leading to MENARCHE, the onset of MENSTRUATION.

The SYSTEM OF CHANNELS (INTERNAL ORGANS) consists of the FALLOPIAN TUBES, the UTERUS and the VAGINA. The fallopian tubes convey the ova from the OVARIES to the UTERUS. They are also the site of FERTILIZATION (sperm travel from the UTERUS toward the ovaries). The UTERUS is the site of zygote implantation; it also provides nourishment, protects and sustains an embryo and fetus during pregnancy. The internal channel system has an active role in PARTURITION (birth of the baby), as the muscular layer (MYOMETRIUM) produces powerful rhythmic contractions. The uterine mucosal lining (the ENDOMETRIUM) undergoes changes associated with the MENSTRUAL CYCLE. The VAGINA houses the erect penis (and semen) during copulation (and ejaculation). It is the lower part of the BIRTH CANAL and the excretory duct for uterine secretions and menstrual flow.

The EXTERNAL STRUCTURES comprising the VULVA (PUDENUM) are the MONS PUBIS, LABIA MAJORA and MINORA, the CLITORIS and the VESTIBULE. The vulva form margins and protective barriers for the vagina and urethra. The clitoris is associated with sexual pleasure (stimulation of sensory nerve endings).

The ACCESSORY GLANDS produce secretions that moisten and lubricate the vestibule and vaginal opening during intercourse (vestibular glands). They produce milk secretions for infant nourishment (mammary).

PUBERTY IN THE FEMALE

Girls attain puberty six months to a year earlier than boys (usually between the age of 12 and14, but sometimes even as early as 9 years or as late as 17 years). The transition to puberty is more abrupt than in boys, due to the ONSET of the MENSTRUAL CYCLE (MENSTRUATION), called MENARCHE. This changes the female child into a woman able to bear children. Its average onset is at 12.5 years of age.

As in the male, maturational changes in the HYPOTHALAMUS produce an increase in GnRH (formerly known as LRH, the LUTEINIZING-RELEASING HORMONE) and its secretion into the HYPOTHALAMO-HYPOPHYSEAL PORTAL VESSELS. GnRH (gonadotrophic-releasing hormone) stimulates release of FSH and LH from the ANTERIOR LOBE of the PITUITARY GLAND directly into the blood stream. FSH and LH stimulate secretion of SEX STEROID HORMONES (ESTROGENS) inside the OVARIES. (A decrease in NEGATIVE FEEDBACK INHIBITION BY THE SEX STEROIDS to the sensitivity of the HYPOTHALAMUS and PITUITARY occurs, thus increasing FSH and LH release.) These SEX STEROIDS (especially ESTRADIOL and PROGESTERONE) cause the appearance and maintenance of SECONDARY SEX CHARACTERISTICS and SECONDARY SEX ORGAN DEVELOPMENT and FUNCTIONING . The FALLOPIAN TUBES lengthen, the VAGINA epithelium thickens, the UTERINE muscle layer (MYOMETRIUM) enlarges and the uterine inner lining (ENDOMETRIUM) proliferates.

MENARCHE (under the control of the HYPOTHALAMUS and PITUITARY) signals the onset of the regulation of all female reproductive activities in a CONTINUAL CYCLICAL TIME PATTERN (an average of every 28 days), ending about 36 years later in MENOPAUSE. This FEMALE HORMONAL CYCLE manifests itself in two ways, effected by the same hormonal changes: the OVARIAN CYCLE focuses on the changes occurring in the OVARIES, and the menstrual cycle focuses on the flow or nonflow of blood from the VAGINA.

(For more on the 28-day MENSTRUAL CYCLE, see WWW. MCMURTRIESANATOMY.COM.)

Primary Sex Organs or Gonads

A | Paired ovaries | (singular: ovary)

Secondary sex organs

INTERNAL ORGANS:
SYSTEM OF CHANNELS

B | 2 Uterine (fallopian) tubes | (oviducts)

C | 1 Uterus |

D | 1 vagina |

EXTERNAL ORGANS:
VULVA (PUDENDUM)

E | Mons pubis | (veneris)

F | Labia majora |

G | Labia minora |

H | Clitoris |

I | Vestibule |

ACCESSORY GLANDS:

J | Paraurethral glands | (Skene's glands)

K | Vestibular glands | (Bartholin's glands)

L | Mammary glands | (Breasts)

COLOR GUIDELINES:
A = reddish brown;
B = light green;
C = light red; D = pink;
E = yellow; F = flesh-tone;
G = light orange; H = orange.

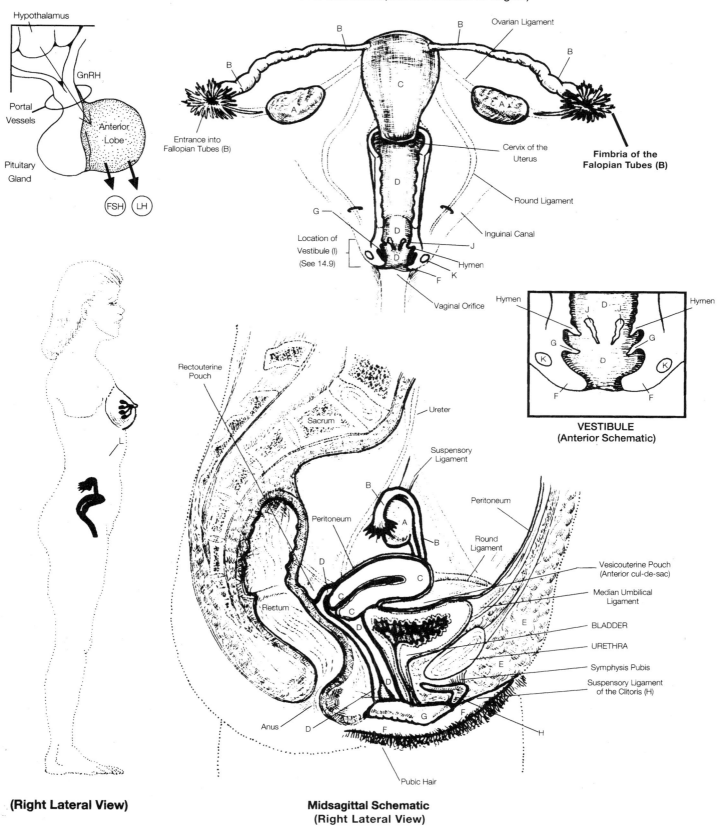

Female Reproductive Organs
Anterior Schematic(Coronal section of Vagina)

Hypothalamus

GnRH

Portal Vessels

Anterior Lobe

Pituitary Gland

FSH   LH

Ovarian Ligament

B   B   B   B

B

A   C   A

Entrance into Fallopian Tubes (B)

Cervix of the Uterus

Fimbria of the Falopian Tubes (B)

Round Ligament

Inguinal Canal

G

Location of Vestibule (I) (See 14.9)

D

D

D

J

Hymen

K

F

Vaginal Orifice

Hymen   Hymen

J   D

G   G

K   D   K

F   F

VESTIBULE
(Anterior Schematic)

Rectouterine Pouch

Sacrum

Ureter

Suspensory Ligament

B

A

B

Peritoneum

Round Ligament

Peritoneum

D

C

C

C

Rectum

D

Vesicouterine Pouch (Anterior cul-de-sac)

Median Umbilical Ligament

E

BLADDER

URETHRA

E

Symphysis Pubis

Suspensory Ligament of the Clitoris (H)

Anus   D

G   F

F

H

L

Pubic Hair

(Right Lateral View)

Midsagittal Schematic
(Right Lateral View)

A Left Ovary

HISTOLOGICAL GENERAL STRUCTURE OF
EACH OVARY: FOUR LAYERS

1 Cuboidal epithelium

2 Tunica albuginea

Stroma

3 Medulla Vascular

4 Cortex Contains ovarian follicles

BLOOD CIRCULATION

5 Ovarian artery

6 Ovarian vein

7 Uterine artery Ovarian branches

8 Uterine vein

9 Capillaries in medulla

10 Mesovarium

Anchors to the mesosalpinx of the
broad ligament

11 Ovarian ligament

Anchors to the uterus

12 Suspensory ligament

Anchors to the posterolateral pelvic wall

SUPPORTIVE CONNECTING STRUCTURES
ATTACHING TO EACH OVARY

13 Broad ligament

14 Mesosalpinx of the
broad ligament

15 Round ligament

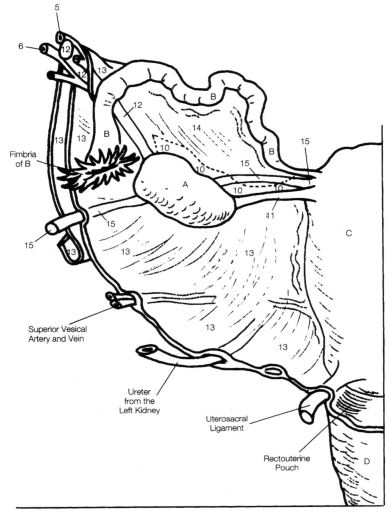

SUPPORTIVE CONNECTING STRUCTURES of LEFT OVARY
(Dorsal-Posterior View)
(LOOKING FROM THE REAR)

Figure labels: Fimbria of B; Superior Vesical Artery and Vein; Ureter from the Left Kidney; Uterosacral Ligament; Rectouterine Pouch

The FEMALE PRIMARY SEX ORGANS consist of the OVARIES and SUPPORTING STRUCTURES. The ovaries are paired glands, almond-shaped, positioned on each side of the UTERUS in the UPPER PELVIC CAVITY, and held in position by several ligaments. Each ovary is nestled in a depression of the posterior abdominal wall called the OVARIAN FOSSA. On the MEDIAL PORTION of each OVARY is an entranceway called the HILUM. All vessels and nerves to each ovary enter only through the HILUM, which is supported by the OVARIAN LIGAMENT. The LATERAL PORTION of each ovary is in close contact with the open ends of each fallopian tube through which the released OVUM from a ruptured GRAAFIAN FOLLICLE enters for possible fertilization from roving spermatozoa.

Regarding their HISTOLOGY, each ovary consists of the CUBODIAL EPITHELIUM, the TUNICA ALBUGINEA and the STROMA (which consists of an inner central vascular MEDULLA and an outer CORTEX area). The cubodial epithelium covers the free surface. It is one layer of simple cubodial epithelium. The TUNICA ALBUGINEA is directly beneath the GERMINAL EPITHELIUM. It is a collagenous connective tissue capsule. The STROMA is an interior principal substance consisting of the CORTEX (outer, dense layer), containing the OVARIAN FOLLICLES (in various stages of development) and the MEDULLA (the inner, loose vascular layer).

A series of ligaments MAINTAIN THE OVARIAN POSITION. The BROAD LIGAMENT is the principal supporting membrane of the female reproductive tract. The MESOVARIUM is the portion of the double-layered fold of peritoneum that connects the anterior border of the ovary to the posterior layer of the broad ligament. It surrounds the ovary and the OVARIAN LIGAMENT. The medial position

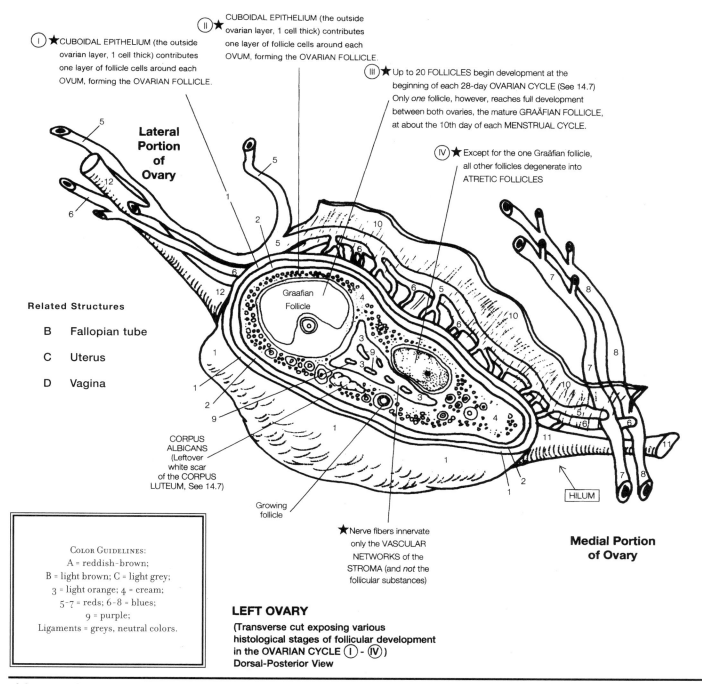

I ★ CUBOIDAL EPITHELIUM (the outside ovarian layer, 1 cell thick) contributes one layer of follicle cells around each OVUM, forming the OVARIAN FOLLICLE.

II ★ CUBOIDAL EPITHELIUM (the outside ovarian layer, 1 cell thick) contributes one layer of follicle cells around each OVUM, forming the OVARIAN FOLLICLE.

III ★ Up to 20 FOLLICLES begin development at the beginning of each 28-day OVARIAN CYCLE (See 14.7) Only *one* follicle, however, reaches full development between both ovaries, the mature GRAÄFIAN FOLLICLE, at about the 10th day of each MENSTRUAL CYCLE.

IV ★ Except for the one Graäfian follicle, all other follicles degenerate into ATRETIC FOLLICLES

Lateral Portion of Ovary

5

Graafian Follicle

Related Structures

B   Fallopian tube

C   Uterus

D   Vagina

CORPUS ALBICANS (Leftover white scar of the CORPUS LUTEUM, See 14.7)

Growing follicle

★ Nerve fibers innervate only the VASCULAR NETWORKS of the STROMA (and *not* the follicular substances)

HILUM

Medial Portion of Ovary

COLOR GUIDELINES:
A = reddish-brown;
B = light brown; C = light grey;
3 = light orange; 4 = cream;
5-7 = reds; 6-8 = blues;
9 = purple;
Ligaments = greys, neutral colors.

LEFT OVARY
(Transverse cut exposing various histological stages of follicular development in the OVARIAN CYCLE Ⓘ - Ⓘⓥ)
Dorsal-Posterior View

of the ovary closest to the uterus is attached to the OVARIAN LIGAMENT of the UTERUS; the LATERAL POSITION of the ovary is attached to the SUSPENSORY LIGAMENT. BLOOD SUPPLY comes from the ovarian arteries arising from the lateral sides of the ABDOMINAL AORTA just below the origin of the RENAL ARTERIES. The OVARIAN ARTERIES anastomoses with OVARIAN BRANCHES of the UTERINE ARTERIES. BLOOD DRAINAGE is through the OVARIAN VEINS. The RIGHT OVARIAN VEIN empties directly into the INFERIOR VENA CAVA. The LEFT OVARIAN VEIN empties into the LEFT RENAL VEIN.

The BROAD LIGAMENT consists of two leaves of a part of the PARIETAL PERITONEUM *between* which are found the remnants of the MESONEPHRIC DUCT, cellular tissues, major blood vessels of the PELVIS, the URETER from the KIDNEY, the UTEROSACRAL LIGAMENT and the ROUND LIGAMENT. It is attached to lateral borders of the

uterus from the insertion of the FALLOPIAN TUBE above to the PELVIC WALL below. The MESOSALPINX of the BROAD LIGAMENT is the free margin of its upper division, within which lies the fallopian tube (oviduct) The MESOVARIUM is that portion of the peritoneal fold that connects the anterior border of the ovary to the posterior layer of the broad ligament. It anchors to the mesosalpinx of the broad ligament.

The ROUND LIGAMENT is not directly attached to the ovary. It is one of four principal supporting ligaments of the UTERUS. It is anchored immediately below and in front of the entrance of the FALLOPIAN TUBE into the UTERUS. Each round ligament extends laterally inside the BROAD LIGAMENT to the PELVIC WALL, where it passes through the INGUINAL RING. Each terminates in a LABIA MAJOR.

The ovarian cycle is a 28-day cycle in which hormonal fluctuations effect changes in the OVARIES. OÖGENESIS (OVIGENESIS) is the process by which in each cycle one mature ovum is prepared (from a set of up to 20 immature eggs) for fertilization. There are 6–7 million OÖGONIA (immature epithelial germ egg cells) produced by MITOSIS in the FETAL OVARIES by the end of the 5th month of gestation, then stopped. Production of the new oogonia never begins again. At the end of gestation (approximately the 9th month) the OÖGONIA are called PRIMARY OÖCYTES (IMMATURE OVA) as they begin MEIOSIS. (As in the male, genesis of the gonadal cells is arrested at PROPHASE I of the 1ST MEIOTIC DIVISION.) Then, the PRIMARY OÖCYTES are surrounded by one layer of FOLLICULAR CELLS from the CUBOIDAL EPITH-ELIUM, forming the tiny PRIMORDIAL FOLLICLES. In response to FSH stimulation from the anterior lobe of the pituitary gland, some primordial follicles get larger (including the inner OÖCYTES). The surrounding FOLLICULAR CELLS divide to produce many small GRANU-LOSA CELLS. They fill the follicle and surround the oocyte, thus forming a PRIMARY FOLLICLE.

Following OVULATION (under the influence of LH), the empty RUPTURED FOLLICLE undergoes both structural and biochemical changes in transforming into a CORPUS LUTEUM, an endocrine structure. The structure of the secondary (growing) follicle changes under FSH stimulation. The antrum fills with liquid folliculi fluid; the granu-lose cells form the cumulus oophorus (a mound that supports the ovum), the corona radiate (a ring encircling the ovum) and the granulose ring (a ring around the circum-ference of the follicle). The zona pellucida, a thin gel-like layer of proteins and polysac-charides, forms between the oöcyte and the corona radiata.

Through further FSH stimulation, some PRIMARY FOLLICLES develop a fluid-filled ANTRUM, and are now called GROWING SECONDARY FOLLICLES. The primary oocyte completes its first meiotic division in a SECONDARY FOLLICLE: two OÖCYTES result, one SECONDARY OÖCYTE (gets all the cytoplasm) and one POLAR BODY, which degenerates. The SECONDARY OÖCYTE in a SECONDARY FOLLICLE now enters the 2nd MEIOTIC DIVISION. It, too, is arrested at METAPHASE II, and is not completed unless fertilization occurs. By the 10th to the 14th day after the onset of MENSTRUATION, only one follicle has continued to grow into a MATURE GRAÄFIAN FOLLICLE (under further FSH stimulation). All other SECONDARY FOLLI-CLES degenerate into ATRETIC FOLLICLES.

In the process of OVULATION, the graäfian follicle containing the MATURE OVUM (sec-ondary oöcyte) bulges from the surface, ruptures and releases the mature ovum, which finds its way into the opening of a FALLOPIAN TUBE, to await a roving spermatozoan.

During the first half of the menstrual cycle the secondary sex organs stimulate duct development in the mammary glands, induce the proliferation and contraction of the fal-lopian tubes, regenerate the glands and fibrous tissue in endometrium in the uterus, and cornify the vagina. During the 2nd half of the menstrual cycle, the secondary sex organs stimulate alveolar development in the mammary glands, mucify and relax the fallopian tubes, effect secretions of glands of the endometrium in the uterus, and mucify the vagina.

Color Guidelines:
A = yellow; b, B = orange; G = purple;
9 = red; 10–12 = yellow-orange, ochres;
c, d, e = light cool colors;
f = reddish-brown; g = light brown;
h = cream; i = light pink

FOLLICULAR DEVELOPMENT IN THE OÖGENIC EPITHELIUM OF THE CORTEX

Gestating OÖGONIA

(a) PRIMARY OÖCYTE (immature ovum)

2 Primordial follicle

3 Primary follicle

4 Secondary (growing) follicle

(5) Mature (Graäfian) follicle

6 Atretic follicles

OVULATION

7 Ruptured follicle

(8) Released secondary oocyte

CORPUS LUTEUM

9 Corpus Hemorragicum
Huge blood clot formed in cavity

10 Young corpus luteum

11 Mature corpus luteum

12 Corpus Albicans
Replaces regressing corpus luteum

TYPES OF OVA

a Primary oöcyte
(immature ovum)

b B Secondary oöcyte
(mature ovum)

FOLLICULAR CAVITIES

c Zona pellucida

d Follicular fluid cavities

e Antrum (fluid-filled cavity)

FOLLICULAR CELL GROUPS

f Granulosa (follicular) cells

g Theca folliculi
(theca interna and externa)

h Cumulus Oöphorus

I Corona radiata

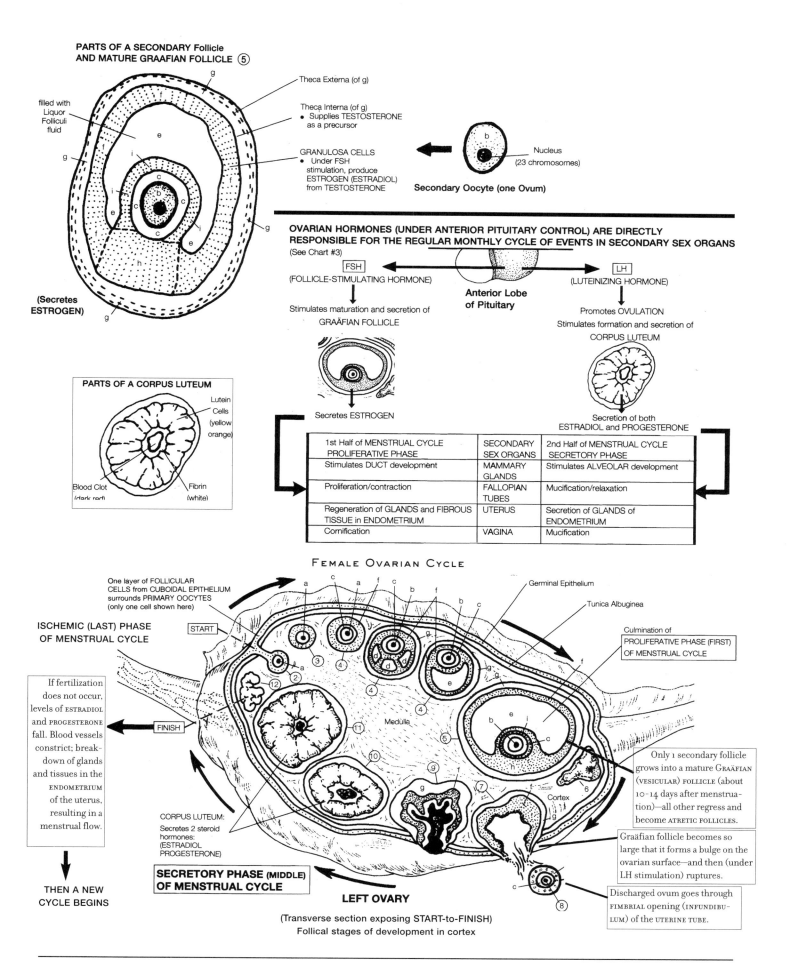

PARTS OF A SECONDARY Follicle AND MATURE GRAAFIAN FOLLICLE ⑤

filled with Liquor Folliculi fluid

Theca Externa (of g)

Theca Interna (of g)
- Supplies TESTOSTERONE as a precursor

GRANULOSA CELLS
- Under FSH stimulation, produce ESTROGEN (ESTRADIOL) from TESTOSTERONE

(Secretes ESTROGEN)

Nucleus (23 chromosomes)

b

Secondary Oocyte (one Ovum)

OVARIAN HORMONES (UNDER ANTERIOR PITUITARY CONTROL) ARE DIRECTLY RESPONSIBLE FOR THE REGULAR MONTHLY CYCLE OF EVENTS IN SECONDARY SEX ORGANS

(See Chart #3)

FSH (FOLLICLE-STIMULATING HORMONE)

Stimulates maturation and secretion of GRAÄFIAN FOLLICLE

Secretes ESTROGEN

Anterior Lobe of Pituitary

LH (LUTEINIZING HORMONE)

Promotes OVULATION

Stimulates formation and secretion of CORPUS LUTEUM

Secretion of both ESTRADIOL and PROGESTERONE

PARTS OF A CORPUS LUTEUM

Lutein Cells (yellow orange)

Blood Clot (dark red)

Fibrin (white)

1st Half of MENSTRUAL CYCLE PROLIFERATIVE PHASE	SECONDARY SEX ORGANS	2nd Half of MENSTRUAL CYCLE SECRETORY PHASE
Stimulates DUCT development	MAMMARY GLANDS	Stimulates ALVEOLAR development
Proliferation/contraction	FALLOPIAN TUBES	Mucification/relaxation
Regeneration of GLANDS and FIBROUS TISSUE in ENDOMETRIUM	UTERUS	Secretion of GLANDS of ENDOMETRIUM
Cornification	VAGINA	Mucification

FEMALE OVARIAN CYCLE

One layer of FOLLICULAR CELLS from CUBOIDAL EPITHELIUM surrounds PRIMARY OOCYTES (only one cell shown here)

Germinal Epithelium

Tunica Albuginea

START

ISCHEMIC (LAST) PHASE OF MENSTRUAL CYCLE

Culmination of PROLIFERATIVE PHASE (FIRST) OF MENSTRUAL CYCLE

If fertilization does not occur, levels of ESTRADIOL and PROGESTERONE fall. Blood vessels constrict; break-down of glands and tissues in the ENDOMETRIUM of the uterus, resulting in a menstrual flow.

FINISH

Medulla

Only 1 secondary follicle grows into a mature GRAÄFIAN (VESICULAR) FOLLICLE (about 10-14 days after menstruation)—all other regress and become ATRETIC FOLLICLES.

CORPUS LUTEUM: Secretes 2 steroid hormones: (ESTRADIOL PROGESTERONE)

Cortex

Graäfian follicle becomes so large that it forms a bulge on the ovarian surface—and then (under LH stimulation) ruptures.

SECRETORY PHASE (MIDDLE) OF MENSTRUAL CYCLE

THEN A NEW CYCLE BEGINS

LEFT OVARY

Discharged ovum goes through FIMBRIAL opening (INFUNDIBU-LUM) of the uterine tube.

(Transverse section exposing START-to-FINISH) Follical stages of development in cortex

The unfertilized OVUM enters one of the FALLOPIAN TUBES (OVIDUCTS) between a finger-like FIMBRIA (sometimes touching the ovary with a single OVARIAN FIMBRIA). CILIARY ACTION and PERISTALSIS from the tube's inner mucosa facing the inner lumen cavity, conduct the OVUM into the wide INFUNDIBULUM of the tube. After SPERM travels up the UTERUS and into the FALLOPIAN TUBES toward the OVARY, FERTILIZATION takes place with the OVUM in the FALLOPIAN TUBE. Ciliary action and peristalsis move the ZYGOTE (if FERTILIZATION occurs) to the uterus.

The uterus is supported by four pairs of ligaments derived from serosa. The thick three-layered muscular MYOMETRIUM is responsible for the muscular contractions needed during LABOR and PARTURITION (delivery). The zygote enters the UTERUS from the FALLOPIAN TUBE and implants in the two-layered mucosal ENDOMETRIUM which faces the inner lumen cavity of the uterus. If fertilization does not occur, the STRATUM FUNCTIONALE, a superficial layer composed of columnar epithelium and secretory glands, is shed as MENSES during menstruation, and is built up again by steroid ovarian hormones. The STRATUM BASALE is a regenerative deeper layer of the endometrium that replenishes STRATUM FUNCTIONALE after each menstruation.

The VAGINA is the copulation passageway. It delivers sperm from intromission of the PENIS to the cervix of uterus. It is also a canal for the discharge of menstrual flow and the birth canal for delivery. The anterior, posterior and two lateral spaces (the four FORNICES OF THE VAGINA—the FORNIX) into which the upper vagina is divided are recesses formed by protrusion of the CERVIX UTERI into the vagina (forming the upper horizontal crescent-shaped walls). The BLADDER is situated adjacent to the anterior vaginal wall. The RECTUM is situated behind the posterior wall. The VAGINA represents a *potential space*, the walls of which are in contact with each other (forming a vertical slit close to the VULVA) The three layers of the vagina consist of an inner mucosa (consisting of stratified squamous epithelium—mucous membrane that forms a series of transverse folds called VAGINA RUGAE), a middle muscularis and an outer fibrous layer.

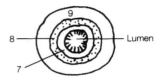

CROSS-SECTION of a UTERINE TUBE

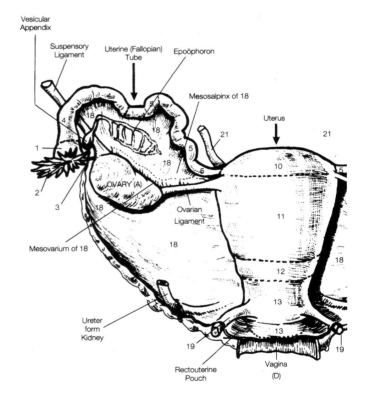

Dorsal-Posterior View

(VIEWED FROM THE REAR)

COLORING NOTES:
B = light greens and blues;
C = light reds; D = pink;
Ligaments = grays and light colors;
Histological layers = warm colors.

B: Fallopian (uterine) tubes

1	Infundibulum
2	Fimbria
3	Ovarian Fimbria
4	Ampulla (Fertilization site)
5	Isthmus
6	Intramural

Three fallopian layers

7	Mucosa (ciliated columnar)
8	Muscularis
9	Serosa

Part of the visceral petoneum

C: Uterus

STRUCTURE

10	Fundus
11	Body
12	Isthmus
13	Cervix (Cervix uteri [neck])

INNER CAVITIES (AND THEIR OPENINGS)

14	Uterine cavity
15	Internal os
16	Cervical canal
17	External os

SUPPORT OF THE UTERUS

Four paired LIGAMENTS

18	Broad ligaments Peritoneum
19	Uterosacral ligaments
20	Cardinal ligaments

Lateral cervical

21	ROUND LIGAMENTS

Three Uterine layers

22	Endometrium (Two-layered mucosa)

A superficial, shedding layer,
stratumbasale, is a deeper regenetive layer

23	Myometrium

Three-layered, thick muscularis

24	Perimetrium (serosa)

D: Vagina

25	Fornix of vagina (4 fornices)

Three vaginal layers:

Mucosa

Muscularis

Fibrous layer

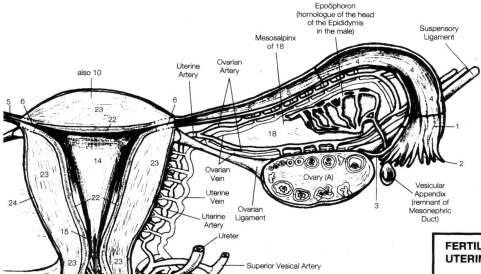

**Coronal Section
Anterior View**

(VAGINA SHOWN
EXPANDED OUT)

THE VAGINA IS HIGHLY VASCULARIZED
BY A COMPLEX, EXTENSIVE PLEXUS
SURROUNDING IT, CALLED THE
VAGINAL AZYGOS ARTERIES (UTERINE,
MIDDLE RECTAL, PUDENDAL AND VAGI-
NAL BRANCHES OF THE INTERNAL
ILIAC ARTERY).

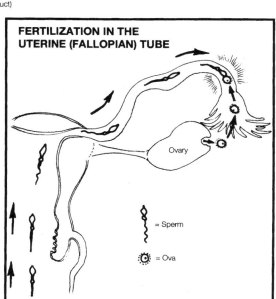

**FERTILIZATION IN THE
UTERINE (FALLOPIAN) TUBE**

= Sperm

= Ova

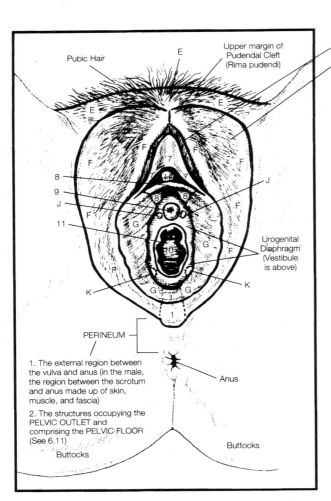

MEDIAL SURFACES
OF THE LABIA
MAJORA (F) UNITE
TO FORM THE ANTE-
RIOR COMMISSURE

THE ROUND
LIGAMENTS
END IN THE
LABIA MAJORA

Pubic Hair

Upper margin of
Pudendal Cleft
(Rima pudendi)

Urogenital
Diaphragm
(Vestibule
is above)

PERINEUM

1. The external region between
the vulva and anus (in the male,
the region between the scrotum
and anus made up of skin,
muscle, and fascia)

2. The structures occupying the
PELVIC OUTLET and
comprising the PELVIC FLOOR
(See 6.11)

Buttocks

Anus

Buttocks

**Inferior View
(LABIA MAJORA and MINORA
spread and folded out to
expose VESTIBULE)**

E	**Mons pubis**	
	Mons veneris (pubic eminence)	

THE VULVA (PUDENDUM)

F	**2 Labia majora**	
	Lateral borders of the vulva	
G	**2 Labia minora**	
H	**Clitoris**	
I	**Vestibule**	
1	**Fourchette**	Frenulum of both labia
2	**Perineal body**	
	"Central tendon" of perineal muscles	

3 PORTIONS OF THE CLITORIS (H)

3	**Glans**	
	(Exposed portion composed of erectile tissue)	
4	**Body**	
	Consists of 2 fused crura (continuations of 5)	
5	**2 Corpora cavernosa**	
6	**Suspensory ligament**	

UNITING CLOAKS OF
THE LABIA MINORA (G)

7	**Prepuce of clitoris**	
	Anterior preputium clitoridis	
8	**Frenulum of clitoris**	
	Posterior frenulum clitoridis	

VESTIBULE: SPATIAL CLEFT BETWEEN
ATTACHMENTS OF LABIA MINORA
(RECEPTACLE OF ORIFICES)

External urethral orifice

10	**Vaginal orifice**	
11	**Hymen**	Carunculae
J	**Paraurethral glands**	
	Skene's glands	
K	**Vestibular glands**	
	Bartholin's glands	
12	**Vestibular bulbs**	

The MONS PUBIS is the pad of adipose tissue and coarse skin that cushions the SYMPH-
YSIS PUBIS and VULVA during copulation. It is hairy after puberty. The EXTERNAL GENI-
TALIA (the VULVA or PUDENDUM) serve to form margins and to enclose and protect the
VAGINAL ORIFICE (the opening of the VAGINA) and other external reproductive organs. The
VULVA lies posterior to the MONS PUBIS and consists of the two LABIA MAJORA, two LABIA
MINORA, a CLITORIS, and a VESTIBULE (including the VAGINAL and URETHRAL ORIFICES and the
BULBS and GLANDS of the VESTIBULE).

The LABIA MAJORA are two hairy, thickened longitudinal folds enclosing and protecting the VULVA (homologous to the SCROTUM). They are separated by a spatial cleft called
the RIMA PUDENI into which the URETHRA and VAGINA opens. Their medial surfaces unite
above the clitoris to form the ANTERIOR COMMISSURE. The LABIA MINORA are two small,
hairless longitudinal folds that protect the VAGINAL and URETHRAL openings and enclose
the vestibule. They lie between the LABIA MAJORA and the HYMEN. Anteriorly, they split to
form the PREPUCE and FRENULUM of the CLITORIS.

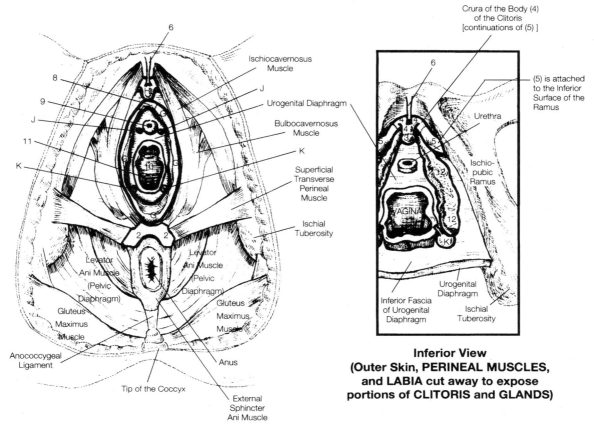

Crura of the Body (4)
of the Clitoris
[continuations of (5)]

(5) is attached
to the Inferior
Surface of the
Ramus

Ischiocavernosus
Muscle

Urogenital Diaphragm

Bulbocavernosus
Muscle

Superficial
Transverse
Perineal
Muscle

Ischial
Tuberosity

Urethra

Ischio-
pubic
Ramus

Urogenital
Diaphragm

Ischial
Tuberosity

Inferior Fascia
of Urogenital
Diaphragm

Levator
Ani Muscle
(Pelvic
Diaphragm)

Levator
Ani Muscle
(Pelvic
Diaphragm)

Gluteus
Maximus
Muscle

Gluteus
Maximus
Muscle

Anococcygeal
Ligament

Tip of the Coccyx

Anus

External
Sphincter
Ani Muscle

Inferior View
♀ PERINEUM
(Outer Skin Layer and LABIA cut away)

Inferior View
(Outer Skin, PERINEAL MUSCLES,
and LABIA cut away to expose
portions of CLITORIS and GLANDS)

COLOR GUIDELINES:
E = light flesh; F = flesh; G = light orange;
H = yellow-orange; 1 = orange; 2 = light grey,;
3 = yellow; 4 = red; 5 = red-orange;
6 = light brown; 7 = light purple; 8 = purple;
9 = light pink; 10 = pink; 11 = light red;
J, K, 12 = greens and blues.
(The vestibule [I] is a space and thus
cannot be colored.)

The CLITORIS is a small, erectile organ. It corresponds to the origin and structure of the penis, although it does not have a urethra, because the female anatomy is more specialized. It is richly supplied with sensory nerve endings that enhance pleasure during sexual stimulation. It is located beneath the anterior commissure of the labia majora and partially hidden by the cloaks of the labia minora. It measures approximately 2 cm (0.8 in) long by 0.5 cm (0.2 in) wide. In young girls, the medial surfaces of the labia majora are in contact with each other, concealing the LABIA MINORA and vesti-bule. In older women, the labia minora may protrude between the labia majora.

The vestibule is a longitudinal spatial cleft enclosed by the LABIA MINORA. It contains the openings of the VAGINA and URETHRA.

During sexual excitement, the VAGINAL OPENING is lubricated by secretions from a pair of VESTIBULAR GLANDS located within the wall just inside the vaginal orifice. The lateral walls of the VESTIBULE are formed inwardly by vascular, erectile tissue (the VESTIBULAR BULBS).

The HYMEN (CARUNCULAE) is a fold of mucous membrane that partially covers the entrance to the vagina. (Contrary to folklore, rupture or absence of the hymen cannot be used to prove or disprove virginity or history of sexual intercourse. Pregnancy can occur with the hymen intact.)

Parasympathetic nervous stimulation causes a dilation of arterioles of the genital erectile tissue and (as in the male) compresses venous return, thus causing swelling or "erection."

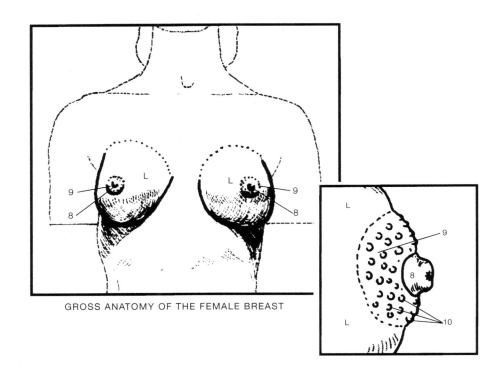

GROSS ANATOMY OF THE FEMALE BREAST

L   Breasts (Left and Right)

Mammary Glands
(GLANDULAR STRUCTURE)

1   Lobe

2   Lobules

3   Glandular alveoli (secrete milk)

4   Secondary tubules

5   Mammary ducts

6   Lactiferous ampullae (sinuses)

7   Lactiferous ducts

(GROSS ANATOMY)

8   Nipple   Primary areola

9   Areola

10   Areolar glands

(LYMPHATIC DRAINAGE)

11   Axillary nodes

12   Axillary-apical nodes

13   Parasternal nodes

14   Nodes to opposite breast

15   Nodes to rectus sheath and diaphragm

RELATED STRUCTURES

16   Suspensory ligaments of Cooper
Supports breasts

17   Superficial fascia
Adipose tissue

18   Pectoralis major muscle
And minor

19   Deep fascia of pectoralis major muscle

20   Intercostal muscles

21   Serratus anterior muscle

22   Clavicle

23   Ribs

Two compound MAMMARY GLANDS are located within the BREASTS, one in each breast. Structurally, the breasts are part of the INTEGUMENTARY SYSTEM, being modified sweat glands embedded in superficial fascia (fat). The amount of adipose tissue determines breast size and shape only. Functionally, they are part of the REPRODUCTIVE SYSTEM, since they secrete MILK for the nourishment of the infant (triggered by the hormones PROLACTIN and OXYTOCIN). At PUBERTY, the OVARIES secrete ESTROGEN, which stimulates the growth and development of the duct system, mammary glands (alveoli) and breast adipose tissue.

Each MAMMARY GLAND is composed of between 15–20 lobes. Each lobe has its external own external drainage pathway. Each LOBE is subdivided into LOBULES, which contain the GLANDULAR ALVEOLI. The ALVEOLAR clusters secrete MILK only in response to secretions of LUTEOTROPHIN and oxytocin HORMONES. Milk is channeled through a series of SECONDARY TUBULES and MAMMARY DUCTS, being stored in LACTIFEROUS AMPULLAE (SINUSES) before draining at the nipple. The NIPPLE remains pliable due to secretions of AREOLAR GLANDS within the pigmented circular AREOLA surrounding it.

Support of the BREASTS comes from the SUSPENSORY LIGAMENTS of COOPER, which run between the LOBULES extending from the skin to the deep fascia overlying the PECTORALIS MUSCLE. The lateral margin of the breast is alongside the anterior border of the axilla (armpit). The axillary tail of the breast comes in close contact with the axillary vessels, and this region is clinically associated with a high incidence of breast cancer within the lymphatic drainage.

During pregnancy, the progesterone secreted by the CORPUS LUTEUM in the OVARY and PLACENTA acts synergistically with ESTROGENS to bring the ALVEOLI to complete development.

The breasts develop during pregnancy. During the first 6–12 weeks they exhibit fullness and tenderness. Erectile tissues develop in the nipples and pigment is deposited around the nipple (the primary areola). During weeks 16–20, the secondary areola

LYMPHATIC DRAINAGE CHANNELS OF THE BREAST
(PARTIALLY SECTIONED)

shows small whitish spots in pigmentation due to hypertrophy of the SEBACEOUS GLANDS (GLANDS OF MONTGOMERY) present in the areola surrounding the nipple.

In the first two or three days after birth (and before the onset of true lactation) the breasts secrete colostrum, a thin, yellowish fluid containing large quantities of proteins and calories (in addition to antibodies and lymphocytes important for protective immunity against infection).

After parturition (CHILDBIRTH, DELIVERY), PROLACTIN (LUTEOTROPHIN) in conjunction with ADRENAL CORTICOIDS initiates LACTATION (MILK SECRETION). OXYTOCIN from the POSTERIOR PITUITARY GLAND induces the ejection of milk. The sucking and MILKING REFLEX restimulates milk secretion and discharge.

FEMALE BREAST—SAGITTAL SCHEMATIC (SIMPLIFIED)

BIBLIOGRAPHY

ATLASES AND TEXTS

Albertine, K.H., *Anatomica,* Global Book Company, 2000.

Dorland's Illustrated Medical Dictionary. 25th edition, Philadelphia: W. B. Saunders & Co., 1974.

Grant, J. C., *An Atlas of Anatomy*, 4th edition, Baltimore: Williams & Wilkens Co., 1956.

Hale, R. B., edition, *Artistic Anatomy*, P. Richer, New York: Watson-Guptill Publications, 1971.

Hole, John W., Jr., *Human Anatomy and Physiology*, 4th edition, Dubuque, IA: Wm. C. Brown Publishers, 1987.

McClintic, J. R., *Basic Anatomy and Physiology of the Human Body*, 2nd edition, New York: John Wiley & Sons, 1980.

McClintic, J. R. *Human Anatomy*. St. Louis: C. V. Mosby Co., 1983.

McNaught, A. B., and R. Callander. *Illustrated Physiology*. 3rd editon, New York: Churchill Livingstone, 1975.

Moore, K.L., *Clinical Anatomy,* Lippincott Williams & Wilkins, 2002.

Netter, F. H. *Atlas of Human Anatomy,* 3rd edition, ICON Learning Systems, 2003.

Netter, F. H. *Ciba Collection of Medical Illustrations*. Volumes I-III. Summit, NJ: Ciba Pharmaceutical Co., 1962.

Peck, S. R. *Atlas of Human Anatomy for the Artist*. 11th edition, New York: Oxford University Press, 1968.

Romanes, G. J., editor. *Cunningham's Textbook of Anatomy*. 10th edition, New York: Oxford University Press, 1964.

Spence, A. P., and E. B. Mason. *Human Anatomy and Physiology*. 2nd edition, Reading, MA: Benjamin and Cummings, 1983.

Stedman's Medical Dictionary. 25th ed. Phila.: W. B. Saunders Co., 1974.

Taber's Cyclopedic Medical Dictionary. 15th edition, Philadelphia: F. A. Davis Co., 1985.

Tortora, G. J. *Principles of Human Anatomy*. 3rd edition, Harper & Row, 1983.

Truex, R. C., and M. B. Carpenter. *Human Neuroanatomy*. 8th edition, Baltimore: Williams & Wilkens Co., 1982.

Van De Graaff, K. M., and S. Fox. *Concepts of Human Anatomy and Physiology*. Dubuque, IA: Wm. C. Brown Publishers, 1989.

Van Dc Graaff, K. M., and R. W. Rhees. *Schaum's Outline of Theory and Problems of Human Anatomy and Physiology*. New York: McGraw-Hill Book Co., 1987.

Vannini, V., and G. Pogliani, ed. *The Color Atlas of Human Anatomy*. New York: Harmony Books, 1980.

GENERAL WORKS

Ackerman, D., *A Natural History of the Senses,* Peter Smith, 2002.

Angier, N., *Woman: An Intimate Geography,* Anchor, 2000.

Barash, D.P., *The Mammal in the Mirror,* St. Martin's, 1999.

Bondeson, J., *A Cabinet of Medical Curiosities,* Norton, 1999.

Damasio, A., *Looking for Spinoza: Joy Sorrow and the Feeling Brain,* Harcourt, 2003.

Eckstein, G., *The Body Has a Head,* Harper & Row, 1969.

Godwin, G., *Heart: A Personal Journey through its Myths and Meaning,* Wm. Morrow, 2001.

LeDoux, J. *The Synaptic Self,* Viking, 2002.

Leroi, A.M., *Mutants: On Genetic Variety and the Human Body,* Viking, 2003.

Miller, J. *The Body in Question,* Randon House, 1978.

McNeill, D., *The Face: A Natural History,* Little, Brown, 1998.

Nuland, S., *How We Die,* Knopf, 1994.

Nuland, S., *How We Live: The Wisdom of the Body,* Knopf, 1997.

Nuland, S., *The Mysteries Within,* Simon & Schuster, 2000.

Ornstein, R., *The Amazing Brain,* Houghton-Mifflin, 1984.

Restak, R., *The Secret Life of the Brain,* National Academy Press, 2001.

Rifkin, B.A., *Human Anatomy from the Renaissance to the Digital Age,* Abrams, 2006.

Rose, K.J., *The Body in Time,* John Wiley & Sons, 1988.

Starr, D., *Blood,* Knopf, 1998

Tsiaris, A., *The Architecture and Design of Man and Woman,* Doubleday, 2004.

Vanderlinden, K. *Foot: A Playful Biography,* Greystone, 2003.

Vogel, S., *Vital Circuits: On Pumps, Pipes, and the Working of Circulatory Systems,* Oxford UP, 1992.

Vogel, S., *Prime Mover: A Natural History of Muscle,* Diane, 2001.

Wilson, F.R., *The Hand,* Pantheon, 1998.

Yalom, M., *A History of the Breast,* Knopf, 1997.

Zimmer, C. *Soul Made Flesh: The Discovery of the Brain,* Free Press, 2004.

INDEX

A

abdomen 18, 20, 94, 96
abdominopelvic cavity 22–23
abducens 150
abductor 95, 100, 104, 100–101, 106–108, 110, 114, 120, 122–125
acetabulum 64, 67–68
acid 249, 261, 264, 266–268, 274
acid-base balance 280, 282, 286
acidophils 180
acromial 20
adductor 93, 100, 102, 104, 106, 108, 112, 122–125
adenohypophysis 178–180
adipose 36, 284
adrenal 18, 21, 174, 176–177, 180, 186–187, 208, 218
adrenaline 161
adrenocorticotrophin 180
adrenoglomerulotropin 177
adventitia 194, 262
afferent system 126–127, 130
air-blood barrier 17
alar nasalis 84
aldosterone 177, 187
alimentary 248, 250, 252, 256, 262, 264, 266, 268, 270, 272–273
alkaline 249, 253, 270, 274
alveolar-capillary membrane 17
alveoli 17, 235–236, 310
amino acids 249–250, 253, 270, 274
ammonia 280
ampulla 170, 252, 268, 272, 274
amygdaloid 138–140
amylase 188, 252–253, 261, 274
anabolism 24
anaphase 28–29
anastomoses 194, 204, 206, 214, 216
anconeus 102, 106–107
androgens 175, 187
angle of Louis 58
annulus fibrosus 57
ans 126–127, 130–131, 134, 158–161, 269, 273
antebrachium 18, 20
anteroposteriorly 58
antibodies 177
antigens 177
anvil 168
aorta 21, 177, 187, 192–193, 196, 198–200, 202, 208–210, 212–214, 218, 247, 255, 296, 303
apocrine 34
appendicular skeleton 44–45, 60, 62, 64, 66, 68, 70, 80–81, 102, 104, 106, 108, 110, 112, 114, 116, 118, 120, 122
appendix 17, 21
aqueous humor 164–165
arachnoid 145–149
arbor vitae 142–143
archicortex 137

B

archipallium 136–137
areola 310–311
argentaffen 266–267
arterioles 39, 194, 226–227, 235, 245, 270, 286
arteriovenous anastomoses 194
articular system 14–15
articulating bones 78–79
aryepiglotticus 86
arytenoid 86, 242–243
ascending rami of the ischia 67
aster 28–29
atria 198–200
atrioventricular 198–200
atrium 192–193, 196, 198, 200, 220
atroventricular (cuspid) valves 199
Attrahens Aurem 84
Auricular 64, 67, 154, 168, 198, 202
Auricularis 84
autonomic nervous system (ANS) 14, 126–127, 130, 158, 160
axial skeleton 44–45, 48, 50, 52, 54, 56, 58, 60
axial skeleton muscles 80, 84, 86, 88, 92–93, 100
axilla 155
axillary 20, 228
axis 44, 56
axolemma 128–129
axons 83, 129–130, 132–133, 136–138, 143–144, 152–153, 158–161, 166, 172–173
axoplasm 128–129
azygos 190, 218, 220

B

B-lymphocytes 17
basal 244
basement membrane 33–34
basophils 180
biceps 102–103, 112–113, 125
bicuspid (mitral) valve 198
bile 188–189, 252–253, 268, 274, 278
biliary system 17
Billroth's cors 230
bladder 17–18, 21, 280, 282, 288–289
blastocoele 32
blind spot 164, 166
body planes 12–15
bolus 252, 256, 262–265
bony orbit 162–164
bony thorax 44, 58
Bowman capsule 286
Bowman's glandular goblet 172
brachial 206, 216
brachial nerve 88
brachialis 102
brachii 102–103, 125, 206
brachiocephalic 182, 184, 202, 206, 216, 218, 224

C

brachioradialis 102–103, 106–107
brachium 18, 20, 60–61, 92–93, 100, 102–103, 216
brain 16, 18, 22–23, 48–49, 51, 53, 56, 190, 202, 204–205, 218
brain sand 176
breast 20, 180, 290, 310
brevis 106, 108–109, 112, 116, 118, 120–123, 125
Broca's speech 136
bronchial 33, 208, 218, 234–236, 244–245
Bronchioles 236, 244–245
bronchomediastinal 228
bronchopulmonary 234, 244–245
bronchus 244–246
Brunner's glands 268, 270
brush-border enzymes 250, 253, 271, 274
Buccal 20, 23, 202, 250, 252, 256
Buccinator 84, 256
bulbar 151, 162–164
Bulbourethral glands 292, 298–299
burns 40
bursa 78
buttocks 18, 20, 156

C

calcaneal 20
calcaneus 70
calcaneus heelbone 70
calyces 284–286, 289
calyx 284–285
canal cavity 23
canine 258
canthus 162–163
capillary 16–17, 39, 190–194, 200, 218, 220
capitate 62
capitulum 61
caput 18
carbohydrates 190–191, 249–250, 252, 261, 264, 270
carcinomas 28
cardiac 80–81
cardiac muscle 33, 32, 80
cardiac notch 247
cardiovascular 44, 190–193, 195, 197, 199–207, 209, 211–224, 234, 236, 249
carina 245
carotid 88, 202, 204, 206
carotid canal 53
carpal 18, 20, 62, 104, 206
cartilage 15, 30–31, 33, 36, 37, 44, 46, 54, 58, 65, 168, 242–243
caruncle 162–163, 260
catabolism 24, 280
catecholamines 174, 187
Cauda Equina 144–145, 152
caudad (inferior) 15